CARBANION CHEMISTRY

Structures and Mechanisms

ERWIN BUNCEL

JULIAN M. DUST

AMERICAN CHEMICAL SOCIETY
Washington, DC

2003

OXFORD
UNIVERSITY PRESS

Oxford New York
Auckland Bangkok Buenos Aires Cape Town Chennai
Dar es Salaam Delhi Hong Kong Istanbul Karachi Kolkata
Kuala Lumpur Madrid Melbourne Mexico City Mumbai Nairobi
São Paulo Shanghai Taipei Tokyo Toronto

Copublished by the American Chemical Society
1155 16th Street, NW
Washington, DC 20036
and Oxford University Press, Inc.
198 Madison Avenue
New York, NY 10016

www.oup.com

Developed and distributed in partnership by the
American Chemical Society and Oxford University Press

Library of Congress Cataloging-in-Publication Data
Buncel, E.
Carbanion chemistry : structures and mechanisms / Erwin Buncel, Julian Michael Dust.
p. cm.
Includes bibliographic references and index.
ISBN 0-8412-3556-2
1. Carbanions.
I. Dust, Julian Michael. II. Title
QD305.C3 .B86 2001
547′.1372—dc21 00-60593

9 8 7 6 5 4 3 2 1

Printed in the United States of America
on acid-free paper

To Penny, Jacquie, and Irene E.B.

To My Parents J.M.D.

Preface

Carbanions have a central role in modern chemistry and the factors governing their structure, stability, and reactivity are recognized as being among the principal foundations of organic chemistry. Carbanions have an undisputed role in modern organic synthesis. The importance of understanding the theoretical principles that direct their behavior is thus self-evident. The last 25 years of the 20th century have witnessed important advances in investigative techniques, such as multinuclear (heteronuclear) nuclear magnetic resonance spectroscopy for solution and solid-state investigations into carbanion structure, and ion cyclotron resonance for gas-phase investigations of intrinsic carbanion reactivity. Alongside the experimental developments have come important advances in theoretical understanding as a result of newer calculational methods being applied for ever larger molecular systems.

In this way, the organic chemist today is able to garner insight not only into the structure and behavior of isolated carbanions, but also into the vital role played by the counterion, the nature of the solvent medium, and the state of aggregation, on carbanion structure and reactivity.

Historically, the 1965 seminal text *Fundamentals of Carbanion Chemistry* by Donald J. Cram[1] was undoubtedly a watershed, and formed the basis for all subsequent work in this area. The monograph, *Carbanions: Isotopic and Mechanistic Aspects* by Erwin Buncel, published in 1975,[2] updated some of the topics in Cram's work and introduced others, such as nonclassical carbanions and orbital symmetry control in carbanion rearrangements. J. R. Jones's book, *The Ionization of Carbon Acids*, illuminating this facet of the subject, had appeared two years earlier.[3] Other books by J. C. Stowell[4] and R. B. Bates and C. A. Ogle[5] covered different aspects of the subject. There followed a series of edited monographs, three volumes in the series *Comprehensive Carbanion Chemistry*, which appeared in the 1980s,[6-8] and two volumes entitled *Advances in Carbanion Chemistry* that appeared in the early 1990s.[9,10] In each case, active investigators in the field contributed chapters outlining advances in their area of expertise. These edited monographs, then, were directed mainly at researchers in the field. A work on organometallic reagents also appeared in 1994;[11] this mainly concerned synthetic aspects.

It appeared to us that the time was ripe for another general text in the area of carbanion chemistry that would update many of the developments that had arisen over the last 25 years of the 20th century. Thus, this book is not intended as a comprehensive account of the subject, which we believe is, in any case, not feasible in view of the explosive growth in this area. Regardless, we hope that our text will be useful for advanced students and researchers alike, both as an introduction and as a source of reference to recent developments in carbanion chemistry.

As the title implies, our work addresses structural and mechanistic aspects of carbanion chemistry, though it is also our aim to point out applications to synthesis. As to the order of presentation of topics, we believe that the order summarized below can be justified on historical grounds and as a logical development of the subject, though, admittedly, an argument could be put forward for a different presentation order.

The first three chapters evolve in a straightforward way. In Chapter 1, the concept of carbon acidity—thermodynamic, kinetic, and solution as compared with gas-phase—leads in Chapter 2 to factors governing the stabilities of carbanions and their reactivities in proton transfer processes, namely, Brønsted correlations, transition-state structure, and imbalances. In Chapter 3, we consider the stereochemistry of carbanions, a parallel facet of their structure, which includes the chirality of carbanions adjacent to sulfur and phosphorus centers. Chapter 4 highlights spectroscopy (NMR, UV–vis, etc.) as the most direct and informative methodology for study of carbanion structure and bonding in solution, providing evidence of ion pairing and association phenomena, as pioneered by Michael Szwarc and others. Chapter 5 deals with carbanion reactions catalyzed by solid bases, which in the past have been treated only in specialized texts but are included here since they put into perspective and traverse much of carbanion chemistry, as influenced by surface and inclusion phenomena. The treatment includes industrially important aspects, such as catalysis of alkanol dehydration, alkene double-bond rearrangements, and catalysis of various condensation reactions by basic zeolites and superbasic solids. Chapter 6 reviews keto-enol tautomerization followed by consideration of transition-state structure in the aldol condensation and selected enolate rearrangements. Chapter 7, on non-enolate carbanion rearrangements, features again the principles of orbital symmetry control in electrocyclic and sigmatropic rearrangements, while also critically examining the evidence for the intervention of radical-ion pathways for a balanced presentation. The chapter closes with some of the synthetically valuable "name rearrangements" of non-enolate carbanions; these have been in the forefront of organic chemistry during much of the 20th century.

To sum up, in this book we have sought to provide an updated overview of carbanion chemistry from a mechanistic viewpoint, in the tradition set by Cram. In so doing, we hope to stimulate further mechanistic investigations in this rapidly expanding field of carbanion chemistry.

Finally, we wish to express our gratitude to colleagues who have read and commented on our manuscript. Parts of the manuscript were read by professors C. A. Bernasconi, R. S. Brown, U. Edlund, G. W. Rayner-Canham, V. Snieckus, and F. Terrier, and their comments are highly appreciated, as are those of the reviewers. However, we accept full responsibility for any errors or inaccuracies in the text. The cooperation of ACS staff is also acknowledged. We thank our families for their patience, support, and other contributions in the preparation of this text. Grateful acknowledgement is made to the Natural Sciences and Engineering Research Council of Canada and the Principal's Research Fund of Sir Wilfred Grenfell College for supporting our research in carbanion chemistry, and to the many students and other co-workers who actively participated in research in this area, for their efforts, and for the many illuminating discussions that made this work possible.

Kingston, Ontario — E. B.
Corner Brook, Newfoundland — J. D.

References

1. Cram, D.J. *Fundamentals of Carbanion Chemistry*; Academic Press: New York, 1965.
2. Buncel, E. *Carbanions. Mechanistic and Isotopic Aspects*; Elsevier: Amsterdam, 1975.
3. Jones, J.R. *The Ionization of Carbon Acids*; Academic Press: London, 1973.
4. Stowell, J.C. *Carbanions in Organic Synthesis*; Wiley: New York, 1979.
5. Bates, R.B.; Ogle, C.A. *Carbanion Chemistry*; Springer-Verlag: Berlin, 1983.
6. Buncel, E.; Durst, T., Eds. *Comprehensive Carbanion Chemistry. Part A. Structure and Reactivity*; Elsevier: Amsterdam, 1980.
7. Buncel, E.; Durst, T., Eds. *Comprehensive Carbanion Chemistry. Part B. Selectivity in Carbon–Carbon Bond Forming Reactions*; Elsevier: Amsterdam, 1984.
8. Buncel E.; Durst, T., Eds. *Comprehensive Carbanion Chemistry. Part C. Ground and Excited State Reactivity*; Elsevier: Amsterdam, 1987.
9. Snieckus, V., Ed. *Advances in Carbanion Chemistry*; JAI Press: Greenwich, CT, 1992; Vol. 1.
10. Snieckus, V., Ed. *Advances in Carbanion Chemistry*; JAI Press: Greenwich, CT, 1994; Vol. 2.
11. Schlosser, M. *Organometallics in Synthesis*; Wiley: New York, 1994.

Contents

CARBANION CHEMISTRY

1

Carbon Acidity and Carbanions

1.1 STRUCTURE AND PROPERTIES OF CARBANIONS

Carbanion chemistry is rich and diverse.[1–6] Carbanions are important intermediates in numerous carbon–carbon bond-forming reactions.[6–8] Thus, carbanion chemistry encompasses both the standard name reactions of introductory organic chemistry, such as the Claisen, Knoevenagel, and Dieckmann condensations, and the chiroselective extensions of these reactions in modern synthetic schemes.[6–16]

As its name indicates, a carbanion is an anionic species in which a carbon atom bears unit negative charge; the carbanionic center is surrounded by a Lewis octet of electrons, including those that are involved in covalent bonding to three other groups. Some examples of carbanions (or carbanides), drawn as standard Lewis electron dot structures, are shown in Chart 1.1.

Unlike other common ions—chloride or hydroxide, for example—most carbanions are not present in high concentration in the free state or in aqueous solution. Of course, those exceptional situations in which carbanions are present in high concentration provide particular insight into the factors that stabilize these intermediates, which are so useful in synthesis, either as "free carbanions" or as disguised carbanionic synthons.[6–16]

The carbanides shown in Chart 1.1 form one general class: carbanions derived from "unsubstituted" hydrocarbons. Another class, which has become increasingly important in synthetic organic chemistry, comprises carbanions in which the anionic center is stabilized by a heteroatom or heteroatom-containing group. Some of the carbanions falling under this category are shown in Chart 1.2.

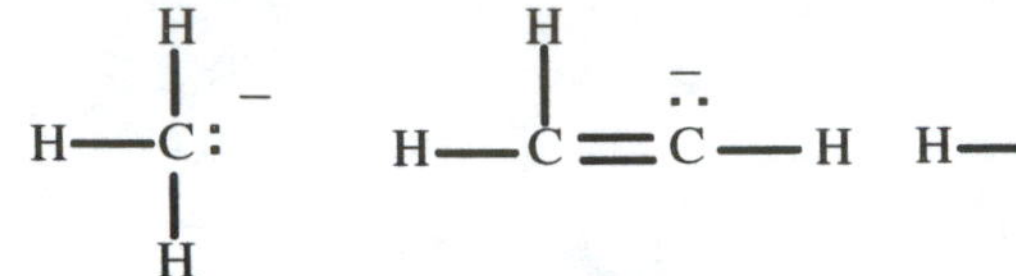

methyl anion (methide)

vinyl anion

acetylide anion (ethynide)

allyl anion

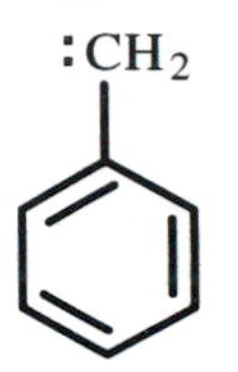

benzyl anion

phenyl anion

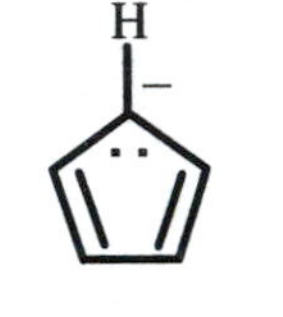

cyclopentadienyl anion (cyclopentadienide)

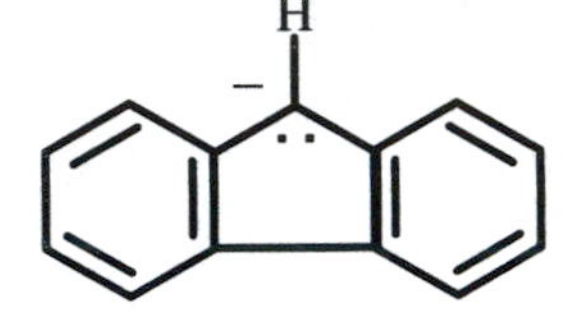

fluorenyl anion (fluorenide)

CHART 1.1

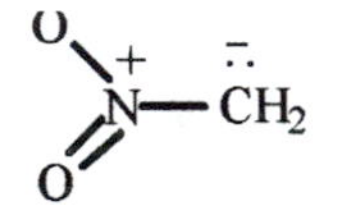

nitromethide anion

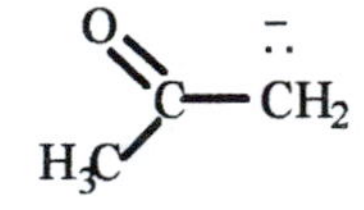

acetone enolate (acetonate) anion

cyanomethide anion

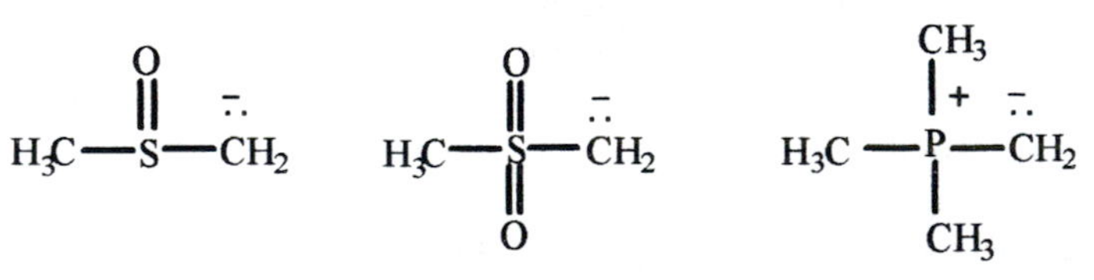

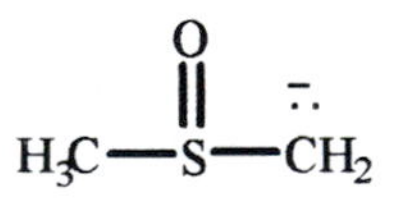

methylsulfinyl methide anion (dimsyl anion)

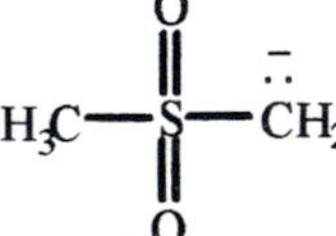

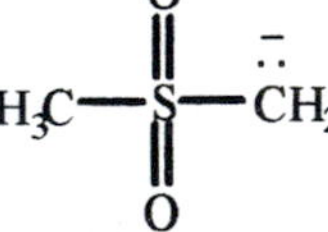

methylsulfonyl methide anion

trimethylphosphonium methide anion (tetramethylphosphonium ylide)

CHART 1.2

Conceptually, the simplest route to carbanions is by loss of a proton from the corresponding hydrocarbon or hydrocarbon derivative. In principle, then, all compounds that contain C–H bonds may potentially be *carbon acids*.[17–20] In this introductory chapter we will explore the link between the acidity of carbon acids and the stability of the resultant carbanions. This chapter is organized, therefore, in a historical fashion. After having considered various definitions of acidity and how these apply to carbon acids, we will examine the acidities (pK_a values) measured for these acids. Those acidities measured first were, of course, those accessible in aqueous solution. We then examine various acidity scales, including those tethered to water as the standard state (via acidity functions), ion-pair acidities in various ethereal solvents, notably the extensive current work of the Streitwieser group[21] in tetrahydrofuran (THF), and acidities determined in dimethyl sulfoxide (DMSO). These are all measures of equilibrium or approximate equilibrium acidities of carbon acids. Finally, the chapter concludes with a consideration of gas-phase acidities, the most recent class of acidity measurements to be undertaken, and draws the comparison between these acidities and those determined in solution.

1.2 DEFINITIONS OF ACIDITY

1.2.1 Brønsted Acidity

One of the most common and useful definitions of acidity is attributed to Brønsted[22] and to Lowry.[23] A proton donor is termed a Brønsted acid, whereas a proton acceptor is a Brønsted base.

$$HA_{(solv)} \overset{K_a}{\rightleftharpoons} H^+_{(solv)} + A^-_{(solv)} \tag{1.1}$$

Acid strength is then reflected in the magnitude of the equilibrium constant, K_a, for the acid dissociation shown in equation 1.1. Acid strengths are frequently listed according to their pK_a values (where $pK_a = -\log K_a$); strong acids are associated with small positive or, alternatively, negative pK_a values.

At first glance, the Brønsted–Lowry (hereafter referred to as Brønsted, for simplicity) definition for acids appears to be similar to the earlier definition by Arrhenius. Only protic species can be acids according to either view. However, in the Brønsted definition there is a specific and significant role for the solvent, as indicated by the subscript "(solv)" in equation 1.1. Where the solvent is water, equation 1.1 is often rendered as

$$\underset{\text{acid}}{HA} + \underset{\text{base (solvent)}}{H_2O} \rightleftharpoons \underset{\text{conjugate acid}}{H_3O^+} + \underset{\text{conjugate base}}{A^-} \tag{1.2}$$

The role of solvent as the species that accepts the proton from the acid of interest is seen more readily in equation 1.3, where the proton that is transferred is set in boldface to highlight it.

$$\mathrm{H{-}A + H_2O \rightleftharpoons H{-}\overset{+}{O}H_2 + A^-} \tag{1.3}$$

Consider the equilibrium derived from inversion of equation 1.3; in this form, the former conjugate base, A^-, is seen to accept a proton from "protonated water" (i.e., the hydronium ion, H_3O^+) to give water and the acid, HA. Here, the Brønsted definition can be seen to extend that of Arrhenius. While both definitions would recognize $HCl_{(aq)}$ as an acid, only the Brønsted rubric recognizes chloride ion as a base, albeit a weak one in aqueous solution. Equation 1.3 also emphasizes the generalization that strong acids will have weak conjugate bases and, conversely, that weak acids (including most carbon acids) will have strong conjugate bases.

In water, the pK_a of acids that range from moderately weak to reasonably strong may be determined from the pH (operationally defined in terms of concentration as $-\log\,[H^+]$) of the aqueous solution and a determination, typically through spectrophotometric measurement, of the buffer ratio, $[HA]/[A^-]$, by use of the Henderson–Hasselbalch equation:

$$pK_a = pH - \log[HA]/[A^-] \tag{1.4}$$

Selected pK_a values for some of these moderately strong carbon acids, determined in water, are listed in Table 1.1. Note how nitro and cyano group substitution enhances the acidity of these methane derivatives.

As can be seen, equation 1.2 is to be preferred over equation 1.1 because it more clearly designates the role of the solvent as a base and that of protonated water as the conjugate acid of HA. The reality of hydronium ion, formulated as H_3O^+, has been a subject for debate.[30–32] However, Giguere[32] has mustered infrared spectrophotometric, X-ray, and neutron diffraction data, as well as thermodynamic evidence, to support the view that the proton in aqueous solution is covalently bonded to a single molecule of water, while a first hydration sphere (equivalent to $H_3O^+ \cdot 3H_3O$ or $H_9O_4^+$) is strongly associated with the hydronium ion.[33,34] In this regard, studies of the binding energies linked to the successive aquation of the proton *in the gas phase* show that up to eight molecules of water may bind to the proton and the process is still exothermic at each step.[35–37] This high degree of hydrogen-bonding and organization about the proton in water is an important feature of the solvent properties of water, and it could be argued that the protonic species in water is best designated as $H^+(H_2O)_n$. Similarly, the proton is likely to be well solvated in most common, hydrogen-bonding organic solvents,[38,39] although actual covalent bonding may occur only between a single molecule of solvent and the proton.[40] For simplicity's sake, wherever the proton, H^+, appears in any subsequent equation or equilibrium in this work, the proton will be understood to be suitably solvated.

Two major points arise from consideration of the Brønsted definition of acidity. First, Brønsted acidity provides the link between our simple concept of the carbon acid and the carbanion; the carbanion is the conjugate base of the carbon acid. Secondly, the role of solvent in determining the acidity of carbon acids merits further careful scrutiny. Therefore, in comparing a series of organic acids in *a single solvent* according to equation 1.1, the degree of solvation of the proton is

Table 1.1. Acidities of some carbon acids in water (298 K)

Acid[a]	pK_a
$\mathbf{H}C(CN)_3$	0[b]
$\mathbf{H}C(NO_2)_3$	0.2[c]
$CH_2(NO_2)_2$	3.6[c]
$CH_2(COCF_3)_2$	6.8[d]
$CH_2(COCH_3)_2$	9.0[e]
$\mathbf{H}CN$	9.2[e]
CH_3NO_2	10.2[b]
$CH_2(CN)_2$	11.2[b]

[a]The acidic proton of interest is set in boldface type in the carbon acid formulae.
SOURCES: [b]Data taken from reference 24, but also see references 25 and 26; pK_a of tricyanomethane is determined to be −5.1 in these last two references; [c]Data taken from reference 27; [d]Data taken from reference 28; [e]Data taken from reference 29.

constant, unlike the situation in considering the acidity of a single carbon acid in going from water, for example, to another solvent such as DMSO.[41] Moreover, undissociated neutral acids, HA, would be expected to be solvated only to a small degree relative to the charged conjugate base, A^-, at least in high dielectric constant (polar) solvents. Hence, an order of Brønsted acidity in a given polar medium will be determined primarily by the changes wrought in the stability of A^- by substituents, structure, and hybridization in a given solvent system.

1.2.2 The Leveling Effect

The role of the medium on acidity imposes a restriction on our ability to measure the acidity of carbon acids (or other weak acids). This limitation is termed the *leveling effect.* In essence, the leveling effect recognizes that there can be no stronger acid in a given solvent than the conjugate acid of that solvent. Similarly, no base stronger than the conjugate base of the medium can exist in that medium. As a consequence, the relative strengths of bases that are stronger than the conjugate base of the solvent cannot be measured in that solvent. The same sentence would be equally true if the words "bases" and "base" were replaced by "acids" and "acid", respectively.

The leveling effect is particularly relevant in the consideration of the acidity of carbon acids, most of which do not ionize appreciably in water. Consider equilibrium 1.2 again. If an acid HA, such as tricyanomethane (Table 1.1), is a very strong acid it will dissociate completely, equilibrium 1.2 will lie far to the right and, in fact, the only acid in the system will be the hydronium ion. The strength of the acid, HA, is "leveled" to that of the conjugate acid of the medium. Therefore, tricyanomethane *appears* to have a pK_a of 0 in water[24] when, in fact, a better estimate is −5.1.[25,26] More to the point, a very weak carbon acid in water would not be dissociated significantly, equilibrium 1.2 will lie far to the left, and the concentration of H_3O^+ *will not be measurably different from that produced due to the autoprotolysis of water itself.* In this context, it would be impossible to measure the acidity of the simplest carbon acid, methane, in water; it has a pK_a value estimated to fall somewhere between ca. 56 (based on extrapolation from measurements in DMSO)[41] and 49[42] (based partly on an ab initio calculation of the pK_a of ethane as 50.6[43]), while the pK_a of water is 15.75. Note that the effective pK_a of methane in water could be even lower, as a result of hydrogen-bonding interactions not present in DMSO and clearly absent in the gas phase. Regardless, methane is much too weak an acid for its pK_a value to be measured by direct methods in water.

Nevertheless, water is generally accepted as the standard solvent for establishing any acidity scale, not only because of its high dielectric constant, which makes it a strong ionizing solvent, and its ability to act as both a hydrogen-bond donor and acceptor, but also because of its historical precedent as one of the first solvents in which acidities were determined (cf. Table 1.1). Several methods will be discussed for measuring the pK_a values of weak carbon acids in more basic solvents. Here, the solvent (:Solv) is a more powerful proton acceptor than water and, therefore, equilibrium 1.5 would be expected to lie farther to the right-hand side than would equilibrium 1.2, using the same carbon acid. These solvents include dimethyl sulfoxide (DMSO), acetonitrile (MeCN), and cyclohexylamine (CHA). The problem then becomes one of tethering these acidity measurements to the standard solvent, water (at infinite dilution at 298 K).

$$HA + :Solv \rightleftharpoons H\text{–}Solv^+ + A^- \quad (1.5)$$

1.2.3 Lewis Acidity

An alternative approach to acidity is broader than the Brønsted definition. Thus, Lewis[44] defined an acid as an electron-pair acceptor and a base as an electron-pair donor. A key portion of this definition is that an adduct is formed between the Lewis acid and the Lewis base. Rather than a single equation (cf. equation 1.1), a series of adduct-forming reactions is needed to illustrate Lewis acidity–basicity.

$$A + :B \rightleftharpoons AB \quad (1.6)$$

$$A + A_1B \rightleftharpoons AB + A_1 \quad (1.7)$$

$$B: + AB_1 \rightleftharpoons AB + :B_1 \quad (1.8)$$

$$AB + A_1B_1 \rightleftharpoons AB_1 + A_1B \quad (1.9)$$

If the pair of electrons is donated from a Lewis base such as a chloride ion, to a proton with concomitant bond formation, then the formal equivalency of equation 1.6 and the inverse of equation 1.2 can be seen; namely, $Cl^- + H^+ \rightarrow HCl$. Numerous reagents that would not be recognized as acidic according to the Brønsted definition, such as $AlCl_3$, BF_3, as well as metal ions such as Ag^+, are considered to be Lewis acids. This definition also expands the category of bases. Alkenes, carbonyl compounds, and ethers would all be viewed as Lewis bases. Some illustrative reactions are given in equations 1.10–1.12.

$$BF_3 + H_3C\ddot{O}CH_3 \rightleftharpoons F_3\overset{-}{B}-\overset{+}{O}(CH_3)_2 \qquad (1.11)$$

$$R_2C{=}O + RMgX \rightleftharpoons R_3C{-}OMgX \qquad (1.12)$$

As can be seen from these equations, the Lewis description is often an apt one for considering the reactions of carbanions and carbanion-like species. Equation 1.10 represents the formation of a cyanohydrin, under basic conditions, while equation 1.12 represents the addition step of the Grignard reaction. These equations fall into the general Lewis acid–base category represented by equation 1.6. Alkylation reactions that involve attack of a carbanion on a haloalkane with concomitant displacement of the halide would follow the Lewis acid–base reaction described by the general equation 1.8 above.

The strength of the Lewis acid may be assigned in theory in a manner strictly analogous to that for Brønsted acids (which, after all, form a subset of the Lewis acids), namely, by determination of the relative magnitude of the equilibrium constants associated with equations 1.6–1.9. For metal ions as Lewis acids, which may be involved in complexation with one or more ligands, the binding constant can often be determined. Quantities such as pL (i.e., pL = – log[Ligand]) and pM (– log[Metal ion]) are measurable in many systems by a variety of methods that include potentiometry, as well as NMR and ultraviolet–visible (UV–vis) spectroscopic techniques.[45] Other approaches to quantifying the Lewis acid–base description include the donor/acceptor number approach of Gutmann[46,47] and the S_a scale of cationic Lewis acid strengths, as defined from inorganic crystal structures by Brown.[48,49]

Extensive research into quantifying Lewis acidity has also been published by Drago and co-workers, who correlated the reaction enthalpy for Lewis adduct formation (according to equation 1.6) in terms of electrostatic (E) and covalent interactions (C) between the Lewis acid and base components.[50–52] More recently, the Drago E–C correlation equation has been extended to permit correlation of any physicochemical property (χ):

$$\Delta_{\chi} = E_{A}E_{B} + C_{A}C_{B} + W \tag{1.13}$$

where either a standard acid (acceptor) or base (donor) is held constant and the test series of bases or acids is varied as the specific property (χ) is measured. In its extended form the Drago E–C equation has been used to correlate 43 reaction-related properties of phosphines interacting with various organometallic compounds and haloalkanes, using properties such as heats of reaction, logarithm of rate constants, and ^{13}C NMR chemical shift data, as well as the pK_a of the phosphines.[53–55] Further work in this area may extend to the reactions of carbanions and carbanion-like species.

Conventional criticism of Lewis acid–base theory holds that it is so general that virtually no reaction in standard organic chemistry would be left unrepresented in the defining equations 1.6–1.9. However, it could be argued that the definition is *not general enough*, in that it is tied to the Lewis concept that reactions occur solely by transfer of electrons in discrete pairs, the "curvy arrow formalism".[56] Many organic reactions are now being redefined in terms of single electron transfer[57–64] or in terms of transition states that have varying degrees of radical or radical-ion character.[65–68] Further, there is a growing body of work that quantitatively links Brønsted acidity, as measured by pK_a, and ease of free-radical formation, as measured by bond dissociation enthalpies (BDE).[69–71] An alternative approach, that implicitly includes single electron transfer (SET)[72–74] and has utility in explaining the ambident behavior of enolates, is the hard–soft acid–base (HSAB) theory, which we will consider in the next section.

1.2.4 Hard–Soft Acidity–Basicity (HSAB)

Hard–soft acid–base theory has its roots in the problem of comparing various families of Lewis acids and bases. As an example of this difficulty, amines as Lewis bases tend to bind more strongly with the proton as the Lewis acid than do phosphines, and phosphines bind more strongly than do arsines; however, with silver ion as the acid, amines now bind more weakly than do phosphines, although the phosphine to arsine order is preserved.[75]

To account for these differences in Lewis adduct formation, Lewis acids and bases were classified as *hard* or *soft* by R. G. Pearson.[76–80] Pearson used kinetic, solution-, and gas-phase equilibrium data (including donor–acceptor equilibria), as well as gas-phase heat of reaction data, to formulate these classifications. In general, *soft bases* are characterized by large central (donor) atoms that are not as electronegative as the donor atoms of the analogous *hard bases*. The soft bases contain donor atoms that are more readily oxidized than those of hard bases, a result of the fact that the central donor atoms of the soft bases usually have low-lying empty orbitals; the soft donor atoms are more polarizable than those of their hard counterparts. Similarly, soft acids contain large acceptor atoms that bear only a small positive charge (i.e., the charge may be delocalized) and are polarizable, whereas hard acids are composed of small acceptor atoms that are less polarizable and are difficult to reduce. Because redox data are used in the assignment of degree of hardness or softness of various acids and bases, single-electron transfer processes are implicit in the HSAB concept.

The prime argument of the HSAB theory is that hard acids prefer to bind with hard bases and soft acids interact preferentially with soft bases. Originally, these preferences were presumed by Pearson to be an added factor in the determination of Lewis acid–base interactions, rather than a synonym for acid–base *strength*. Inasmuch as acid strength is often measured by pK_a or other equilibrium constant determination, acid strength can be related to Gibbs free energy change ($\Delta G^{\circ}{}_a$), whereas HSAB interactions *as originally defined* are *extrathermodynamic*.

The utility of the HSAB description in explaining carbanion reactions can be illustrated with an appropriate example. Enolate anions generally react via their carbon centers, particularly with carbonyl substrates. There are cases, nonetheless, in which reaction via the oxygen center is found,[81,82] notably in alkylation reactions. Thus, the enolate ion can act as an *ambident nucleophile* to give products of oxygen attack (arrow from oxygen) or products of carbon attack (arrow from carbon) as shown in Reaction 1.1.

On the basis of the resonance forms that represent enolate anion, it might be expected that the majority of reactions would occur at the oxygen center. The argument here is that the negative charge would preferentially reside on the more electronegative oxygen center and this canonical form would, therefore, make a greater contribution to the overall hybrid. Recent molecular orbital calculations concur with the valence bond view that the majority of the negative charge resides on the oxygen center in enolate ions.[83–85] Nevertheless, reactions of enolates often result in predominant C-alkylation.

Significantly, the degree of C- versus O-alkylation is modified by the reaction conditions, as well as by the choice of enolate ion. Thus, the enolate formed by the sterically hindered diphenylacetophenone has been found to react with methyl iodide to yield 50% of the O-methylated product in DMSO as compared with only 4% of the methyl ether in acetone.[86] Note that the C/O alkylation ratio decreases as the size of the counterion of the enolate decreases.[87,88] Solvents in which anions are desolvated, such as DMSO[89] or, more notably, hexamethylphosphortriamide (HMPA),[90] also increase the relative amount of O-alkylation. Insulating the enolate anion from the counterion with crown ethers also stimulates O-alkylation.

In HSAB terms, the O-center of an enolate anion is hard compared with the C-site. Therefore, the enolate O-center would be expected to interact preferentially

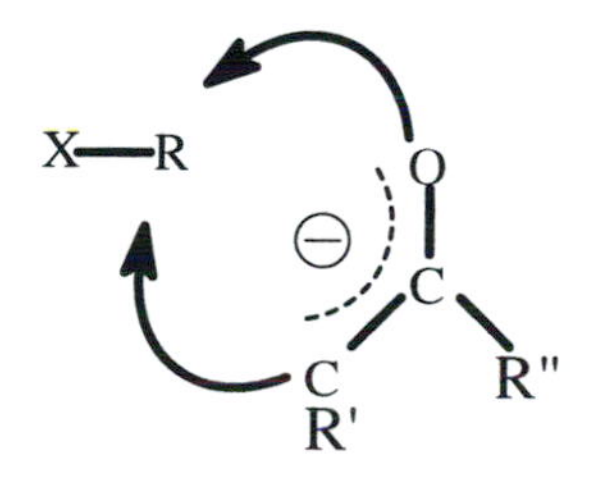

REACTION 1.1

with the hard metal cation (usually Li^+). The soft C-center, on the other hand, would be expected to react primarily with the C-center of the alkylating agent. In this context, a larger, more polarizable counterion (e.g., Cs^+ or Rb^+) would decrease ion pairing between the oxyanion center and the counterion. The result would be a larger percentage of O-alkylation. Conversely, a smaller, harder ion would form a tighter ion pair involving the O-center. The result would be a larger percentage of C-alkylation.

The HSAB theory can also accommodate some of the other observed reactivity modifications. Solvents that desolvate the anion or crown ethers that efficiently complex the counterion would free the O-center of the enolate and encourage reaction via the oxygen. Alcohols contain hard oxygen centers and the hard proton bonded to the oxygen would be expected to coordinate more strongly with the hard oxygen center of enolate anions, and, therefore, C-alkylation would be favored in alkanols. But DMSO with a soft sulfur center would not coordinate to the same degree with the oxygen and O-alkylation would be favored. O-alkylation would also be favored when a harder fluoroalkane alkylating agent was used instead of a softer iodoalkane.

Given the strong connection between the O- versus C-reactivity of ambident nucleophiles like enolate anions, and the nature of the counterion and solvent, it is useful to consider enolate reactivity in the gas phase. Determination of the relative propensity for attack via oxygen versus carbon has been hampered by the fact that while the products of the two pathways are different, the various gas-phase techniques [ion cyclotron resonance (ICR), flowing afterglow, and high-pressure mass spectrometry] detect only charged products. Careful choice of systems,[91,92] as well as modified sampling techniques,[93] have permitted some data to be collected. For example, Brauman and co-workers[94,95] determined the O:C reactivity ratio (Table 1.2) from the ratio of negative product ions formed ultimately from attack by a series of enolate ions on 2,2,2-trifluoroacetyl chloride (see Scheme 1.1 for acetophenone enolate).[95] The logarithm of the O:C ratio found in these studies correlated well with the keto–enol energy differences (ΔH_{k-e}) for the neutral

Table 1.2. Selected gas-phase O:C reactivity ratios determined for reaction of a series of enolates ($R–COCH_2^-$) with 2,2,2-trifluoroacetyl chloride (cf. Scheme 1.1)

$R–COCH_2^-$	O:C ratio
H	5.9
CH_3	4.0
$(CH_3)_3C$	4.1
C_6H_5	6.0
F	0.9

SOURCE: Data taken from references 94 and 95.

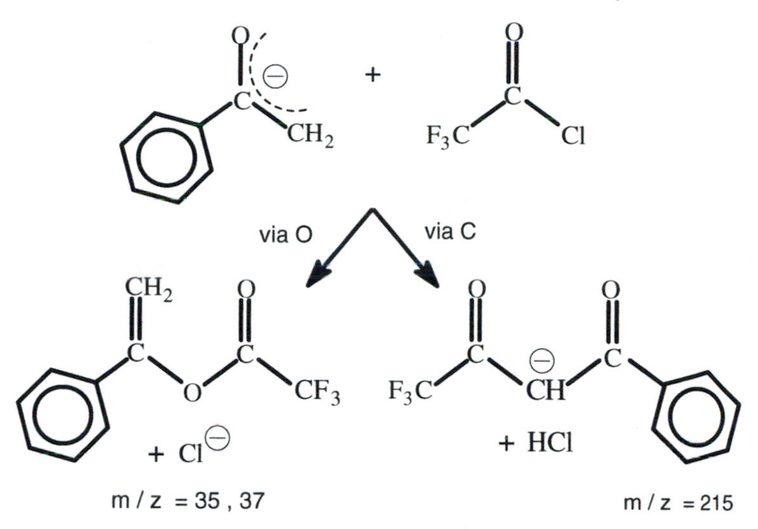

SCHEME 1.1

parent carbonyl compounds, following an earlier proposal of Brickhouse and Squires.[96] Other gas-phase systems, however, have been satisfactorily explained by invoking HOMO–LUMO interactions (HOMO, highest occupied molecular orbital; LUMO, lowest unoccupied molecular orbital);[97] the connection between such HOMO–LUMO interactions and the HSAB description of Lewis acids and bases will be considered further below. A selection of inorganic and organic Lewis acids and bases are divided into hard, soft, and borderline cases[98] in Table 1.3. As can be seen from the table, carbanions are generally considered to be soft bases, and from the foregoing discussion the carbon center of enolate anions would be expected to be softer than the oxygen center. Similarly, Table 1.3 classifies 1,3,5-trinitrobenzene (TNB) as a soft acid. The interaction between the soft C-site of an enolate and TNB[99] should be favored over reaction via the O-center according to HSAB theory. In fact, reaction of TNB with a wide range of nucleophiles, including carbanions, gives rise to anionic σ-bonded adducts, termed Meisenheimer complexes.[100–102] The Lewis acid–base equilibrium 1.6 provides a valid representation of Meisenheimer complexation.

Until recently, only the Meisenheimer adducts formed by attack via the carbon center of carbanions had been observed,[103–107] although, where the structure permits, 1,3-diadducts form as a result of intramolecular cyclization involving another methylene functionality α to the carbonyl group.[106,107] However, at low temperature (−50 °C) in acetonitrile:1,2-dimethoxyethane [MeCN-d_3:DME, 1:1 v/v), the oxygen-bonded adduct of acetophenone enolate (**1**, Scheme 1.2) has been observed by ^{1}H NMR spectroscopy.[108] As the temperature was raised to close to ambient (20 °C), the O-adduct gave way and was replaced by the carbon-centered enolate adduct (**2**, Scheme 1.2).

This result is consistent with the foregoing discussion, in that 18-crown-6 polyether was used to complex the potassium counterion of the enolate and, so,

Table 1.3. Classification of Lewis acids and bases as *hard* or *soft*

Hard	Borderline	Soft
	Acids	
H^+, Li^+, Na^+, Be^{2+} Mg^{2+}, Ca^{2+}, Sr^{2+}, Mn^{2+} Al^{3+}, Sc^{3+}, Cr^{3+}, Co^{3+}, Fe^{3+} BF_3, $B(OR)_3$ $Al(CH_3)_3$, $AlCl_3$, AlH_3 RPO_2^+, $ROPO^{2+}$ RSO^{2+}, $ROSO^{2+}$, SO_3 RCO^+, CO_2, NC^+, HX	Fe^{2+}, Co^{2+}, Ni^{2+}, Cu^{2+}, Zn^{2+} Pb^{2+}, Sn^{2+}, $B(CH_3)_3$, SO_2 NO^+, R_3C^+	Cu^+, Ag^+, Hg^+, Hg^{2+}, BH_3, RS^+ I^+, Br^+, HO^+, RO^+, I_2, Br_2 Trinitrobenzene, tricyanobenzene Chloranil, quinones, etc. Tetracyanoethylene, etc. $:CH_2$ and other carbenes, M^0 (metal atoms)
	Bases	
H_2O, OH^-, CH_3COO^- PO_4^{3-}, SO_4^{2-}, F^-, Cl^- CO_3^{2-}, ClO_4^-, ROH, RO^- R_2O, NH_3, RNH_2	$C_6H_5NH_2$, C_5H_5N, N_3^-, Br^- NO_2^-, SO_3^{2-}, N_2	R_2S, RSH, RS^-, I^-, SCN^- $S_2O_3^{2-}$, R_3P, $(RO)_3P$, R_3As RNC, CO, C_2H_4, C_6H_6, H^- R^-, CN^-

[a] SOURCE: Data taken from reference 98.

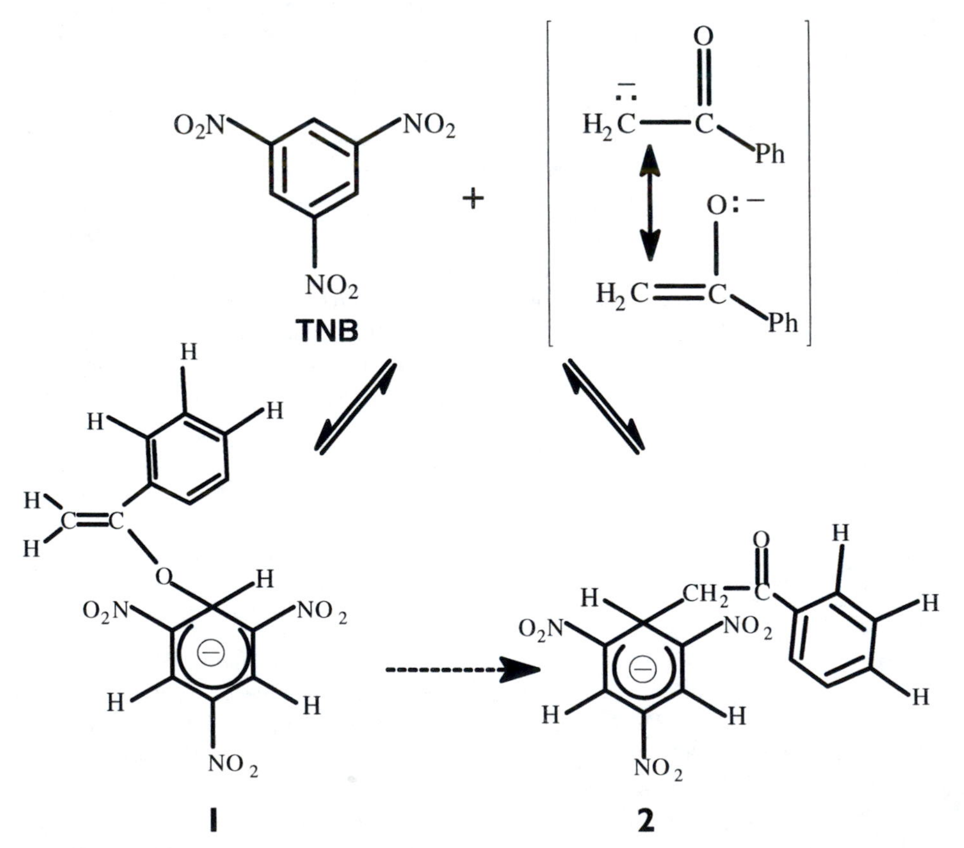

SCHEME 1.2

solubilize the acetophenone enolate; this would also be expected to free the oxygen center of the enolate and make it susceptible to electrophilic attack. The interesting observation that the O-adduct (**1**) is the kinetically preferred product and the C-adduct (**2**) is the product of thermodynamic control is in agreement with calculations by Houk and Paddon-Row[109] that show that the activation energy for O-attack is lower than that for C-attack in the gas-phase alkylation of a number of enolates.

Interestingly, the pathways for rearrangement of the TNB O-adduct **1** include not only dissociation back to TNB and enolate, which may then attack TNB via the carbon site to give **2**, but also a [3,3] sigmatropic shift (dashed arrow in Scheme 1.2), akin to the Claisen rearrangement[110,111] of allyl vinyl ethers.[112,113] Rearrangements of carbanions will be considered further in Chapter 6 and especially in Chapter 7.

Clearly, while HSAB theory provides a useful shorthand to explain, for example, the ambident behavior of enolate ions, other factors, including the relative stability of the parent keto and enol forms of the carbonyl compounds, must be considered to obtain a full picture of this reactivity.

The underlying justification for the HSAB classification was developed first by Klopman[114] on the basis of *frontier molecular orbital* (FMO) theory. Hard bases and acids have large HOMO–LUMO gaps. As a result, in considering a reference base it is unlikely that there will be much transfer of electron density from the HOMO of a reference base to the LUMO of the acid. Similarly, there will be little transfer of electron density from the HOMO of a hard base to the LUMO of a reference hard acid. Overall, then, there should also be little electron density transfer between hard bases and hard acids and little covalent bonding between these species. However, because of the concentrated negative and positive charges of the hard bases and acids, respectively, an electrostatic interaction would be favorable; the process would be *charge-controlled*. Conversely, the small HOMO–LUMO gaps of the soft acids and bases would enhance HOMO–LUMO electron density transfer and concomitant covalent bonding. Qualitative HOMO–LUMO diagrams to illustrate this concept are shown in Figure 1.1.

In recognizing the link between HOMO–LUMO interactions and soft–soft interactions and the further connection between hard–hard interactions and charge control, it is natural to see a similarity between the qualitative HSAB theory and Drago's E–C scheme,[50–55] which explicitly divides Lewis acid–base adduct formation into covalent and electrostatic components. However, it is important to recognize that the HOMO–LUMO interaction scheme shown in Figure 1.1 is not tied to the Lewis concept of donation of *pairs* of electrons.[56] Again, the HSAB description implicitly recognizes that electrons in an acid–base interaction may be transferred singly, a consequence of utilizing redox potentials in this classification scheme.

Nonetheless, the HSAB modification of the Lewis description of acidity and basicity could be criticized on the grounds that it is essentially qualitative in nature.[76–80] Consequently, the current thrust of work in HSAB theory has been the assignment of *absolute hardnesses*.[76,115,116] In this analysis (supported by density functional theory),[117,118] each chemical species (whether molecule, ion, or

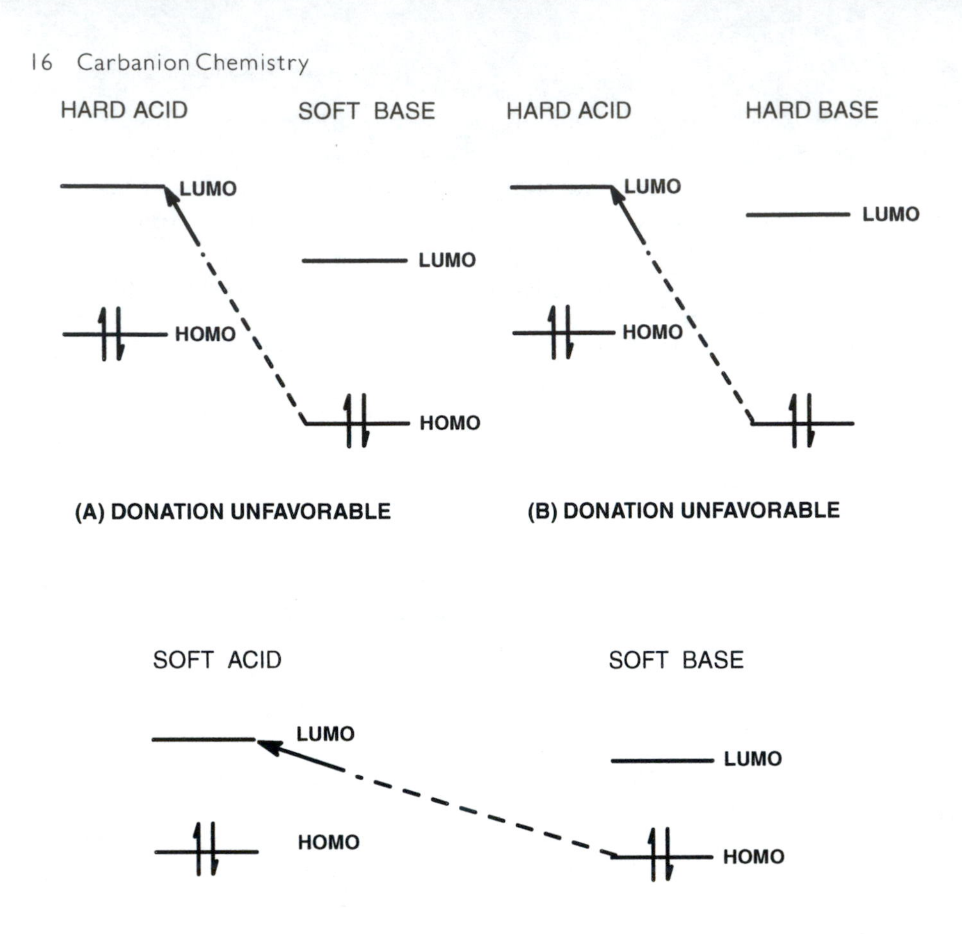

Figure 1.1. Qualitative HOMO–LUMO interactions for (A) a hard acid with a soft base, (B) a hard acid with a hard base, and (C) a soft acid with a soft base.

radical) is associated with an electronic chemical potential, μ, and an absolute hardness, η, defined in equations 1.14 and 1.15:

$$\mu = (\partial E/\partial N)_{\nu} \tag{1.14}$$

$$\eta = 1/2(\partial \mu/\partial N)_{\nu} \tag{1.15}$$

where N refers to the total number of electrons in the species, E is the energy, and ν refers to the potential due to the nuclei plus any external potential. These differential equations can be rewritten in the approximate but more readily accessible forms given as equations 1.16 and 1.17.

$$-\mu = (\mathrm{IP} + \mathrm{EA})/2 = \chi_{\mathrm{M}} \tag{1.16}$$

$$\eta = (\mathrm{IP} - \mathrm{EA})/2 \tag{1.17}$$

where IP and EA represent the ionization potential and electron affinity of the species, respectively, in electron volts (eV). Note that the negative chemical potential is numerically equivalent to the Mulliken electronegativity, χ_M;[119] this has been renamed the *absolute electronegativity*, χ.[120] In solution, IP and EA may be replaced by the *reversible* redox potentials of the chemical species.[121,122] (Also see reference 123 for an alternative definition of absolute hardness.)

As two systems representing a typical Lewis acid, A, and a common Lewis base, B, are brought together, electron density would be expected to flow from the Lewis base, which is characterized by its lower electronegativity, χ, to the Lewis acid, which generally has a higher electronegativity. This flow continues until the chemical potentials equalize, a proposition similar to Sanderson's statement of electronegativity equalization.[124] Thus, within the limits of the approximations used, including the fact that covalency is not considered, the fraction of electrons transferred in the process of Lewis acid–base adduct formation (cf. equation 1.6), is given as equation 1.18:

$$\Delta N = \chi_B - \chi_A / 2(\eta_B + \eta_A) \qquad (1.18)$$

Hence, as described by Liu and Parr[125]: "Hardness measures resistance to the charge flow that a difference in electronegativities commands." It follows that absolute softness, σ, is simply defined as the inverse of the absolute hardness; that is, $\sigma = 1/\eta$.

Unfortunately, the approximate equations 1.16 and 1.17 are not valid for anions,[116] and so the utility of this approach to quantify HSAB theory is impaired, as far as consideration of carbanion chemistry in concerned. For anions, including carbanions, only local hardnesses have thus far been defined.[125] While these local hardnesses may be applicable to the problem of the ambident (O vs C) behavior of enolate and related carbanions (Reaction 1.1), HSAB theory will likely remain a qualitative, although useful, way of examining the reactivity of carbanions.

Although the Lewis and other definitions of acidity may be called upon as appropriate, the majority of this work will rely on the Brønsted view of carbon acids as proton donors.

1.3 THERMODYNAMIC (EQUILIBRIUM) ACIDITY

Thermodynamic acidity is determined by measuring equilibria such as equation 1.1 or an equilibrium related to equation 1.1. In many cases, this measurement must be accomplished indirectly.

It is important to note that the acid dissociation constant, K_a, which is usually calculated from concentrations, should be written using the activities, a, of the species involved. For equation 1.1 the thermodynamic expression for K_a is given by equation 1.19:

$$K_a = (a_{(H^+)} \times a_{(A^-)}) / a_{(HA)} \qquad (1.19)$$

where $a_{(H^+)}$ represents the activity of the proton, and the other activities are similarly denoted by subscripts. The activities are related to the concentrations by a proportionality constant, the activity coefficient, f. Hence, equation 1.19 may be rendered as equation 1.20:

$$K_a = [H^+][A^-](f_{(H^+)} \times f_{(A-)})/[HA]f_{(HA)} \quad (1.20)$$

Upon extrapolation to infinite dilution, the activity coefficients approach unity and equation 1.1 and 1.20 become equivalent. Water (at 298 K) is typically taken as the standard state for acidity measurements, but when measurements are made in nonaqueous solvents the K_a values can theoretically be related if the various activity coefficients are known in the nonaqueous solvent, as shown in equation 1.21.

$$K_{a(H_2O)} = K_{a(solv)}(f_{(H^+)} \times f_{(A^-)}/f_{(HA)}) \quad (1.21)$$

The significance of equation 1.21 will become clear in the following discussion of acidity scales (Section 1.3.1). However, where a pK_a value is estimated solely on the basis of concentrations, the pK_a measured is an "apparent" one rather than a thermodynamic one.

Several methods are used to determine pK_a values of carbon acids. Potentiometry is particularly useful for relatively acidic carbon acids (pK_a of between 4 and 10). In this method, the carbon acid is partially deprotonated by the addition of aqueous NaOH solution; the pH is measured, and, assuming a suitably low concentration so that equation 1.1 applies (i.e., equation 1.20 collapses to equation 1.1 as explained above), the pK_a may be calculated from the Henderson–Hasselbalch equation (equation 1.4). In these titrations, the glass electrode was originally used because it functions reversibly at low concentrations.[126–129] In standard titrations of acids of known pK_a, the accuracy of modern versions of this system is better than 0.1 pK_a unit; higher errors would be expected for carbon acids outside the 4–10 pK_a unit perimeter. (Titrations of carbon acids in nonaqueous media are also frequently used; a study of the acidity of $(CF_3)_3CH$ used titration of the carbon acid in dimethyl sulfoxide with a 0.01 M tetrabutylammonium hydroxide solution in a 1:4 (v/v) mixture of isopropyl alcohol and benzene to determine a pK_a value whose errors fell within 0.02–0.04 pK_a unit per titration.[129]) For equilibria that are achieved relatively rapidly, relaxation techniques, such as pH-jump, can be used to determine the K_a from the forward and reverse rate constants. In this case, reported pK_a values have error limits of ca. 0.05 pK_a unit.[130] The data presented in Table 1.1, therefore, should be reviewed in light of these estimates of uncertainty. Conductance methods that depend on long extrapolations back to infinite conductance, on the other hand, cannot be used for any but the most acidic carbon acids. A "back-titration" method, in which the carbon acid is first largely converted to the sodium salt and is then titrated with acid, successfully uses conductivity as the method of measurement.[131] (The electrochemical approach of Breslow, which relies on a thermodynamic cycle, is discussed in Section 1.3.2.)

Spectrophotometric methods are also widely used. Again with the assumption that the activity coefficients may be neglected, K_a can be calculated from the carbanion/carbon acid concentration ratio. This ratio can be obtained spectrophotometrically if the carbon acid–base pair have sufficiently different absorption spectroscopic characteristics and if the Beer–Lambert law relating absorbance and concentration is obeyed. (Typically, if the solutions are dilute enough to meet the requirement that concentrations can be used without correction, then the solutions are likely dilute enough to permit use of the Beer–Lambert relationship.) Not all acids undergo significant spectroscopic change upon ionization (e.g., the ionization of H_2),[132,133] but an indirect spectrophotometric method can still be used in some of these cases.[132–135] Some specific examples in which the spectrophotometric method has been used will be discussed in the following, as appropriate.

1.3.1 Acidity Scales

The earliest attempts to estimate hydrocarbon acidities[136,137] involved equilibration of the weak carbon acid with the potassium or sodium salt of another carbon acid whose pK_a was known:

$$\text{R–H} + (\text{R}')^-\text{K}^+ \rightleftharpoons (\text{R})^-\text{K}^+ + \text{R}'\text{–H} \tag{1.22}$$

The pK_a of the unknown carbon acid is then obtained by difference, according to equation 1.23.

$$pK_a(\text{R}-\text{H}) - pK_a(\text{R}'-\text{H}) = \log\left\{\frac{[\text{R}-\text{H}][(\text{R}')^-\text{K}^+]}{[\text{R}'-\text{H}][(\text{R})^-\text{K}^+]}\right\} \tag{1.23}$$

In this way a series of carbon acid pK_a values can be built up, one at a time, by comparison with a carbon acid of known pK_a. Once a series of "indicator hydrocarbons" has been established, other pK_a values can be determined by interpolation between these indicators. To anchor the series to water, the initial equilibration would use a carbon acid whose aqueous pK_a had been determined (or estimated) by other means.

Experimentally, the concentrations of the species involved were determined spectrophotometrically or, alternatively, by quenching the carbanions with carbon dioxide and subsequent isolation of the corresponding carboxylic acids.

These studies were undertaken in ether or benzene as solvent. In these media of low dielectric constant, the "salts" would be expected to exist as ion pairs. It is relevant to note here that organometals such as *n*-butyllithium form polymeric aggregates; consequently, ether or benzene solutions of such compounds are nonconducting.[138,139] There is considerable current effort being put into assessing the type and degree of aggregation of organolithium compounds including enolates.[140–147] While varying degrees of aggregation have been found, including formation of dimers and tetramers, the nature of the reactive species is still uncertain and may vary from system to system. For example, the methylation of lithium isobutyrophenone (the lithium salt of 2-methyl-1-phenylpropanone) was believed to involve only the aggregates observed by NMR spectroscopy,[145,146] but recent evidence presented by Streitwieser demonstrates that alkylation of the related lithium 1-(4-biphenylyl)-2-methyl-1-propanone enolate proceeds through the simple monomeric ion pair in THF.[147] Further, it has been shown that addition of lithium salts, such as lithium perchlorate, that can enhance the rate of reaction of enolates,[148] as well as affect selectivity and solubility in chiral alkylation and aldol reactions,[149,150] causes formation of mixed aggregates in ethereal solvents (e.g., diethyl ether, THF, etc.) with organolithiums and lithium dialkylamide salts.[151–153] Both experimental and calculational results indicate that the structure of these mixed aggregates influences the degree of stereochemical induction in enolate reactions;[154] no doubt, definitive analysis of the factors that affect the degree of aggregation and the nature of the reactive species as a function of solvent, added salts, and carbanion structure will emerge from the current application of multinuclear NMR spectroscopic techniques,[155,156] X-ray structural determination,[157,158] and theoretical calculations.[159,160]

Clearly, the variable degree of ion pairing and aggregation makes comparison of the pK_a values of carbon acids determined in ethereal solvents with those measured in water a procedure that should be approached with caution. Although the organometallic compounds (or metal carbanides) formed from delocalized carbanions such as fluorenyl- or polyarylalkylmetallics may approach true salts in the sense that the carbon–metal bond is significantly ionic, such salts are essentially undissociated in ether or benzene. In these low dielectric solvents, all of the "salts" are likely to exist as ion pairs[161,162] to the extent that the salts dissociate at all. The consequence of this for acidities determined in low-polarity solvents is that the acidities found by the McEwen[137] and like methods must be considered as "ion-pair acidities" that only approximate true acidities if the ion-pair dissociation constants of the two organometallics $(R)^-K^+$ and $(R')^-K^+$ (equation 1.22) are approximately equivalent. This amounts to a requirement that RH and R′H be structurally similar.

Streitwieser and his collaborators[163–167] have done extensive work on the acidity of carbon acids in cyclohexylamine ($C_6H_{11}NH_2$, CHA) as the solvent. In this solvent, ion-pair dissociation constants for lithium salts are low; for fluorenyllithium (cf. Chart 1.1) in CHA at room temperature the ion pairs are described as solvent separated $(R^- \| L^+)$.[168] The Streitwieser group has also studied structurally similar hydrocarbons, most of which would yield charge-delocalized carbanions as the conjugate bases, using the larger (softer) cesium cation. Fluorenylcesium in cyclohexylamine is present almost exclusively as contact or tight ion pairs (R^-,Cs^+). These observations concerning the variation in types of ion pairs with the counterion in cyclohexylamine are consistent with the HSAB principle. Consequently, equations 1.24 and 1.25 will be applicable for measurements made with cesium cyclohexylamide and lithium cyclohexylamide, respectively:

$$R'\text{–}H + Cs^+CHA^- \rightleftharpoons (R')^-, Cs^+ + CHA \tag{1.24}$$

$$R'\text{–}H + Li^+CHA^- \rightleftharpoons (R')^- \| Li^+ + CHA \tag{1.25}$$

Addition of a second hydrocarbon of known pK_a, R″H, to the system would establish an equilibrium equivalent to that of equation 1.22. Because the carbanions formed were delocalized in the Streitwieser systems, significant spectral changes attend their formation and the equilibrium concentrations could be determined spectrophotometrically. Calculation of the unknown pK_a value is then accomplished using the corresponding equation related to equation 1.23.

Since there is greater charge separation in the $R^- \| Li^+$ species of equation 1.25 than in the R^-,Cs^+ species of equation 1.24, the extent of ionization of the carbon acid will generally be smaller for the LiCHA case. Thus, triphenylmethane is known to be more acidic toward CsCHA than it is toward LiCHA. It may be concluded that the CsCHA system provides a better measure of carbon acidity than the LiCHA system (though both systems afford internal self-consistency). Cesium carbanide ion pairs should be valid models for carbanions and the nature of the ion pair should remain unchanged within a series of structurally related compounds. On the other hand, the use of organolithium reagents in synthesis in THF solvent argues in favor of the use of the lithium ion-pair scale in THF as a guide to carbanion acidity in these systems and the nature of aggregation in this solvent (see below). Thus, the pK_a of many weak carbon acids that were not accessible in water or ether could be measured in CHA because the pK_a of CHA is about 42. Much of this work, including ion-pair pK_a values for numerous carbon acids, has been previously compiled and reviewed by Streitwieser[169] (also see references 21 and 135). Selected acidities determined using CsCHA in CHA (i.e., p$K_{a(CsCHA–CHA)}$), and statistically corrected for numbers of acidic protons, are given in Table 1.4.

The pK_a values determined in CHA were originally linked to water, as the standard state, through cyclopentadiene, whose pK_a (= 16)[170] could be measured by the acidity function method and, so, ultimately anchored to water. However, this was later modified to 9-phenylfluorene with a reference value of 18.49.[169]

It is pertinent to note that CsCHA can cause a double deprotonation in certain cases in which both negative charges can be delocalized. An example is provided by 9-benzylfluorene (Scheme 1.3).[171]

In examining Table 1.4 several conclusions can be made. Thus, triphenylmethane is more acidic than diphenylmethane, which, in turn, is more acidic than the estimated pK_a of toluene indicates. Just as we previously noted the enhancement of acidity that accompanies substitution of a methane by electron-withdrawing groups such as nitro, cyano, and alkanoyl groups (Table 1.1) and further noted that acidity increases with the number of these substituents on the methyl carbon, it is also apparent that increasing phenyl substitution increases the carbon acidity. Careful examination of both types of substitution will show that the increase in acidity that accompanies attachment of each acidifying group is not monotonic. Thus, addition of a single phenyl substituent to toluene decreases the pK_a value by almost 9 units (diphenylmethane compared with toluene), while addition of a further phenyl ring only lowers the pK_a by another 2 units (triphenylmethane compared with diphenylmethane). This "saturation

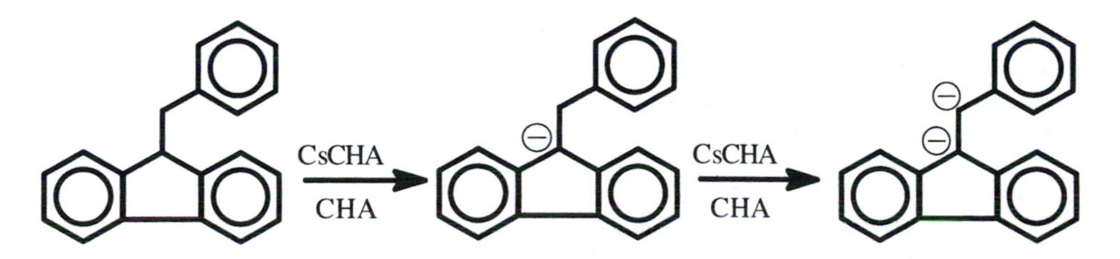

SCHEME 1.3

effect" is usually considered to arise from steric inhibition of resonance in the resultant carbanions. Cross-comparisons between Tables 1.1 and 1.4, of course, are hampered by the differences in solvent and ion pairing. The greater efficacy of fluorine relative to chlorine in augmenting the acidity of the benzene ring protons is apparent in the pK_a difference for pentafluorobenzene as compared with pentachlorobenzene, that is, $\Delta pK_a = 4.1$ favoring pentafluorobenzene and, at first glance, seems explicable solely on the basis of the higher electronegativity of fluorine as compared with chlorine. However, the effect of fluorine on aliphatic carbanions, particularly for α- and β-F-substituted carbanions, is clearly different and depends not only upon the high electronegativity of fluorine but also upon the relative tendency of fluorine and chlorine to support hyperconjugation in the β-substituted carbanions,[129] as well as π-backbonding;[172–174] these conflicting factors will be explored further in Chapter 2. Groups that contain third-period elements also increase the acidity of the C–H group to which they are attached, as evidenced partly from the acidity of phenyl methyl sulfone and benzyl diphenyl phosphine oxide.[175–178] (In this regard, the ^{13}C NMR method of determining pK_a values for lithium carbanides in THF, developed by the Fraser group,[179–183] also shows that α-silicon and -tin substituted carbon acids have virtually the same acid-enhancing effects, but both are less acidifying than an α-thio substituent.[180,181])

Streitwieser's group has determined pK_a values in tetrahydrofuran (THF). The cesium ion-pair acidities previously reported in THF using fluorene as the anchor,[184–186] with a pK_a of 22.9 in dimethyl sulfoxide (DMSO),[41] have been superseded by a cesium ion scale in THF that uses more hydrocarbon indicators,[187] and which has even more recently been corrected for impurities in three of the indicators.[188] The correspondence between the two THF scales is reasonable. Acidities for the more acidic carbon acids (pK_a values of 25 or less) were adjusted by between 0 and 0.1 pK_a units in going to the new scale, while the less acidic carbon acids had values adjusted by 0.2–0.3 pK_a units. More striking was the

Table 1.4. Selected carbon acidities determined in cyclohexylamine by the cesium cyclohexylamide equilibrium method (equation 1.25)

Compound	pK_a	Compound	pK_a
Toluene	(42.2)[a,b]	Phenyl ethyl sulfone	29.3[e]
1,3,5-Trichlorobenzene	33.6[c]	Phenyl methyl sulfone	29.1[e]
o-Benzylbiphenyl	33.51[b]	Pentafluorobenzene	25.8[f]
Diphenylmethane	33.38[d]	*t*-Butylacetylene	25.52[g]
p-Benzylbiphenyl	31.82[e]	Benzyl diphenyl phosphine oxide	24.89[e]
Triphenylmethane	31.45[d]	Phenylacetylene	23.24[g]
Pentachlorobenzene	29.9[f]	Benzyl methyl sulfone	22.6[e]
Phenyl isopropyl sulfone	29.3[e]	Tris(pentafluorophenyl)methane	21.43[h]

[a]Value obtained by extrapolation.
SOURCES: [b]Data taken from reference 175; [c]Data taken from reference 177; [d]Data taken from reference 166; [e]Data taken from reference 178; [f]Data taken from reference 169; [g]Data taken from reference 165; [h]Data taken from reference 164.

excellent linear correlation[186] between the acidity scale determined in cyclohexylamine with cesium cyclohexylamide (Table 1.4) versus THF; the regression equation ($r^2 = 0.999$) is given as equation 1.26.

$$pK_{a(CsCHA-CHA)} = 1.0084pK_{a(Cs-THF)} - 0.13 \quad (1.26)$$

The correspondence suggests that the variation of dissociation constants for the cesium carbanide ion pairs as a function of structure is essentially the same in both solvent systems. However, there is no doubt that this agreement reflects the charge delocalized nature of the carbon acid indicators used in the establishment of both acidity scales.

Selected cesium ion-pair acidities[187] (on a per proton basis), according to the modified scale in tetrahydrofuran, are listed in Table 1.5. Structures of the 9-substituted fluorenes are shown in Chart 1.3.

Cursory comparison of Tables 1.4 and 1.5 bears out the correlation between the two scales reported. In this regard, the differences in acidity between toluene and diphenylmethane and those between diphenylmethane and triphenylmethane from Table 1.4 are 8.8 and 1.9, respectively; however, from Table 1.5 the same changes in pK_a become 7.7 and 2.0, respectively. The main change is in the value of toluene, a value that is derived from extrapolation of a Brønsted plot (see Section 1.4.) and, consequently, subject to greater error than the pK_a values measured directly.

While it is clear that in some systems the various kinds of ion pairs and higher-order aggregates can be differentiated,[144,147,188] it is less clear whether these results will require the introduction of different ion-pair acidity scales dependent upon the degree of aggregation, as well as on the solvent. Examination of the lithium and cesium ion-pair acidities[189] of a set of acetylenes (4-ethynylbiphenyl (4-biphenylylacetylene, **3**), 3,3,3-triphenylpropyne, **4**, and 1-ethynyladamantane, **5**) yielded the order of pK_a values and average aggregation number for ion pairs as shown in Chart 1.4.

Table 1.5. Selected carbon acidities determined in tetrahydrofuran by the cesium fluorenide equilibrium method

Compound	pK_a	Compound	pK_a
9-(4-Biphenylyl)fluorene	17.72	Triphenylmethane	31.26
9-Phenylfluorene	18.15	*p*-Benzylbiphenyl	31.70
9-(4-Dimethylaminophenyl) fluorene	19.23	Diphenylmethane	33.25
9-Benzylfluorene	21.30	Benzyl methyl sulfide	34.68
9-Methylfluorene	22.32	Bis(2,4-dimethylphenyl)methane	35.96
Fluorene	(22.90)[a]	Phenyl *p*-tolyl sulfide	37.69
9-*tert*-Butylfluorene	24.39	Toluene	(40.9)[b]

SOURCE: Data from reference 187.
[a]Tethered to the DMSO scale; pK_a value set to 22.90.
[b]Value obtained by extrapolation.

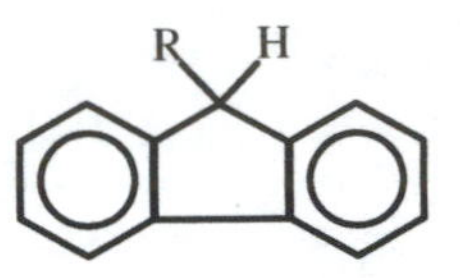

$R = H$ = fluorene $R = t\text{-}C_4H_9$ = 9-*t*-butylfluorene
$R = CH_3$ = 9-methylfluorene $R = C_7H_7$ = 9-benzylfluorene
$R = C_6H_5$ = 9-phenylfluorene $R = C_6H_5{-}C_6H_4$ = 9-(4-biphenylyl)fluorene

CHART 1.3

In this case the pK_a values were rationalized on the basis that with these essentially localized carbanions, the ion pairs, whether with lithium or cesium, were all contact ion pairs. The results show that aggregation, as well as solvent and counterion, are important in assessing carbanion acidity. Comparisons with acidities determined in DMSO[41] also suggest that, at least in some systems, ion pairing is also significant in DMSO.

Now that the various Streitwieser scales are tethered to the pK_a of fluorene in DMSO rather than to an aqueous pK_a value, and noting the extent of the set of carbon acid acidities measured thus far in DMSO,[41,190] it could be argued that DMSO will supplant water as the common standard state for acidity, at least for carbon acids. In this context, it should be noted that, unlike cyclohexylamine ($\epsilon = 4.7$),[191] DMSO is a relatively high dielectric constant ($\epsilon = 47$)[191] solvent in which ion pairing would be less likely *for sufficiently dilute solution* than is the case[189] in THF ($\epsilon = 7.6$).[191] On the other hand, the autoprotolysis constant of DMSO (35.1) imposes a stricter limitation of the pK_a values of carbon acids accessible in this solvent as compared with CHA (41), although the range of measurable pK_a values is clearly much greater than that possible in water. Obviously, the body of work done by the Bordwell group[41,169,190] serves as an important database with which to make structure–acidity (and structure–reactivity) generalizations.

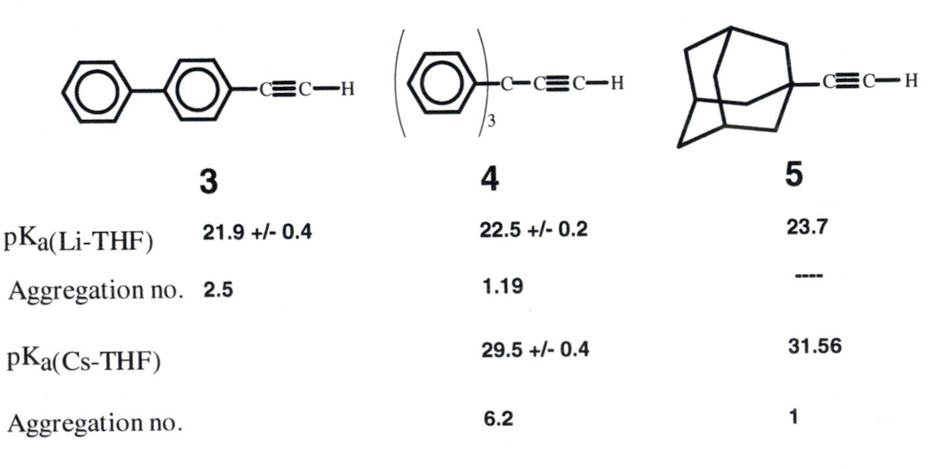

CHART 1.4

The Bordwell approach rests on the titration of the weak carbon acid with the dimsyl anion ($CH_3SOCH_2^-$, Chart 1.2) in the presence of a suitable indicator.[192,193] The indicators commonly used are substituted anilines that have overlapping pK_a values. The actual pK_a values of these indicators (and their purity) are then significant in assessing the pK_a of a given unknown carbon acid,[194] as we have seen above in the realignment of the cesium ion-pair acidity scale in THF.[186,187] As with the scales defined by Streitwieser for CHA and THF solvent media, the pK_a of an unknown hydrocarbon in DMSO can be determined by equilibration of the unknown carbon acid with two or more secondary potassium carbanide indicators, whose acidities would be expected to be close to that of the unknown acid.

Many of the carbon acid acidities in DMSO (hereafter $pK_{a(DMSO)}$) have been tabulated and critically reviewed,[41,190,169] and, consequently, Tables 1.6 and 1.7 contain only a limited selection of $pK_{a(DMSO)}$ values; we will discuss limited sets of these acidities to determine what structural features enlarge or diminish acidity. These data and the relationships between the various acidity scales and gas-phase acidity will be discussed further in later sections.

It is useful to reiterate that DMSO pK_a values are absolute values equivalent to pK_a values in water in that each set of acidities defines its standard state internally: the DMSO set to DMSO and the aqueous set to water. However, conversion between the scales (cf. equation 1.21) is problematic. Normally, acidity is modified nonlinearly with solvent; conversion of the pK_a in one solvent into a pK_a in another solvent is not simply a matter of multiplying through by a constant factor. Thus, picric acid (2,4,6-trinitrophenol) has virtually the same acidity in water or DMSO,[195] a result of the fact that the negative charge in the aryloxide

Table 1.6. Acidities in DMSO ($pK_{a(DMSO)}$, 298 K) of selected lactones, esters, and diesters

Compound	Class	$pK_{a(DMSO)}$
Lactones		
Meldrum's acid		7.32[a]
Methyl Meldrum's acid		7.42[a]
Ethyl Meldrum's acid		7.52[a]
Esters and diesters		
Ethyl 1-phenylacetate		22.6[b]
Ethyl acetate		27.45[a]
Dimethyl malonate (dimethyl 1,3-propanedioate)		15.87[a]
Diethyl malonate		16.37[b]
Di-*tert*-butyl malonate		18.4[b]
Diethyl ethylmalonate (diethyl 2-ethyl-1,3-propanedioate)		19.1[c]
Diethyl methylmalonate		18.7[c]
Dimethyl methylmalonate		18.04[c]
Diethyl phenylmalonate		16.28[d]

SOURCES: [a]Data taken from reference 195; [b]Data taken from reference 41; [c]Data taken from reference 196; [d]Data taken from reference 199.

Table 1.7. Acidities in DMSO ($pK_{a(DMSO)}$ at 298 K) of selected ketones, alkanediones, cycloalkanones, and cycloalkanediones

Compound	$pK_{a(DMSO)}$	Compound	$pK_{a(DMSO)}$
Ketones and alkanediones		*Cycloalkanones and cycloalkanediones*	
Dibenzyl ketone	18.7[a]	Cyclobutanone	25.05[a]
Phenylacetone	19.9[a]	Cyclopentanone	25.8[a]
Methyl ethyl ketone (butanone)	24.4[a]	3,3-Dimethylcyclohexanone	25.84[a]
Acetophenone	24.72[b]	Cyclohexanone	26.26[c]
p-Methoxyacetophenone	25.66[b]	Cyclodecanone	26.8[a]
o-Methoxyacetophenone	25.66[b]	1,3-Cyclohexanedione	10.3[c]
Acetone	26.5[a]	Dimedone	11.24[c]
Acetylacetone (2,4-pentanedione)	13.32[c]		
Dibenzoylmethane	13.36[b]		
3-Methyl-2,4-pentanedione	15.05[a]		

SOURCES: [a]Data taken from reference 41; [b]Data taken from reference 199; [c]Data taken from reference 195.

conjugate base is highly delocalized into the aromatic ring and onto the nitro groups and, so, the stability of the anion is relatively insensitive to the nature of the solvent. Similarly, we have already seen that acidity scales designed for THF or CHA media correlate well with the $pK_{a(DMSO)}$ scale, where these scales are all defined using carbon acids that yield charge-delocalized (and, therefore, stabilized) anions. On the other hand, it can be expected that the acidity of hydrocarbons that give localized carbanions will differ significantly from scale to scale and will be affected more by the degree of ion pairing or aggregation found in one medium as compared with those in another. Thus, nitromethane, whose anion would be expected to be delocalized, has a pK_a of 10.2 in water[196] while its value in DMSO is 17.2.[197] Here, of course, hydrogen-bonding of water to the nitro group will assist stabilization in water relative to DMSO.

From the viewpoint of synthetic utility, it is possible to take advantage of the differences in pK_a caused by the change in solvent. In this regard, in a preparation of a series of alkynols by reaction of a range of acetylide anions with a number of ketones, the acetylide anions were generated by reaction of the parent terminal alkyne with potassium *tert*-butoxide in DMSO; the pK_a of *t*-butyl alcohol *in DMSO* (32.2) was found to be greater than that of a typical terminal alkyne, such as phenylacetylene (29.7), unlike the situation in water.[198] Therefore, the acetylide anions could be prepared in higher concentration in DMSO with *tert*-butoxide than they could be in *t*-butyl alcohol.

One synthetically important class of carbanions is that formed from carbon acids in which the acidic hydrogen is attached to a carbon that is α to at least one carbonyl group. The structures of some of the more esoteric, though mechanistically important, of these carbon acids are given in Chart 1.5 and the acidic protons are highlighted. The $pK_{a(DMSO)}$ values for the carbonyl compounds shown in Chart 1.5, as well as a number of common esters and diesters, are given in Table 1.6.

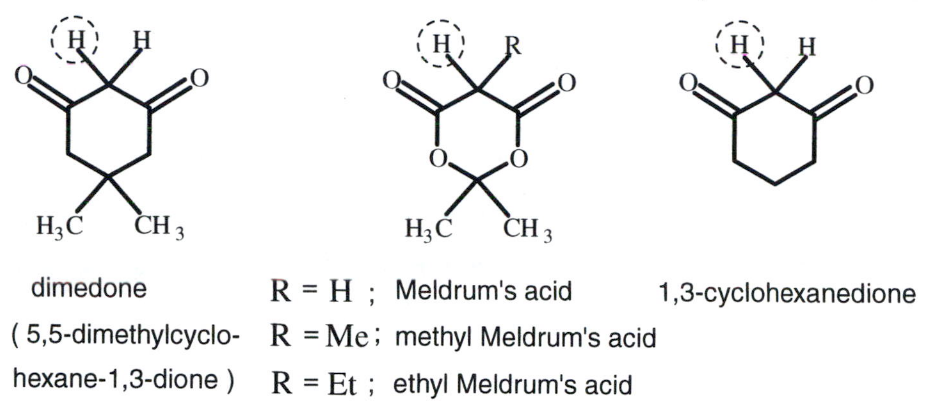

CHART 1.5

The acidifying influence of two α-carbonyl groups can be seen in the low pK_a of Meldrum's acid (7.32). Comparison can also be made between diethyl malonate (pK_a = 16.37) and its monoester analogue, ethyl acetate (pK_a = 27.45); a second α-ester group increases the acidity in this case by ca. 11 orders of magnitude. The high acidity of malonic acid esters[197,199] lays the foundation of the malonic ester synthesis[200] in which, in the presence of excess alkylating agent, the ester is dialkylated. Note that while monoalkylated malonic ester (e.g., diethyl methylmalonate) is less acidic than its precursor malonic ester, the monoalkylated esters are still about 8 orders of magnitude more acidic than ethyl acetate and, therefore, can be further alkylated in the synthesis.

Ketones and similar carbonyl compounds are also synthetically useful in aldol and related condensation reactions. The acidities of these compounds, determined in DMSO, are listed in Table 1.7.

Again, the enhancement in acidity that accompanies bis-carbonyl substitution of a carbon acid site can be seen in the comparison of the pK_a values for acetone and acetylacetone (2,4-pentanedione); the pK_a difference ($\Delta\Delta$pK_a) between the two carbon acids is 13.2; the acidity of acetylacetone increases by 13 orders of magnitude relative to acetone. The addition of a benzoyl group to acetophenone also enhances the acidity of the carbon acid. However, reiterating the importance of steric congestion on *α*-group substituent effects, note for comparison that dibenzoylmethane is only about 11 pK_a units more acidic than acetophenone.

Note that comparisons between the current DMSO data for the dicarbonyl compounds (Tables 1.6 and 1.7) and data obtained in THF or water would be hampered by the varying effect of metal ion chelation in these systems. Thus, the metal counterion of a dicarbonyl carbanion is likely ion paired to both oxyanionic centers[196] and the degree of this ion pairing would presumably be solvent dependent, though always of a higher degree for the dicarbonyl enolates than for less acidic and more conventional enolates.

Comment may also be made about the effect of benzene ring substituents; both *para* and *ortho* methoxy-substituted acetophenones are less acidic than the parent acetophenone, an observation consistent with standard physical organic analysis of the effect on a center of remote methoxy groups as electron-donating substituents.[56] In fact, Bordwell's group has established good-to-excellent Hammett substituent (σ) constant[201] correlations with more than 20 families of aryl-substituted or aromatic compounds.[41,190] Such correlations permit interpolation of many acidities for compounds that have not been directly measured. In the case of substituted toluenes, electron-withdrawing groups tend to increase the acidity of toluene substantially. Thus, the $pK_{a(DMSO)}$ values for toluenes[202,203] substituted with electron-withdrawing groups in the *para* position decrease in the following order:

	4-H	>	4-CN	>	4-$PhSO_2$	>	4-PhCO	>	4-SO_2CF_3	>	4-NO_2
pK_a	43		30.8		29.8		26.9		24.0		20.4

Again, however, the powerful electron-withdrawing ability of the nitro group, whether remote as in the case of the substituted toluenes or α to the carbon acid center as in the nitromethane series (Table 1.1), is manifest. This acidity-enhancing effect of nitro groups was again confirmed in a study by Bordwell and Zhao[204] of a series of nitromethane derivatives, where 1-nitro-1-(3,5-dinitrophenyl)ethane was found to have a $pK_{a(DMSO)}$ of 9.9 as compared with the 1-(4-nitrophenyl) and the parent 1-phenyl 1-nitroethane derivatives that had pK_a values of 10.3 and 14.0, respectively.

Third-period and later-period elements can also amplify the acidity of carbon acids, particularly when these atoms are α to the acidic C–H center. Thus, replacement of one of the C-9 protons of fluorene (Chart 1.3) with a thiomethyl group (SMe) increases the acidity by almost 5 $pK_{a(DMSO)}$ units; that is, $pK_{a(DMSO)}$fluorene–H − $pK_{a(DMSO)}$fluorene-9–SCH_3 = 22.9 − 18.0 = 4.9.[41] Other thioalkyl groups have a similar effect.[179–183] The thiophenyl group enhances fluorene acidity by 7.5 pK_a units. Increasing the oxidation number of the sulfur center increases the acidifying ability of the group attached at the 9-position; 9-phenylsulfonylfluorene is more acidic than 9-thiophenylfluorene by ca. 4 $pK_{a(DMSO)}$ units.[41]

The thiophenyl group (PhS) is apparently more effective at augmenting the acidity of an adjacent CH_2 group in 1-thiophenylacetophenone than is the corresponding selenophenyl group (PhSe) in 1-selenophenylacetophenone; the pK_a in DMSO of the former is 17.1 while that of the latter is 18.6.[205] This result was interpreted as indicating that stabilization of the respective carbanions requires a conjugative interaction with the thio or seleno group and not just polarization.[206] The contentious mode of operation of such substituents will be explored further in this book (Chapter 2).

An alternative to the relative approach of McEwen and Streitwieser (as far as it is linked to a known pK_a in water or DMSO) and the absolute method of Bordwell (which uses DMSO at 298 K as its standard state) is the *acidity function* (AF) method. This method, inaugurated by Hammett and Deyrup[207] for the determination of the acidity of very strong acids, has been reviewed by Rochester,[208] and by Buncel and Wilson[89] for the determination of the acidity of weak acids in mixed water–aprotic polar solvent media.

In brief, for weak acids, including most carbon acids, the basicity (i.e., Brønsted proton abstraction ability) of a very basic medium must be measured.

A series of acids is chosen whose UV–vis spectra differ significantly upon deprotonation; hence, the concentration of the acid (HA) and its conjugate base (A^-) can be monitored spectrophotometrically. (An equivalent NMR method has also come into use.[179–183,209]) The initial acid of the series must dissociate appreciably in basic water. The next acid may dissociate to some degree in pure water but must dissociate to a greater degree in a medium that contains DMSO, for example, and added tetramethylammonium hydroxide. Each successively weaker acid would dissociate with increasing amounts of the aprotic polar solvent. Thus, in practice, one measures spectrophotometrically the extent of ionization of the acid under examination in media of constant [OH^-] but as a function of an increasing proportion of the aprotic polar component.

Each dissociation could be described by equation 1.20; the ratio of the acid dissociation constant determined in water to the constant determined in the mixed solvent medium, $K_a(HA_1)/K_a(HA_2)$, is given by equation 1.27.

$$\frac{K_a(HA_1)}{K_a(HA_2)} = \frac{[HA_1][A_2^-]}{[HA_2][A_1^-]} \times \frac{f(HA_1)f(A_2^-)}{f(HA_2)f(A_1^-)} \tag{1.27}$$

Because the indicator ratio, $[HA_1]/[A_1^-]$, as well as $K_a(HA_1)$, determined in water, are known, only the ratio of the activity coefficients is required to evaluate $K_a(HA_2)$. The concentrations of HA_2 and A_2^- can be measured by spectroscopic means, of course. If the structure of the indicator acid is sufficiently similar to the structure of the unknown acid, and if the reaction involves the same charge types as shown in equation 1.27, then $f(HA_1)/f(A_1^-)$ would equal the activity coefficient ratio for HA_2/A_2^- and the overall ratio of the activity coefficients in equation 1.27 would be equal to unity. If the logarithm of each side of equation 1.27 is taken, a procedure suggested by the magnitudes of the quantities usually involved, the activity coefficient term becomes nil and the equation takes the form of equation 1.28.

$$pK_a(HA_1) - pK_a(HA_2) = \log\left\{\frac{[HA_1][A_1^-]}{[HA_2][A_2^-]}\right\} \tag{1.28}$$

Note the similarity between equation 1.28 and equation 1.23.

Having determined the pK_a of the unknown acid, HA_2, from equations 1.27 and 1.28, the pK_a of the next indicator acid may be determined in the same medium and in a medium that contains more DMSO until the pK_a values of a range of acids have been determined in a range of aqueous-DMSO systems. These pK_a values are anchored to water as the standard state and can be used to define the basicity of the medium in a scale labeled H_- according to equation 1.29.[210]

$$H_- = pK_a - \log[HA]/[A^-] \tag{1.29}$$

Note the similarity between equations 1.29 and 1.4; in pure water the acidity function H_- is replaced by pH.

Acidity functions permit extrapolation of pK_a from pure water to a very basic medium, in this case a DMSO-rich medium. The key to this mixed solvent approach is that when DMSO or a comparable aprotic polar solvent is added to an aqueous or alcoholic solution containing hydroxide ion (or alkoxide ion, as required), the reactivity of the base in proton abstraction processes is greatly enhanced.[89,211–214] Dimethyl sulfoxide, unlike water, is capable of forming only weak hydrogen-bonds with hydroxide ion, and as a result the thermodynamic activity of the latter increases tremendously as the proportion of DMSO is continuously increased. A further consideration is the effect of adding increasingly larger amounts of DMSO on the structure of what was initially a highly ordered water solution.[214] Thus, a 50% DMSO solution containing 0.01 M hydroxide has a basicity that is 10^5 times greater than that of a pure aqueous medium.

The H_- scale outlined above was defined for mixtures of water and DMSO containing 0.011 M tetramethylammonium hydroxide. However, H_- AF scales may similarly be defined for tetramethylsulfone (sulfolane)[215] and dimethylformamide (DMF).[216,217] Selected H_- data for aqueous sulfolane, DMSO, and DMF are listed in Table 1.8.

The utility of AF scales such as H_- lies in the ability to evaluate the pK_a values of unknown weak acids and link these acidities to water as the standard state, so that rigorous comparisons can be made between structural effects on acidity in water (Table 1.1) and in binary aqueous media. With the H_- value defined for a given solvent composition, and the ratio of $[HA_{unknown}]$ to $[A^-_{unknown}]$ for the unknown acid calculated from the UV–vis or other spectroscopic data, the pK_a of the unknown acid can be estimated.

The key assumption in this process is that the activity coefficient ratios remain close to unity. For this assumption to be valid, the structures of the unknown acids should be similar to those of the indicator acids used to construct the acidity function. In the case of H_- the indicators used are substituted anilines. Generally, the requirement for structural similarity is met only approximately[207] and, although modifications have been introduced,[219–224] AF methods remain approximate ones, albeit AF approaches do link the acidity of weak carbon acids to water as the standard state.

The acidities of some substituted fluorenes are listed[210] in Table 1.9; these values were obtained using the H_- AF method. Comparison can be made between

Table 1.8. Selected H_- data for aqueous binary mixtures with several aprotic polar solvents (0.011 M tetramethylammonium hydroxide)

Mol % aprotic polar component	H_- values		
	Sulfolane	DMF	DMSO
20	13.22	14.20	14.48
40	14.25	15.75	16.50
60	15.56	17.34	18.50
80.78			20.68
90.07			21.98
99.59			26.59

SOURCE: Data taken from references 215–218.

Table 1.9. Acidity of selected substituted fluorenes as derived from the H_ acidity function for aqueous-DMSO binary mixtures

Fluorene	pK_a	Fluorene	pK_a
2-nitro	17.96	9-(4-biphenylyl)	18.21
2-cyano	18.96	9-(4-methoxyphenyl)	19.01
2-bromo	20.56	9-(4-dimethylaminophenyl)	19.61
2-chloro	20.59	9-benzyl	21.20
2-methoxy	22.36	9-methyl	21.80
9-(3-chlorophenyl)	17.66	9-ethyl	22.20
9-(3-trifluorophenyl)	17.69	9-isopropyl	22.70
9-(4-chlorophenyl)	18.10	9-*tert*-butyl	23.41

SOURCE: Data taken from reference 210.

the acidities given in Table 1.9 and those in Table 1.5 (cesium ion-pair acidities in THF). For example, 9-benzylfluorene has a pK_a of 21.20 according to the H_ AF and a pK_a of 21.30 as determined by the cesium ion-pair method in THF (standard state linked to DMSO; see above). This merely re-emphasizes the fact that acidities of carbon acids that yield highly delocalized carbanions are found to be similar no matter what the method of measurement.

1.3.2 Heats of Protonation

Equilibrium measurements are of value because they can be related directly to the thermodynamics involved in the deprotonation. Thus, the Gibbs free energy, ΔG_a, where the subscript "a" indicates acid dissociation, is related to the equilibrium constant for acid dissociation, K_a, in accord with equation 1.30.

$$\Delta G_a = -RT \ln K_a \tag{1.30}$$

Naturally, equation 1.30 applies for thermodynamically valid acid dissociation constants.

To the extent that both the entropy change and enthalpy change associated with acid dissociation, ΔS_a and ΔH_a, respectively, vary linearly with ΔG_a, or that ΔS_a may be small enough to neglect, the Gibbs free energy change of acid dissociation, ΔG_a, will correspond linearly with the heat of acid dissociation, ΔH_a (i.e., the heat of protonation). This linear correlation has been shown by Arnett[225] for 30 weak acids in DMSO over a range of 20 pK_a units. In this case, the relative order of the heat of protonation values for the weak acids would be a guide to their thermodynamic acidity. The method of determining heats of protonation is titration with potassium dimsyl in DMSO, with the heat of neutralization indicative of the end point.

This method has the advantage of utilizing straightforward calorimetric measurement but, in common with the Bordwell treatment, the range of pK_a values that are accessible is, of course, limited by the acidity of DMSO itself. In this context, it should be noted that heats of protonation have more recently been determined in THF solvent[199] [the strong base lithium bis(trimethylsilyl)amide was used in these calorimetric titrations] and the heats of protonation were found to be insensitive to the degree of aggregation, as measured by vapor-phase osmometry. Further, the heats of protonation in THF were found to correlate with the pK_a values for the same carbon acids, determined by the Bordwell method in DMSO. Thus, for diethyl malonate (p$K_{a(DMSO)}$ = 16.28) the heat of protonation was found to be exothermic (-25.9 ± 0.3 kcal/mol), while the more acidic dibenzoylmethane (p$K_{a(DMSO)}$ = 13.36) was found to be more highly exothermic (with ΔH_a of -28.9 ± 0.1 kcal/mol).[199]

On the other hand, a method that has the possibility of determining a very large range of acidities has been developed by Breslow and co-workers.[226–230] The method relies on a thermodynamic cycle that incorporates bond dissociation enthalpy (BDE) data. To determine the free energy for dissociation of a hydrocarbon, R–H, the free energy associated with the electrochemical reduction of an alkyl halide, R–X, is measured. This reduction step, however, can be resolved into the BDE for homolytic dissociation of the hydrocarbon and the energy for reduction of the radical, $R^{\cdot}$, to the carbanion (which takes place spontaneously at the electrode surface when the alkyl halide is reduced electrochemically). A further step is required—the reduction of hydrogen atom to the proton. This can be determined for triphenylmethane, whose radical is relatively stable and persistent in solution, by measuring the electrochemical reduction of this radical followed by combination of this energy with the known (accurate) pK_a of triphenylmethane. Thus, the scale is tied to the value chosen for the pK_a of triphenylmethane. Finally, with this information for triphenylmethane, Ph_3CH, the energy of the ($R^{\cdot} + H^{\cdot}$) to ($R^- + H^+$) step can be estimated relative to Ph_3CH, whose BDE is also well known.

Selected pK_a values determined by thermodynamic cycle method[230] are given in Table 1.10. While the method yields *measured* values of much weaker carbon acids than those obtained by other means (e.g., trimethylcyclopropene), the standard state of the pK_a values is ill-defined. Specifically, the reduction potentials for the alkyl halides are usually not reversible and, therefore, not thermodynamic

Table 1.10. Acidities of selected hydrocarbons as determined by the electrochemical thermodynamic cycle method

Compound	pK_a[a]
Tris(4-chlorophenyl)methane	28.9
Tris(4-methoxyphenyl)methane	34.6
Tris(4-dimethylaminophenyl)methane	40.3
Cycloheptatriene	36
Triphenylcyclopropene	50
Toluene	41
Trimethylcyclopropene	62

SOURCE: Data taken from reference 230.
[a]Errors are estimated to be 5–10%, typically.

values. The electrode processes are governed by kinetic considerations. In this regard, the solution in the vicinity of the electrode has been compared to a fused salt. This is a thermodynamically less well defined state than the infinitely dilute aqueous state at 298 K. For these reasons, the pK_a values determined in this way must be considered to be apparent pK_a values.

It is interesting to note the value found for toluene by this method, 41. Extrapolated values from the various ion-pair acidity scales range from 42.2 (cf. Table 1.4)[169] to 40.9 (cf. Table 1.5),[187] while the $pK_{a(DMSO)}$ for toluene has been estimated to be 43.[41] It seems that there is an approach to consensus between all of the methods; an average value for the pK_a of toluene is 42. Similarly, an electrochemical determination[231] of the pK_a of methane affords a value of 58.7 that is in reasonable agreement with the extrapolated $pK_{a(DMSO)}$ value of 56.[41] Of course, the effective pK_a of methane in water is likely to be lower. One more recent estimate assigns a value of 49 for the pK_a of methane.[42]

As we can see, the Breslow electrochemical approach sought to measure elusive pK_a values using accepted bond dissociation enthalpies. More recently, the Bordwell group,[41,190,196,204–206] among others,[232] have used the $pK_{a(DMSO)}$ database in thermodynamic cycles to determine the bond dissociation enthalpies of hydrocarbons, where these BDE values are unknown! Thus, we have come full circle.

1.4 KINETIC ACIDITY

An alternative to the thermodynamic method of assessing acidity is the kinetic approach. At one extreme, such kinetic assessment may be purely qualitative, involving determination of relative rates of proton abstraction from competitive experiments. At the other extreme, valuable rankings of acidity (in terms of estimated pK_a values) can be determined, particularly from isotopic exchange studies.

As we have seen, equilibrium methods for the measurement of the acidity of carbon acids rely, in the majority of cases, on color differences between the carbon acid and the carbanide. Obviously, this accounts for the determination of the pK_a values of hydrocarbons primarily whose anions form extended conjugated systems and, as a result, absorb strongly in the visible or ultraviolet region of the spectrum. (But, see also references 179–183 and 209 that deal with equilibrium pK_a measurements using NMR techniques.) The kinetic acidity method is, in principle, independent of color change and does not have the limitations of medium polarity or dielectric constant that are inherent in potentiometric and conductivity-type measurements.

While a number of kinetic methods exist to estimate hydrocarbon acidities,[233–236] we will focus our discussion on determination of kinetic acidity based on isotopic exchange. In this approach, exchange occurs between a normal substrate, H–R, and a deuterium- or tritium-labeled medium (i.e., D-Solv or T-Solv) or between an isotopically labeled substrate, D-R or T-R, and the normal medium. In either case, the rate of exchange is followed. The exchange is base catalyzed, usually by the conjugate base of the medium (Solv:$^-$), and in the case of exchange between a normal substrate and a deuterium-labeled medium this can be rendered as shown in equations 1.31a and 1.31b.

$$\mathrm{HA + Solv{:}^- \xrightarrow{\text{slow}} A{:}^- + H\text{–}Solv} \tag{1.31a}$$

$$\mathrm{A{:}^- + D\text{–}Solv \underset{\text{slow}}{\overset{\text{fast}}{\rightleftharpoons}} DA + Solv{:}^-} \tag{1.31b}$$

Thus, deprotonation by the conjugate base of the solvent, Solv:$^-$, is normally the slow, rate-limiting step. The carbanion, A:$^-$, that is formed is then protonated by D-Solv, which forms the deuterium-containing solvent pool, in a rapidly achieved equilibrium process. Other types of isotopic exchange[237,238] may be represented in an analogous fashion by equations similar to 1.31a and 1.31b.

What is the relationship between the acid dissociation equilibrium constant, K_a, and the rate of proton exchange for the given acid, HA? Intuition prompts us to suggest that the *stronger* an acid is according to its K_a, the *faster* the rate of proton transfer in a kinetic process and, hence, the greater the rate constant for deprotonation. Thus, there should be a direct correspondence between the magnitude of K_a and a proton transfer rate constant, k_a. In fact, this relationship is called the Brønsted catalysis law[239] and in its linear form appears as equation 1.32,

$$\log k_a = \alpha \log K_a + \log C \tag{1.32}$$

where k_a is the rate constant for proton transfer from HA to Solv:$^-$ in equation 1.31a, K_a is the thermodynamic acid dissociation constant for HA, and α, the Brønsted exponent, is traditionally taken as a measure of the degree of proton transfer[240] in the transition state (i.e., α varies between the limits of 0 and 1; cf. Chapter 2, Sections 2.2.1.1 and 2.2.1.2). The Brønsted catalysis law implies that for a given base and a series of increasingly strong acids (or, alternatively, a given acid with a series of increasingly strong bases, where β replaces α in equation 1.32), the rate of protonation of the base will increase in proportion to thermodynamic acid strengths. Furthermore, as originally conceived Brønsted plots were expected to show a continuous variation in slope from 0 to 1 (or vice versa) as the strength of the acid/base was varied,[239,241] α (equation 1.32) would approach 1 for strong acid catalysts (i.e., weak conjugate bases where β would tend toward 0) and α would approach 0 (and β approach 1) for weak catalytic acids (strong bases). In fact, few reactions were expected to span a wide enough range of acidities to result in Brønsted plot curvature,[242–244] and such curvature, where found, was generally considered to be the result of scatter and poor construction of the plot[245] or indicative of a change in mechanism.[246] [However, also see references 247–249 concerning the reactivity-selectivity principle (RSP).]

More recently, curvature in Brønsted plots, even those constructed using a limited range of pK_a values, has been explained as a natural consequence of the principle of nonperfect synchronization (PNS).[250–252] Although an apparently separate issue is the observation of "abnormal" values of α and β in the literature (i.e., values outside the range set by 0 and 1),[253–260] these values have been interpreted by Bernasconi in terms of the PNS as well.[252] Of course, examples explained by the PNS must be drawn from those Brønsted plots that were properly constructed, that is, where the investigators refrained from use of bases from different families, with markedly different steric and stereoelectronic requirements.[245]

A novel alternative to the classical method of constructing Brønsted plots has been presented[261,262] in which a single base is used; the pK_a of the base changes in a continuous manner as the solvent composition is varied from pure water to 90 mol % DMSO. Since it avoids the problems inherent in the use of different families of structurally different bases in the preparation of the plot, the method holds promise for the detection of actual curvature in Brønsted plots.[263–265]

The foregoing discussion stresses the connection between kinetic and thermodynamic acidity. In Chapter 2, the Brønsted relationship and the PNS will be used to probe the reactivity of carbanions and to aid in explaining such apparent contradictions in our intuitive understanding of this reactivity as what Kresge has dubbed the "nitroalkane anomaly".[260,266–270]

In the context of our discussion of determination of acidity via kinetics, however, the Brønsted catalysis law should be treated as an empirical relationship. It allows us to estimate thermodynamic acidities kinetically. Note that even if the relationship between pK_a and $\log k$ were nonlinear, it could still be used as a calibration plot so long as the function relating acidity and reactivity remains a smooth and continuous one. Thus, application of the Brønsted equation (1.32) to determine the acidity of an unknown weak carbon acid by the isotopic exchange method proceeds as follows. A standard base is chosen that can deprotonate a given series of weak carbon acids at rates that can be followed. The rates of deprotonation of the carbon acids, whether isotopically labeled with deuterium or tritium (in exchange with a normal medium) or unlabeled (in exchange with a labeled medium), are followed and the rate constants for the exchange correlated with the pK_a values, according to a known Brønsted function.

It is clear, then, that the pK_a values may be determined for the carbon acids if (1) the rate of exchange can be shown to be related directly to the rate of proton transfer and (2) the slope of the relevant Brønsted line, that is, β in this case, is known. The first condition is usually satisfied if the magnitude of the kinetic isotope effect is large,[271,272] indicative of significant C–H (or C–D or C–T) bond scission in the rate-determining step of the reaction.[177] The second prerequisite is met by first determining the Brønsted slope from previously measured pK_a values for compounds *that are structurally related to the unknown carbon acid.* For example, in DMSO the K_a values of a number of substituted toluenes are known.[202,203] The pK_a of 4-ethoxytoluene, say, could be determined if the rates of proton transfer from isotopic exchange were known for the toluene series, since a Brønsted plot could then be constructed. Moreover, for localized carbanions the slope of the Brønsted plot often approaches unity[177,273] and so an unknown pK_a could be determined from the proton transfer kinetics alone.

The method is hampered in some respects by the process of *internal return.*[274] Scheme 1.4 illustrates the process of internal return. Here, the carbanion, formed by proton abstraction from the carbon acid (k_1) by the conjugate base of the solvent ($^-$Solv), is hydrogen-bonded to the solvent (H-Solv). This hydrogen-bonded carbanion, carbanion···H-Solv, may partition between reprotonation

(k_{-1}) or proceed to products (k_2). In principle, the k_2 step could represent any number of processes but, as shown in Scheme 1.4, k_2 depicts the overall transfer of a deuteron from deuterated solvent (D-Solv, as in equation 1.31b). Consequently, k_2 would likely comprise a number of microscopic steps including the exchange of D-Solv for H-Solv in the carbanion···H-Solv encounter complex, the actual transfer of the deuteron from D-Solv to the carbanion to give the deuterio-carbon acid (R′R″R‴C–D), and dissociation of the deuterio-carbon acid····Solv encounter complex to yield free R′R″R‴C–D.

Nonetheless, if an initially formed carbanion of a protio-carbon acid is merely reprotonated in the k_{-1} step (rather than deuterated in the k_2 step of Scheme 1.4) immediately after formation, then the process of isotopic exchange will go unrecorded. In fact, determination of kinetic deuterium-hydrogen isotope effects (k^H/k^D) of close to 1 is taken to be strong experimental evidence for significant internal return in these systems. However, in these cases the observed isotopic effect is also small and the Brønsted extrapolation would be invalid.[274]

The extent of systems that have been examined by the isotopic exchange method is large and includes ketones (with OD^-/D_2O),[275,276] polynitrobenzenes (with OD^-/D_2O/DMF),[237,238,277–279] triphenyl- and diphenylmethanes (with MeO^-/MeOT),[280] toluenes and xylenes (with *t*-BuO^-/tritiated DMSO),[281,282] and cyclopropane and other cycloalkanes (with $C_6H_{11}ND^-$/*N*,*N*-dideuteriocyclohexylamine).[283] A variety of tritium-labeled carbon acids have been detritiated in OH^-/H_2O, MeO^-/MeOH, and EtO^-/EtOH and the rate constant ratio for detritiation with hydroxide and methoxide (k^T_{OH}/k^T_{MeO-}) has been proposed as an additional probe for the classification of carbon acids.[284] The detritiation of tritiated diethyl malonate by a number of heterocyclic bases has also been examined.[285]

The Kresge group has examined the process of detritiation of a series of terminal acetylenes[286,287] and concluded, in agreement with equilibrium acidity measurements,[189] that the corresponding acetylide anions were essentially localized carbanions and so their conjugate acids acted as "normal" (i.e., like typical oxy-

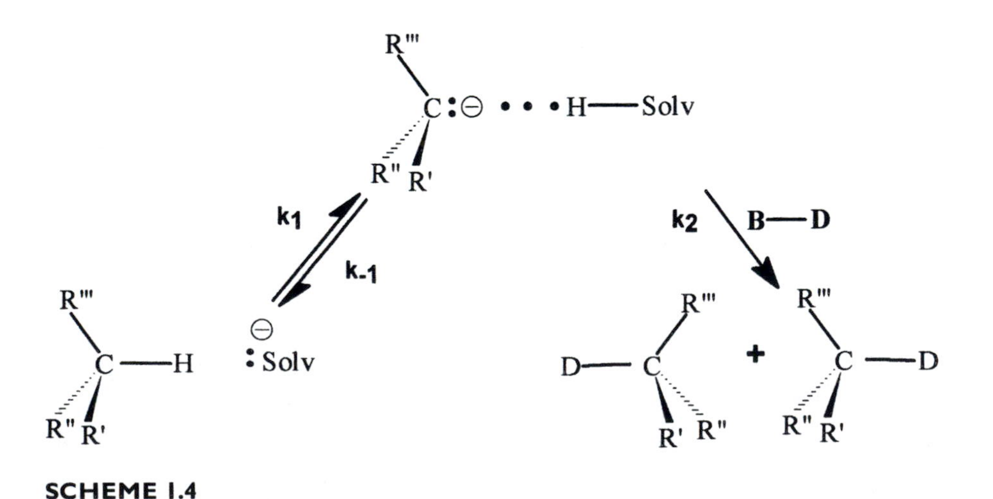

SCHEME 1.4

gen- and nitrogen-centered) acids.[240] Following the definitions of Eigen,[240] "normal" acids are those for which rate constants for proton transfer are on the order expected for diffusion control, that is, between 10^{10} and 10^{11} $M^{-1}{\cdot}s^{-1}$. To meet this kinetic criterion, a carbon acid and its conjugate base must not differ significantly in molecular or electronic structure or, in other words, proton transfer must not require significant solvational and structural reorganization. An interesting comparison is that put forward by Albery:[288] while ethanoic (acetic) acid and 1,1,1-trifluoro-2,4-pentanedione (trifluoroacetylacetone) have approximately the same K_a (1.7×10^{-5} M for the former and 2.0×10^{-5} M for the latter), the proton transfer constants (forward and reverse) that make up the acid dissociation equilibrium constant are significantly different. Proton transfer from ethanoic acid ($k_f = 9 \times 10^5\ s^{-1}$) is much faster than proton loss from trifluoroacetylacetone ($k_f = 1.5 \times 10^{-2}\ s^{-1}$), and on these grounds it appears odd that deprotonation of trifluoroacetylacetone is so slow. On the other hand, the protonation of ethanoate ion ($k_b = 5 \times 10^{10}\ M^{-1}{\cdot}s^{-1}$) is also much faster than protonation of the enolate ion of trifluoroacetylacetone ($k_b = 7.5 \times 10^2\ M^{-1}{\cdot}s^{-1}$), and it is this slow protonation of the conjugate base that combined with the relatively small deprotonation rate constant yields the coincidentally similar K_a value for trifluoroacetylacetone as compared to that of ethanoic acid. Thus, the protonation and deprotonation processes for trifluoroacetylacetone are anomalous compared with those for ethanoic acid, and trifluoroacetylacetone could be termed a "pseudoacid"[240,288] whereas ethanoic acid would be considered a "normal acid". Returning to our original example, terminal acetylenes were found to be "normal" acids, where major solvent and structural reorganization does not accompany protonation/deprotonation.[286,287]

To catalog the utility of kinetic acidity methods in but one area of research, kinetic H/D/T exchange has been used to probe the effect of transition metals on carbon acidity (i.e., the *metal activating factor*) in both labile[289–291] and stable[292,293] complexes of biologically significant heterocycles, including thiazole,[294] imidazole,[295] 1-methylimidazole,[296–298] histidine,[299] and purines.[300,301] Extensive and complementary studies of the *proton activating factor* (PAF) in carbon acidity of nitrogen-containing heterocycles have been undertaken by the Stewart group;[302–305] here, protonation (sometimes modeled by alkylation)[304] of a ring nitrogen activates the adjacent carbon acid center(s). It may be concluded that the kinetic approach to the measurement of acidity represents a powerful method to determine the acidities of very weak carbon acids and, in turn, to provide insight into the factors that influence the acidities of these acids.

Kinetic acidity measurements provide access to pK_a values of acids that could not be measured in DMSO or cyclohexylamine by equilibrium methods. In this context, drastic experimental conditions are often required to bring about exchange in very weak acids. For example, cyclohexane with 1 M potassium amide in deuterated liquid ammonia exchanges on average only one of its protons with the medium *after 180 h at 120 °C*![306]

A further limitation arises from the fact that equation 1.32 contains no specific terms to allow for variation in solvent or for significant "nonlinear" structural variation of the carbon acid (though recall that so long as a continuous function exists relating acidity and rate data, a kinetic acidity determination can be made).

By using the substituted toluene data,[202,203] it would be valid to consider the kinetic acidity of 4-ethoxytoluene in DMSO but not in HMPA, or that of any given 4- or 3-substituted toluene but not that of an *ortho*- or otherwise sterically hindered toluene. Regardless of the limitations outlined in this and previous paragraphs, kinetic measurements of acidity represent one of the few methods of estimating the acidity of weak carbon acids that avoids the problems associated with the leveling effect in solvents.[275–308]

A comparable, though more limited, approach uses the Hammett equation,[201]

$$\log k_X/k_O \quad \text{or} \quad \log K_X/K_O = \rho\sigma \text{ (or } \rho\,\sigma^-) \tag{1.33}$$

another linear free-energy relationship (LFER) where ρ normally reflects the sensitivity of the reaction or equilibrium to the substituent effect and σ (or σ^-)[309] is a substituent constant that may be correlated with either kinetic or equilibrium data for a substituent in the *meta* or *para* positions of a benzene ring. Because the substituent constant, σ, was defined originally by the acidity of substituted benzoic acids (relative to the unsubstituted benzoic acid) in *water*, it might be presumed that data determined in other solvents would not correlate. However, this is not always the case, as we have previously noted.[41] Bordwell, for instance, has correlated the acidities of several families of compounds, including the equilibrium acidities of 23 *meta*- and *para*-substituted acetophenones in DMSO, with the Hammett σ-constant and thereby redefined the constants for use in DMSO.[310] In theory, any set of *meta*- and *para*-substituted aryl carbon acids could be correlated with the new DMSO-corrected constants. Again, kinetic data including isotopic exchange reactions could be used. The pK_a values of unknown compounds that should fit the resultant Hammett plot would then be interpolated or extrapolated from the Hammett line.

In fact, the Hammett equation can be viewed as a subsidiary equation of the Brønsted equation.[190] A relationship has been noted between the Brønsted coefficient and the Hammett substituent constant for a series of 4-substituted (aryloxy)-tetrahydrofurans; it has been suggested that this connection may offer a "shortcut" to the Brønsted exponent.[311]

At this juncture, it should be apparent that if rate data can be used, via application of the Brønsted equation (1.32), to obtain the acidities of weak carbon acids, then, conversely, equation 1.32 may be used in combination with pK_a data to predict the reactivity of carbanions. In fact, the discussion of curvature in Brønsted plots is an example of the use of pK_a data to probe rate behavior. In Chapter 2 we will examine the reactivity of carbanions in further detail and draw further on the Brønsted analysis in making our comparisons.

1.5 GAS-PHASE ACIDITY: IS IT INTRINSIC ACIDITY?

Gas-phase acidities are determined by a number of methods. First, the gas-phase acid ionization constant, K_a, may be determined from the relative concentrations of the two carbanions (equation 1.34); these concentrations may be measured by a variety of techniques including high-pressure mass spectrometry (MS)[312–315] and ion cyclotron resonance (ICR)[316,317] and, more recently, Fourier transform ICR (FT-ICR),[129,318,319] as well as flowing afterglow.[320–324] (These techniques are described in detail in the references cited, as well as in references 325–327, which consider the techniques in the context of carbon acid gas-phase acidity. Consequently, the results obtained from these methods but not the measurement techniques themselves will be discussed further.)

$$R^- + R'\text{–}H \rightleftharpoons R\text{–}H + R'^- \qquad (1.34)$$

The overall equilibrium constant for equation 1.34 may also be determined from measurement of the rate constants for the forward ($R^- + R'\text{–}H \rightarrow R\text{–}H + R'^-$) and reverse ($R'^- + R\text{–}H \rightarrow R^- + R'\text{–}H$) kinetic approaches to equilibrium.[326,327] In some cases,[328,329] only the ratio of $[R^-]/[R'^-]$ can be obtained experimentally and, then, the acidity of R′–H, for example, can be determined only if the acidity of R–H has already been determined. For a series of carbon acids whose relative acidity is known, a "ladder" can be constructed from which the acidities of all of the carbon acids can be determined from the anchoring pK_a value of the first R–H in the series.[129] More recently, the extensive gas-phase acidity results determined in four different laboratories (i.e., those of Kebarle,[330–332] Taft,[333] Meot-Ner and Sieck,[334,335] and Szuljecko and McMahon[336]) have been critically evaluated,[326] particularly in light of the scale of proton affinity values determined by Smith and Radom[337] among others[338] from ab initio calculations. In the critical evaluation, inconsistencies caused by different experimental temperatures were assessed. The result is a self-consistent and comprehensive compilation of gas-phase acidities (and basicities),[326] the most recent version of which is available on the Internet.[339]

Where direct measurements of proton transfer equilibrium constants are difficult or otherwise experimentally inaccessible, backeting experiments may provide acidity/basicity estimates. In bracketing techniques, a given carbanion, R^-, is allowed to react with each member of a series of reference acids, in turn. Acids that react with R^- are presumably more acidic than R–H, whereas those that do not react are less acidic than the conjugate acid of the carbanion of interest. Given the pK_a values for the series of acids, the pK_a of the unknown acid, R–H, can be bracketed by the two reference acids of closest pK_a; that is, the pK_a of the unknown R–H falls between the pK_a of the acid that reacts with R^- and that of the nearest reference acid that does not react. Generally, results of such bracketing experiments have greater intrinsic errors compared with direct equilibrium measurements.[326] These errors arise from reactions that may compete with the proton transfer, as well as possible proton transfer from isomeric carbanionic structures.

Finally, gas-phase acidities may be calculated from a combination of known bond dissociation enthalpies (BDE), electron affinities (EA), and the ionization potential of the hydrogen atom [IP(H$^\cdot$)] through the use of a thermochemical cycle, derived from equations 1.35a–1.35c.

$$R\text{–}H \longrightarrow R^\bullet + H^\bullet \qquad (1.35a)$$

$$\mathrm{R^{\bullet} + e^- \longrightarrow R^-} \tag{1.35b}$$

$$\mathrm{H^{\bullet} \longrightarrow H^+ + e^-} \tag{1.35c}$$

The first equation represents the BDE of the carbon acid, the second equation is the inverse of the EA, and the final equation shows the IP(H·) for the hydrogen atom. Addition of equations 1.35a to 1.35c yields equation 1.36, which we recognize as the gas-phase acid ionization of the carbon acid, R–H.

$$\mathrm{R{-}H \longrightarrow H^+ + R^-} \tag{1.36}$$

The enthalpy change for acid ionization (ΔH_a) of the carbon acid, R–H, is then given by equation 1.37.

$$\Delta H_a = \mathrm{BDE(R-H) - EA(R^{\cdot}) + IP(H^{\cdot})} \tag{1.37}$$

where the ionization potential of the hydrogen atom is a recurring constant (313.6 kcal/mol). Two comments should be made about equation 1.36. First, in its derivation we have ignored the fact that the electron affinity and ionization potential terms (equations 1.35b and 1.35c, respectively) are defined at 0 K, while the bond dissociation energy (equation 1.35a) and the final heat of acid dissociation (equation 1.36) are 298 K values. Although corrections can be made for these temperature differences, the errors in the measurement of the individual values is often greater than the absolute value of the correction factors and, therefore, these minor adjustments are rarely undertaken.[340,341] Secondly, the entropy term in the dissociation of R–H has been found to be relatively constant and small (ca. 2–10 eu, entropy units) and, therefore, gas-phase acidities are usually given as the enthalpy changes for acid dissociation (ΔH_a, equation 1.37) or, alternatively, as the *proton affinity* of the anions.[312–315,326]

The proton affinity is given by the reverse of equation 1.36, which generally corresponds to equation 1.38, where M represents a gas-phase base.

$$\mathrm{H^+ + M \longrightarrow MH^+} \tag{1.38}$$

More specific to our topic of carbanions, equation 1.38 may be rendered as 1.39,

$$\mathrm{H^+ + R^- \longrightarrow R{-}H} \tag{1.39}$$

for which the negative of the reaction enthalpy change is defined as the proton affinity.

Clearly, gas-phase acidities cannot be influenced by solvent and, consequently, acidities of species determined in the gas phase are frequently taken to be intrinsic acidities. Specific solvent effects such as hydrogen-bonding of a substituent with the solvent[333,342,343] or, conversely, steric inhibition of solvation, as apparently occurs in acetonitrile solvent,[344,345] can conceivably mask the inherent effect on acidity due to substituents, for example, and cause considerable differences in acidity to appear in going from one solvent to another.[346] In fact, it now appears that gas-phase data will be used as the standard for the evaluation of the various factors, including resonance, polarizability, and field/inductive effects, whereby a substituent interacts with a reactive center.[347,348]

A frequently cited example of the differences in acidity found in going from protic solvents to the gas phase is that reported by Brauman and Blair[328,329] where the order of the acidity of a series of alkanols from most to least acidic was as follows: neopentyl > *t*-butyl > isopropyl > ethyl > methyl > water. This order is virtually the inverse of that found for the alcohols in neat solvent. Thus, in the gas phase, *tert*-butyl alcohol is more acidic than methanol, whereas the opposite is true in the parent alcoholic solutions. In this case, the observed gas-phase order was taken as indicative of the polarizable nature of alkyl groups that can stabilize negative charge as well as positive charge. Further, the apparent enhanced basicity of the *tert*-butoxide ion (i.e., the decreased acidity of *tert*-butyl alcohol) was presumed to result from decreased efficacy of hydrogen-bonding and a shielding of the negative charge from solvent in solution, both because of the steric bulk of the *t*-butyl group as compared with that of the methyl group.[328,329]

In fact, it should be surprising that there is any correlation between the acidity of carbon acids in the gas phase and their acidity in any solvent. After all, it appears to be necessary to invoke many separate terms (e.g., for hydrogen-donating and -accepting ability of both solute and solvent, for cavity formation to dissolve the solute, etc.) to correlate the effect of solvent on reaction rates and on spectroscopic measurements.[349,350]

The gas-phase acidities (heats of acid dissociation) for a selected set of carbon acids are listed in Table 1.11. Also, a few gas-phase values for carboxylic acids are included for sake of comparison. In examining the table it should be mentioned first that gas-phase enthalpies of acid dissociation (ΔH_a, equations 1.36 and 1.37) are generally positive (endothermic) values. Therefore, the most acidic carbon acids in Table 1.11 are those with the smallest ΔH_a values. These values are taken from the recent critical evaluation of these values,[339] but the original reference sources are also cited.

It is important to note that for groups of structurally similar carbon acids, gas-phase and solution data show the same broad trends, although there are some striking anomalies. For example, the order of the acidity of substituted methanes increases with the order of electron-withdrawing power of the substituent: NO_2 > SO_2Ph > COPh > SO_2CH_3 > $COCH_3$ > CN > $SOCH_3$. Bordwell's order, determined in DMSO, is broadly similar, except that the positions of the substituent pairs $COCH_3$ and SO_2CH_3, and COPh and SO_2Ph, are interchanged.[202,203,353] In some cases, these anomalies are readily explicable by considering some of the specific solvation effects mentioned above.[333,342–348,355] Furthermore, anomalies due to discrepancies between sets of measurements from different laboratories have been eliminated in the current critically assessed gas-phase acidities.[339] Where solution acidities are known, and particularly where such measurements can aid in assessing the effect of the solvent environment (solvation, aggregation, etc.), solution pK_a values are to be preferred in determining the reactivity of carbanions. However, where solution acidities are not known, gas-phase determinations are particularly valuable.

1.6 COMPARISON OF SOLUTION- AND GAS-PHASE ACIDITIES

Although it is clear that there are a number of specific solvent–acid–conjugate base interactions that may prevent comparison of solution- and gas-phase data, DMSO, the solvent for which the largest number of pK_a values of carbon acids have been determined,[41] is unique. It is well known that DMSO solvates anions (e.g., conjugate bases of uncharged acids) only poorly.[89,209] The main difference in solvation for the acid dissociation equilibrium (equation 1.1) lies in solvation of the various anions. Therefore, the low degree of solvation by DMSO (or a similar solvent) of any given anion results in acid–base equilibria in DMSO that correlate with the same equilibria in the gas phase, to a first approximation.[355–357]

Table 1.11. Gas-phase acidities (enthalpies and/or free energies) for selected acids as measured by high-pressure mass spectrometry

Compound	$\Delta H°$ (kcal/mol)	$\Delta G°$ (kcal/mol)[a]
Nitriles		
$CH_2(CN)_2$	336.0 ± 2.6[b]	328.2
$PhCH_2CN$	351.5 ± 3.2[b]	344.9
CH_3CN	374.5 ± 2.1[e]	373.1
Carbonyl compounds		
$CF_3COCH_2COCH_3$	328.5 ± 4.1[b]	322.1
Dimedone	338.8 ± 4.1[b]	331.1
Diethyl malonate	348.3 ± 2.3[b]	342.3
CF_3COCH_3	350.3 ± 3.5[b]	343.2
$PhCH_2COCH_3$	351.1 ± 3.5[b]	345.4
CH_3COCH_3	369.6 ± 2.6[d]	362.4
Nitroalkanes		
CH_3NO_2	357.4 ± 2.9[d]	350.7
$CH_3CH_2NO_2$	357.5 ± 2.9[e]	351.0
Sulfoxides and sulfones		
$PhSO_2CH_3$	362.7 ± 2.3[f]	355.3[f]
$CH_3SO_2CH_3$	366.4 ± 2.0[f]	358.9[f]
CH_3SOCH_3	374.6 ± 2.3[f]	367.2[f]
Carboxylic acids		
CF_3COOH	324.4 ± 2.9[g]	317.4[g]
PhCOOH	340.1 ± 2.2[g]	333.1[g]
CH_3COOH	348.5 ± 2.9[g]	341.5[g]

SOURCE: Data are taken from reference 339.
[a]Errors for all Gibbs free energies are ±2.0 kcal/mol.
ORIGINAL SOURCES: [b]Reference 312; [c]Reference 351; [d]Reference 314; [e]Reference 352; [f]Reference 315; [g]Reference 354.

For a series of sulfoxides, sulfones, ketones, nitroalkanes, and cyanoalkanes—all of which could be viewed as structural derivatives of methane—a fair correlation was found between the gas-phase acidities [ΔG_a (gas)] and the DMSO solution acidities [rendered as ΔG_a(DMSO)].[358] However, the nitroalkanes show a significantly higher acidity with increase in alkyl substitution in water, as expected from the trend to localization of charge in carbanions with alkyl substitution in the gas phase and the ease of hydrogen-bonding between the nitro substituent and water in the aqueous phase.[344–348] Such correlations are not general[190] but clearly are valid for limited sets of structurally similar compounds.[342,358]

Moreover, pK_a values for 10 compounds that would yield highly conjugated carbanions, representing a range of 15 pK_a units as determined in CHA with cesium as the counterion, have been found to correlate well with the acidities of the same carbon acids in DMSO.[169] The same acids that would yield delocalized carbanions show a pK_a–pK_a correlation between 1,2-dimethoxyethane (with Cs^+ counterion) and CHA (with the same counterion). Because these carbon acidities are all cross-correlated between the various solvents and the gas phase, it may be expected that individual solvent-to-gas-phase correlations will also be valid. Naturally, these cross-correlations tend to break down for carbon acids that yield more localized carbanions.

There is a further value in the gas-phase determinations and that is that these acidities can be directly compared with values determined by various calculational methods.[83–85,109,337,359–361] (However, also see reference 362.) Thus, in the future it may be expected that calculational studies will probe the intrinsic electronic and structural effects on acidity, that these studies will be confirmed by gas-phase acidity determinations, and that, finally, these results will be applied to the solution phase, for those systems where gas-phase and condensed-phase acidities follow the same trends.

In this chapter, we have examined the methods[363] of determining acidities of carbon acids. We have drawn a number of conclusions concerning the effect of substituents on these acidities. In the next chapter, we will examine the structure and reactivity of the carbanions that are derived from carbon acids.

Notes and References

1. Cram, D.J. *Fundamentals of Carbanion Chemistry*; Academic Press: New York, 1965.
2. Buncel, E. *Carbanions: Mechanistic and Isotopic Aspects*; Elsevier: Amsterdam, 1975.
3. *Comprehensive Carbanion Chemistry. Part A: Structure and Reactivity*; Buncel, E.; Durst, T., Eds.; Elsevier: Amsterdam, 1980.
4. *Comprehensive Carbanion Chemistry. Part B: Selectivity in Carbon–Carbon Bond Forming Reactions*; Buncel, E.; Durst, T., Eds.; Elsevier: Amsterdam, 1984.
5. *Comprehensive Carbanion Chemistry. Part C: Ground and Excited State Reactivity*; Buncel, E.; Durst, T., Eds.; Elsevier: Amsterdam, 1987.
6. Snieckus, V.A. *Advances in Carbanion Chemistry*; JAI Press: Greenwich, CT, 1992; Vol. 1.
7. Trost, B.M., Ed.; *Comprehensive Organic Synthesis, Vol. 3, Carbon–Carbon σ-Bond Formation*; Pergamon Press: Oxford, 1991.
8. Heathcock, C.H.; Oare, D.A. *Top. Stereochem.* **1991**, *20*, 87.

9. Evans, D.A.; Andrews, G.C. *Acc. Chem. Res.* **1974**, *7*, 147.
10. Seebach, D. *Angew. Chem., Int. Ed.. Engl.* **1979**, *18*, 239.
11. Seebach, D.; Miller, D.D.; Müller, S.; Weber, T. *Helv. Chim. Acta* **1985**, *68*, 949.
12. Blank, S.; Seebach, D. *Angew. Chem., Int. Ed. Engl.* **1993**, *32*, 1765.
13. Juaristi, E.; Anzorena, J.L.; Boog, A.; Madrigal, D.; Seebach D.; Garcia-Baez, E.V.; Garcia-Barradas, O.; Gordillo, B.; Kramer, A.; Steiner, I.; Zürcher, S. *J. Org. Chem.* **1995**, *60*, 6408.
14. Denmark, S.E.; Cramer, C.J. *J. Org. Chem.* **1990**, *55*, 1806.
15. Cramer, C.J.; Denmark, S.E.; Miller, P.C.; Dorow, R.L.; Swiss, K.A.; Wilson, S.R. *J. Am. Chem. Soc.* **1994**, *116*, 2437.
16. Denmark, S.E.; Kim, J.H. *J. Org. Chem.* **1995**, *60*, 7535.
17. Bell, R.P. *The Proton in Chemistry*, 2nd ed.; Cornell University Press: Ithaca, NY, 1973.
18. Jones, J.R. *The Ionization of Carbon Acids*; Academic Press: London, 1973.
19. Stewart, R. *The Proton: Applications to Organic Chemistry*; Academic Press: Orlando, FL, 1985.
20. Jones, J.R. *Annu. Rep. Prog. Chem. C.* **1987**, *83*, 197.
21. Streitwieser, A.; Wang, D.Z.; Stratakis, M.; Facchetti, A.; Gareyev, R.; Abbotto, A.; Krom, J.A.; Kilway, K.V. *Can. J. Chem.* **1998**, *76*, 765.
22. Brønsted, J.N. *Recl. Trav. Chim. Pays-Bas* **1923**, *42*, 718.
23. Lowry, T.M. *Chem. Ind. (London)* **1923**, *42*, 43.
24. Pearson, R.G.; Dillon, R.L. *J. Am. Chem. Soc.* **1953**, *75*, 2493.
25. Boyd, R.H. *J. Phys. Chem.* **1963**, *67*, 737.
26. Boyd, R.H. *J. Am. Chem. Soc.* **1961**, *83*, 4288.
27. Slovetskii, V.I.; Shevelev, S.A.; Fainzil-berg, A.A.; Novikov, S.S. *Zh. Khim. Obshch. Mendeleeva* **1961**, *6*, 599 and 707.
28. Jones, J.R. *Prog. Phys. Org. Chem.* **1972**, *9*, 241.
29. Schwarzenbach, G.; Felder, E. *Helv. Chim. Acta* **1944**, *27*, 1701.
30. Zundel, G. *Hydration and Intermolecular Interaction*; Academic Press: New York, 1969.
31. Giguere, P.A.; Turrell, S. *Can. J. Chem.* **1976**, *54*, 3466.
32. Giguere, P.A. *J. Chem. Educ.* **1979**, *56*, 571.
33. Eigen, M.; De Maeyer, L. *Proc. R. Soc. London* **1958**, *A247*, 505.
34. Newton, M.D. *J. Chem. Phys.* **1977**, *67*, 5535.
35. Kebarle, P.; Searles, S.K.; Zolla, A.; Scarborough, J.; Arshadi, M. *J. Am. Chem. Soc.* **1967**, *89*, 6393.
36. Kebarle, P. *J. Phys. Chem.* **1970**, *53*, 2129.
37. Keesee, R.G.; Castleman, Jr., A.W. *J. Phys. Chem. Ref. Data* **1986**, *15*, 1011.
38. Grimsrud, E.P.; Kebarle, P. *J. Am. Chem. Soc.* **1973**, *95*, 7939.
39. Meot-Ner, M. *J. Am. Chem. Soc.* **1978**, *100*, 4694.
40. Giguere, P.A.; Martel, C.; Turrell, S. *Chem. Phys. Lett.* **1978**, *56*, 231.
41. Bordwell, F.G. *Acc. Chem. Res.* **1988**, *21*, 456.
42. Richard, J.P.; Williams, G.; Gao. J. *J. Am. Chem. Soc.* **1999**, *121*, 715.
43. Jorgensen, W.L.; Briggs, J.M.; Gao, J. *J. Am. Chem. Soc.* **1987**, *109*, 6857.
44. Lewis, G.N. *Valence and Structure of Atoms and Molecules, ACS Monograph*; Chemical Catalogue Company: New York, 1923.
45. Connors, K.A. *Binding Constants. The Measurement of Molecular Complex Stability*; Wiley-Interscience: New York, 1987, p 243.
46. Gutmann, V. *Coord. Chem. Rev.* **1975**, *15*, 207.
47. Gutmann, V. *The Donor–Acceptor Approach to Molecular Interactions*; Plenum: New York, 1978.

48. Brown, I.D. *Acta Crystallogr., Sect. B*, **1988**, *44*, 545.
49. Brown, I.D.; Skowron, A. *J. Am. Chem. Soc.* **1990**, *112*, 3401.
50. Drago, R.S.; Wayland, B.B. *J. Am. Chem. Soc.* **1965**, *87*, 3571.
51. Drago, R.S.; Parr, L.B.; Chamberlain, C.S. *J. Am. Chem. Soc.* **1977**, *99*, 3203.
52. Drago, R.S. *Applications of Electrostatic–Covalent Models in Chemistry*; Surfside Scientific Publishers: Gainesville, FL, 1994.
53. Drago, R.S.; Joerg, S. *J. Am. Chem. Soc.* **1996**, *118*, 2654.
54. Drago, R.S. *Organometallics* **1995**, *14*, 3408.
55. Drago, R.S.; Zoltewicz, J.A. *J. Org. Chem.* **1994**, *59*, 2824.
56. Ingold, C.K. *Structure and Mechanism in Organic Chemistry*, 2nd ed.; Cornell University Press: Ithaca, NY, 1969.
57. Sawyer, D.T.; Roberts, J.L. Jr., *Acc. Chem. Res.* **1988**, *21*, 469.
58. Bacaloglu, R.; Blasko, A.; Bunton, C.A.; Ortega, F.; Zucco, C. *J. Am. Chem. Soc.* **1992**, *114*, 7708.
59. Bunnett, J.F.; Kim, J.K. *J. Am. Chem. Soc.* **1970**, *92*, 7463.
60. Bunnett, J.F. *J. Chem. Educ.* **1974**, *51*, 312.
61. Ashby, E.C. *Acc. Chem. Res.* **1988**, *21*, 414.
62. Ashby, E.C.; Mehdizadeh, A.; Deshpande, A.K. *J. Org. Chem.* **1996**, *61*, 1322.
63. Ashby, E.C.; Deshpande, A.K.; Doktorovich, F. *J. Org. Chem.* **1994**, *59*, 6223.
64. Ashby, E.C. *Pure Appl. Chem.* **1980**, *52*, 545.
65. Pross, A. *Acc. Chem. Res.* **1985**, *18*, 212.
66. Pross, A.; Shaik, S.S. *Acc. Chem. Res.* **1983**, *16*, 363.
67. Shaik, S.S.; Schlegel, H.B.; Wolfe, S. *Theoretical Aspects of Physical Organic Chemistry. The S_N2 Mechanism*; Wiley-Interscience: New York, 1992.
68. Hoz, S. *Acc. Chem. Res.* **1993**, *26*, 69.
69. Bordwell, F.G.; Bausch, M.J. *J. Am. Chem. Soc.* **1986**, *108*, 1979.
70. Sim, B.A.; Milner, G.H.; Griller, D.; Wayner, D.D.M. *J. Am. Chem. Soc.* **1990**, *112*, 6635.
71. Venimadhavan, S.; Amarnath, K.; Harvey, N.G.; Cheng, J.P.; Arnett, E.M. *J. Am. Chem. Soc.* **1992**, *114*, 221.
72. A theory that explicitly includes electron transfer in the definition of acids and bases is that attributed to Usanovich, references 73 and 74.
73. Usanovich, M. *Zh. Obshch. Khim.* **1939**, *9*, 182.
74. Gehlen, H. *Z. Phys. Chem.* **1954**, *203*, 125.
75. Ahrland, S.; Chatt, J.; Davies, N.R. *Q. Rev. (London)* **1958**, *12*, 265.
76. Pearson, R.G. *J. Am. Chem. Soc.* **1963**, *85*, 3533.
77. Pearson, R.G.; Songstad, J. *J. Am. Chem. Soc.* **1967**, *89*, 1827.
78. Parr, R.G.; Pearson, R.G. *J. Am. Chem. Soc.* **1983**, *105*, 7512.
79. Pearson, R.G. *J. Am. Chem. Soc.* **1985**, 107, 6801.
80. Pearson, R.G. *J. Chem. Educ.* **1987**, 64, 561.
81. March, J. *Advanced Organic Chemistry*, 3rd ed.; Wiley: New York, 1985; pp 322–325.
82. House, H.O. *Modern Synthetic Reactions*, 2nd ed.; Benjamin: Menlo Park, CA, 1972; p 492.
83. Wiberg, K.B.; Breneman, C.M.; LePage, T.J. *J. Am. Chem. Soc.* **1990**, *112*, 61.
84. Wiberg, K.B. *J. Am. Chem. Soc.* **1990**, *112*, 4177.
85. Wiberg, K.B.; Castejon, H. *J. Org. Chem.* **1996**, *60*, 6327.
86. Zook, H.D.; Russo, T.J.; Ferrand, E.F.; Stotz, D.S. *J. Org. Chem.* **1968**, *33*, 2222.
87. Kornblum, N.; Seltzer, R.; Haberfield, P. *J. Am. Chem. Soc.* **1963**, *85*, 1148.
88. Sarthou, P.; Bram, G.; Guibé, F. *Can. J. Chem.* **1980**, *58*, 786.
89. Buncel, E.; Wilson, H.A. *Adv. Phys. Org. Chem.* **1977**, *14*, 133.
90. LeNoble, W.J.; Morris, H.F. *J. Org. Chem.* **1969**, *34*, 1969.

91. Brickhouse, M.D.; Squires, R.R. *J. Am. Chem. Soc.* **1988**, *110*, 2706.
92. Hayes, R.N.; Grese, R.P.; Gross, M.L. *J. Am. Chem. Soc.* **1989**, *111*, 8336.
93. Jones, M.E.; Kass, S.R.; Filley, J.; Barkley, R.M.; Ellison, G.B. *J. Am. Chem. Soc.* **1985**, *107*, 109.
94. Wladkowski, B.D.; Wilbur, J.L.; Zhong, M.; Brauman, J.I. *J. Am. Chem. Soc.* **1993**, *115*, 8833.
95. Zhong, M.; Brauman, J.I. *J. Am. Chem. Soc.* **1996**, *118*, 636.
96. Brickhouse, M.D.; Squires, R.R. *J. Phys. Org. Chem.* **1989**, *2*, 389.
97. Freriks, I.L.; deKoning, L.J.; Nibbering, N.M.M. *J. Am. Chem. Soc.* **1991**, *113*, 9119.
98. Pearson, R.G. *Surv. Prog. Chem.* **1969**, *5*, 1.
99. Bacaloglu, R.; Bunton, C.A.; Ortega, F. *J. Am. Chem. Soc.* **1989**, *111*, 1041.
100. Buncel, E.; Crampton, M.R.; Strauss, M.J.; Terrier, F. *Electron Deficient Aromatic- and Heteroaromatic-Base Interactions. The Chemistry of Anionic Sigma Complexes*; Elsevier: Amsterdam, 1984; pp 244–250.
101. Buncel, E.; Dust, J.M.; Terrier, F. *Chem. Rev.* **1995**, *95*, 2261.
102. Terrier, F. *Chem. Rev.* **1982**, 82, 77.
103. Renfrow, R.A.; Strauss, M.J.; Terrier, F. *J. Org. Chem.* **1980**, *45*, 471.
104. Murphy, R.M.; Wulff, C.A.; Strauss, M.J. *J. Am. Chem. Soc.* **1974**, *96*, 2678.
105. Fendler, J.H.; Hinze, W.L.; Liu, L. *J. Chem. Soc., Perkin Trans. 2* **1975**, 1768.
106. Strauss, M.J. *Acc. Chem. Res.* **1974**, *7*, 181.
107. Strauss, M.J.; Schran, H. *J. Am. Chem. Soc.* **1969**, *111*, 3974.
108. Buncel, E.; Dust, J.M.; Manderville, R.A. *J. Am. Chem. Soc.* **1996**, *118*, 2072.
109. Houk, K.N.; Paddon-Row, M.N. *J. Am. Chem. Soc.* **1988**, *110*, 2706.
110. Wunderli, A.; Winkler, T.; Hansen, H.J. *Helv. Chim. Acta* **1977**, *60*, 2436.
111. Dao, L.H.; Dust, J.M.; Mackay, D.; Watson, K.N. *Can. J. Chem.* **1979**, *57*, 1712.
112. Burrows, C.J.; Carpenter, B.K. *J. Am. Chem. Soc.* **1981**, *103*, 6983.
113. Burrows, C.J.; Carpenter, B.K. *J. Am. Chem. Soc.* **1981**, *103*, 6984.
114. Klopman, G. *J. Am. Chem. Soc.* **1968**, *90*, 223.
115. Pearson, R.G. *Acc. Chem. Res.* **1990**, *23*, 1.
116. Pearson, R.G. *J. Org. Chem.* **1989**, *54*, 1423.
117. Parr, R.G.; Yang, W. *Density Functional Theory in Atoms and Molecules*; Oxford University Press: London, 1989.
118. Perdew, J.P.; Parr, R.G.; Levy, M.; Balduz, J.L. Jr., *J. Phys. Chem.* **1982**, *49*, 1691.
119. Mulliken, R.S. *J. Phys. Chem.* **1934**, *2*, 782.
120. Parr, R.G.; Donnelly, R.A.; Levy, M.; Palke, W.E. *J. Chem. Phys.* **1978**, *68*, 3801.
121. Pearson, R.G. *Acc. Chem. Res.* **1993**, *26*, 250.
122. Parr, R.G. *Acc. Chem. Res.* **1993**, *26*, 256.
123. Orsky, A.R.; Whitehead, M.A. *Can. J. Chem.* **1987**, *65*, 1970.
124. Sanderson, R.T. *Science* **1955**, *121*, 207.
125. Liu, G.H.; Parr, R.G. *J. Am. Chem. Soc.* **1995**, *117*, 3179.
126. Ritchie, C.D.; Uschold, R.E. *J. Am. Chem. Soc.* **1968**, *90*, 2821.
127. For titrations in nonaqueous media see the following and references 128 and 129: Koppel, I.A.; Koppel, J.B.; Degerbeek, F.; Grehn, L.; Ragnarsson, U. *J. Org. Chem.* **1991**, *56*, 7172.
128. Gyenes, I. *Titration in Non-aqueous Media*; Van Nostrand: Princeton, NJ, 1967.
129. Koppel, I.A.; Pihl, V.; Koppel, J.; Anvia, F.; Taft, R.W. *J. Org. Chem.* **1994**, *116*, 8654.
130. Terrier, F.; Croisat, D.; Chatrousse, A.P.; Pouet, M.J.; Hallé, J.C.; Jacob, G. *J. Org. Chem.* **1992**, *57*, 3684.
131. Ballinger, P.; Long, F.A. *J. Am. Chem. Soc.* **1959**, *81*, 1050.
132. Buncel, E.; Menon, B.C. *J. Am. Chem. Soc.* **1977**, *99*, 4457.

133. Buncel, E.; Menon, B.C. *J. Chem. Soc., Chem. Commun.* **1976,** 648.
134. Solov'yanov, A.A.; Beletskaya, I.P.; Reutov, O.A. *Zh. Org. Khim.* **1983**, *19*, 1822 (Engl. Transl., p 1593).
135. Streitwieser, A.; Kim, Y.J. *J. Am. Chem. Soc.* **2000**, *122*, 11783. See reference 186.
136. Conant, J.B.; Wheland, G.W. *J. Am. Chem. Soc.* **1932**, *54*, 1212.
137. McEwen, W.K. *J. Am. Chem. Soc. Chem.* **1936**, *58*, 1124.
138. Brown, T.L. *Adv. Organomet. Chem.* **1965**, *3*, 365.
139. Sovell, V.M.; Kimura, B.Y.; Spiro, T.G. *J. Coord. Chem.* **1971**, *1*, 107.
140. Boche, G. *Angew. Chem., Int. Ed. Engl.* **1989**, *28*, 277.
141. Arnett, E.M.; Palmer, C.A. *J. Am. Chem. Soc.* **1990**, *112*, 7354.
142. Arnett, E.M.; Fisher, F.J.; Nichols, M.A.; Ribeiro, A.A. *J. Am. Chem. Soc.* **1990**, *112*, 801.
143. Facchetti, A.; Kim, Y.J.; Streitwieser, A. *J. Org. Chem.* **2000**, *65*, 4195.
144. Ciula, J.C.; Streitwieser, A. *J. Org. Chem.* **1992**, *57*, 431.
145. Jackman, L.M.; Lange, B.C. *J. Am. Chem. Soc.* **1981**, *103*, 4494.
146. Jackman, L.M.; Dunne, T.S. *J. Am. Chem. Soc.* **1985**, *107*, 2805.
147. Abbotto, A.; Streitwieser, A. *J. Am. Chem. Soc.* **1995**, *117*, 6358.
148. Ashby, E.C.; Nodding, S.A. *J. Org. Chem.* **1979**, *44*, 4371.
149. Seebach, D.; Beck, A.K.; Studer, A. *Mod. Synth. Methods* **1995**, *7*,1.
150. Juaristi, E.; Beck, A.K.; Hansen, J.; Matt, T.; Mukhopadhyay, T.; Simson, M.; Seebach, D. *Synthesis* **1993**, 1271.
151. Novak, D.P.; Brown, T.L. *J. Am. Chem. Soc.* **1972**, *94*, 3793.
152. Sun, X.; Winemiller, M.D.; Xiang, B.; Collum, D.B. *J. Am. Chem. Soc.* **2001**, *123*, 8039.
153. Hall, P.L.; Gilchrist, J.H.; Harrison, A.T.; Fuller, D.J.; Collum, D.B. *J. Am. Chem. Soc.* **1991**, *113*, 9575.
154. Henderson, K.W.; Dorigo, A.E.; Liu, Q.Y.; Williard, P.G.; Schleyer, P. von R.; Bernstein, P.R. *J. Am. Chem. Soc.* **1996**, *118*, 1339.
155. Collum, D.B. *Acc. Chem. Res.* **1993**, *26*, 227.
156. Parsons, R.L., Jr.; Fortunak, J.M.; Darow, R.L.; Harris, G.D.; Kauffman, G.S.; Nugent, W.A.; Winemiller, M.D.; Briggs, T.F.; Xiang, B.; Collum, D.B. *J. Am. Chem. Soc.* **2001**, *123*, 9135.
157. Amstutz, R.; Schweizer, W.B.; Seebach, D.; Dunitz, J.D. *Helv. Chim. Acta* **1981**, *64*, 2617.
158. Williard, P.G.; Liu, Q.Y. *J. Am. Chem. Soc.* **1993**, *115*, 3380.
159. Li, Y.; Paddon-Row, M.N.; Houk, K.N. *J. Org. Chem.* **1990**, *55*, 481.
160. Leung-Toung, R.; Tidwell, T.T. *J. Am. Chem. Soc.* **1990**, *112*, 1042.
161. Szwarc, M. *Ions and Ion-pairs in Organic Chemistry. 1*; Interscience: New York, 1972.
162. Szwarc, M. *Ions and Ion-pairs in Organic Chemistry. 2*; Interscience: New York, 1974.
163. Streitwieser, A.; Brauman, J.I.; Hammons, J.H.; Pudjaatmaka, A.H. *J. Am. Chem. Soc.* **1967**, *87*, 384.
164. Streitwieser, A.; Ciuffarin, E.; Hammons, J.H. *J. Am. Chem. Soc.* **1967**, *89*, 63.
165. Streitwieser, A.; Reuben, D.M.E. *J. Am. Chem. Soc.* **1971**, *93*, 1794.
166. Streitwieser, A.; Murdoch, J.R.; Hafelinger, G.; Chang, J.J. *J. Am. Chem. Soc.* **1973**, *95*, 4248.
167. Streitwieser, A.; Guibé, F. *J. Am. Chem. Soc.* **1978**, *100*, 4532.
168. Streitwieser, A.; Padgett, W.M., III; Schwager, I. *J. Phys. Chem.* **1964**, *68*, 2922.
169. Streitwieser, A.; Juaristi, E.; Nebenzahl, L.L. In *Comprehensive Carbanion Chemistry. Part A. Structure and Reactivity*; Buncel, E.; Durst, T., Eds.; Elsevier: Amsterdam,

1980; pp 323–381. This reference also reviews critically some data from the Bordwell group.
170. Streitwieser, A.; Nebenzahl, L.L. *J. Am. Chem. Soc.* **1976**, *94*, 5730.
171. Streitwieser, A.; Chang, C.J.; Reuben, D.M.E. *J. Am. Chem. Soc.* **1972**, *94*, 5730.
172. Silvester, M.J. *Aldrichimica Acta* **1995**, *28*, 45.
173. Chambers, R.D.; Bryce, M.R. In *Comprehensive Carbanion Chemistry. Part. C. Ground and Excited State Reactivity*; Buncel E.; Durst, T., Eds.; Elsevier: Amsterdam, 1987, pp 271–321.
174. Gutsev, G.L.; Ziegler, T. *Can. J. Chem.* **1991**, *69*, 993.
175. Juaristi, E.; Streitwieser, A. *J. Org. Chem.* **1978**, *43*, 2704.
176. Belikov, V.M.; Mairanovskii, S.G.; Korchemuaya, Ts. B.; Novikov, S.S.; Klimova, V.A. *Izv. Akad. Nauk. SSSR, Otd., Khim. Nauk.* **1960**, 1787.
177. Streitwieser, A.; Holtz, D.; Ziegler, J.D.; Brokaw, M.L., Guibé, F. *J. Am. Chem. Soc.* **1976**, *98*, 5229.
178. Terekhova, M.I.; Petrov, E.S.; Mesyats, S.P.; Shatenshtein, A.I. *Zh. Obshch. Khim.* **1975**, 2351.
179. Fraser, R.R. *Tetrahedron Lett.* **1982**, *23*, 4195.
180. Miah, M.A.; Fraser, R.R. *J. Bangladesh. Chem. Soc.* **1989**, *2*, 77.
181. Miah, M.A.; Fraser, R.R. *Indian J. Chem. Sect. A* **1990**, *29A*, 588.
182. Fraser, R.R.; Savard, S.; Mansour, T.S. *Can. J. Chem.* **1985**, *63*, 3505.
183. Fraser, R.R.; Mansour, T.S.; Savard, S. *J. Org. Chem.* **1985**, *50*, 3232.
184. Bors, D.A.; Kaufman, M.J.; Streitwieser, A. *J. Am. Chem. Soc.* **1985**, *107*, 6975.
185. Kaufman, M.J.; Gronert, S.; Bors, D.A. *J. Am. Chem. Soc.* **1987**, *109*, 602.
186. Kaufman, M.J.; Gronert, S.; Streitwieser, A. *J. Am. Chem. Soc.* **1988**, *110*, 2829.
187. Streitwieser, A.; Ciula, J.A.; Krom, J.A.; Thiele, G. *J. Org. Chem.* **1991**, *56*, 1074.
188. Xie, L.; Streitwieser, A. *J. Org. Chem.* **1995**, *60*, 1339.
189. Gareyev, R.; Streitwieser, A. *J. Org. Chem.* **1996**, *61*, 1742.
190. Taft, R.W.; Bordwell, F.G. *Acc. Chem. Res.* **1988**, *21*, 463.
191. Riddick, J.A.; Bunger, W.B.; Sakano, T.K. *Techniques of Chemistry, Vol. II. Organic Solvents. Physical Properties and Methods of Purification*; 4th ed.; Wiley: New York, 1986; pp 310, 609, 685.
192. Steiner, E.C.; Gilbert, J.M. *J. Am. Chem. Soc.* **1965**, *87*, 382.
193. Steiner, E.C.; Starkey, J.D. *J. Am. Chem. Soc.* **1967**, *89*, 2651.
194. Bordwell, F.G.; Algrim, D.; Fried, H.E. *J. Chem. Soc., Perkin Trans. 2* **1979**, 726.
195. Arnett, E.M.; Harrelson, J.A., Jr. *J. Am. Chem. Soc.* **1987**, *109*, 809.
196. Zhang, X.M.; Bordwell, F.G. *J. Phys. Org. Chem.* **1994**, *7*, 751.
197. Arnett, E.M.; Maroldo, S.G.; Schilling, S.L.; Harrelson, J.A., Jr. *J. Am. Chem. Soc.* **1984**, *106*, 6759.
198. Babler, J.H.; Liptak, V.P.; Phan, N. *J. Org. Chem.* **1996**, *61*, 416.
199. Arnett, E.M.; Moe, K.D. *J. Am. Chem. Soc.* **1991**, *113*, 7288.
200. Streitwieser, A.; Heathcock, C.H.; Kosower, E. *Introduction to Organic Chemistry*, 4th ed.; Maxwell-MacMillan: Toronto, 1992; pp 889–891.
201. Hammett, L.P. *J. Am. Chem. Soc.* **1937**, *59*, 96.
202. Bordwell, F.G.; Bares, J.E.; Bartmess, J.E.; Drucker, G.E.; Gerhold, J.; McCollum, G.J.; Van Der Puy, M.; Vanier, N.R.; Mathews, W.S. *J. Org. Chem.* **1977**, *42*, 326.
203. Bordwell, F.G.; Bares, J.E.; Bartmess, J.E.; McCollum, G.J.; Van Der Puy, M.; Vanier, N.R.; Mathews, W.S. *J. Org. Chem.* **1977**, *42*, 321.
204. Bordwell, F.G.; Zhao, Y. *J. Org. Chem.* **1995**, *60*, 6348.
205. Bordwell, F.G.; Van Der Puy, M.; Vanier, N.R. *J. Org. Chem.* **1976**, *41*, 1885.
206. Bordwell, F.G.; Van Der Puy, M.; Vanier, N.R. *J. Org. Chem.* **1976**, *41*, 1883.
207. Hammett, L.P.; Deyrup, A.J. *J. Am. Chem. Soc.* **1932**, *54*, 2721.

208. Rochester, C.H. *Acidity Functions*; Academic Press: London, 1970.
209. Arnett, E.M.; Scorrano, G. *Adv. Phys. Org. Chem.* **1976**, *13*, 83.
210. Cox, R.A.; Stewart, R. *J. Am. Chem. Soc.* **1976**, *98*, 488.
211. Cram, D.J.; Mateos, J.L.; Hauck, F.; Langemann, A.; Kopecky, K.R.; Nielsen, W.D.; Allinger, J. *J. Am. Chem. Soc.* **1959**, *81*, 5774.
212. Cram, D.J.; Rickborn, B.; Knox, G.R. *J. Am. Chem. Soc.* **1960**, *82*, 6412.
213. Parker, A.J. *Chem. Rev.* **1969**, *69*, 1.
214. Stewart, R. *The Proton: Applications to Organic Chemistry*; Academic Press: Orlando, FL, 1985; pp 65–67.
215. Langford, C.H.; Burwell, R.L. *J. Am. Chem. Soc.* **1960**, *82*, 1503.
216. Buncel, E.; Symons, E.A.; Stewart, R.; Dolman, D. *Can. J. Chem.* **1970**, *48*, 3354.
217. Bethel, D.; Cockerill, A.F. *J. Chem. Soc. B* **1966**, 913.
218. Dolman, D.; Stewart, R. *Can. J. Chem.* **1967**, *45*, 911.
219. Garcia, B.; Leal, J.M. *J. Phys. Org. Chem.* **1991**, *4*, 413.
220. Garcia, B.; Leal, J.M.; Herrero, A.; Palacios, J.C. *J. Chem. Soc., Perkin Trans. 2* **1988**, 1742.
221. Cox, R.A.; Yates, K. *Can. J. Chem.* **1983**, *61*, 22.
222. Cox, R.A. *Acc. Chem. Res.* **1987**, 20, 27.
223. Bunnett, J.F.; Olsen, F.P. *Can. J. Chem.* **1966**, *44*, 1899.
224. Marziano, N.C.; Cimino, G.M.; Passerini, R.C. *J. Chem. Soc., Perkin Trans. 2* **1973**, 1915.
225. Arnett, E.M.; Moriarty, T.C.; Small, L.E.; Rudolph, J.P.; Quirk, R.P. *J. Am. Chem. Soc.* **1973**, *95*, 5229.
226. Breslow, R.; Chu, W. *J. Am. Chem. Soc.* **1973**, *95*, 411.
227. Wasielewski, M.R.; Breslow, R. *J. Am. Chem. Soc.* **1976**, *98*, 4222.
228. Breslow, R.; Grant, J.L. *J. Am. Chem. Soc.* **1977**, *99*, 7745.
229. Fox, M.A. In *Comprehensive Carbanion Chemistry. Part C. Ground and Excited State Reactivity*; Buncel, E.; Durst, T., Eds.; Elsevier: Amsterdam, 1987; pp 107–116.
230. Breslow, R. *Pure Appl. Chem.* **1974**, *40*, 493.
231. Schiffrin, D. J. *Discuss Faraday Soc.* **1973**, *56*, 1975.
232. Chattaraj, P.K.; Cedillo, A.; Parr, R.G.; Arnett, E.M. *J. Org. Chem.* **1995**, *60*, 4707.
233. Guthrie, J.P.; Cossar, J.; Klym, A. *J. Am. Chem. Soc.* **1982**, *104*, 895.
234. Guthrie, J.P.; Cossar, J.; Klym, A. *J. Am. Chem. Soc.* **1984**, *106*, 1351.
235. Guthrie, J.P.; Cossar, J.; Klym, A. *Can. J. Chem.* **1987**, *65*, 2154.
236. Buncel, E.; Chuaqui, C. *Can. J. Chem.* **1976**, *54*, 673.
237. Buncel, E.; Elvidge, J.A.; Jones, J.R.; Walkin, K.T. *J. Chem. Res., Synop.* **1980**, 272.
238. Buncel, E.; Norris, A.R.; Elvidge, J.A.; Jones, J.R.; Walkin, K.T. *J. Chem. Res., Synop.* **1980**, 326.
239. Brønsted, J.N.; Pedersen, K.J. *Z. Phys. Chem.* **1924**, *108*, 185.
240. Eigen, M. *Angew. Chem., Int. Ed. Engl.* **1964**, *3*, 1.
241. Brønsted, J.N.; Guggenheim, E.A. *J. Am. Chem. Soc.* **1927**, *49*, 2554.
242. Ahrens, M.L.; Eigen, M.; Druse, W.; Maas, G. *Ber. Bunsen-Ges. Phys. Chem.* **1970**, *74*, 380.
243. Freter, R.; Pohl, E.R.; Wilson, J.M.; Hupe, D.J. *J. Org. Chem.* **1979**, *50*, 1771.
244. Hupe, D.J.; Wu, D. *J. Am. Chem. Soc.* **1977**, *99*, 7648.
245. Bordwell, F.G.; Cripe, T.A.; Hughes, D.L. In *Nucleophilicity*; Harris, J.M.; McManus, S.P., Eds.; Advances in Chemistry Series 215; American Chemical Society: Washington, DC, 1985; pp 137–153.
246. For example, Harris, J.M.; Sedaghat-Herati, M.R.; McManus, S.P. *J. Org. Chem.* **1988**, *47*, 3224.

247. An issue of the *Israel Journal of Chemistry* is devoted to consideration of the RSP: *Isr. J. Chem.* **1985**, *26*.
248. Exner, O. *J. Chem. Soc., Perkin Trans. 2* **1993**, 973.
249. Buncel, E.; Wilson, H. *J. Chem. Educ.* **1987**, *64*, 475.
250. Bernasconi, C.F. *Acc. Chem. Res.* **1987**, *20*, 301.
251. Bernasconi, C.F.; Wenzel, P.J. *J. Am. Chem. Soc.* **2001**, *123*, 7146.
252. Bernasconi, C.F. *Adv. Phys. Org. Chem.* **1992**, *27*, 119.
253. Bordwell, F.G.; Boyle, W.J., Jr.; Hautala, J.A.; Lee, K.C. *J. Am. Chem. Soc.* **1969**, *91*, 4002.
254. Bordwell, F.G. *Acc. Chem. Res.* **1970**, *3*, 456.
255. Blandamer, M.J.; Robertson, R.E.; Scott, J.M.W. *J. Am. Chem. Soc.* **1982**, *104*, 1136.
256. For examples where the Brønsted catalysis law is well behaved see references 240, 257, and 258.
257. Bell, R.P.; Lidwell, O.M. *Proc. R. Soc. London, Ser. A* **1940**, *176*, 88.
258. Bell, R.P. *The Proton in Chemistry*, 2nd Ed.; Cornell University Press: Ithaca, NY, 1973; p 203.
259. Lin, A.C.; Chiang, Y.; Dahlberg, D.B.; Kresge, A.J. *J. Am. Chem. Soc.* **1983**, *105*, 5380.
260. Kresge, A.J. *Chem. Soc. Rev.* **1973**, *2*, 475.
261. Buncel, E.; Um, I.H.; Hoz, S. *J. Am. Chem. Soc.* **1989**, *11*, 971.
262. Buncel, E.; Tarkka, R.M., Hoz, S. *J. Chem. Soc., Chem. Commun.* **1993**, 109.
263. Tarkka, R.M.; Park, W.K.C.; Liu, P.; Buncel, E.; Hoz, S. *J. Chem. Soc., Perkin Trans. 2*, **1994**, 2439.
264. Tarkka, R.M.; Buncel, E. *J. Am. Chem. Soc.* **1995**, *117*, 1503.
265. Terrier, F.; Moutiers, G.; Xiao, L.; Le Guével, E.; Guir, F. *J. Org. Chem.* **1995**, *60*, 1748.
266. Kresge, A.J. *Can. J. Chem.* **1974**, *52*, 1897.
267. Kresge, A.J. *Acc. Chem. Res.* **1975**, *8*, 354.
268. Kresge, A.J. In *Proton Transfer Reactions*; Caldin, E.F.; Gold, V., Eds.; Chapman & Hall, London, 1975; pp 140–200.
269. Kresge, A.J.; Leibovitch, M. *Can. J. Chem.* **1990**, *68*, 1786.
270. Yamataka, H.; Mustanir, A.; Mishima, M. *J. Am. Chem. Soc.* **1999**, *121*, 10223.
271. Melander, L.; Saunders, W.H., Jr. *Reaction Rates of Isotopic Molecules*; Wiley: New York, 1980.
272. Shiner, V.H., Jr.; Wilgis, F.P. *Isotopes in Organic Chemistry*; Elsevier: Amsterdam, 1992.
273. Margolin, Z.; Long, F.A. *J. Am. Chem. Soc.* **1973**, *95*, 2757.
274. Koch, H.F. *Acc. Chem. Res.* **1984**, *17*, 137.
275. Jones, J. R.; Stewart, R. *J. Chem. Soc. B* **1967**, 1173.
276. Warkentin, J.; Barnett, C. *J. Am. Chem. Soc.* **1968**, *90*, 4629.
277. Buncel, E.; Zabel, A.W. *J. Am. Chem. Soc.* **1967**, *89*, 3082.
278. Buncel, E.; Symons, E.A. *J. Org. Chem.* **1973**, *38*, 1201.
279. Buncel, E.; Zabel, A.W. *Can. J. Chem.* **1981**, *59*, 3177.
280. Cram, D.J.; Kollmeyer, W.D. *J. Am. Chem. Soc.* **1968**, *90*, 1791.
281. Hofman, J.E.; Muller, R.J.; Schriesheim, A. *J. Am. Chem. Soc.* **1963**, *85*, 3002.
282. Hofman, J.E.; Schriesheim, A.; Nickols, R.E. *Tetrahedron Lett.* **1965**, *22*, 1725.
283. Streitwieser, A.; Taylor, D.R. *J. Chem. Soc., Chem. Commun.* **1970**, 1248.
284. Jones, J.R.; Walkin, K.T.; Davey, J.P.; Buncel, E. *J. Phys. Chem.* **1989**, *93*, 1362.
285. Davey, J.P.; Jones, J.R.; Buncel, E. *Can. J. Chem.* **1986**, *64*, 1246.
286. Kresge, A.J.; Powell, M.F. *J. Org. Chem.* **1986**, *51*, 822.
287. Kresge, A.J.; Powell, M.F. *J. Org. Chem.* **1986**, *51*, 819.

288. Albery, W.J. *Prog. React. Kinet.* **1967**, *4*, 355.
289. Jones, J.R.; Taylor, S.E. *J. Chem. Soc., Perkin Trans. 2* **1979**, 1587.
290. Jones, J.R.; Taylor, S.E. *Chem. Soc. Rev.* **1981**, *10*, 329.
291. Brodsky, N.; Nguyen, N.M.; Rowan, N.S.; Storm, C.D.; Butcher, R.J.; Sinn, E. *Inorg. Chem.* **1984**, *23*, 891.
292. Jones, J.R.; Taylor, S.E. *J. Chem. Soc., Perkin Trans. 2* **1979**, 1773.
293. Martin, R.B.; Scheller-Krattiger, V. *J. Am. Chem. Soc.* **1982**, *104*, 1078.
294. Buncel, E.; Clement, O. *J. Chem. Soc., Perkin Trans. 2* **1995**, 1333.
295. Buncel, E.; Clement, O.; Onyido, I. *J. Am. Chem. Soc.* **1994**, *116*, 2679.
296. Buncel, E.; Joly, H.A.; Jones, J.R. *Can. J. Chem.* **1986**, *64*, 1240.
297. Buncel, E.; Yang, F.; Moir, R.Y.; Onyido, I. *Can. J. Chem.* **1995**, *73*, 772.
298. Clement, O.; Roszak, A.W.; Buncel, E. *J. Am. Chem. Soc.* **1996**, *118*, 612.
299. Buncel, E.; Joly, H.A.; Yee, D.C. *Can. J. Chem.* **1989**, *67*, 1426.
300. Buncel, E.; Boone, C.; Joly, H.A. *Inorg. Chim. Acta* **1986**, *125*, 167.
301. Buncel, E.; Kumar, A.; Norris, A.R. *J. Inorg. Biochem.* **1986**, *64*, 442.
302. Stewart, R. *The Proton: Application to Organic Chemistry*; Academic Press: Orlando, FL, 1985; pp 291–294.
303. Stewart, R.; Srinivasan, R. *Acc. Chem. Res.* **1978**, *11*, 271.
304. Srinivasan, R.; Gumbley, S.; Stewart, R. *Tetrahedron* **1979**, *35*, 1257.
305. Lee, T.W.S.; Rettig, S.J.; Stewart, R.; Trotter, J. *Can. J. Chem.* **1984**, *62*, 1194.
306. Buncel, E. *Carbanions: Mechanistic and Isotopic Aspects*; Elsevier: Amsterdam, 1975; p 19.
307. Brown, A.L.; Chiang, Y.; Kresge, A.J.; Tang, Y.S.; Wang, W.H. *J. Am. Chem. Soc.* **1989**, *111*, 4918.
308. Chiang, Y.; Jones, J, Jr.; Kresge, A.J. *J. Am. Chem. Soc.* **1994**, *116*, 8358.
309. Jaffe, H.H. *Chem. Rev.* **1953**, *53*, 191.
310. Bordwell, F.G.; Cornforth, F.J. *J. Org. Chem.* **1978**, *43*, 1763.
311. Lahti, M.; Lindstrom, R.; Lonnberg, H. *J. Chem. Soc., Perkin Trans. 2* **1989**, 603.
312. McMahon, T.B.; Kebarle, P. *J. Am. Chem. Soc.* **1976**, *98*, 3399.
313. Cumming, J.B.; Kebarle, P. *J. Am. Chem. Soc.* **1977**, *99*, 5818.
314. Cumming, J.B.; Magnera, T.F.; Kebarle, P. *Can. J. Chem.* **1977**, *55*, 3474.
315. Cumming, J.B.; Kebarle, P. *J. Am. Chem. Soc.* **1978**, *100*, 1835.
316. Baldeschweiler, J.D.; Woodgate, S.G. *Acc. Chem. Res.* **1971**, *4*, 114.
317. McIver, R.T. Jr., *Rev. Sci. Instrum.* **1978**, *49*, 111.
318. Fujio, M.; McIver, R.T. Jr.; Taft, R.W. *J. Am. Chem. Soc.* **1981**, *103*, 4017.
319. Koppel, I.A.; Taft, R.W.; Anvia, F.; Zhu, S.Z.; Hu, L.Q.; Sung, K.S.; DesMarteau, D.D.; Yagupolskii, L.M.; Yagupolskii, Y.L.; Ignatev, N.V.; Kondratenko, N.V.; Volkonskii, A.Y.; Vlasov, V.M.; Notario, R.; Maria, P.C. *J. Am. Chem. Soc.* **1994**, *116*, 3047.
320. Schmeltekopf, A.L.; Fehsenfeld, F.C. *J. Chem. Phys.* **1970**, *53*, 3173.
321. Fehsenfeld, F.C. *Int. J. Mass Spectrom. Ion Phys.* **1975,** *16*, 151.
322. Bohme, D.K.; Young, L.B. *J. Am. Chem. Soc.* **1970**, *92*, 3301.
323. Bohme D.K.; Lee-Ruff, E.; Young, L.B. *J. Am. Chem. Soc.* **1971**, *93*, 4608.
324. Bohme, D.K.; Lee-Ruff, E.; Young, L.B. *J. Am. Chem. Soc.* **1972**, *94*, 5153.
325. Pellerite, M.J.; Brauman, J.I. In *Comprehensive Carbanion Chemistry. Part A. Structure and Reactivity*; Buncel, E.; Durst, T., Eds.; Elsevier: Amsterdam, 1980; pp. 55–97.
326. Hunter, E.P.; Lias, S.G. *J. Phys. Chem. Ref. Data* **1998**, *27*, 413; and references therein.
327. Lias, S.G.; Bartmess, J.E.; Liebman, J.F.; Holmes, J.L.; Levin, D.; Mallard, G.W. *J. Phys. Chem. Ref. Data* **1988**, *17* (Suppl. 1).

328. Brauman, J.I.; Blair, L.K. *J. Am. Chem. Soc.* **1968**, *90*, 5636.
329. Brauman, J.I.; Blair, L.K. *J. Am. Chem. Soc.* **1968**, *90*, 6561.
330. Kebarle, P. *Annu. Rev. Phys. Chem.* **1977**, *28*, 445.
331. Lau, Y.K.; Saluja, P.P.S. ; Kebarle, P.; Alder, R.W. *J. Am. Chem. Soc.* **1978**, *100*, 7328; and references therein.
332. Lau, Y.K. Ph.D. Thesis; University of Alberta: Edmonton, AB, 1979.
333. Taft, R.W. *Prog. Phys. Org. Chem.* **1983**, *14*, 248.
334. Meot-Ner, M.; Sieck, L.W. *J. Am. Chem. Soc.* **1991**, *113*, 4320.
335. Sieck, L.W. *J. Phys. Chem.* **1997**, *101*, 8140.
336. Szuljecko, J.; McMahon, T.B. *J. Am. Chem. Soc.* **1993**, *115*, 7841; and references therein.
337. Smith, B.J.; Radom, L. *J. Am. Chem. Soc.* **1993**, *115*, 4885; and references therein.
338. Bernasconi, C.F.; Wenzel, P.J. *J. Org. Chem.* **2001**, *66*, 968.
339. For the searchable database see: http//webbook.nist.gov/chemistry.
340. Brinkman, E.A.; Berger, S.; Brauman, J.I. *J. Am. Chem. Soc.* **1994**, *116*, 8304.
341. Damrauer, R.; Hankin, J.A. *Chem. Rev.* **1995**, *95*, 1137.
342. Mishima, M.; McIver, R.T. Jr.; Taft, R.W.; Bordwell, F.G.; Olmstead, W.N. *J. Am. Chem. Soc.* **1984**, *106*, 2717.
343. Taft, R.W.; Abboud, J.L.; Anvia, F.; Berthelot, M.; Fujio, M.; Gal, J.F.; Headley, A.D.; Henderson, W.G.; Koppel, I.; Qian, J.H.; Mishima, M.; Taagepera, M.; Ueji, S. *J. Am. Chem. Soc.* **1988**, *110*, 1797.
344. Coetzee, J.F.; Padmanabhan, G.R. *J. Am. Chem. Soc.* **1965**, *87*, 5005.
345. Leffek, K.T.; Preuszynski, P.; Thanapaalasingham, K. *Can. J. Chem.* **1989**, *67*, 590.
346. Benoit, T.L.; Baulet, D.; Frechette, M. *Can. J. Chem.* **1988**, *66*, 3038.
347. Taft, R.W.; Topsom, R.D. *Prog. Phys. Org. Chem.* **1987**, *16*, 1.
348. Catalán, J.; Fabero, J.; Sánchez-Cabezudo, M.; De Paz, J.L.G.; Taft, R.W. *J. Phys. Org. Chem.* **1996**, *9*, 87.
349. Abraham, M.H.; Grellier, P.L.; Abboud, J.L.M.; Doherty, R.M.; Taft, R.W. *Can. J. Chem.* **1988**, *66*, 2673.
350. Abraham, M.H.; Grellier, P.L.; Prior, D.; Morris, J.J.; Taylor, P.J.; Maria, P.C.; Gal, J.F. *J. Phys. Org. Chem.* **1989**, *2*, 243.
351. Matimba, H.E.K.; Crabbendan, A.M.; Ingemann, S.; Nibbering, N.M.M. *Int. J. Mass. Spectrom. Ion Processes.* **1992**, *114*, 85.
352. Cumming, J.B.; Kebarle, P. *Can. J. Chem.* **1978**, *56*, 1.
353. Bordwell, F.G.; Imes, R.H.; Steiner, E.C. *J. Am. Chem. Soc.* **1967**, *89*, 3905.
354. Caldwell, G.; Renneboog, R.; Kebarle, P. *Can. J. Chem.* **1989**, *67*, 611.
355. Bartmess, J.E.; McIver, R.T. In *Gas Phase Ion Chemistry, Vol. 2.*; Bowers, M.T., Ed.; Academic Press: New York, 1979; Chapter 11.
356. Arnett, E.M.; Johnson, D.E.; Small, L.E. *J. Am. Chem. Soc.* **1975**, *97*, 5598.
357. Gal, J.F.; Maria, J.C. *Prog. Phys. Org. Chem.* **1990**, *17*, 159.
358. Bordwell, F.G.; Bartmess, J.E.; Hautala, J.A. *J. Org. Chem.* **1978**, *43*, 3095.
359. Bernasconi, C.F.; Wenzel, P.J. *J. Am. Chem. Soc.* **2001**, *123*, 2430.
360. Werstiuk, N.H. *Can. J. Chem.* **1988**, *66*, 2958.
361. Pugh, J.K.; Streitwieser, A. *J. Org. Chem.* **2001**, *66*, 1334.
362. Merrill, G.N.; Kass, S.R. *J. Phys. Chem.* **1996**, *100*, 17465.
363. Chiang, Y.; Kresge, A. J. *Can. J. Chem.* **2000**, *78*, 1627.

2

Stability of Carbanions

2.1 KINETIC ISOTOPE EFFECTS AND THE STUDY OF CARBON ACIDS

In Chapter 1 we considered the acidity of carbon acids and noted a number of the factors that enhance or diminish the acidity of these acids. Among the methods used to determine the pK_a values of carbon acids, we examined kinetic methods that, generally, rely upon application of the Brønsted catalysis law (Section 1.4). Further, we noted that isotopic exchange methods specifically require a clear linkage of the rate of isotopic exchange to the rate of proton transfer and that this is usually met by finding a large kinetic isotope effect (KIE), inasmuch as a large KIE indicates significant C–H, C–D, or C–T bond breakage in the rate-determining step of the reaction under investigation. Clearly, the KIE is not only an adjunct to the measurement of kinetic acidities but also a valuable probe of carbon acidity in its own right.

In this section we shall briefly explore the KIE, and specifically its origin, with isotopes of hydrogen (i.e., deuterium and tritium). More extensive discussions of the KIE, including kinetic isotope effects with other elements,[1] may be found in the mongraphs by Melander[2] and Saunders[3] and Shiner and Wilgis.[4] A number of other works are also extant[5–11] that focus on both the theory[10] and various applications of isotopes to organic chemistry.

2.1.1 Simple Zero-Point Energy (Two-Atom) Model

In our simplified discussion, the rupture of a carbon–hydrogen (or carbon–deuterium or –tritium) covalent bond to yield a carbanion and the conjugate acid of the base, B, can be represented by an energy-reaction coordinate profile (Figure 2.1). In this profile the *zero-point energy* for the relevant C–H bond (ZPE^{C-H}) and the corresponding C–D bond (ZPE^{C-D}) are shown. The zero-point energy for the bond is defined as the vibrational energy of the bond that exists even at a temperature of absolute zero (0 K) and, therefore, the C–H bond (equation 2.1) is located in the vibrational energy level assigned the vibrational quantum number, n, of zero. Of course, higher vibrational states are assigned their own values of n, increasing from the zero point in increments of 1. Therefore, in this quantum mechanical treatment the various vibrational energy levels have energies given by equation 2.2 and for the ZPE^{C-H} energies given by equation 2.3.

$$\text{—}\overset{|}{\underset{|}{C}}\text{—H (D,T)} + B^{-} \longrightarrow \text{—}\overset{|}{\underset{|}{C}}^{-} + B\text{—H} \tag{2.1}$$

$$E = (n + 1/2)\, h\nu \tag{2.2}$$

$$ZPE^{C-H} = 1/2\, h\nu \tag{2.3}$$

In the simplest classical model the carbon moiety (equation 2.1) and the attached hydrogen are treated as two point masses or "two balls connected by a spring". Hook's law then defines the frequency of the vibration, ν, by equation 2.4, where μ is the reduced mass according to equation 2.5.

$$\nu = 1/2\pi\sqrt{k/\mu} \tag{2.4}$$

$$\mu = (m_1 m_2)/(m_1 + m_2) \tag{2.5}$$

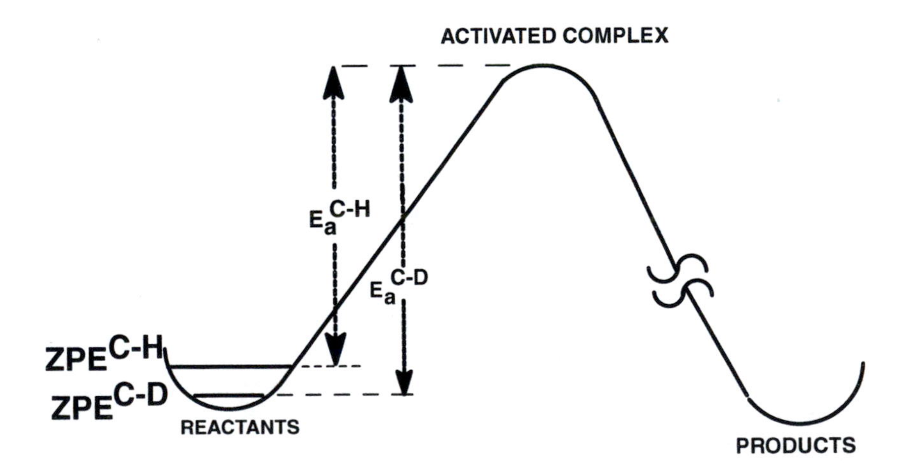

Figure 2.1. Energy profile that illustrates the zero-point energy origin of the kinetic isotope effect for carbon–hydrogen versus carbon–deuterium bond scission in a primary isotope effect.

When one of the masses (m_2, say) is much greater than the other, as in the case where the carbon moiety (equation 2.1) is much more massive than the hydrogen (or deuterium), μ is approximated by the smaller mass. Thus, in the case of equation 2.1, μ is replaced by m_1 in equation 2.4; here, m_1 is the mass of hydrogen. In equation 2.4, k represents the "spring force constant", which we recognize as the bond strength.

From equation 2.4 it is clear that the frequency of the vibration is directly proportional to the reciprocal of the square root of the mass; that is, $\nu \propto 1/(m)^{1/2}$. Thus, the frequency of vibration for the C–D bond will be lower than that for the C–H bond because the mass of deuterium is double that of hydrogen. While our model envisages the C–D frequency as being smaller than the C–H frequency by a factor of $1/\sqrt{2}$, that is, ca. 1/1.41, the maximum observed value, in fact, is approximately 1/1.35.[12] Further, from equation 2.2 we can see that a lower frequency of vibration must translate into a lower vibrational energy and, in this case, a lower ZPE. Consequently, the $ZPE^{C\text{–}D}$ is less than the $ZPE^{C\text{–}H}$, as is illustrated in Figure 2.1.

An important general principle is implicit in the energy profile of Figure 2.1, namely, that the energy profile does not vary with isotopic exchange. This arises from the understanding that a change in mass in the nucleus of an atom (as a result of changing the number of neutrons in the nucleus) should have negligible effect on the electronic distribution of the atom (wave function) and, consequently, little effect on the electron distribution in the molecule. Of course, it is the change in electronic distribution that constitutes bond breaking and bond making and defines the energy-reaction coordinate profile. Therefore, whether a C–H or α C–D bond is being broken (equation 2.1), the energy profile is, to a good approximation, unchanged.[13]

As shown in Figure 2.1, then, the KIE arises from the differences in zero-point energy between the relevant C–H and C–D bonds that are broken in the rate-determining step. In other words, the difference in activation energy between the C–H and C–D reactions (equation 2.1) is wholly reflected in the difference in zero-point energy, and the kinetic isotope effect (i.e., given as the ratio of the rate constant for the C–H reaction as compared with that for the C–D reaction) can be determined from this zero-point energy difference alone. Thus, equation 2.6,

$$\mathrm{KIE} = k_H/k_D = e^{-(E_a^{C\text{–}H} - E_a^{C\text{–}D})/RT} = e^{(ZPE^{C\text{–}D} - ZPE^{C\text{–}H})/RT} = e^{\Delta ZPE/RT} \tag{2.6}$$

may be derived from the Arrhenius treatment of activation energy. Combination of typical values for the vibrational frequency of C–H versus C–D bonds with equations 2.2 through 2.6 leads to values of k_H/k_D of 8.3, 6.9, and 4.3 at temperatures of 0 °C, 25 °C, and 100 °C, respectively. A comparable treatment that relies on transition-state theory, and incorporates a correction for the *observed* maximum differential in C–H and C–D vibration frequencies,[12] provides a KIE value of 6.5 at 25 °C, using the Eyring equation.[14] In a similar way, a typical hydrogen–tritium KIE (k_H/k_T) can be calculated to be 20 at 0 °C.[15] Equations that relate the hydrogen–tritium and hydrogen–deuterium KIE have been reported[16] and, more recently, confirmed for the H–D and H–T KIE in the hydroxide-ion-promoted enolization of acetone.[17]

2.1.2 Extended ZPE (Three-Atom) Model

Although useful, our simplified view of the KIE ignores a number of factors pertinent to the study of carbon acids. First, in our zero-point energy analysis we have considered only the stretching mode of vibration of the relevant C–H and C–D bonds; contributions from the bending modes may also increase the value of the KIE.[15] Secondly, the zero-point energy analysis implies that the C–H and C–D bonds break essentially without participation of the base that accepts the proton or deuteron. In fact, the transition state for this *primary isotope effect* plausibly includes the base; the proton (or deuteron) being transferred is partially bonded to both the original carbon and to the base in the transition state (TS), as shown in Figure 2.2. (Note that the TS depicted in Figure 2.2 illustrates *linear* transfer of the proton from the carbon acid to the base and is, therefore, an example of a linear transition state, as examined below).

Since in the transition state the proton is partly bonded both to the carbon acid and to the abstracting base, the transition state will also have vibrational modes and, more importantly, zero-point energies associated with these vibrational quantum levels. Therefore, the KIE should reflect the energy difference between the zero-point energy of the C–H bond in the reactant ground state and the zero-point energy of the same bond in the transition state as compared with the same difference for the C–D bond in a C–H/C–D primary deuterium isotope effect. (The C–H and C–D stretching modes, taken to approximate the reaction coordinate for C–H/C–D bond breaking in the present case, may be coupled to other vibrational modes; these modes lead to *secondary isotope effects*.[2–4,11,18,19] Such effects lie outside the scope of the current discussion.) We will explore the effect of TS structure on the KIE later.

All of our discussion of the KIE, thus far, has been tied to the implicit idea that in breaking the C–H and C–D bonds the reaction follows the path of minimum energy indicated by the energy profile (Figure 2.1). In short, the molecules "go over the hill" from reactants to products. However, because of the small mass of the proton, it undergoes *quantum mechanical tunneling*[20,21] more readily than does the deuteron. Thus, rather than proceeding "over the hill", the proton "tunnels through" to the product side. The salient point is that the rate constant for proton transfer thereby becomes exaggerated relative to the rate constant for deuteron transfer, and very large KIE (on the order of 25–30 for hydrogen–deuterium isotope effects) are measured.[22–24] Characteristic of tunneling is the expectation that Arrhenius-type plots (log k_H/k_D vs $1/T$) will show curvature.[25,26]

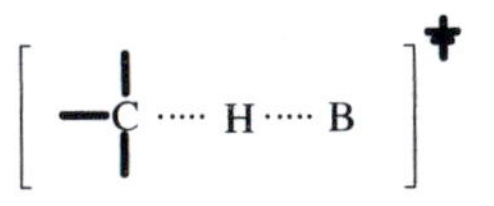

Figure 2.2. Transition state (TS) for transfer of a proton from a carbon acid to a base, B. Here, the symbol ‡ indicates a transition state structure.

2.1.3 Linear versus Nonlinear Transition States in Proton Transfer

With the proviso that the KIE in a given system adheres to the qualitative energy profile given above (Figure 2.1), the isotope effect can be used to probe the nature of the transition state for proton transfer. Two main transition-state structures may be envisaged for proton transfer: (1) a linear transition-state (as shown in Figure 2.2) and (2) various nonlinear TS structures. In this regard, the seminal analysis of degree of proton transfer and transition-state structure provided by Westheimer should be highlighted;[27] it is fair to say that prior to this work there were no universally acceptable explanations for the observation of primary KIE values (at 25 °C) smaller than 6–7.[28]

Let us consider the linear TS first. The model for this TS has already been given in Section 2.1.2, where we briefly considered TS vibrational modes. In examining the TS again it may be noted that the proton being transferred is drawn as being equidistant between the carbon atom of the acid and the abstracting base. Two possible stretching modes would be associated with this linear TS, as shown in Figure 2.3.

Clearly, in Figure 2.3 the asymmetric stretch illustrated in TS-a (by the arrows) represents progress along the reaction coordinate, and at the transition state itself (i.e., TS-a) it *is* the reaction coordinate. Therefore, it does not contribute to the zero-point energy of the transition state and, in this case, only the reactant-state zero-point energy difference (ΔZPE) need be considered. Under these circumstances the ground-state zero-point energy treatment that leads to the derivation of equation 2.6 is valid and the KIE values cited above would obtain. However, the transition state would also exhibit a symmetric stretch, as shown by TS-s, which is a mode of vibration that does not exist in the reactants. (Note that bending modes in the transition state are often taken to be similar to reactant C–H/C–D bending and that these effects are expected to cancel, as far as the primary isotope effect is concerned, but *not* the secondary KIE.[29]) More important, as drawn, TS-s is highly symmetrical and here the hydrogen atom is motionless. Since substitution of the motionless hydrogen for a motionless deuterium would have no effect on this mode of vibration, there would be no contribution of this mode to the ZPE of the transition state and, again, equation 2.6 would hold. In summary, for a highly symmetrical linear TS the KIE should be approximately that described by the simple zero-point energy analysis; that is, k_H/k_D ca. 6–7 at 25 °C.

Figure 2.3. Vibrational modes in linear proton transfer transition states. TS-a shows the antisymmetric stretching mode, while TS-s shows the symmetric stretching mode. In both TS structures the proton is positioned equidistant between carbon acid and the base.

Is it reasonable to expect the transition state for proton transfer from a carbon acid to a base to be so highly symmetrical? It is likely that only when the carbon acid and the Brønsted base are equal in base (or, conversely, acid) strength, and have similar structures, would the transition state for proton transfer approach the model we have discussed. In fact, maximum KIE values have been found when the difference in pK_a values between proton donor and acceptor have approached zero.[30–33] Conversely, small KIE could be taken to indicate that the transition state, although linear, is not symmetrical; the hydrogen (deuterium) is closer either to the carbon center or to the base center, but is not equidistant from the two sites. In this case, hydrogen or deuterium move along with the symmetric stretch of TS-s. Now, the symmetric stretching mode will contribute to the transition-state zero-point energies and partly cancel the difference in ground-state ZPE, leading to a small KIE. (Of course, the magnitude of the isotope effect will also be dependent on the position of the transition state along the reaction coordinate).

It should be emphasized that the foregoing discussion ignores the potential role of tunneling in these systems,[34] as well as the problem of internal return (Section 1.4.).[35] Focusing on the latter for a moment, if, instead of equation 2.1, equation 2.7 were to apply, a small KIE could result from rapid return of the initially formed hydrogen-bonded carbanion to the carbon acid; that is, $k_{-1} \gg k_2$. In this case, simple proton abstraction is no longer rate-determining and arguments based on the nature of the rate-determining transition state would be invalid. Thus, the low KIE ($k_H/k_D = 3$) found in the reaction of 2-phenylbutane and its deuterio analogue, 2-phenylbutane-2*d*, with the strong base, potassium *t*-butoxide in DMSO/*t*-butyl alcohol,[36] as well as the small k_H/k_D of 0.6 found in deprotonation of toluene (benzylic position),[37] have been attributed to the intervention of internal return.[38]

$$\text{—C—H(D,T)} + {}^{-}\text{B} \underset{k_{-1}}{\overset{k_1}{\rightleftharpoons}} \left[\text{—C}\text{------H(D,T)B}\right] \xrightarrow{k_2} \text{—C}^{-} + \text{H(D,T)B} \qquad (2.7)$$

However, small kinetic isotope effects may also arise from a nonlinear transition state. In this regard, the symmetric stretching mode that appears as TS-s in the linear example (Figure 2.3) could be approximated by NLTS-s in Figure 2.4. The magnitude of the arrows emphasizes that this stretching mode is symmetrical.

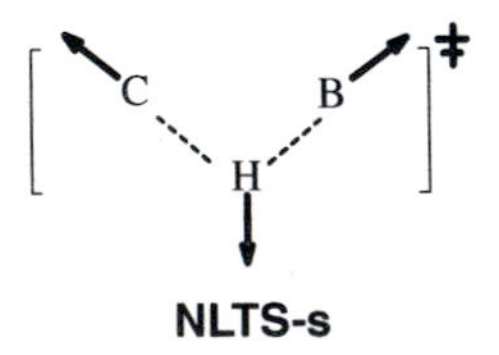

Figure 2.4. A non-linear transition state structure (NLTS-s) that illustrates the symmetric stretching vibration. Compare to TS-s depicted in Figure 2.3.

It should be recognized that NLTS-s is only one possible structure that could be postulated for a nonlinear TS and that there is a continuum of TS with varying degrees of nonlinearity for proton transfer. Nonetheless, a large contribution would be made to the TS zero-point energy by the symmetrical stretch in NLTS-s, and this contribution would largely cancel out the reactant-state ZPE differences for C–H versus C–D. A small KIE would be predicted. Further, it has been argued that for nonlinear TS the KIE should not only be small, but also independent of temperature.[39] Although this temperature independence may be characteristic of nonlinear proton transfer transition states, the use of this characteristic to assign TS structure is still controversial.[25,40–42] In this regard, it is noteworthy that Stewart has concluded,[43] primarily on the basis of hydrogen-bonding studies,[44–50] that while a linear TS is preferred in proton transfers, nonlinear TS structures are possible and may not involve an undue energy cost relative to the linear analogue.[50]

From these considerations it follows that evaluation of kinetic isotope effects requires analysis of the *summation* of all pertinent vibrational modes, both in the ground state and in the transition state (or activated complex). Thus, Figure 2.5 shows a general energy-reaction coordinate diagram that illustrates the differences in ZPE for both the ground and transition states for a light isotope (subscripted *l,i*, where *i* denotes the various vibrational modes) and the corresponding heavy isotope (subscripted *h,i*). Not only is consideration of the sum of the vibrational modes ($\nu = 1, 2, 3, \ldots, i$) in both ground and transition state important in deter-

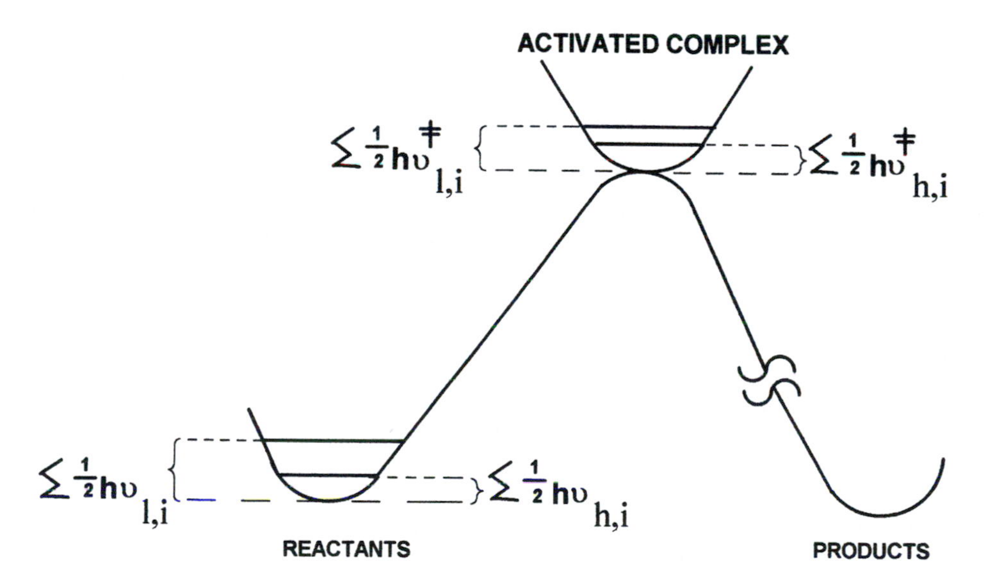

Figure 2.5. A general energy profile that illustrates the origin of the kinetic isotope effect in terms of the zero-point energies of both ground and transition states. The ZPE for the heavy isotope (subscripted *h*) and the light isotope (subscripted ℓ) arise from the summation of the various vibrational modes, *i*.

mining primary KIE, but also it is the necessary prerequisite to describe the origin of secondary KIE.

Clearly, determination of the kinetic isotope effect complements evaluation of the pK_a values of carbon acids. The magnitude of the primary KIE can provide a clue to the transition-state structure for proton transfer, although the analysis is hampered by tunneling, which on the one hand may lead to abnormally high KIE, and by internal return, which on the other hand is one source of low KIE.[35] Thus, in a recent study of D/T (and, by the Swain–Schaad relationship,[16,17] the H/D) kinetic isotope effects for proton abstraction from isotopically labeled 1,1,1-triphenylmethane by cesium cyclohexylamide (CsCHA) in cyclohexylamine, Dixon and Streitwieser ascribed the observed low KIE ($k_D/k_T = 1.81$ and $k_H/k_D = 3.83–4.03$) to either an unsymmetrical (presumably linear) TS for proton transfer or to internal return.[51]

Among other physical organic tools, KIE, has been used by Koch and coworkers[35,52–56] to probe the variation in mechanism for dehydrohalogenation. These studies have also been significant in delineating the impact of internal return on the mechanistic interpretation of primary KIE. Among the mechanisms proposed for elimination reactions, elimination in a system where a strong base is present, and the leaving group is not particularly good, might be expected to proceed via the E_{1cb} (elimination unimolecular via the conjugate base of the carbon acid) mechanism, as shown generally by equation 2.8,

$$\mathrm{LG{-}CRR'{-}C(R''R'')'{-}H + B:^{-} \rightarrow LG{-}CRR'{-}(R''R''')C:^{-} \rightarrow CRR'{=}CR''R'''} \tag{2.8}$$

where the leaving group is designated LG and the strong base is labeled $B:^-$.

In their study of the dehydrohalogenation of 1-X-1,1-difluoro-2-chloro-2-phenylethane (X = Cl or F) by ethoxide, the Koch group[55] found a very large kinetic element effect (i.e., k for elimination of HCl was ca. 10^5-fold larger than k for elimination of HF), which, accompanied by primary KIE and substituent (Hammett σ–ρ) studies, supported the view that hydrogen-bonded carbanions were central intermediates in these eliminations. The temperature dependence (i.e., Arrhenius behavior) of the primary KIE for the related system (where X = Br or Cl) was shown to be inconsistent with the concerted E2 mechanism.[56] In considering eliminations of HX promoted by ethoxide with the substrates $C_6H_5CHClCF_2Cl$ and $C_6H_5CHClCF_3$, loss of chloride was shown to proceed from a hydrogen-bonded carbanion (HBC in Scheme 2.1), whereas loss of fluoride from the second substrate occurs via a free cabanion (FC in Scheme 2.1), and this difference gives rise to the large element effect. On the other hand, exchange of the benzylic hydrogen (or isotope) of $C_6H_5CHClCF_3$ occurs 13-fold faster than elimination of HF and the close-to-unity KIE indicates both internal return and the loss of the fluoride ion from the free carbanion (Scheme 2.1). These results have more recently been extended to a wide range of polyhalogenated phenylethane systems[54] and the interpretation of the E_{1cb} dehydrohalogenation mechanism, involving hydrogen-bonded versus free carbanions, has been bolstered by quantum chemical calculations

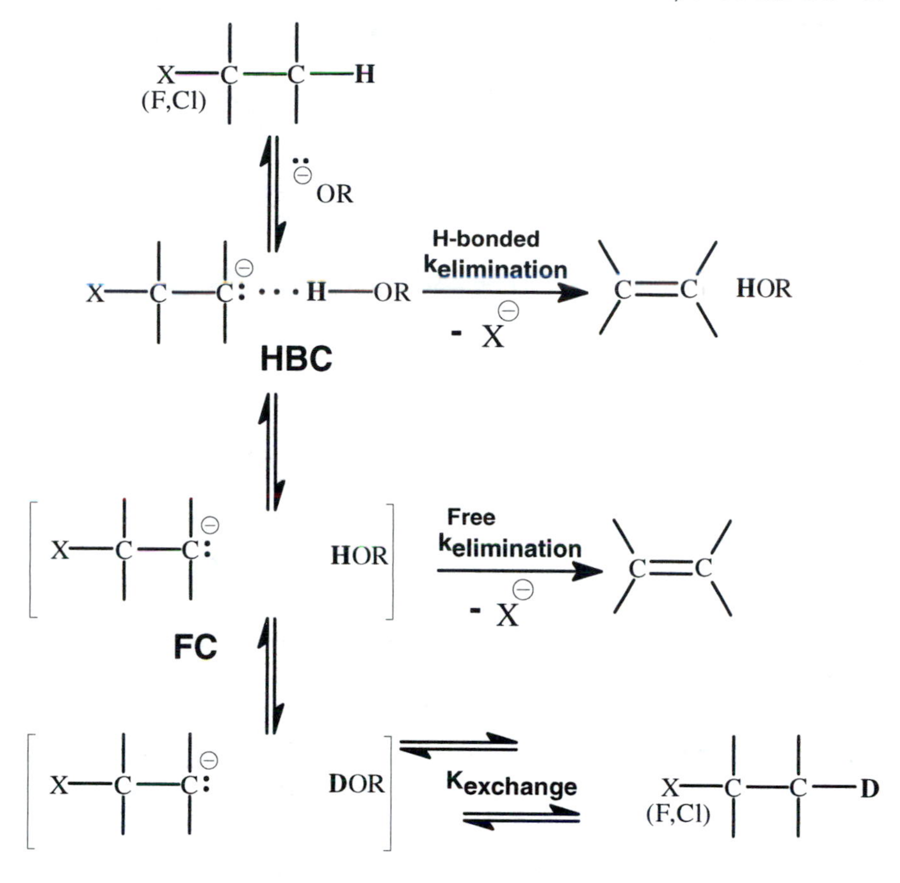

SCHEME 2.1

(PM3, both gas phase and in a dielectic medium).[57] Clearly, interpretation of even primary KIE, particularly where the effects are small, requires careful consideration of all putative intermediates. Further, the results underscore the importance of hydrogen-bonding in proton transfer, generally.[58]

For a carbon acid that demonstrates a "normal" primary H/D isotope effect of 6–7 at 25 °C (and which shows a typical temperature dependence), it appears reasonable to conclude that the proton transfer is rate determining, that it involves a linear TS, and that neither tunneling nor internal return intervene to any large extent.

Kinetic isotope effects, therefore, can permit a glimpse into the transition state for proton transfer from a carbon acid, while the pK_a measurements detailed in Chapter 1 provide insight into the thermodynamics of carbon acidity. In the following section we will consider the connection between acidity and the reactivity of the carbanion, the species formed in the ionization of a carbon acid.

2.2 RELATION OF ACIDITY TO REACTIVITY

In Chapter 1, the relationship between acidity and rate constant was considered from the viewpoint of determining the acidity (pK_a values) of families of carbon acids from kinetic data via the Brønsted catalysis law (cf. equation 1.32).[59] In this section, we will consider the reactivity of carbanions, the conjugate bases of carbon acids, using the pK_a values as a guide, that is, we will approach the Brønsted relationship in the inverse manner.

2.2.1 Brønsted Correlations

2.2.1.1 Methodologies of Constructing Brønsted Plots

In principle, the Brønsted experiment may be conducted in two different ways. First, the kinetic acidity of a given carbon acid (**CH**) may be probed by studying the rate of its reaction *with a series of bases* ($:B_i$) to give the carbanion (C^-) and the conjugate acid of the base in question, according to equation 2.9:

$$\mathbf{CH} + :\mathbf{B}_i \underset{k_p^{BH}}{\overset{k_p^B}{\rightleftharpoons}} \mathbf{C}^- + \mathbf{B}_i\mathbf{H}^+ \tag{2.9}$$

The Brønsted plot [log k_p^B vs log K_a^{BH}, where k_p^B is the rate constant for proton transfer from **CH** to $:B_i$ and K_a^{BH} is the acid dissociation constant of the conjugate acids of the series of bases used in the deprotonation of the carbon acid **CH,** (equation 2.10)]

$$\log k_p^B = -\beta^B \log K_a^{BH} + C \tag{2.10}$$

yields a slope, β^B, which is taken as a measure of the degree of proton transfer in the transition state. Ideally, β^B approaches 0 for proton transfer to strong bases at rates approaching diffusion control, where proton transfer is very exothermic and, so, the extent of proton transfer at the transition state (TS) is very low; β^B approaches 1 for very low rates of proton transfer to weak bases and proton transfer is virtually complete at the TS. For a limited series of bases that are not very strong or very weak relative to the carbon acid, the value of β^B would lie somewhere between the limits of 1 and 0 and its value can, again, be taken as a measure of the degree of proton transfer occurring at the TS. Alternatively, the Brønsted plot could be constructed by considering the inverse of equation 2.9, where the acids, $\mathbf{B}_i\mathbf{H}^+$ (subscript *i* designates the different acids, $i = 1, 2, 3$, etc.), are varied in strength and the rate of proton transfer from these acids to the carbanion, C^-, is measured. Here, the plot (log k_p^{BH} vs log K_a^{BH}, where K_a^{BH} is the acid dissociation constant of the acids) yields a slope, α^{BH}, that varies in the limits 0 to 1 and reflects the degree of proton transfer at the TS (equation 2.11).

$$\log k_p^{BH} = \alpha^{BH} \log K_a^{BH} + C' \tag{2.11}$$

Thus, α^{BH} and β^B are the relevant Brønsted coefficients that give a measure of the position of the TS. The rationale behind this idea will be outlined below.

Where anomalous values of α^{BH} or β^{B} (i.e., values greater than 1 or less than 0) determined in this first method (i.e., equations 2.9–2.11) have been found,[60–62] the anomalies have been attributed to the use of acids (or bases) of very different structural type.[63,64] Consequently, these structurally different acids/bases are solvated differently or have different steric environments, and this variation in solvation or steric demand accounts for the abnormal Brønsted behavior.

In the second type of Brønsted experiment, the acidity of a series of carbon acids, XC_iH, is varied, ideally through the use of *remote* substituents, so that the problems due to differential solvation or steric demand at the reaction site are reduced. The Brønsted plot is now constructed on the basis of equation 2.12.

$$XC_iH + :B \underset{k_p^{C}}{\overset{k_p^{CH}}{\rightleftharpoons}} XC_i^- + BH^+ \tag{2.12}$$

In the forward direction (left to right), equation 2.13 defines the plot with the slope, α^{CH}, and in the opposite direction (right to left) equation 2.14 yields the slope, β^{C} (equations 2.12 and 2.13). Of course, the term "$\log K_a$" in equations 2.10, 2.11 and 2.13, 2.14 is equivalent to $-pK_a$.

$$\log k_p^{CH} = \alpha^{CH} \log K_a^{CH} + C'' \tag{2.13}$$

$$\log k_p^{C} = \beta^{C} \log K_a^{CH} + C''' \tag{2.14}$$

To be perfectly clear, the first method (equations 2.9–2.11) yields the "normal" Brønsted coefficient, β^{B}, for proton transfer from a carbon acid to a series of bases, B, and gives the "normal" Brønsted coefficient, α^{BH}, for proton transfer from a series of acids, BH, to a carbanion. According to convention, we have transposed α and β in the second method; α^{CH} is the new Brønsted coefficient for proton transfer from a series of carbon acids to a single base, while β^{C} represents the protonation of a series of carbanions (conjugate bases of CH) by a given acid. *It is through this latter method that "genuinely abnormal" Brønsted coefficients may be detected* that are particularly informative. These abnormal coefficients will be examined in detail in subsequent sections.

2.2.1.2 Relationship between Brønsted Coefficients and Transition-State Position

At this juncture, it is appropriate to consider the interpretation of the Brønsted slopes, α and β, as measures of the position of the TS for proton transfer and the limits that this imposes on the theoretical values that these slopes can have. It must be recognized, first of all, that the Brønsted correlation, in common with other linear free-energy relationships (LFER), relates kinetics and thermodynamics. Following the derivation of Leffler and Grunwald,[65] it may be postulated that changes δ in the free energy of activation for a reaction, $G^{\ddagger}$ (where the superscript $\ddagger$ signifies the TS), are a linear combination of changes in the standard Gibbs free energy of products and reactants, G_P^o and G_R^o, respectively, as shown in equation 2.15.

$$\delta G^{\ddagger} = a\delta G_{\mathrm{P}}^{\mathrm{o}} + b\delta G_{\mathrm{R}}^{\mathrm{o}} \tag{2.15}$$

The premise here is that the transition state is intermediate between the reactants and products and that a change wrought in the reactants and/or products (indicated by the *Leffler–Grunwald operator*, δ) that affects the energy of the reactants and/or products should also influence the transition-state energy. If the operator affects the TS entirely, as it does the products, and not at all, as it does the reactants, then the constant a should be at some maximum value and the constant b at some minimum value. In this case, the TS must resemble the products in energy and, presumably, in structure.

Conversely, if the TS responds to the operator in exactly the same manner as the reactants and not at all as the products do, then the constant b should have some maximum value, the constant a some minimum value, and the TS should resemble the reactants in energy and structure. Note that the operator represents the incremental energy change concomitant with change of substituents or modification of the medium, and so on.

What if the TS is equidistant along the reaction coordinate from products and reactants? In this situation, we would expect the constants to be equal, that is, $a = b$. It seems reasonable at this point to suggest that the maximum value should be 1 for 100% effect of the operator and 0 for nil effect.[66] Furthermore, the constants should be inversely related. Simply put, if the TS is closer along the reaction coordinate to the reactants then it should be farther from the products, and vice versa. This leads to the reformulation of equation 2.15 as equation 2.16.

$$\delta G^{\ddagger} = a\delta G_{\mathrm{P}}^{\mathrm{o}} + (1 - a)\delta G_{\mathrm{R}}^{\mathrm{o}} \tag{2.16}$$

It is apparent, then, that the constant a has become an "index of progress" along the reaction coordinate. This progression of the transition state along the reaction coordinate is illustrated graphically in Figure 2.6.

As shown in the figure, the progress index a may be replaced by the Brønsted coefficients α^{CH} and β^{B}, while index b may be substituted for by β^{C} and α^{BH}. The proof of this follows directly from equation 2.16. Thus, from standard thermodynamics we know that equation 2.17 may be applied.

$$\Delta G^{\mathrm{o}} = -RT\ 2.303 \log K \tag{2.17}$$

$\Delta G^{\mathrm{o}} = G_{\mathrm{P}}^{\mathrm{o}} - G_{\mathrm{R}}^{\mathrm{o}}$ here and rearrangement followed by substitution in equation 2.16 gives rise to equation 2.18.

$$\delta G^{\ddagger} = -a\delta RT\ 2.303 \log K + \delta G_{\mathrm{R}}^{\mathrm{o}} \tag{2.18}$$

However, $\Delta G^{\ddagger}$ can be written as the difference in energy between the reactant(s) and the TS; namely, $\Delta G^{\ddagger} = G^{\ddagger} - G_{\mathrm{R}}^{\mathrm{o}}$. Substitution into equation 2.18 then leads to equation 2.19.

$$\delta\Delta G^{\ddagger} = -a\delta RT\ 2.303 \log K \tag{2.19}$$

From the linear form of the Eyring equation that defines transition-state theory,[67] a rate constant, k, can be expressed in terms of the difference in the free energy of activation ($\Delta G^{\ddagger}$), as given in equation 2.20.

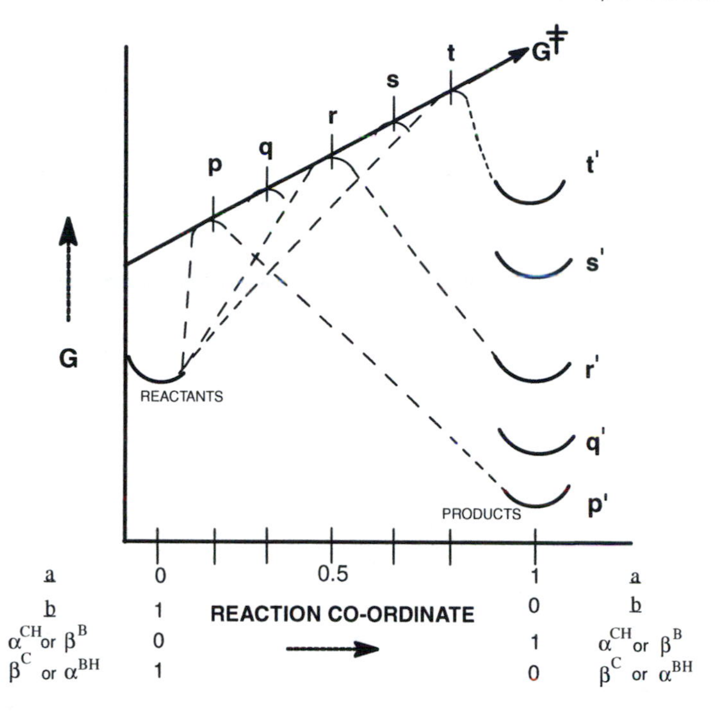

Figure 2.6. Free-energy-reaction coordinate diagram that illustrates the connection between kinetics and thermodynamics. As the products become progessively more stable relative to the reactants (t′⟶p′) the TS shifts (t⟶p) *earlier* on the reaction coordinate. The constants, *a* and *b*, or the Brønsted coefficients, α, and β, are indices of reaction progress and indicate the position of the TS.

$$\Delta G^{\ddagger} = -RT\ 2.303 \log k + RT\ 2.303 \log \boldsymbol{k}T/\mathbf{h} \qquad (2.20)$$

Multiplication of equation 2.20 by the Leffler–Grunwald operator and rearranging gives equation 2.21.

$$\delta\,\Delta G^{\ddagger} = -\delta RT\ 2.303 \log k + \delta RT\ 2.303 \log \boldsymbol{k}T/\mathbf{h} \qquad (2.21)$$

Since the left-hand sides of equations 2.19 and 2.21 are the same, the right-hand sides of the equations may be set equal to one another to give equation 2.22.

$$-\delta RT\ 2.303 \log k + \delta RT\ 2.303 \log \boldsymbol{k}T/\mathbf{h} = -a\delta RT\ 2.303 \log K \qquad (2.22)$$

Rearrangement and division through by the quantity ($-RT$ 2.303) yields equation 2.23.

$$\delta \log k = a\delta \log K - \delta \log \boldsymbol{k}T/\mathbf{h} \qquad (2.23)$$

Replacement of the last term with a symbol for the constant, *C*, yields an equation that we recognize as a form of the Brønsted equation, where the rate constant *k* refers to the process of proton transfer and the equilibrium constant *K* is replaced by an acid dissociation constant, K_a to give equation 2.24.

$$\delta \log k = a\delta \log K_a - C \quad (2.24)$$

It is now clear that the Brønsted slope parameters, α (and β), may supplant a (and b) and that the Brønsted coefficients are, therefore, also measures of transition-state position (Figure 2.6).

Figure 2.6 and the analysis of TS position and energy has been approached in a number of different ways. The contributions of many eminent scientists have been recognized in the acronym *Bema Hapothle* (*B*ell[68]–*E*vans[69]–*Ma*rcus[70]–*Ha*mmond[71]–*P*olyani[69]–*Tho*rnton[72]–*Le*ffler[65,66]) suggested by Jencks.[73] In accord with the Hammond postulate,[71] Figure 2.6 shows the TS moving progressively *earlier* on the reaction coordinate as the products become increasingly more stable than the reactants (or, alternatively, the TS becomes progressively *later* as the products become increasingly less stable than the reactants).

In our treatment thus far, we have *assumed* that the relationship between kinetic and thermodynamic properties is a linear one. However, any kinetic process attains a practical maximum of diffusion- or encounter-control; that is, a reaction rate reaches a maximum when each microscopic collision leads to formation of product. For a bimolecular process, the diffusion-controlled rate constant is limited by solvent viscosity (which is temperature dependent) and at room temperature has a typical value of 10^{10} $M^{-1} \cdot s^{-1}$. Thus, one would expect Brønsted constants α^{CH} and β^{B} to have a minimum value of 0 (β^{C} and $\alpha^{BH} = 1$) when the proton transfer rate constant is ca. 10^{10} $M^{-1} \cdot s^{-1}$ and to tend toward a value of 1 (α^{CH} and $\beta^{B} = 0$) for very slow rates of proton transfer. Over this large range of kinetic and thermodynamic characteristics, the value of the Brønsted coefficients would be expected to vary continuously and the resultant plot should be curved.[74,75] (The "progress indices" may be redefined as *selectivities* and, so, this discussion also may be framed in terms of the reactivity-selectivity principle, that is, RSP.)[76]

More to the point, proton transfer can be viewed as a competition for the proton between the carbon acid and the base. In principle, a diffusion-controlled process will arise whenever ΔpK_a (i.e., $\Delta pK_a = pK_a^{CH} - pK_a^{BH}$) is very large, and it follows that the Brønsted plot (log k vs pK_a) is *necessarily* curved if the pK_a range studied is very large. Of course, for carbon acids whose proton transfers involve very high activation barriers it is unreasonable to think that diffusion-control will be approached even under a large thermodynamic driving force, that is, even if a very large acidity difference (ΔpK_a) could be achieved practically. Focusing on those carbon acids where proton transfer could be very fast, then, if the carbon acid is a much stronger acid than the conjugate acid of the base, the base will take the proton at a diffusion-controlled rate; that is, the forward reaction in equation 2.9 (or equation 2.12) has a diffusion-controlled rate constant, k_p^B, ca. 10^{10} $M^{-1} \cdot s^{-1} (= k_p^{CH})$ and β^B (or α^{CH}) equals zero. Similarly, if the conjugate acid of the base (the Brønsted acid catalyst) is a much stronger acid than the carbon acid is, then the reverse reaction (k_p^{BH} in equation 2.9 and k_p^{C} in equation 2.12) becomes encounter-controlled and α^{BH} (or β^{C}) will equal zero. Naturally, when the forward reaction in equations 2.9 and 2.12 is diffusion-controlled, the reverse reactions represent base catalysis where proton transfer depends linearly on the strength of the bases and the respective Brønsted slopes have a value of one, that is when β^B and α^{CH} are zero, α^{BH} and β^C have values of 1. Where the carbon acid and conjugate acid of the base (equations. 2.9 and 2.12) are similar in acidity, the Brønsted coefficients will have values intermediate between 1 and 0. In traversing a *wide* range of ΔpK_a values, the coefficients must span the full range of 0 to 1 and the Brønsted plot must curve.

Our foregoing discussion is clearly tied to transition-state theory[67] and the assumption that the transition state is intermediate not only in terms of energy but also in structure between the reactants and products. Simply, we have taken the two-dimensional reaction profile (Figure 2.6) as an adequate representation of a generalized reaction. However, in proton transfer processes we recognize that quantum mechanical tunneling of the proton, for example, does not follow the "minimum energy pathway" suggested by the two-dimensional profile. Nor have we explicitly considered possible motion of the TS perpendicular to the reaction coordinate, although this movement has been shown to be important in assessing the nature of the TS in a range of systems.[73,76–78] Such perpendicular or *anti-Hammond* movement may be appropriately analyzed in terms of three-dimensional reaction profiles termed More O'Ferrall–Jencks diagrams.[73,76–78]

In addition, the possibility that the transition state may include contributions from excited states of the reactants and/or products, as advanced by the configuration mixing model (CMM), has been ignored at this stage in the discussion.[79,80] In this case, the TS may have a structure that is not a weighted linear combination of the energies (and structures) of the reactants and products, as required in our development of equations 2.15–2.24.

While recognizing these aspects of the problem of relating kinetics and thermodynamics, we shall proceed initially to examine proton transfer reactions of carbon acids in terms of the Brønsted correlation and with the assumption that the Brønsted coefficients represent normal TS motion along the reaction coordinate. Importantly, with this assumption we can analyze those situations where the Brønsted coefficients are abnormal, that is, where they are outside the range of 1 to 0 expected on the basis of Figure 2.6 and equations 2.15–2.24.

2.2.1.3 The Nitroalkane Anomaly

Having set the stage by exploring the methodologies of constructing Brønsted correlations, we now turn our attention to an interesting application of these strategies, the "nitroalkane anomaly".[81] Instead of obtaining Brønsted coefficients (equations 2.12 and 2.13) with values falling between 1 and 0, the Brønsted coefficients are significantly larger than unity; for example, an α^{CH} value of 1.29 was obtained in the deprotonation of a family of 1-aryl-1-nitromethanes with morpholine as base.[82] Note that in this example *the Brønsted plots were devised according to the second method* outlined above (Section 2.2.1.1., equations 2.12–2.14). However, Brønsted coefficients β^B determined in the "normal" way (method 1, equations 2.9–2.11) had conventional values clustered about 0.55 in the deprotonation of the 1-arylnitroethanes with a series of secondary amines; a given 1-arylnitroethane was deprotonated by the series of amines to yield a Brønsted β^B, then another arylnitroethane was studied, and so forth.[82] Kinetic deprotonation of 3-X-nitropropanes with hydroxide yielded an α^{CH} value of 1.67, while deprotonation by pyridine gave an α^{CH} of 1.89.[83] It is apparent, however, on the basis of our discussion in section 2.2.1.2 that these Brønsted coefficients are unreasonable since they fall outside the range of 1 to 0.[65,66,68,69,75]

Another aspect of the nitroalkane anomaly is the apparent inversion of the parallelism between the effect of a substituent on rates and equilibria. Thus, Kresge has shown that while the pK_a values of a series of nitroalkanes (nitromethane, nitroethane, and 2-nitropropane) decrease (from 10.2 to 8.6 and 7.74), the rate constants for proton abstraction by hydroxide also *decline* (from 27.6 to 5.19 to 0.316 $M^{-1} \cdot s^{-1}$), that is, the stronger the acid becomes, the slower the rate of proton transfer becomes.[81] This lack of correlation of kinetic and thermodynamic acidity is the nitroalkane anomaly.

2.2.1.4 Transition-State Imbalances

The most widely accepted explanation for these observations is that the transition state for the proton transfer in these cases is highly *imbalanced.*[73] In the nitroalkane example discussed above, proton transfer is highly advanced relative to delocalization of the negative carbanion charge into the nitro group and so proton transfer does not benefit from this delocalization, while the thermodynamic acidity fully reflects the delocalization of charge into the nitro function.[79,84] Therefore, the nitroalkane anomaly reveals two important features in assessing carbanion reactivity: (1) that transition states for proton transfer that lead to carbanion formation may be imbalanced and (2) that reactivity may not always parallel acidity (or basicity).

It should be emphasized that transition-state imbalances are a more general phenomenon than the limited discussion here may suggest. In fact, Bernasconi, in laying the groundwork for his principle of nonperfect synchronization (PNS), observed that whenever two or more processes occur nonsynchronously at the transition state, a TS imbalance will result.[84–87] These "coupled" processes include bond formation–bond rupture, charge delocalization–localization, solvation–desolvation, and development–loss of resonance.[73,74,88–92] Solvational effects have been found to be particularly important; ab initio molecular orbital (MO) and density functional theory (DFT) calculations (HF/6-31 + G^*, B3LYP/6-31 + G^*, MP2/6-31 + G^*, and MP2/6-311 + G^{**} levels of theory) suggest that the nitroalkane anomaly does not exist in the gas phase.[93]

Clearly, anomalous Brønsted coefficients can be taken as diagnostic of TS imbalances, as in the case of the nitroalkanes. However, the TS imbalance in the case of nitroalkane deprotonation (and nitrocarbanide protonation) is generally considered to be large. What of those situations where the imbalance is small, where the relevant Brønsted coefficients (even determined by method 2), in fact, may still fall within the traditional range?

In many of these situations, the imbalance may be assessed, according to Bernasconi,[84] from the difference between the Brønsted coefficients and, really, between the two types of Brønsted coefficient. Consider the following situation. The α^{CH} value for deprotonation of a series of carbon acids, XC_iH, by a given base, :B, (method 2 coefficient) has been determined. Similarly, β^B for deprotonation of a carbon acid, XCH, by a series of bases, :B_i, has been measured (i.e., by method 1). It follows that if the base :B used in method 2 is a member of the set of bases $:B_i$ and if the carbon acid **XCH** used in method 1 is a member of the set of carbon acids $\mathbf{XC_iH}$, then the values of α^{CH} and β^B should be equal; both coefficients are measuring the same degree of proton transfer in the TS for the deprotonation of a $\mathbf{XC_iH}$ acid by a $:B_i$ base. The quantity $\alpha^{CH} - \beta^B$ is one measure of transition-state charge imbalance, and when it is greater than zero the imbalance is termed "positive"; conversely, if the difference is less than zero, the imbalance is "negative".

In general, positive imbalances arise when a substituent is closer to the negative charge center in the TS but farther away in the product. (The inverse situation accounts for negative imbalances). Thus, in deprotonation of a series of 1-aryl-1-nitromethanes with benzoate ion in DMSO (25 °C)[83] α^{CH} is 0.92, while the Brønsted β^B value for the deprotonation of 1-nitro-1-phenylmethane by a series of substituted benzoate ion bases in DMSO (25 °C) is 0.55.[94] The positive imbalance ($\alpha^{CH} - \beta^B = 0.37$) arises from the fact that the substituent, X (placed in the *para* position here), is closer to the partial negative charge in the transition state (shown on the left in Figure 2.7 as a linear proton transfer TS, cf. Section 2.1.3) than it is in the resultant carbanion (shown on the right in Figure 2.7). This is tantamount to saying that in the TS, charge is more highly localized on the carbon center than it is in the final nitrocarbanide. In the carbanion, the resonance form that places the negative charge on the more electronegative oxygen center would be expected to make a more significant contribution to the overall hybrid.

Further, recall that specific solvent interactions, notably in the nitrocarbanides where hydrogen-bonding solvents may further stabilize a negative charge that is delocalized out onto the NO_2 oxygens, can lead to enhanced thermodynamic acidities. Thus, the solvent and the structural reorganization are important in assessing transition-state imbalances. It is important to note, in this context, that even when nitro (and other strong electron-withdrawing) substituents are placed in sites remote from the carbon acidic site where hydrogen-bonding effects would be expected to be less important, strong TS imbalances have been found by the Terrier group.[95–98] An interesting example is provided by the *P*-(formyl-

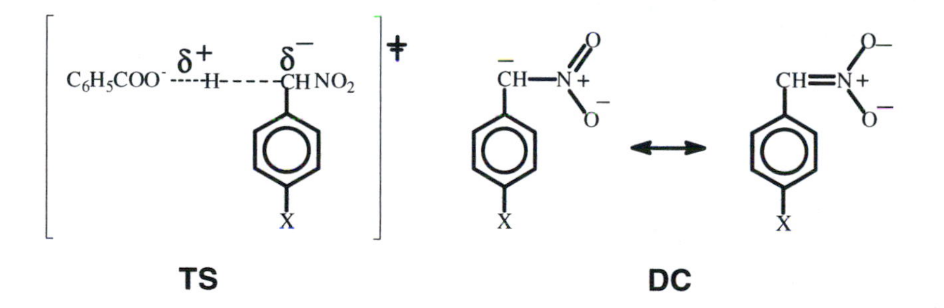

Figure 2.7. The structure of the transition state (TS) for transfer of a proton from a series of 4-X substituted nitrotoluenes, assuming a linear TS structure, is shown on the left-hand side. Two of the canonical forms for the delocalized carbanion (DC) are depicted on the right-hand side.

methyl)triphenyl phosphonium cation [$(C_6H_5)_3P^+{-}CH_2COH$], which, as a carbon acid, deprotonates to give the corresponding phosphonium ylide [$(C_6H_5)_3P^+$ ^-CHCOR, where R = H, CH_3, or C_6H_5]; this carbon acid displays the lowest Marcus intrinsic reactivity reported thus far (70% water:30% DMSO, amine bases) and this result has been attributed to covalent hydration of the carbonyl function, that is, deprotonation of the carbon acid proceeds via the geminal diol rather than via the free carbonyl compound.[99]

It should be emphasized here that the difference ($\alpha^{CH} - \beta^{B}$) may indicate transition-state imbalance, but the magnitude of the difference does not correlate directly with the "degree of lag". That is, a large difference would appear to imply that there is a large lag in the development of resonance behind proton transfer and, conversely, a small difference would appear to imply a small difference in development of resonance stabilization for the carbanion and its formation by deprotonation of the carbon acid. In fact, a large $\alpha^{CH} - \beta^{B}$ value may reflect only much greater resonance stabilization of the carbanion, rather than greater lag. Bernasconi has further developed the quantitative application of PNS through the introduction of another measure of lag, the $\lambda_{res} - \beta^{B}$ difference.[84,85] The interested reader is directed to these thorough reviews of the quantitative measure of these effects. Our discussion will remain more qualitative, with the understanding that while the $\alpha^{CH} - \beta^{B}$ difference may indicate TS imbalance it does not quantify the degree of lag.

Since TS imbalances in proton transfer reactions that lead to delocalized carbanions are common, it is reasonable to ask whether one can separate the effect of delocalization on the stability of the carbanion from the rate for its formation from the corresponding carbon acid. In short, can we separate the thermodynamic driving force for formation of a given carbanion (or series of carbanions) from some *intrinsic kinetic factor* for the proton transfer process?

The process of carbon acid ionization is represented by equation 2.25,

$$\text{X–CH}_2\text{–H} + {}^-\text{B} \underset{k_{-1}}{\overset{k_1}{\rightleftharpoons}} \text{X–}\bar{\text{C}}\text{H}_2 + \text{HB} \qquad (2.25)$$

where k_1/k_{-1} is equal to the equilibrium constant for acid dissociation, K_a^{CH}/K_a^{BH}. Of course, K_a^{CH}/K_a^{BH} is directly related to the incremental standard Gibbs free energy change, $\Delta\Delta G^\circ$, and when K_a^{CH}/K_a^{BH} is equal to 1, $\Delta\Delta G^\circ$ is zero; that is, there is no driving force in either the forward (k_1) or reverse (k_{-1}) directions. As we have seen, a Brønsted plot (equation 1.32) relates the rate constant, k_p^{CH} ($= k_1$), to the acid ionization constant, K_a^{CH}. Therefore, extrapolation of a Brønsted plot to the point [i.e., $(x, y) = (\log k_1, \log K_a^{CH})$] where K_a^{CH} is unity ($-\log K_a^{CH} = pK_a^{CH} = 0$) should yield a value of $k_p^{CH} (= k_1)$ that represents the *intrinsic rate constant* (hereafter k_o) for the forward proton transfer process. From k_o, the intrinsic kinetic barrier, $\Delta G_o^\ddagger$ (where $\Delta G_o^\ddagger = \Delta G_1^\ddagger$), free of any thermodynamic driving force, can be determined. (A similar argument may be developed from the product side of equilibrium 2.25, which would yield the intrinsic rate constant for protonation of the carbon acid.) While determination of any kinetic parameter in a reaction is valuable, determination of the intrinsic barrier can provide particular insights into chemical reactivity; the intrinsic barrier is a measure of reactivity for a *class* of compounds.

The meaning of the intrinsic barrier is tied to Marcus theory and the Marcus analysis of proton transfer reactions for carbon acids will be discussed in Section 2.2.2.

2.2.1.5 The Principle of Nonperfect Synchronization

While we have, thus far, considered only TS imbalances that arise from resonance stabilization preceding or trailing behind proton transfer in the formation of a carbanion from a carbon acid, Bernasconi has also discussed imbalances due to differences in solvation, as well as steric effects and the effect of intramolecular hydrogen-bonding.[84–86,100] In general, analysis of these and other imbalances have led to the enunciation of the principle of nonperfect synchronization (PNS).[84,85] The PNS states that when a factor that stabilizes the product (e.g., the carbanion) occurs *late* along the reaction coordinate, this factor will result in a lowering of the intrinsic reactivity, k_o, and, conversely, *early* development of a factor that stabilizes the product will result in an increase in the intrinsic reactivity. By similar reasoning, elimination of a factor that stabilizes the reactant *late* on the reaction coordinate raises the intrinsic reactivity, while elimination of the reactant-stabilizing factor *early* on the reaction coordinate results in decreased intrinsic reactivity.[84,85]

Interestingly, in analyzing the limited number of systems in which the kinetics of proton transfer between two carbon centers (i.e., between a carbon acid and another carbanion; cf. equation 1.22) has been studied in terms of the PNS,[84,100] Bernasconi found that the intrinsic rate constants, k_o, for these deprotonations were significantly lower than those found for the same carbon acids when piperidine or morpholine were used as the bases. These observations were attributed to a "dual imbalance" in the TS brought about by the fact that both the carbanions and the carbon acids used in the deprotonation studies were remotely ring-substituted; resonance stabilization would both develop late for the new carbanion derived from the original carbon acid and would be lost early as far as the original carbanion was concerned. (See reference 101 for the 9-alkylfluorene/9-alkylfluorenide anion systems where anomalous Brønsted coefficients also suggest a similar interpretation.)

In drawing together our discussion of kinetic isotope effects (KIE) and the PNS, it is tempting to speculate that in deprotonation of a series of carbon acids by a series of carbanions of similar structural type (and consequent relative closeness in pK_a for the respective carbon acids), one could find a high KIE, indicative of a highly symmetrical linear transition state (presuming no appreciable internal return), and, simultaneously, find relatively low intrinsic rates for the deprotonations.

In fact, Bernasconi has concluded that in carbanion formation, resonance stabilization frequently trails the charge transfer that accompanies proton transfer. Further, solvation, intermolecular hydrogen-bonding, and π-overlap that involves remote aryl groups, all generally lag behind charge transfer and/or any accompanying bond modifications.[84–87,100] These all lead to lower intrinsic reactivity, namely, lower k_o values, than are found for systems lacking these factors. (However, also see reference 84 for the development of a quantitative measure of lag, $\lambda_{res} - \beta^B$, and see reference 102 for the use of ab initio calculations in determining TS imbalances.)

A number of reactions of carbanions, other than simple proton transfer, are amenable to analysis using intrinsic rate constants obtained from this Brønsted approach. Thus, cleavage of diaryl disulfides by a range of carbanions, including carbanions stabilized by bis-carbonyl substitution (e.g., carbanion of acetylacetone), cyano-, and nitro-substituted carbanides, showed the following order of decreasing intrinsic rate constant values: k_o (cyano-substituted carbanions) > k_o (bis-carbonyl-substituted carbanions) $\gg$ k_o (nitro-substituted carbanions).[103]

The reaction of carbanions with electron-deficient aromatics leads to the formation of anionic σ-bonded adducts, termed Meisenheimer complexes (e.g., Scheme 1.2).[104–108] Where the carbanions are suitably substituted with good leaving groups (i.e., L in equation 2.26), attack at an unsubstituted position of the ring of the electron-deficient aromatic leads to elimination of a ring proton and the leaving group from the erstwhile carbanion (as HL) in an overall process (equation 2.26) that has been named vicarious nucleophilic substitution (VNS) of hydrogen; acidic workup yields an arylmethane.[109–111] In equation 2.26, the electron-deficient aromatic is typically a polynitrobenzene.

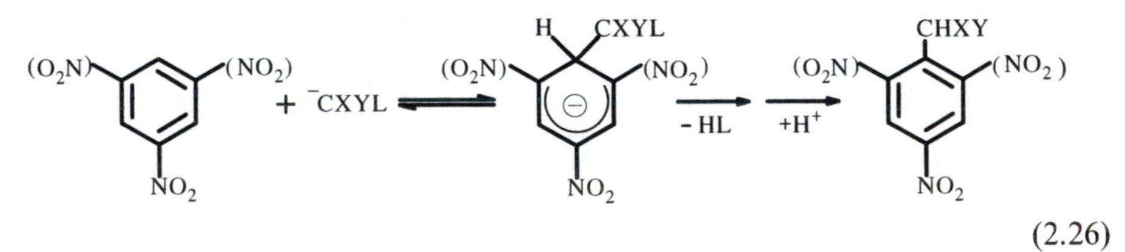

(2.26)

Note that if the electron-deficient aromatic bears a good leaving group (e.g., 1-chloro-2,4,6- trinitrobenzene, shown more generally as 1-L-2,4,6-trinitrobenzene in equation 2.27),

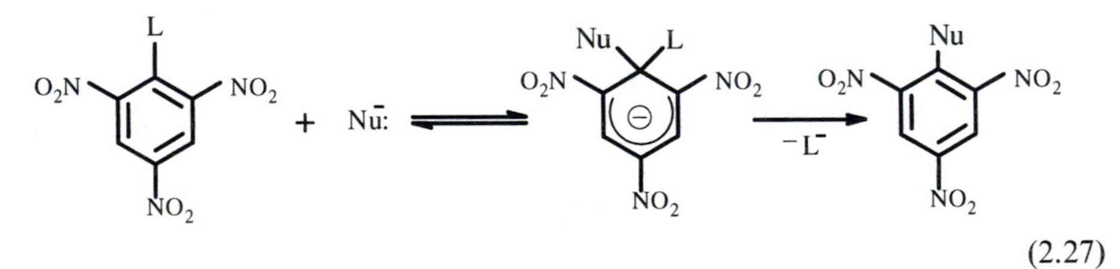

(2.27)

attack of a nucleophile (Nu:) may occur competitively with, or subsequent to, attack at the unsubstituted position. Expulsion of the leaving group leads to a substitution product in accord with the S_NAr mechanism of nucleophilic aromatic substitution.[104,105,112] The propensity for attack at an unsubstituted as compared with a substituted position, as well as the stability of the respective adducts, depends on a number of factors, including the steric bulk of the nucleophile and the opportunity for stereoelectronic stabilization of the Meisenheimer complex.[106,109,113]

Furthermore, attack of a carbanion *that lacks suitable leaving groups* on an electron-deficient aromatic *that lacks a suitable leaving group*, for example, 1,3,5-

trinitrobenzene (TNB), may still lead to formation of a stable Meisenheimer complex. An example (equation 2.28) is found in the reaction of aryloxides with various electron-deficient aromatics, including TNB;[114–117] the aryloxides react with these substrates via a ring carbon site of the ambident aryloxide.[114–117]

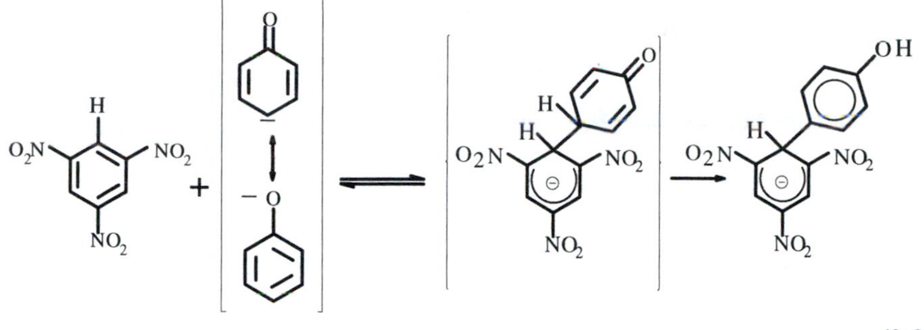

(2.28)

Re-aromatization of the attached aryloxyl group in the initially formed (and unobserved) quinoidal adduct results in effectively irreversible formation of the final Meisenheimer adduct (equation 2.28 for the reaction of TNB with phenoxide ion). Oxidation of the carbon-centered Meisenheimer adduct affords a convenient preparation of substituted biphenyl derivatives (equation 2.29). Similar results have been observed in the reaction of indolide and pyrrolide anions with TNB and related electron-poor substrates.[118,119]

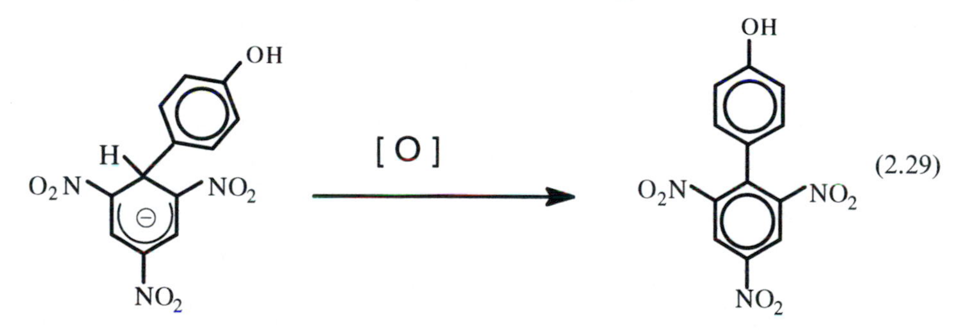

(2.29)

Finally, carbanions that are stabilized by resonance or by relatively strong inductive/polarizability effects may also attack electron-poor aromatic systems, leading to stable Meisenheimer adducts (cf. Scheme 1.2). Crampton and co-workers have examined the reactions of some of these delocalized anions, such as malononitrile anion [dicyanomethide, $CH(CN)_2^-$], with 1-X-trinitrobenzenes.[120] The order of intrinsic rate constant values (log k_o) for malononitrile anion to nitromethide anion (methanol solvent with TNB and 2,4,6-trinitrotoluene) is consistent with the order of those found in the study cited above: 4.40 (malononitrile) $\gg -0.66$ (nitroethide) > -0.74 (nitromethide anion).[120] It is also notable that the intrinsic rate constants for these delocalized carbanions were significantly smaller than those obtained for methoxide, as a typical localized O-nucleophile, in the same Meisenheimer complexation. Similar results have been obtained by the same group in related Meisenheimer complexation studies[121,122] and by the Terrier group in their study of the Meisenheimer complexation of nitrocarbanides[123] with the superelectrophile, 4,6-dinitrobenzofuroxan (DNBF).[124,125] Interestingly, although elimination of HNO_2 is possible from the Meisenheimer complex formed by attack of the dinitromethide ion on DNBF (equation 2.30), as in the VNS mechanism, such elimination does not occur in this case.[123] This further confirms the great stability associated with Meisenheimer complexes of DNBF relative to those derived from TNB.[104,124,125]

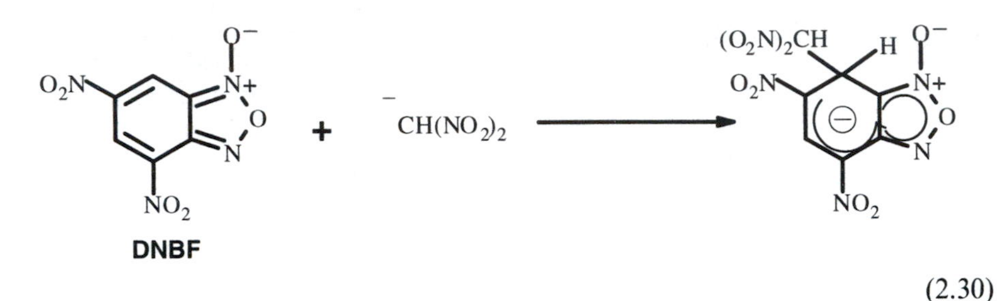

(2.30)

Finally, the S_N2-type displacement of chloride from 1-chlorobutane in DMSO by a series of arylphenylsulfonylmethide ions [$Ar(PhSO_2)CH^-$] was studied by the Bordwell group,[126] who found a good Brønsted correlation except for the carbanions in which the aryl ring was *para*-substituted by strong resonance electron-withdrawing groups such as NO_2 and COPh (which may also interact specifically with the solvent to enhance their electron-withdrawing ability further). These deviations from the Brønsted line could be attributed[84] to the "resonance lag" described by the PNS for proton transfers involving carbon acids or to greatly enhanced resonance stabilization of the carbanion.

Clearly, the Brønsted analysis described will become an ever more important tool for the assessment of the inherent reactivity of carbanions.[84,95–99,120–123,126] When Brønsted plots are constructed in a novel methodology using a single carbon acid, whose pK_a is varied by varying the solvent medium from water through DMSO–water mixtures to almost pure DMSO,[127–131] curvature in the plots has been found consistent with the large variation in ΔpK_a attainable through this modulation of the medium. In other cases, related *though different* sets of Brønsted lines have been found using the solvent variation approach;[132] these results may also be interpretable according to the PNS.

It should be noted that other approaches to the analysis of carbon acidity and carbanion reactivity, particularly attempts to reconcile the nitroalkane anomaly,[133,134] have been suggested.[133–139] Among these, the configuration mixing model (CMM)[79,80,134,137–139] offers a persuasive alternative theoretical framework for the understanding of nitroalkane carbon acidity. Here, the transition state for proton transfer arises from the mixing of a number of valence bond configurations. In the case of proton abstraction from a nitroalkane, the TS is a mixture of configurations for the reactants, the products (where the negative charge of the carbanion is fully delocalized on the NO_2 group oxygens), and a configuration in which the negative charge is localized on the carbanionic center.[137] It is the relatively high contribution of this localized carbanion configuration to the TS that leads to the nitroalkane anomaly, namely, the lack of correlation of kinetic and equilibrium acidities.

Nonetheless, the relative experimental simplicity of the PNS analysis, based as it is on the Brønsted methodology, commends it to more general use. In the next section, we will compare this approach to that of Marcus.

2.2.2 *Marcus Analysis*

As we have seen, the Brønsted equation relates a thermodynamic property of carbon acids (pK_a) to a kinetic property (rate constant) and, by extrapolation, yields the intrinsic reactivity of the carbanion (via k_o, the intrinsic rate constant) from the carbon acid acidity of a family of carbon acids. In point of fact, the concept of an intrinsic kinetic barrier is derived from the work of Marcus, initially on (outer sphere) electron transfer reactions.[140,141] The original Marcus equation related the actual free energy of activation ($\Delta G^{\ddagger}$) to the intrinsic free energy of activation ($\Delta G_o^{\ddagger}$) and to the thermodynamic driving force, the standard Gibbs free energy of reaction (ΔG_o) as given in equation 2.31.

$$\Delta G^{\ddagger} = w_r + \Delta G_o^{\ddagger}[1 + (\Delta G_o - w_r + w_p)/4\Delta G_o^{\ddagger}]^2 \qquad (2.31)$$

This is the Marcus equation for outer-sphere electron transfer, where w_r and w_p represent work terms associated with bringing reactants (w_r) together and pulling products ($-w_p$) apart. These work functions are particularly relevant when dealing with two ionic species. However, if species under consideration are neutral, as would be the case in proton transfer from a carbon acid to a neutral base such as an amine, then it may be possible to neglect these work functions and the form of the Marcus equation that applies is equation 2.32.

$$\Delta G^{\ddagger} = \Delta G_o^{\ddagger}(1 + \Delta G_o/4\Delta G_o^{\ddagger})^2 \qquad (2.32)$$

Equation 2.31 also can be rendered as equation 2.33.

$$\Delta G^{\ddagger} = \Delta G_o^{\ddagger} + \Delta G_o/2 + (\Delta G_o)^2/16\Delta G_o^{\ddagger} \qquad (2.33)$$

These equations or variants of them have been applied not only to proton transfer reactions,[142–144] but also to methyl transfers (i.e., S_N2 displacement at a methyl carbon)[145,146] and to nucleophilic addition reactions.[147]

Examination of equation 2.33 leads us to a further conclusion. If the standard Gibbs free energy for a reaction, ΔG_o, is moderate to small but, more importantly, if it is significantly smaller than the intrinsic free energy of activation, $\Delta G_o^{\ddagger}$, then the second term in equation 2.33 approaches nil and the following approximate equation (2.34) can be written.

$$\Delta G^{\ddagger} = \Delta G_o^{\ddagger} + \Delta G_o/2 \qquad \textit{or} \qquad \Delta G_o^{\ddagger} - \Delta G_o^{\ddagger} = \Delta G_o/2 \qquad (2.34)$$

Obviously, equation 2.34 is a linear equation that relates kinetic and thermodynamic parameters (cf. section 2.2.1.2) in much the same way as the extended Brønsted relationship may be used to determine the intrinsic rate constant, k_o, for a series of reactions.

However, we can also discern the origin of curvature in Brønsted plots. Over limited ranges of pK_a or, more important, for reactions that involve large intrinsic kinetic barriers, the Brønsted plot would be expected to be linear, just as the

Marcus equation collapses to a linear form under the same conditions. On the contrary, when the intrinsic kinetic barrier is small (particularly relative to the thermodynamic driving force) in a reaction, curvature of the Brønsted plot may be expected, as equation 2.33 ceases to be a reasonable approximation for the relationship of kinetics and thermodynamics; the curvilinear equation 2.31 becomes the appropriate description for the system.

In principle, then, a version of the Marcus equation (2.31–2.34) can be used to calculate the intrinsic free energy of activation from a single set of kinetic and thermodynamic data; for example, $\Delta G_0^{\ddagger}$ can be calculated from a rate constant and an equilibrium constant (or, more germane to our discussion, a proton transfer rate constant and a single pK_a value).

J. P. Guthrie has developed a method of predicting proton transfer rate constants for carbon acids,[148] based on a multidimensional Marcus theory.[149] Further work along these lines will no doubt lead to a fuller understanding of the factors that influence the magnitude of the intrinsic barrier in proton transfers, as well as other reactions, at the molecular level.[150]

We will now turn our attention to consideration of the factors that stabilize carbanions.

2.3 FACTORS THAT INFLUENCE CARBANION STABILITY

2.3.1 Stability versus Persistence of Carbanions

We have seen in the preceding sections that carbon acids that exhibit relatively low pK_a values, as a result of the delocalization of charge in the carbanion via resonance, often exhibit low intrinsic reactivities (high kinetic barriers) for the proton transfer reaction that represents ionization of the carbon acid. Frequently, resonance stabilization lags behind proton transfer in these cases. It is clear, however, that the thermodynamic stability of a carbanion is reflected in the pK_a of its conjugate acid and, of course, under standard conditions the free energy can be calculated from the equilibrium constant. The acid dissociation equilibrium (cf. equation 1.1) involves all of the species—carbon acid, solvated proton, and resultant carbanion—and, therefore, indicates the stability of the carbanion and proton *relative to the carbon acid.* In comparing the relative K_a (pK_a) values of a series of structurally similar carbon acids, the trend in the K_a values is largely accounted for by the stabilities of the carbanions.[151] Where solution (e.g., DMSO) and gas-phase acidities correlate, this K_a-stability parallelism is certainly the case.[152]

At this point, we should distinguish clearly between stability and persistence. Although a carbanion may have a long lifetime, for example, as a consequence of steric hindrance about the carbanionic site or because the environment (solvent/structure) provides few decomposition pathways, the carbanion may still be unstable in terms of its conjugate acid pK_a. Persistence is, therefore, a property determined essentially by kinetics, whereas stability is a thermodynamic property, as detailed by Griller and Ingold for the case of another class of reactive intermediates, the free radicals.[153]

Experimentally, carbanions that exhibit low persistence introduce significant difficulties in the determination of the gas-phase acidities of the corresponding carbon acids. Thus, McMahon and Kebarle initially found it difficult to obtain the equilibrium gas-phase acidity of nitromethane, likely owing to nucleophilic displacement of nitrite from nitromethane by the nitromethide anion, acting as nucleophile,[154] that is, the nitromethide ion has a facile pathway for decomposition. Moreover, we have shown that nitroalkanes are associated with a relatively large intrinsic kinetic barrier to proton abstraction; formation of the nitromethide ion is, consequently, more difficult than is formation of another carbanion whose conjugate acid has a comparable pK_a.

In fact, the nitromethide anion is more stable than the cyclopentadienyl anion, based on earlier gas-phase measurements;[155] the solution pK_a values of nitromethane (in H_2O) and cyclopentadiene (anchored to water by an AF method) are 10.2[156] and 16,[157] respectively, which again supports the contention that the nitromethide ion should be stable. We recognize, however, that the pK_a values of nitroalkanes are reduced in water relative to aprotic solvents because of hydrogen-bonding of the water to the nitro group in the nitrocarbanion. Thus, nitroalkane acidity would be preferentially exalted in water as compared with that of cyclopentadiene. Nonetheless, it is reasonable to expect that a nitro group should stabilize a methyl carbanion, in part inductively and in part as a result of resonance delocalization of the charge onto the electronegative oxygens (as shown in 2.35).

$$\overset{-}{C}H_2\text{—}\overset{+}{N}(\text{—}O^-)(\text{=}O) \longleftrightarrow CH_2\text{=}\overset{+}{N}(\text{—}O^-)(\text{—}O^-) \qquad (2.35)$$

In this example, determination of the pK_a of the carbon acid, as a measure of carbanion stability, is hampered by the limited persistence of the nitromethide ion, which, in turn, arises from the relatively high kinetic barrier to formation of the carbanion and the favorable pathway for its decomposition.

A similar difficulty has been noted in the determination of the gas-phase acidity of α-silyl carbon acids by the bracketing technique,[158,159] where alkoxides used to bracket the acidity of bis(trimethylsilyl)methane led preferentially to formation of the trimethylsiloxide ion as shown in equation 2.36.

$$RO^- + [(CH_3)_3Si]_2CH_2 \longrightarrow (CH_3)_3SiO^- + (CH_3)_3SiCH_2R \qquad (2.36)$$

Interestingly, the Bordwell group has also reported[160] interference from desilylation in their attempts to determine the pK_a value of the α-silyl carbon acid, trimethylsilylacetophenone ($PhCOCH_2SiMe_3$), by equilibration with deuterated dimsyl potassium (CD_3SOCD_2K) in perdeuterated DMSO. Again, the problem is one of persistence rather than a reflection on the inherent stability of the respective carbanions.

In the following sections, stability will be discussed on the basis of the equilibrium acidities for the precursor carbon acids. Where possible, the pK_a values used in the comparisons will be drawn from data determined in the same solvent or measured in the gas phase.

2.3.2 Structural Factors in Carbanion Stabilization

2.3.2.1 Hybridization at the Carbanionic Center

In discussing the factors that influence the stability of carbanions, as defined above from the relative acidities of their conjugate carbon acids, two broad classifications can be made. First, there are stabilizing factors that are directly related to structure and bonding at the carbanionic center. These factors may be usefully subdivided into two subfactors: hybridization and Hückel aromaticity of the resultant carbanion. The other main factors that influence carbanion stability fall into the area of substituent effects. These substituent effects may be further classified, as we shall do later on in this chapter (Sections 2.3.3 and 2.3.4).

The standard example of the effect of hybridization on carbanion stability is provided by comparison of the following carbon acids: ethane, ethene (ethylene), and ethyne (acetylene), where the acidity increases in the order given. These hydrocarbons have pK_a values of 49 and 44 for ethane and ethene, respectively, that are extrapolated from the ion-pair acidities, whereas ethyne has a measured ion-pair acidity of 25.[161] In fact, terminal alkynes readily form various metal salts, including silver acetylide; formation of this salt by treatment of the terminal alkyne with silver(diammine) hydroxide (Tollens' reagent) is the basis of a standard laboratory test-tube test that distinguishes such alkynes from their internal isomers.[162] To explain this trend in carbon acidities we shall concentrate on the corresponding carbanions. Thus, in the ethynide anion the anionic electron pair resides in an sp hybrid orbital, in the ethenide anion the pair resides in an sp^2 orbital, and in the ethanide anion the electron pair is located in an sp^3 orbital. The spatial probability distribution function of s-orbitals is nonzero at the nucleus, whereas the p-orbital distribution has a node at the nucleus and, since an sp orbital has 50% s-character while an sp^2 orbital has only 33% s-character, it is clear that the anionic electrons of the ethynide ion located in the sp hybrid orbital should be more tightly held than the corresponding electrons of the ethenide ion that are located in the sp^2 hybrid orbital. The ethynide ion is, therefore, a weaker base than the carbanion of ethene is. Similarly, the ethenide anion is a weaker base than the carbanion of ethane, whose anionic lone pair is situated in an sp^3 hybrid orbital that has only 25% s-character. In a Brønsted equilibrium, the weakest base has the strongest conjugate acid and ethyne is the strongest carbon acid of the series.

Alternatively, we may invoke the concept of orbital electronegativity and note that the electronegativity of a carbon center increases with the degree of s-character assigned to the hybrid orbital.[161] In any case, it should be noted that the trend in solution pK_a data is also supported by gas-phase acidity estimates;[163,164] the gas-phase heat of acid dissociation (ΔH°) of ethene[165] is 409.4 ± 0.6 kcal/mol while 420.1 ± 2.0 kcal/mol is assigned to ethane.[159]

Stabilization through hybridization can also be seen in cases where the degree of s-character of the carbon center of the acid is not as obvious as in the example cited above. For example, the hybridization of the carbons of cyclopropane has been determined to be $sp^{2.28}$ rather than the sp^3 that might be expected (which leads to classical arguments concerning ring strain in such small ring cycloalkanes).[166] Based on our foregoing discussion, we would expect a solution acidity intermediate between that of ethane (pK_a ca. 49) and ethene (pK_a ca. 44) and, in fact, the pK_a of cyclopropane has been estimated to be 46.[167] The structure of cyclopropane itself stabilizes the cyclopropane carbanide.

In summary, the higher the degree of s-character of the carbon center of the carbon acid under consideration, the more readily the corresponding carbanion may be formed; the carbanion is stabilized by this structural feature. The trends considered above are in agreement with later calculational studies.[168] It is also noteworthy that kinetic acidity has been correlated by the Streitwieser group with ^{13}C NMR (^{13}C–H) coupling constants, as a measure of s-character of the relevant C–H bond.[169] These hybridization effects have been confirmed in tritium NMR spectroscopic studies by the same group.[170,171]

2.3.2.2 Hückel Aromaticity

Another factor that arises from the structure of a given carbon acid is the aromaticity that may result from deprotonation. We have selected the cyclopentadienide anion in our comparison of the acidity of nitromethane and cyclopentadiene earlier in this section. At that point, we noted that nitromethane is more acidic than cyclopentadiene. The relatively low pK_a value obtained in solution for cyclopentadiene (16.0 by AF methods)[157] cannot be attributed to its allylic nature; the methyl C–H of propene has a pK_a of 43 determined by kinetic methods[172] or ca. 47 as measured by Breslow's electrochemical method (cf. Section 1.3.2),[173] and by the same electrochemical approach the pK_a of cyclohepatriene has been measured as 36 (Table 1.10).[174] The key is the aromaticity of cyclopentadienide, aromaticity imparted by the placement of the two electrons of the anion into a p-type orbital such that a planar array of overlapping p-orbitals with a "Hückel number" of $4n + 2$ p-electrons is achieved.[175,176] In the case of the cyclopentadienyl anion, n in the Hückel formula is 1 and the total of p-electrons in the π-array is 6, as it is in the prototypical aromatic compound, benzene. Of course, n may be any integer including zero, and both charged and neutral species may meet the Hückel requirement for aromaticity. In the case of cyclopentadiene, the pK_a in DMSO (i.e., 18.0) is 17 pK_a units more acidic than its open-chain analogue, the (*E*)-1,3-pentadienyl anion;[151] this translates into a stabilization energy of 24 kcal/mol. Thus, the special stability associated with aromaticity accounts for the ease of formation of cyclopentadienide ion. Finally, note that pyrrole, which contains the more electronegative nitrogen atom in place of the carbon in cyclopentadiene, is also less acidic than cyclopentadiene; the pK_a of pyrrole in DMSO is 23.05,[151] an indication of the lower aromaticity associated with such heteroaromatic systems.

However, in our comparison we also considered cycloheptatriene. The cycloheptatrienide ion would have a total of 8 p-electrons in its planar array of p-orbitals. With $4n$ π-electrons in the *planar* array, cycloheptatrienide ion could be considered anti-aromatic[177] and, consequently, destabilized relative to its open-chain analogue.[178,179] The high pK_a value observed for cycloheptatriene is, therefore, the result of such anti-aromatic destabilization of the resultant carbanion. (If a system is highly nonplanar, however, it would have to be considered neither aromatic nor anti-aromatic, but *non-aromatic*, regardless of the number of p electrons potentially available to form the π-molecular orbital.)

Consideration of the stability of cyclopentadienide leads to other related aromatic systems, including the indenide ion (10π-electron; Hückel $n = 2$) and the fluorenide ion (14π-electron; Hückel $n = 3$). Other structurally related carbanions include the 9-phenylfluorenide and the fluoradenide ion. The pattern of $pK_{a(DMSO)}$ values is shown in Chart 2.1; the acidic protons have been highlighted.

The $pK_{a(DMSO)}$ values[151,180] increase with benzo-substitution in going from cyclopentadiene to indene and then to fluorene (Chart 2.1). Recall that fluorene has been used as the anchor for current ion-pair acidity scales. However, fluoradene is more acidic than cyclopentadiene; its pK_a could, in fact, be measured in water. Only with 9-phenylfluorene does the acidity of the cyclopentadienyl moiety approximate that of the parent cyclopentadiene. In examining the trend, we recognize the significance of the nature of the resultant carbanion, namely, its potential aromaticity. Importantly, highly delocalized planar carbanions show a good correlation between acidity of the parent carbon acids and the π-energy difference between the hydrocarbons and their carbanions, where the energy difference was determined from a linear combination of atomic orbitals–molecular orbital self-consistent field (LCAO–MO SCF) calculation.[181] However, another factor worth consideration is the role of the group attached to the carbanionic center, apart from its ability to delocalize charge through resonance, for

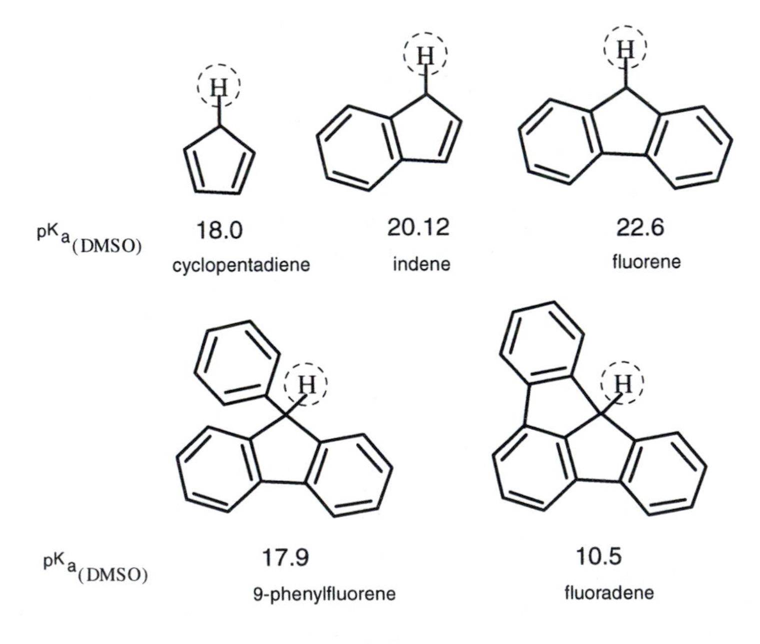

CHART 2.1

example, the acidifying effect of the 9-phenyl group in 9-phenylfluorene. These proximate substituent effects on stabilization of the carbanion will be considered further in the next section.

2.3.3 Effect of Substituents Attached to the Carbanionic Center

2.3.3.1 The Saturation Effect: Steric Inhibition of Resonance

Let us return to consideration of fluorene, 9-phenylfluorene, and fluoradene, but add to this set of hydrocarbons under review the following: toluene, diphenylmethane, and triphenylmethane. The pK_a values of these carbon acids measured (or obtained by extrapolation) in DMSO are 43,[151] 32.6,[182] and 30.8,[151] respectively. Cursory examination of the toluene series illustrates the well-known saturation effect that we have alluded to before (Section 1.1.3, Table 1.4), using ion-pair acidity data. The same conclusions may be reached in consideration of the acidity data determined in DMSO media, namely, that while attachment of a single phenyl ring to the carbanionic center in the methide ion lowers the pK_a (or increases the acidity) by 13 $pK_{a(DMSO)}$ units (where the pK_a of methane is taken to be 56), addition of a further phenyl substituent to the carbanionic center only enhances the acidity of the corresponding hydrocarbon by about 10 pK_a units. Moreover, triphenylmethane is only about 2 pK_a units more acidic than diphenylmethane in DMSO. In short, substitution by a phenyl ring does not lead to a set amount of stabilization of the carbanion; rather, the degree of stabilization decreases with each added phenyl ring, although each ring does lead to some degree of carbanion stabilization, as evidenced by the decrease in pK_a of the parent hydrocarbons.

The saturation effect has been ascribed to the lack of coplanarity of the phenyl rings that arises from the increasing steric congestion about the carbanionic center as phenyl rings are added to the methane moiety. In reality, the phenyl rings of the triphenylmethide ion take up a "propeller-like" conformation as indicated by both calculation[183] studies and X-ray analysis of the triphenylmethyllithium tetramethylenediamine complex[184] and shown in Figure 2.8 (TMEDA and Li omitted). Thus, it is unreasonable in a steric sense to expect all three of the benzene rings of the triphenylmethyl anion to be simultaneously coplanar, and the degree of delocalization found in Ph_3C^- is greater than that in Ph_2CH^- but not as great as expected on the basis of comparison with the $PhCH_2^-$ anion where the single ring is, without doubt, coplanar, with the p-(or hybrid) orbital that contains the negative charge and can delocalize the charge fully. (The geometry at the carbanionic center itself will be discussed further in the next chapter.) Consequently, what we have been calling a saturation effect is really the result of steric inhibition of resonance in the carbanion.

However, steric congestion about the carbanionic center in triarylmethides, for example, would be expected to render such carbanions more persistent in solution. This persistence permits spectroscopic investigation of the structure of arylmethyl carbanions and the nature of their ion pairs in solution.[185–187]

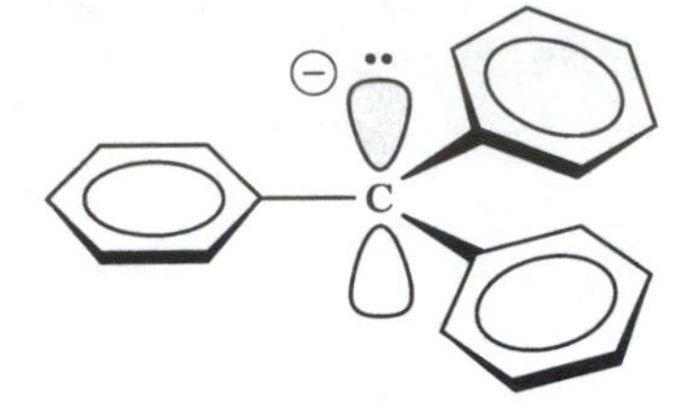

Figure 2.8. Propeller-like structure of triphenylmethide.

Steric inhibition of resonance also accounts for the lack of additivity on carbon acid acidity that follows progressive nitro group substitution of methane; thus, nitromethane, dinitromethane, and trinitromethane have pK_a values in water of 10.2, 3.6, and 0.2, respectively (Table 1.1). Here, the bulky nitro groups sterically interfere with the planarity necessary for full resonance delocalization into each nitro group, though inductive electron-withdrawal remains. Similar effects are expected in the acetyl-substituted methane series, namely, with acetone ("acetylmethane"), 2,4-pentanedione (acetylacetone), and 3-acetyl-2,4-pentanedione [tris(acetyl)methane]. In mitigation of steric inhibition of resonance, chelation of the anions derived from 2,4-pentanedione and tris(acetyl)methane with metal counterions provides a degree of stabilization that is unavailable to the acetone enolate.

In this regard, substitution by linear cyano groups should lead to a monotonic increase in acidity with the number of cyano functions introduced, since in these systems steric repulsion between the groups would be minimized. In proceeding from methane to acetonitrile (cyanomethane), the pK_a decreases from 56 (estimate based on DMSO solution data)[151] to 25,[188] a difference of 31. (Also see reference 189.) Addition of another CN drops the pK_a to 11.2, that is, a further ΔpK_a of 13.8, while tricyanomethane (cyanoform) has a pK_a in water of zero, though acidity function determinations rate the acidity of cyanoform significantly higher (p$K_a = -5.1$; Table 1.1). Therefore, the difference in pK_a between cyanoform and dicyanomethane (malononitrile) is between 11.2 and 16.3. Within the limits of our comparison, then, after introduction of the first CN group into methane, subsequent additions have approximately the same effect on the acidity of the carbon acid and on the stability of the resultant carbanion. It is interesting to note that the first cyano substitution is less effective than the first nitro substitution in enhancing the acidity of the substituted methane. This observation, combined with the observation that the intrinsic rate constant (k_o) for proton transfer from malononitrile (and other cyano-substituted carbon acids, generally) is moderately high,[84] suggests that resonance stabilization is less significant in the case of cyano substitution than in nitro substitution and that the CN group stabilizes a carbanion primarily via inductive and polarizability mechanisms.

Returning to the other members of the cyclopentadienide series, fluorenide, 9-phenylfluorenide, and fluoradenide (Chart 2.1), it is useful to consider these ions as analogues or derivatives not of cyclopentadiene but of triphenyl- and diphenylmethane, respectively. In this context, the fluorenide ion may be considered a "constrained analogue" of diphenylmethane. In the fluorenide ion, the two phenyl rings are forced to approach coplanarity and the pK_a value drops from 32.6 for diphenylmethane to 22.6 for fluorene.[151] Similarly, as a constrained analogue of triphenylmethane, fluoradenide ion has a significantly lower pK_a than triphenylmethane, namely, 10.5 rather than 30.8 for triphenylmethane, an increase in acidity of over 20 orders of magnitude!

Note also that 9-phenylfluorene is more acidic than fluorene, by 4.7 pK_a units. This more modest effect is consistent with the limited coplanarity of the 9-phenyl substituent with the fluorenyl moiety. In fact, Streitwieser and Nebenzahl[190] have performed a more general analysis of such systems and concluded that the twist angle between the fluorenyl portion and the 9-phenyl group is 50°, while on the basis of remote substituent effects Cockerill and Lamper assigned a twist angle of 39° between the two moieties.[191]

Similarly, on the basis of the ion-pair acidities of 10-(4′-phenylphenyl)-9,9-dimethyl-9,10-dihydroanthracene, **6**, and 10-phenyl-9,9-dimethyl-9,10-dihydroanthracene, **7**, (ion-pair pK_a = 27.7 and 28.0, respectively), the Streitwieser group concluded that the phenyl substituent was essentially orthogonal to the carbanionic center.[192] The Bordwell group came to a similar conclusion from their DMSO acidity data for the related 9-phenyl dihydroanthracene and examination of suitable models.[193]

A more recent study of the 9-phenyl dihydroanthracenyl anion by semi-empirical (AM1) calculations and ^{13}C NMR spectroscopy (of the sodium salt of 9-phenyl dihydroanthracene) suggests that the twist angle between the phenyl and the dihydroanthracenyl rings is between 35° and 40° (where 0° represents full phenyl ring coplanarity).[194] However, there is a relatively small energy difference between a phenyl twisted form in which the dihydroanthracenyl anionic moiety is planar and a form in which the dihydroanthracenyl group is folded; for the 9-(4′-cyanophenyl) dihydroanthracenyl anion, this folded form is apparently the major equilibrium form and it permits further delocalization of the negative charge into the aryl ring.[194]

In our discussion of the influence of substitution of the carbanionic center with phenyl groups, we have considered only the possibility of delocalizing the negative charge through resonance. It should also be noted that a phenyl group (or any substituent, for that matter) may stabilize the carbanion via polar/inductive/field

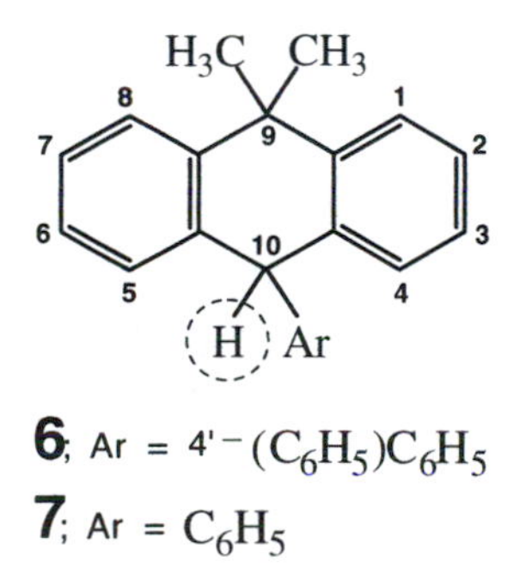

6; Ar = 4′-$(C_6H_5)C_6H_5$

7; Ar = C_6H_5

effects or via polarization of the substituent. In fact, in the Bordwell study of the effect of phenyl substitution on the stability of 9-phenyl dihydroanthracenide, comparison with systems in which a phenyl substituent could fully act as a resonance delocalizer led to the estimation of a ratio of resonance to polar effects of 4:1.[193] In this regard, triptycene, whose phenyl rings are precluded from achieving coplanarity with the carbanionic center, still has an ion-pair pK_a comparable to benzene (ca. 42).[182] Hence, stabilization of carbanions may occur via inductive/field and polarization mechanisms, as well as through resonance delocalization.

2.3.3.2 Polar Mechanisms of Stabilization of Carbanions by Proximate Substituents

(a) Inductive/field and pπ–pπ repulsion effects Attachment of one or more electronegative substituents to the anionic center stabilizes the carbanion via various polar mechanisms including inductive/field and polarization effects. In the first category is the trimethylammonium group, which is one of the most electronegative substituents known. As a positively charged substituent, $(CH_3)_3N^+$ is a powerful electron-withdrawing group in the inductive sense, but, as a result of its second-row central atom, it cannot act as a conjugative or π-acceptor group. Hence, trimethylammonioacetonitrile has a pK_a of 20.6, whereas acetonitrile has a pK_a of 31.3 in DMSO, a difference ($\Delta pK_a = 10.7$) comparable to that caused by attachment of a single benzene ring to methane [i.e., ΔpK_a (methane − toluene) = 13].[151,195] Therefore, stabilization of the carbanion with $(CH_3)_3N^+$ may be achieved that is equal to that provided by C_6H_5, even though the former α-substituent can operate only via the inductive/field mechanism and the latter functions primarily by resonance.

As we have pointed out earlier in our discussion, a nitro substituent would be expected to stabilize the negative charge in the nitromethide ion both inductively and via resonance delocalization of the carbanionic charge onto the oxygens of the nitro group (equation 2.35).[196] Other second-row central atom substituents that lack the possibility of acting in a conjugative fashion to delocalize the negative charge tend to stabilize the anion in accord with their electronegativities, although this effect also depends upon the structure of the carbanion. Exceptions to this generalization are particularly interesting and will be discussed further below.

It might be expected that fluorine would exert a large stabilizing effect on a carbanion to which it is directly attached and that this effect would arise from the large inductive electron withdrawal associated with the high electronegativity of fluorine. In fact, the effect of fluorine substitution on the stability of a carbanion is strongly dependent on the structure of the carbanion. Measurement of the kinetic acidity of a series of fluoroalkanes (CH_3O^-/CH_3OD or CH_3O^-/CH_3OT) yielded the following values for the pK_a: $CF_3\mathbf{H}$, 31; $CF_3(CF_2)_5CF_2\mathbf{H}$, 30; $(CF_3)_2CF\mathbf{H}$, 20; and $(CF_3)_3C\mathbf{H}$, 11.[197] Clearly, the fluorines in trifluoromethane (fluoroform) stabilize the resultant trifluoromethyl anion, though not as effectively as the chlorines of the analogous chloroform ($pK_a = 24.1$)[62,198] stabilize the trichloromethide ion. However, the β-fluorines of 2-(trifluoromethyl)-1,1,1,3,3,3-hexafluoropropane [$(CF_3)_3C\mathbf{H}$, hereafter tris(trifluoromethyl)methane] are even more effective in stabilizing the carbanionic center of its conjugate base than are the α-fluorines of fluoroform and this gives rise to the ΔpK_a of 20 between the two fluoroalkanes $(CF_3)_3CH$ and CF_3H. Another piece of the puzzle is provided by consideration of the effect of substitution in nitromethanes, $XYCHNO_2$, where Y = NO_2, Cl, $CONH_2$, or COOEt. The acidity of these substituted nitromethanes is actually diminished when X is F relative to the systems in which X is H. Conversely, when X is Cl, the acidity is enhanced relative to the nitromethanes where X is H (with but one exception).[198] Here, fluorine substitution of the carbanionic center is destabilizing.

In putting the pieces together, we must carefully consider the structure of two carbanions, namely, the trifluoromethyl anion and the fluoronitromethyl anion, $^-CHFNO_2$. It is reasonable to presume that the high acidity of nitromethane itself requires resonance delocalization of the negative charge into the oxygens of the nitro group and, in turn, argues for a structure that would approach planarity for the nitromethide ion. The planarity of the nitromethide ion is supported by recent ab initio (MP2/6–31 + G*) calculations,[199] though calculations at even higher levels may not bear this out. Notwithstanding the foregoing, if it is assumed that the fluoronitromethide ion is planar, and further assuming that the trifluoromethyl anion is pyramidal in structure, then the acidity data can be rationalized on the basis of two opposing types of inductive effect: an inductive effect involving the σ-bonded framework, I_σ, and an electron-pair repulsion effect, I_π (as shown in Figure 2.9).[200]

In the pyramidal trifluoromethide ion, the lone pair of the carbanion is placed farther away from the lone pairs of the fluorines than would be the case in the approximately planar fluoronitromethide ion. Consequently, electron-pair repul-

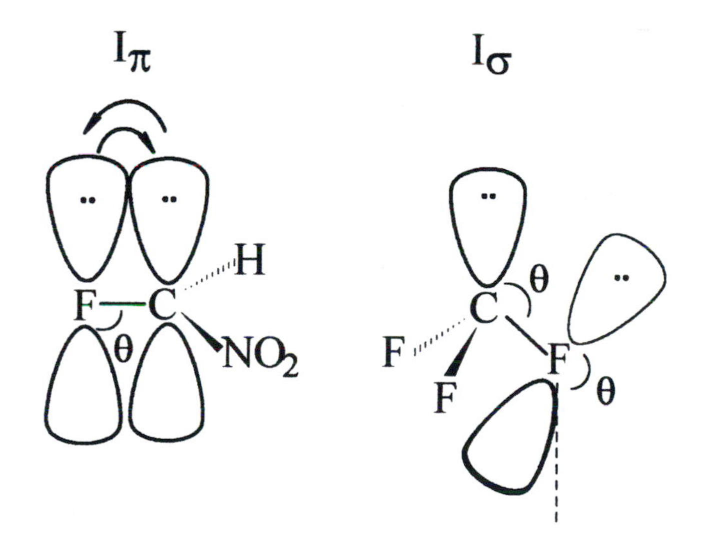

Figure 2.9. Opposing inductive effects in fluorine-substituted carbanions: the I_π electron-pair repulsion and the I_σ inductive effect; which effect is dominant is dependent on the structure of the carbanion, as discussed in the text.

sion, the destabilizing I_π effect, is minimized in the trifluoromethide ion and maximized in the fluoronitromethide ion; fluorine stabilizes the carbanion center in the trifluoromethide ion by the normal inductive effect, I_σ, while in the fluoronitromethide ion the fluorine is destabilizing because of the dominant I_π (predominant pπ–pπ repulsion) effect.[200]

In accord with the view presented, 9-fluorofluorene undergoes proton exchange with methanolic methoxide at a rate that is approximately eight times slower than that of the parent fluorene.[201] Recall that the fluorenide ion is constrained to be planar and, therefore, the I_π electron-pair repulsion mechanism is dominant in 9-fluorofluorenide and, so, the kinetic acidity of its conjugate acid is significantly less than that of the unsubstituted fluorene. (Of course, this assumes that kinetic acidity here follows thermodynamic acidity.) It is also important to recognize that such a pπ–pπ lone-pair repulsion effect[202] or +R-p-orbital feedback destabilization[203] operates in the case of α-oxygen (e.g., alkoxy) substitution of the carbanionic center.[204,205] In fact, the very strong Lochmann–Schlosser base (*n*-butyllithium-potassium *t*-butoxide) is required to significantly deprotonate a series of substituted 1,3-dioxanes.[206]

A preliminary conclusion is that in pyramidal carbanions, α-fluorine substitution stabilizes the carbanion, as indicated by the pK_a values of the relevant conjugate carbon acids. Conversely, in planar carbanions, α-fluorine substitution is destabilizing.

(b) Anionic Hyperconjugation in β-fluorine Substitution We have noted the significantly greater stabilization that β-fluorine substitution exerts on the carbanion relative to α-substitution, even when fluorine substitution at the carbanionic center would be expected to be stabilizing. This β-fluorine effect has been attributed to anionic hyperconjugation, which may be illustrated in terms of double bond–no bond or "sacrificial" resonance forms,[207–209] as shown in (2.37) for the carbanion of tris(trifluoromethyl)methane.

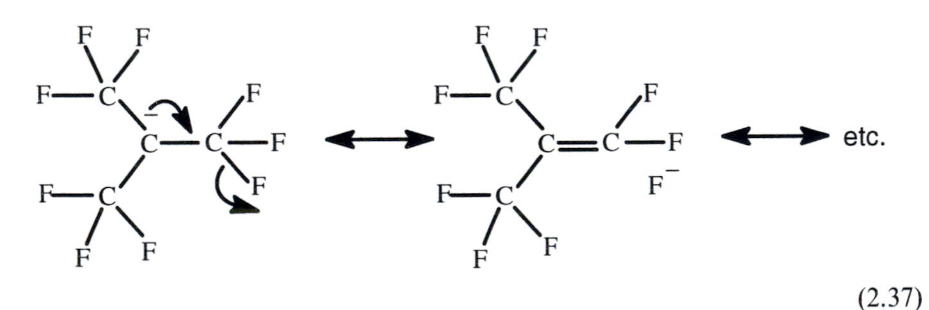

(2.37)

Understandably, if this form of hyperconjugation (2.37) is effective, in which 9 of the 10 possible resonance contributors involve a double bond between the central carbon and an α-trifluoromethyl moiety, it would be expected that the anion would be planar about the carbanionic center, and current evidence supports this surmise.[204,210,211] The valence bond view outlined above may be recast in molecular orbital terms in which the lone-pair electrons of the anionic center interact with the empty antibonding orbitals (σ*) and the filled bonding orbitals

(σ) of the β–C–F bonds.[212–215] The $n \rightarrow \sigma^*$ (or $p \rightarrow \pi^*$) interactions would be stabilizing, while the $n \rightarrow \sigma$ (or $p \rightarrow \pi$) interactions would be destabilizing. The degree of stabilization attendant upon β-fluorine substitution would, therefore, emerge from the balance between these interactions, which, in turn, depend upon the various orbital energies of the interacting fragments.

Alternatively, β-fluorine stabilization could arise from standard inductive/field effects.[199,215] The key to the apparently anomalous degree of β-fluorine stabilization relative to α-fluorine stabilization is the reduction in electron-pair repulsion between the fluorine and the carbanionic center. Here, the I_σ effect becomes dominant because the destabilizing I_π repulsion has been diminished.

In assessing the anionic hyperconjugation mechanism, as compared with the I_σ/I_π approach, which we have already found valuable in considering α-fluorine substitution, a number of groups have targeted bicyclic systems (e.g., **8–10**) in which hyperconjugation would be expected to be impossible.[215–217] For example, in compounds **8–10** the acidity of the bridgehead protons (highlighted) would be expected to be influenced by inductive/field effects as a result of the fluorine substitution, but anionic hyperconjugation is precluded because the necessary planar conformation cannot be attained by the carbanion.

No-bond resonance forms such as those shown earlier for tris(trifluoromethyl)methane cannot be drawn for the carbanions derived from compounds **8–10** without violating Bredt's rule. If the resultant carbanions of **8–10** are obliged to be pyramidal about the carbanionic center then, according to our previous arguments concerning the fluorine I_σ mechanism, maximum inductive stabilization of the carbanion would be predicted. In fact, the bridgehead proton of 1-H-undecafluorobicyclo[2.2.1]heptane, **8**, has been shown to exchange with sodium methoxide in tritiated methanol approximately twice as fast as does the corresponding proton of tris(trifluoromethyl)methane, demonstrating that the kinetic acidity of the former is greater than that of the latter, even though hyperconjugation is not possible for **8**.[217] On the other hand, recent measurements of gas-phase acidity of **8**, **9** (1,4-di-*H*-decafluorobicyclo[2.2.1]heptane), and tris(trifluoromethyl)methane by Taft and co-workers[218] showed that while **8** is slightly more acidic than **9** (in agreement with a small inductive/field effect caused by substitution of one of the bridgehead protons with a fluorine), tris(trifluoromethyl)methane is a significantly stronger acid than the polyfluorobiocycloalkane, **8**; tris(trifluoromethyl)methane was found to have a gas-phase

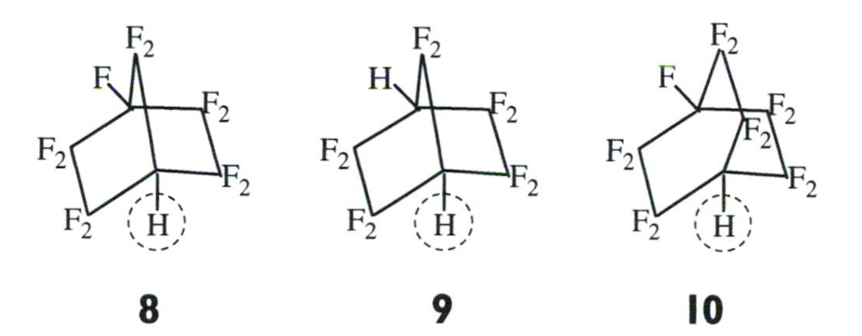

acidity ($\Delta G_a°$) of 326.6 kcal/mol, which makes it a stronger gas-phase acid than **8** by 7.6 kcal/mol (5.6 pK_a units). Interestingly, tris(trifluoromethyl)methane is not only a stronger gas-phase acid than trifluoromethane ($\Delta G_a° = 368.6$ kcal/mol)[219] but also is more acidic than HCl in the gas phase, by 1.7 kcal/mol.[220] These results also agree with competitive deuteration studies involving **8**, **10** (i.e., 1-*H*-trideca-fluorobicyclo[2.2.2]octane), and tris(trifluoromethyl)methane that determined that tris(trifluoromethyl)methane incorporated deuterium about 100-fold faster than **8** and 10-fold faster than **10**.[220]

Finally, the measured p$K_{a(DMSO)}$ of tris(trifluoromethyl)methane (12.6)[218] compares favorably with malononitrile (dicyanomethane, 11.0) and fluoradene (10.5),[151] and demonstrates that it is a moderately strong carbon acid. In conclusion, current evidence supports the view that anionic hyperconjugation is an important mechanism of carbanion stabilization by β-fluorines in carbon acids that can achieve the requisite planarity; it operates in concert with inductive/field electron-withdrawal.

(c) Polarization in Alkyl Group Substitution Alkyl substituents attached directly to the carbanionic center do not act in strict accordance with the electronegativity of these groups. Standard analysis of the effect of alkyl groups would hold that such electron-releasing groups stabilize positive charges and destabilize negative charges.[221] The Taft substituent constants, σ^*, derived from the hydrolyses of esters generally supported an order of decreasing electron-releasing power: *t*-butyl > *i*-propyl > methyl.[222] However, when the Bordwell group examined the DMSO acidity of a series of alkyl-substituted carboxamides that are structurally similar to the Taft esters, little change in pK_a was found as a function of the nature of the alkyl group.[193] A later analysis by Taft and co-workers concluded that alkyl substituents operate by both an inductive/field effect, which depends on the electric dipole of the substituent (related to the group electronegativity of the substituent), and a polarizability effect.[222] The former was found to depend on the first power of the charge and to decrease with the inverse square of the distance between the group and the charge, whereas the latter depends on the square of the charge and decreases according to the inverse fourth power of the distance. By assuming that inductive/field effects stabilize only a positively charged center and that polarizability may stabilize both negative and positive charges, it was possible to factor out the polarizability effect from gas-phase acidity data. The residual inductive/field effect data for alkyl groups correlated with σ_I, a modified polar effect substituent constant.[222] In summary, where inductive/field effects do not dominate, alkyl substituents may stabilize both anionic and cationic centers as a result of their polarizability.

It would be expected on the basis of the foregoing discussion that examples of alkyl-substitution enhancing and diminishing acidity would be found particularly in the gas phase, where solvent effects that could obscure the inherent effects of alkyl substitution would be absent. In this regard, addition of a methyl group to toluene to give ethylbenzene (phenylethane) or to nitromethane to give nitroethane enhances the gas-phase acidity of the ethanes relative to the methanes.[223] Here, methyl substitution stabilizes the carbanion. On the other hand, addition of a methyl substituent to acetonitrile (ethanenitrile) to give propanenitrile or to methyl phenyl sulfone to give ethyl phenyl sulfone results in decreased gas-phase acidities for the methyl-substituted carbon acids.[223] Similarly, propyne (methyl acetylene) is less acidic than the parent ethyne in the gas-phase. However, substitution of a methyl at the carbon acid center for any other larger alkyl group, such as ethyl, propyl, and so on, results in an increase in acidity of the acid; the resultant carbanion is stabilized.[223] These results are consistent with the view that polarizability is the major factor in stabilization of carbanions by alkyl groups in the gas phase. However, variable results accompany methyl substitution and the origin of this variability will be considered next.

The pattern of gas-phase acidities can also be considered within the context of orbital interactions, as laid out by Pross and Radom.[224] In this approach, the interaction of the methyl group with the carbanion center can be viewed as the orbital interaction of the π- and π^*-orbitals of the methyl group with the π-symmetry orbitals of the carbanion. As applied to the ethyne–propyne pair, the interaction of π-CH_3 with the empty π^*-orbitals of the fragments $-C{\equiv}C-H$ and $-C{\equiv}C^-$ results in stabilization of the parent propyne as compared with its conjugate base and, hence, propyne is less acidic than ethyne.

More generally, inductive/field electron release that destabilizes carbanions can be seen to arise from the interaction of the π-orbital(s) of the methyl fragment with the π-orbital(s) of the carbanionic center; however, electron withdrawal via polarizability that stabilizes carbanions results from the interaction of the empty π^*-orbital(s) of the alkyl fragment with the same π-orbital(s) of the acidic site.[224] In principle, both interactions are possible and which mechanism, polarizability or inductive/field effect, will determine the acidity depends on the relative orbital energies and degree of overlap in the specific molecule.

The variability of the effect of methyl substitution is also observed in the DMSO systems examined by Bordwell.[151] This should come as no surprise considering the observed correlation of DMSO solution acidity data with gas-phase measurements, at least for limited series of structurally related carbon acids (cf. Sections 1.5 and 1.6). Therefore, in examining a set of substituted methanes (CH_3–X) and ethanes (CH_3-CH_2-X), where the latter compounds were taken as models for methyl group substitution, the results were understandably mixed.[193,195,225,226] When X was a nitro group, the effect of methyl substitution was to increase the acidity of the carbon acid by 0.7 pK_a unit, but when X was an acetyl function a small decrease in acidity attendant upon methyl substitution was noted (0.3 pK_a unit). When X was a phenylsulfonyl or trifluoromethylsulfonyl group, the decrease in acidity that followed methyl substitution was more significant (1.7 and 1.4 pK_a units, respectively).

It is apparent, then, that whether in the gas phase or DMSO solution, methyl groups may stabilize or destabilize a carbanion, the former by a polarizability mechanism and the latter by a more classical inductive/field mechanism. The effect is, at least partially, also dependent on the nature and, specifically, the electron demand of the other substituents attached to the carbanionic center. On the other hand, in solvents such as water or alcohols the inductive/field destabilization of the carbanion by alkyl groups may be significant.

2.3.3.3 Effect of Third- and Later-Period Central Atoms in the Proximate Substituent on the Carbanion

Another important class of stabilizing substituents that must be considered are those in which an element of the third or a later period is attached to the carbanionic center. It is well recognized that substituents that contain third-period/later-period elements enhance the acidity of the carbon acid center to which they are attached. Thus, base-catalyzed isotopic exchange (H/D) is a facile process with the tetramethylphosphonium ion [$(CH_3)_4P^+$],[227] the trimethylsulfonium ion [$(CH_3)_3S^+$],[227] and with dimethylsulfone [$(CH_3)_2SO_2$],[228] but exchange occurs less readily with dimethyl sulfoxide [DMSO, $(CH_3)_2SO$].[229] Isotopic exchange is similarly enhanced by other phosphorus- and sulfur-containing substituents, with other oxidation states.[230–233] Cursory examination of the tetramethylphosphonium ion or the trimethylsulfonium ion would suggest that the inductive electron-withdrawal due to the positive charge on the α-substituents is important in determining the ease of exchange. Significantly, however, the rate of exchange with the tetramethylphosphonium ion is 2.4×10^6 faster than it is with the tetramethylammonium ion.[227]

Substituents that contain neutral third-period elements attached to the carbon acid site also magnify the kinetic and thermodynamic acidity of carbon acids. For example, examination of the pK_a values determined in DMSO for a selection of 9-substituted fluorenes (Table 2.1) demonstrates clearly the comparative efficacy of substitution by third-period element groups relative to other well-known carbanion stabilizing functions. Thus, substitution of the 9-position by a thiomethyl group lowers the pK_a value from 22.6 (fluorene, uncorrected for the number of acidic protons) to 18.0, while 9-phenyl substitution has a very similar effect and lowers the pK_a to 17.9.[151]

Note, as well, the quite small increase in the acidity of fluorene that follows from substitution of the 9-position with a methoxyl group; $\Delta pK_a = 0.5$. This is consistent with the balance anticipated between the inductive effect due to the electronegativity of oxygen and the destabilizing repulsion of the oxygen lone pairs and the lone pair of the resultant planar carbanion, as previously discussed in detail for fluorine. Also consistent with our previous considerations, alkyl groups have either a slight stabilizing effect or no effect on the acidity of fluorene. Concerning the other third-period-centered substituents, it is apparent that increasing the oxidation state of the attached sulfur further increases the acidity of the fluorene, as in the comparison between the effect of 9-thiomethyl substitution and 9-sulfonylmethyl substitution.[151] Finally, α-substitution with silyl groups stabilizes the planar fluorenide ion, though to a lesser degree than sulfur group substitution.[234]

In a separate study, $pK_{a(DMSO)}$ values support the view that the 9-triphenylphosphonium ion substituent is significantly more stabilizing of the resultant carbanion than substitution at the 9-position by the triphenylsilyl group; the 16 kcal/mol acidifying effect of the former as compared with the 5.6 kcal/mol stabilization of the latter was ascribed to both the greater polarizability of the phosphonium group and the inductive/field electron withdrawal caused by the positive charge on the group.[234]

Table 2.1. Equilibrium acidities of some selected 9-substituted fluorenes, determined in DMSO (298 K)

9-substituent	$pK_{a(DMSO)}$	9-substituent	$pK_{a(DMSO)}$
$COOCH_3$	10.35	$Si(C_6H_5)_3$	18.6
$SO_2C_6H_5$	11.55	OC_6H_5	19.9
$SO_2CH_2CH_3$	12.3	$Si(CH_2CH_3)_3$	21.4
SO_2CH_3	12.75	$Si(CH_3)_3$	21.5
SC_6H_5	15.4	OCH_3	22.1
SCH_2CH_3	17.5	CH_3	22.34
C_6H_5	17.9	CH_2CH_3	22.6
SCH_3	18.0	H	22.6

SOURCE: Data taken from references 151 and 234.

The magnitude of the effect of α-silyl substitution is dependent on structural features of the silyl-functionalized carbon acid. This variability has made assessment of the effective mechanisms of α-silyl stabilization problematic. In this context, substitution in ethyne by either a trimethylsilyl or a triethylsilyl group provided stabilization of only about 1 pK_a unit, based on the ion-pair acidity scale (Li^+ counterion in dimethoxyethane) developed by Petrov, Shatenshtein, and co-workers.[235,236] These effects on a carbanion that would be localized are consistent with a polarizability mechanism for stabilization, given the electropositive nature of silicon relative to carbon. The pK_a determination (equilibrium lithium ion-pair acidity in THF) for benzyltrimethylsilane[237] and phenylthiomethyltrimethylsilane[235] shows that these carbon acids are more acidic than toluene, according to the ^{13}C NMR method developed by the Fraser group.[237–240] Interestingly, while the methylene protons of phenylthiomethyltrimethylsilane ($pK_a = 35.4$) are more acidic than those of thioanisole (phenylthiomethane; $pK_a = 38.7$), phenylthiomethyltrimethylstannane was found to be equally acidic; α-silyl and α-stannyl substitution are equally effective in stabilizing carbanions, even though tin should be more polarizable than silicon.[238] The carbanion stabilization in this case was attributed to C–Si and C–Sn (pπ–dπ) overlap.[238] Importantly, we have seen that substitution at the 9-position of fluorene stabilizes the resultant planar carbanion (Table 2.1), according to $pK_{a(DMSO)}$ measurements, and that in agreement with the results of Miah and Fraser[237,238] α-silyl substitution is less stabilizing than α-functionalization by sulfur-containing substituents. However, the Bordwell group also reported that trimethylsilyl substitution of malonic esters *decreased* the acidity by 2.3 $pK_{a(DMSO)}$ units.[151] Variable degrees of carbanion stabilization of between 1 and 3.4 pK_a units upon α-silyl substitution have been reported by the Streitwieser group, determined by Li/THF (solvent-separated ion pair) and Cs/THF (contact ion pair) acidities of 9-trimethylsilylfluorene, 9-*tert*-butyldimethylsilylfluorene, as well as the ion-pair acidities of benzyltrimethylsilane ($pK_{a(Cs/THF)} = 37.5$), 2-(trimethylsilyl)-1,3-dithiane ($pK_{a(Cs/THF)} = 33.5$), trimethylsilylacetonitrile ($pK_{a(Cs/THF)} = 28.8$), and tris(trimethylsilyl)methane ($pK_{a(Cs/THF)} = 36.8$).[241]

In their examination of α-silyl group effects on the $pK_{a(DMSO)}$ values of 9-X-fluorenes and 1-X-1-phenylsulfonylmethanes (i.e., substituted methyl phenyl sulfones), two of the Bordwell group's results are especially interesting.[234] The first is the observation that the 9-triphenylsilyl group stabilizes the substituted fluorenide ion by 5.9 kcal/mol, an effect that is 2.3 kcal/mol greater than the stabilization afforded the carbanion by 9-triphenylmethyl substitution. While, overall, the stabilization accompanying α-silyl group substitution was attributed to the polarizability of the attached third-period element, the large effect in the case of the 9-triphenylsilyl group indicated that another factor was involved. This extra stabilization of the carbanion *relative to the initial carbon acid* is attributable to relief from steric strain; 9-triphenylsilylfluorene is approximately tetrahedral about the acidic carbon center while deprotonation yields an approximately planar carbanion that is, consequently, less strained.

The second interesting observation concerns the effect of multiple substitution. Thus, introduction of the first α-trimethylsilyl group into methyl phenyl sulfone [to give $(CH_3)_3SiCH_2SO_2Ph$] increases the acidity by 2.8 $pK_{a(DMSO)}$ units, but attachment of a second trimethylsilyl group (to give $[(CH_3)_3Si]_2CHSO_2Ph$) increases the acidity by a further 5.7 $pK_{a(DMSO)}$ units. Here, contrary to the usual saturation effect that accompanies multiple substitution, *the second substituent is more stabilizing than the first!* Again, relief of strain that results from deprotonation of the 1-bis(trimethylsilyl)-1-phenylsulfonylmethane combined with the polarizability of the α-silyl substituents accounts for the exalted acidity of this carbon acid as compared with the monosubstituted methyl phenyl sulfone.[234]

These arguments for relief of steric strain may also explain the earlier observation that tris(trimethylsilyl)methane-*t* undergoes detritiation five-fold to seven-fold faster than tritiated triphenylmethane,[242] whereas gas-phase acidity studies of trimethylsilylmethane have shown that silyl substitution of the anionic center stabilizes the carbanion by 20 kcal/mol, comparable to, but not greater than, the stabilization provided by phenyl functionalization.[159,243] Also consistent with this view is the observation that substitution further from the acidic center, as in β-silyl group substitution, has little stabilizing effect in carbanion formation based on $pK_{a(DMSO)}$ measurements; a 9-$(CH_3)_3SiCH_2$ group stabilizes fluorenide ion by 1.3 $pK_{a(DMSO)}$ units.[244]

Gas-phase acidities permit a comparison of the efficacy of stabilization by neutral phosphorus-, sulfur- and silicon-containing groups (Table 2.2). In this comparison, α-silylmethyl substitution appears to be more effective than α-thiomethyl substitution and may be even more effective than subsititution by the phosphorus-centered group. (Recall that gas-phase acidities are generally endothermic and, therefore, the most favorable deprotonation is that which occurs with the numerically lowest enthalpy change.) Interestingly, α-substitution by phosphorus-, sulfur-, or silicon-centered neutral groups is more stabilizing of the carbanion than substitution by chlorine.

Undoubtedly, substitution of the carbon acidic center by third-period elements stabilizes the resultant carbanion, as shown by the experimental evidence presented above (among the many studies)[252–254] and also by calculational studies.[255–261] However, there is no strict order for the degree of stabilization;

thioalkyl and thioaryl groups are more effective than silylalkyl and silylaryl groups in solution (Table 2.1),[151,231,234,235] but the silylmethyl group is apparently more stabilizing than its thiomethyl counterpart in the gas phase (Table 2.2). Structural features of the systems studied contribute to the variability by introducing other factors, such as relief of ground-state strain, to the acidity measurements.

Although a traditional view[226,237,238] has held that third- (and later-) period elements stabilize carbanions by delocalization of the negative charge into the low-lying d-orbitals of the central atom (pπ–dπ delocalization compared with the pπ–pπ delocalization of phenyl substituents), numerous calculations have stressed the *un*importance of d-orbital participation in the stabilization of carbanions, including carbanions like $HSCH_2^-$, $HSOCH_2^-$, and $HSSO_2CH_2^-$.[261–263]

Consequently, it appears probable that stabilization of the carbanion occurs primarily through a polarization mechanism that is more effective with these more polarizable third-period elements than with second-period-centered groups, as can be seen in the comparison of alkyl group substitution of fluorene with substitution by any of the groups that contain a third-period element as the central atom bonded to the fluorene acidic site (Table 2.1). Ab initio calculational studies of substitution of carbanions by α-silyl groups[264] concluded that beyond polarizability there may also be effective negative hyperconjugation, as represented in (2.38) for trimethylsilylmethide in valence bond resonance forms.

Table 2.2 Gas-phase acidity data for methanes α-substituted by groups that contain third- or later-period central atoms as compared with phenyl groups

Carbon acid	ΔH°_{acid}(kcal/mol)
$ClCH_3$	399.6±2.5[a]
$(CH_3)_2PCH_3$	391.1±2.1[b]
	384.2±3.2[c]
CH_3SCH_3	390.2±1.5[d]
$(CH_3)_3SiCH_3$	387 ±3[e]
	390.2±2[f]
$[(CH_3)_3Si]_2CH_2$	373.6±4.0[e]
$C_6H_5CH_3$	380.6±1.0[g]
$(C_6H_5)_2CH_2$	363.6±2.1[h]
$(C_6H_5)_3CH$	358.7±2.2[i]

SOURCE: All data are taken from reference 164.

ORIGINAL SOURCES: [a]Reference 245; [b]Reference 246; [c]Reference 247; [d]Reference 248; [e]Reference 158; [f]Reference 243; [g]Reference 249; [h]Reference 250; [i]Reference 251.

$$(H_3C)_2\overset{\overset{\displaystyle CH_3}{|}}{Si}-\overset{-}{C}H_2 \longleftrightarrow (H_3C)_2Si{=}CH_2 \; (\overset{-}{C}H_3) \longleftrightarrow (CH_3)_2\overset{\overset{\displaystyle CH_3}{|}}{\overset{-}{Si}}{=}CH_2 \qquad (2.38)$$

As in other calculational studies,[265,266] d-orbital inclusion in the computational basis sets did not lead to improved energies and, therefore, d-orbital participation was considered unimportant for energetics (although necessary for proper description of the geometries of the carbanions examined). In this context, the resonance contributors drawn can, at best, describe *very weak* π-bonding between silicon and carbon. However, the middle canonical form drawn that shifts negative charge from silicon out onto the more electronegative carbon of the methyl substituents may still contribute, albeit to a minor extent. The final structure is a representation of the polarization of silicon that withdraws electron density from the carbon center.

Finally, comment should be made concerning substitution of the carbanionic center by trifluoromethylsulfonyl groups. The trifluoromethylsulfonyl group is a powerful uncharged electron-withdrawing group. In DMSO, trifluoromethylsulfonyl substitution enhances acidity significantly. For example, the difference in $pK_{a(DMSO)}$ between dimethyl sulfone ($pK_{a(DMSO)} = 31.1$) and methyl trifluoromethyl sulfone ($pK_{a(DMSO)} = 18.75$) is 12.3 pK_a units.[151] In the gas-phase, multiple trifluoromethylsulfonyl substitution converts methane into a "CH superacid".[217] While HCl has a gas-phase acidity (ΔG°_{acid}) of 328.3 kcal/mol, bis(trifluoromethylsulfonyl)methane has an acidity of 301.5 kcal/mol and the gas-phase acidity of tris(trifluoromethylsulfonyl)methane has been measured as 280.0 kcal/mol.[217–219]

With the methylsulfonyl methide ion, $CH_3SO_2CH_2^-$, according to ab initio computational studies[265–268] the most stable rotameric conformation places the lone-pair orbital of the anion gauche to both oxygens of the sulfonyl; this arrangement is favorable for a $n \rightarrow \sigma^*$ stereoelectronic interaction involving the antibonding orbital of the CH_3–SO_2 moiety (which could also be described by a double bond–no bond canonical form, as drawn above for the trimethylsilyl methide ion; equation 2.38). For the trifluoromethylsulfonyl methide ion, the same favorable rotameric conformation has been found to be the optimum geometry; the amplified acidity of methyl trifluoromethyl sulfone (and presumably the multiply substituted trifluoromethylsulfonyl methanes) is a result of the same anionic hyperconjugation, whose efficacy is increased because the increased electronegativity of the trifluoromethyl group lowers the energy of the σ*-orbital and, consequently, enhances the stabilizing $n \rightarrow \sigma^*$ (SO_2–CF_3) interaction in the carbanion.[269]

In this section, we have considered in detail the mechanisms of stabilization of carbanions and particularly the effect of substituents directly attached to the carbanionic center. What happens when the stabilizing groups are not in close proximity to the anionic site? We will explore this question in the next section.

2.3.4 *Effect of Remote Substituents on Carbanion Stability*

Substituents far from the reactive site may still interact with that site through the same range of mechanisms that we have already seen operate for functions attached directly to the carbanionic site: inductive/field, polarizability, and resonance delocalization. Some of these processes, however, will be attenuated by the structure that separates the anionic center and the substituent. Thus, with the 9-substituted fluorenes, dimethylamino substitution decreases the acidity by 11.0 $pK_{a(DMSO)}$ units compared with the effect of phenylsulfonyl substitution.[270] However, the difference in these $pK_{a(DMSO)}$ values for the same substituents drops to 8.8 pK_a units in the 4-substituted phenylacetonitrile series[271] and to 7.6 pK_a units in the 4-substituted benzylphenyl sulfone series.[272]

The sensitivity of deprotonation to the effect of remote substituents on aryl rings is often reflected in the Hammett slope parameter, the ρ-value. In this regard, Bordwell lists 11 different families of carbon acids whose acidities correlate with the Hammett substituent constant (σ); the Hammett ρ-value varies from a low of 4.2 for the 3-X-$C_6H_4COCH_3$ family to a high of over 10 for the 9-methylanthracenes.[151]

In these studies, the Hammett plots were generally restricted to correlations with *meta* substituents; *para* substituents such as nitro and acetyl would be expected to have anomalously high effects on the carbon acidity relative to gas-phase results because of specific dipole–dipole interactions with the DMSO solvent, as shown in Figure 2.10 for the 4-acetylbenzyl anion.[223,251,273] More significant enhancements of acidity relative to the gas phase would be expected in water, where hydrogen-bonding is the mode of solvent–substituent interaction. Since these effects operate primarily in resonance-delocalized systems, they have been named *solvent solvation-assisted resonance* (SSAR) effects. For *para*-substituted phenyl systems, SSAR effects increase in the following order with the *para* substituent: $SOCH_3$, SO_2R < CN < SO_2CF_3 < RCO < NO_2 < NO.[251,273]

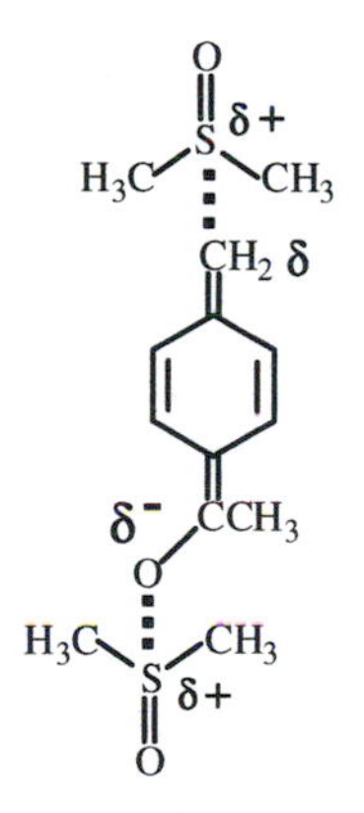

Figure 2.10. 4-Acetylbenzyl anion is depicted here as interacting in a dipole–dipole fashion with the dimethyl sulfoxide solvent. These specific interactions stabilize carbanions such as the 4-acetylbenzyl anion in solution relative to the gas phase and lead to enhanced solution acidity.

Substituents on a phenyl or other aromatic/heteroaromatic ring may interact with the carbanionic center via resonance delocalization, as well as the various polar mechanisms. Multiple substitution of a ring system can lead to significant stabilization of the carbanion and the various substituents may each interact with the carbanionic center via different mechanisms; the most acidic simple substituted fluorene, 2,7-dibromo-9-(methoxycarbonyl)fluorene, has a $pK_{a(DMSO)}$ of 6.5, as a result of such multiple effects.[151]

2.3.4.1 Resonance Stabilization via Remote Substituents: the Nitro Group

Nitro groups are effective carbanion stabilizers that exert a significant polar effect (conveyed by the polarizability of the aromatic nucleus as shown in structure **11** and consistent with the high group electronegativity of the NO_2 group)[274] and are also able to withdraw electron density by delocalization onto the nitro group oxygens (via resonance, as in structures **12** and **13**).

In considering the effect of a substituent like the NO_2 group, other canonical forms that stabilize the anionic charge by bringing it close to the electron-withdrawing functional group should also be recognized (structure **14**).[275,276] Solvents, especially water, also interact with nitro groups to further delocalize charge in the SSAR effect.[251]

The 2,4,6-trinitrophenyl (picryl) group is a powerful acidifying group, even though the nitro groups that confer this ability are remote from the acidic site. Thus, 2,4,6-trinitrophenol (picric acid) has a pK_a of ca. 0 in either water[151] or DMSO.[276] 2,4,6-Trinitrotoluene (TNT) has a pK_a of 13.6 in water (15.6 in methanol).[277] Also, for the sake of comparison, 2,4,6-tris(trifluoromethylsulfonyl)toluene has been found to have a pK_a of 9.46 in methanol.[96] The Terrier group has undertaken a systematic study of the effect of nitro-substituted aryl and heteroaryl groups on carbon acidity.[95,97,98,123,278,279] In the kinetic study of the deprotonation of 2,4,4′-trinitrodiphenyl- and 2,2′,4,4′-tetranitrodiphenylmethane (and the protonation of their conjugate carbanion bases), using a range of amine, carboxylate, and phenoxide bases in DMSO:H_2O (50%:50%), high intrinsic barriers (low k_o values) were found.[95,279] Recall from our discussion of intrinsic reactivity (sections 2.2.1 and 2.2.2) and the principle of non-perfect synchroniza-

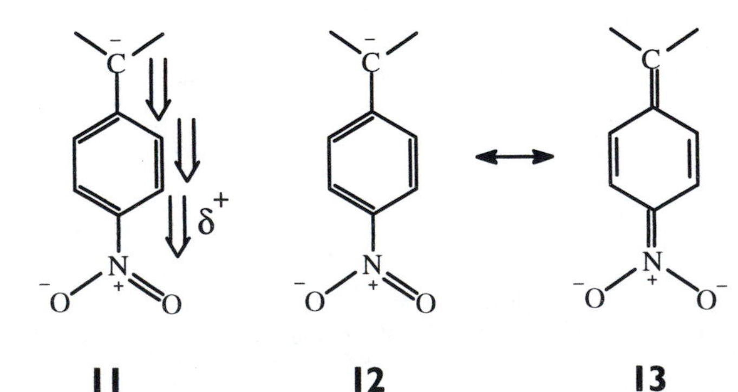

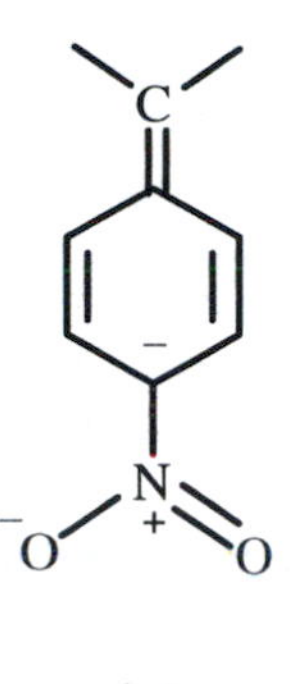

14

tion that these tri- and tetranitrodiphenylmethide ions are resonance stabilized and that such delocalization apparently trails proton transfer here. A separate structural study (NMR and UV–visible spectroscopy) of the carbanions **15a–f** (Chart 2.2) demonstrated that for carbanions **15a–d** the aryl rings are coplanar or approach coplanarity.[279] However, further addition of nitro groups to the *ortho* positions, as in the carbanions derived from 2,2′,4,4′,6-pentanitrodiphenylmethane, **15e**, and those from 2,2′,4,4′,6,6′-hexanitrodiphenylmethane, **15f**, would sterically preclude the simultaneous coplanarity of the two aryl rings and, consequently, inhibit resonance delocalization of charge onto the nitro groups.

Interestingly, the pK_a values for this series of polynitrodiphenylmethanes deviate from expectations. First, steric inhibition of resonance should lead to a significant increase in pK_a for **15e** and **15f** as compared with **15d**, in which the two aryl rings can both delocalize the carbanionic charge simultaneously. Secondly, even when substituents that interact by resonance with the carbanion are remote and are not affected by steric hindrance, the stabilization imparted by these substituents is

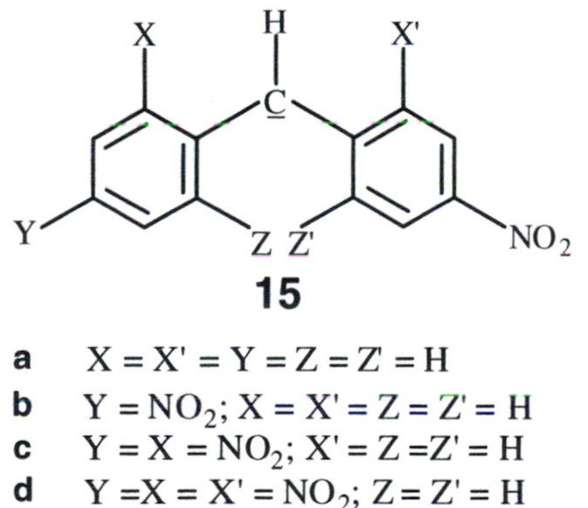

15

a X = X' = Y = Z = Z' = H
b Y = NO_2; X = X' = Z = Z' = H
c Y = X = NO_2; X' = Z = Z' = H
d Y = X = X' = NO_2; Z = Z' = H
e Y = X = X' = Z = NO_2; Z' = H
f Y = X = X' = Z = Z' = NO_2

CHART 2.2

diminished by "resonance saturation". As one example, Bordwell and co-workers found that cyano substitution at C-2 of a fluorene increased acidity by 4.4 pK_a units relative to the parent fluorene, whereas insertion of a second CN at the equivalent C-7 position raised the acidity by only 3.6 pK_a units.[226] The argument here is that the first CN delocalizes the charge by resonance, while the second CN can operate only on the already dispersed charge and, therefore, this CN is less effective. Notwithstanding these reasons, the order of increasing acidity of the polynitrodiphenylmethanes in 50% DMSO:H_2O (298 K) is as follows:

	15a	<	**15b**	<	**15c**	<	**15d**	<	**15e**	<	**15f**
pK_a	≈ 20		14.94		12.19		10.90		7.68		5.01
ΔpK_a		5.06		2.75		1.29		3.22		2.67	

From the differences in the pK_a values, it is clear that the nitrodiphenylmethanes from **15a** to **15d** (i.e., from 4-nitrodiphenylmethane to 2,2′,4,4′-tetranitrodiphenylmethane) behave as might be expected on the basis of resonance saturation and steric inhibition of resonance; each additional nitro group added to either aryl ring is less effective in stabilizing the carbanion than the previous NO_2.[84] However, **15e**, 2,2′,4,4′,6-pentanitrodiphenylmethane is significantly more acidic than would be expected. On the basis of the previous spectroscopic study, it was concluded that **15e** and **15f** should be considered to be 2,4,6-trinitrotoluene derivatives rather than diphenylmethanes. In this case, the acidity of **15f**, for example, arises from the largely inductive/field/polarizability effect of a picryl group on TNT.

It should be noted here that the picryl group is a moderately strong electron-withdrawing group; the σ_p^- substituent constant for the picryl group is 0.48, determined from the pK_a of 4-(2,4,6-trinitrophenyl)phenol, in which the picryl and phenol rings have a twist angle of 59.3° between them, indicative of little overlap of the π-systems.[280]

While steric hindrance was shown to have little effect on the rate of proton transfer in the polynitrodiphenylmethane systems, it is possible that the anomalously high acidity of **15e** and **15f** arises from relief of strain in proceeding from the diarylmethanes to the diarylmethide ions (and this would not be present in **15a–d**), as Bordwell has previously suggested for the 9-triphenylsilylfluorene system among others.[234]

In accord with our expectations concerning the intrinsic reactivity of highly stabilized carbanions, the intrinsic rate constants (log k_o) for the proton transfer reaction of carboxylates with the polynitrodiphenylmethanes range from 0.41 for 2,2′,4,4′,6,6′-hexanitrodiphenylmethane, **15f**, to −1.10 for 2,4,4′-trinitrodiphenylmethane, **15c**.[95] The lower intrinsic barriers to proton transfer for **15e** and **15f** (log $k_o = 0.15$ and 0.41, respectively) are consistent with the inability of these carbanions to delocalize the negative charge into both nitro-substituted phenyl rings simultaneously. Intrinsic rate constants have also been reported by the Terrier group for picrylacetophenone.[127] In fact, the similarity in the intrinsic rate constants for proton transfer from picrylacetophenone and **15f** to aryloxides provides evidence that, as in **15f**, the carbanionic charge is effectively delocalized only into the (single) picryl ring in picrylacetophenone.[127]

The Terrier group has also determined the very high intrinsic barrier to proton transfer from 7-methyl-4,6-dinitrobenzofuroxan, **16**, to carboxylate ions: 20.4 kcal/mol ($\Delta G_0^{\ddagger}$; $\log k_o = -2.15$ at 298 K in water).[98] Again, the low intrinsic reactivity is accompanied by a high equilibrium acidity for this carbon acid; **16** has a pK_a in water of 2.50. Thus, 7-methyl-4,6-dinitrobenzofuroxan is about 10^{11} times more acidic than TNT (pK_a = 13.60 in water).[278] The acidity of **12** ranks it between dinitromethane and trinitromethane (Table 1.1) in the list of strong carbon acids. This high acidity of **16** compared with TNT was determined to arise from the low rate constant of reprotonation of the carbanion, **16a**, versus the same rate constant for reprotonation of 2,4,6-trinitrophenylmethide by water.

The process of deprotonation and reprotonation, as shown in Scheme 2.2, could occur directly via the carbon center or it could proceed through a rapid pre-equilibrium that generates first the tautomeric acinitro form of the carbon acid, **16b.** In fact, Terrier was able to show that under the conditions of the experiment the tautomeric equilibrium (K_T = **16** ⇄ **16b**) is unimportant; the pK_a of **16b** was estimated to be < 0.5 in aqueous solution and K_T has a value < 0.01. Thus, the process of protonation and deprotonation occurs on the carbon; **16** is justifiably a *carbon* acid.[98]

As we have seen, remote substituents may interact with the carbanionic center to delocalize the charge via resonance, as in the case of **16** (i.e., **16a** in Scheme 2.2) or the coplanar polynitrodiphenylmethanes, **15a–d** (Chart 2.2). In effect, the substituents extend the conjugation that already exists between the carbanionic site and the attached phenyl, fluorenyl, or other aromatic or heteroatomic ring systems. Of course, this resonance delocalization may be extended further by SSAR effects, as outlined earlier.[251,273] However, in systems exemplified by **16** the car-

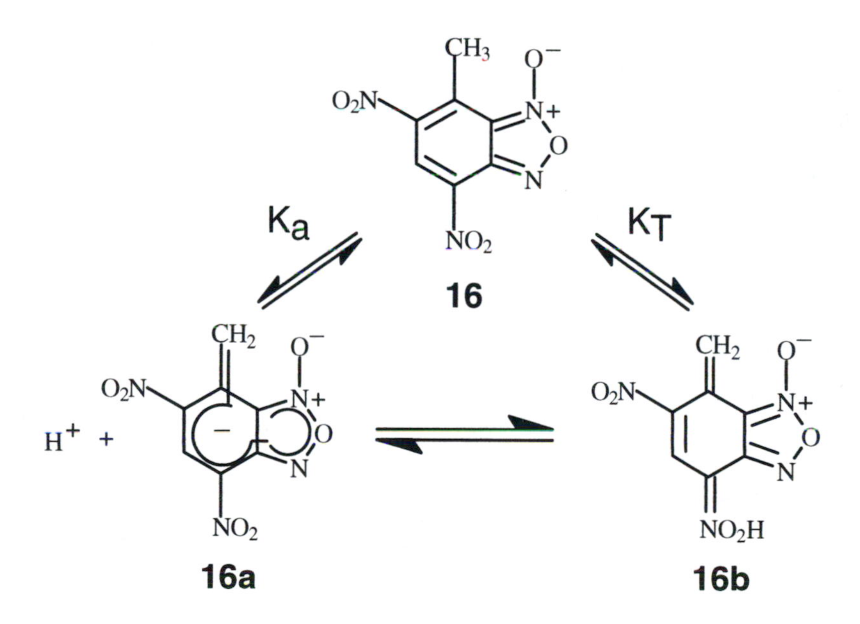

SCHEME 2.2

banionic charge is delocalized into several functions: into each nitro group and into the furoxan ring in **16a**. While this would be expected to lead to resonance saturation,[226] the corollary is that solvent-assisted stabilization of charge by any given group should also be lessened. As a consequence, a truer picture of the α-substituent effect due to the whole 4,6-dinitrobenzofuroxan moiety should emerge.

Substituents also function to enhance the polarizability of the aryl/heteroaryl moiety, as well as directly withdrawing electron density via the sigma-framework or through-space (field effect). The acidity of benzene (ion-pair pK_a 43.0 with Cs^+ counterion in cyclohexylamine solvent) reflects the electronegativity of the sp^2-hybridized carbons of the ring, but also the polarizability of the benzene π-cloud.[181] However, the pentafluorophenyl anion is appreciably more stable than the phenyl carbanion as a result of the multiple fluorine substitution; the ion-pair pK_a (Cs^+/CHA) is 25.8. The fluorines act by withdrawing electron density here by the inductive/field mechanism and by enhancing the polarizability of the phenyl ring.

In the next section, we will probe the concept of conjugation further and its effect on the stability of carbanions.

2.3.5 Conjugative Effects on Carbanion Stability

Early in this work (Table 1.1), we introduced a select group of carbon compounds whose acidity could be measured in water. The high acidity of these carbon acids as compared with the majority of carbon compounds can be attributed to the effect of the substituents adjacent to the acidic center; in most cases, these substituents stabilize the resultant carbanions by resonance. In this context, however, the necessity of attaining planarity about the carbanionic center proved to be paramount. Thus, 2-indanone and 2-tetralone (Chart 2.3, where the acidic proton is highlighted), both of which presumably orient the carbonyl group coplanar with the anion, have acidities ($pK_{a(water)}$) of 12.2 and 12.9,[281] respectively, which are greater than those of benzyl methyl ketone ($pK_a = 15.9$, Chart 2.3),[282] the nonrigid analogue, or acetone ($pK_a = 19.36$ anchored to water).[283]

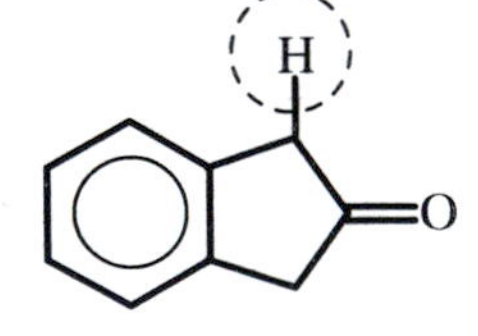

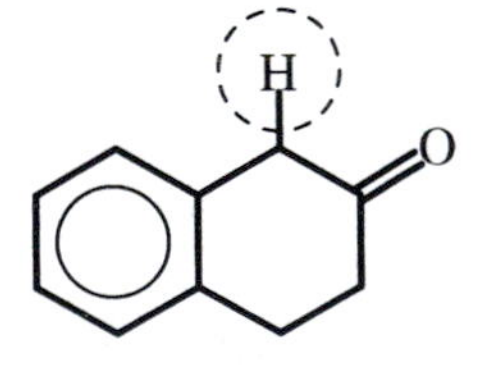

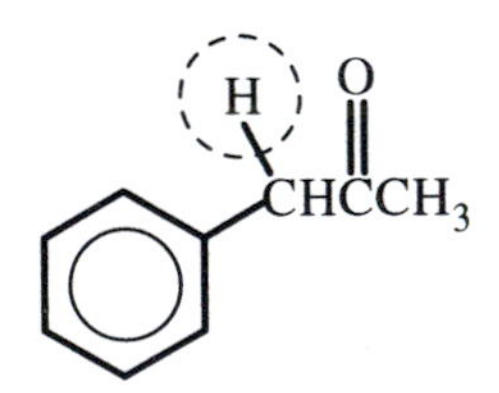

	2-indanone	2-tetralone	benzyl methyl ketone
$pK_{a(H_2O)}$	12.2	12.9	15.9

CHART 2.3

Table 2.3. Acidities of some conjugated hydrocarbons determined in DMSO (acetic acid/sodium acetate) by the H_ AF method (anchored to water as standard state)

Compound	Number	pK_a
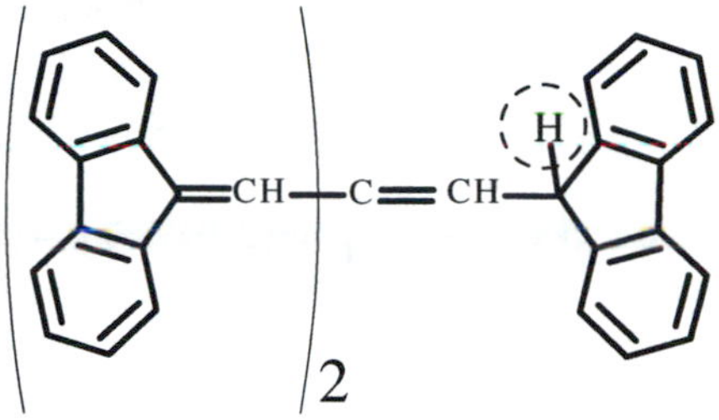 =CH—C=CH— H 2	**17**	10.4
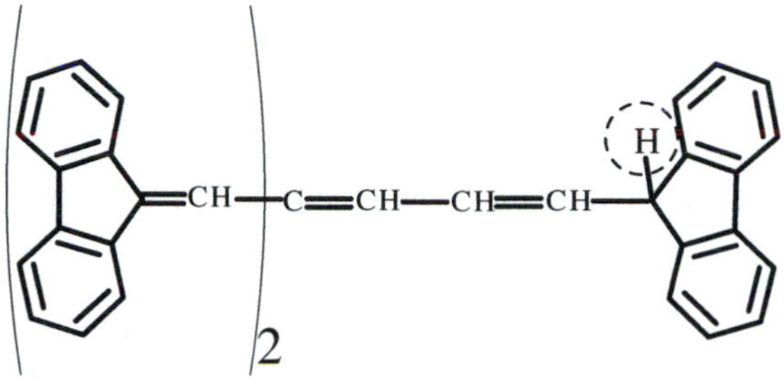 =CH—C=CH—CH=CH— H 2	**18**	11.2
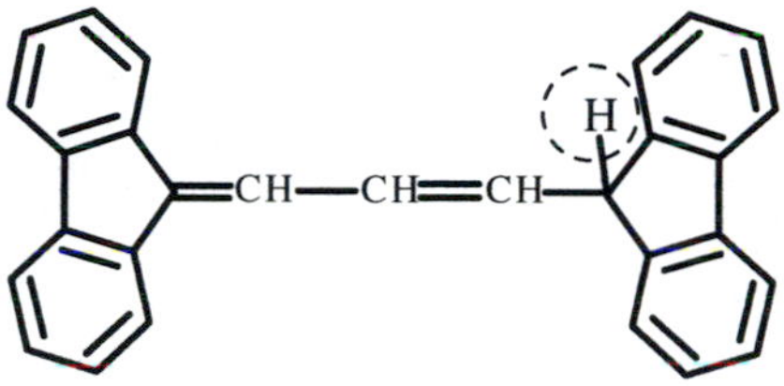 =CH—CH=CH— H	**19**	12.3
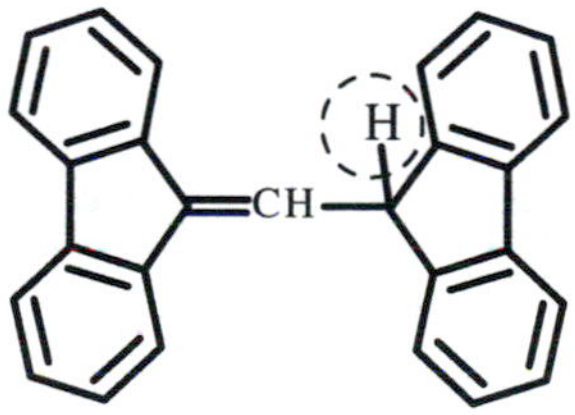 =CH— H	**20**	14.4

Source: Data were taken from reference 265.

Then what of extended conjugated systems, where suitable extensive overlap of p-orbitals is not at issue?

The acidity of a series of hydrocarbons that would yield fluorenyl carbanions with extended conjugation was investigated by Kuhn and Rewicki.[284] Selected acidity data for these compounds, determined by the H_ acidity function (AF) method (which links the pK_a values so determined to water as the standard state), are collected in Table 2.3. As can be readily seen, these are all fairly acidic hydrocarbons and, in general, the pK_a values fall as the degree of conjugation increases, with the exception of compound **18** whose acidity appears somewhat higher than expected. Note, nonetheless, that the carbon acids in Table 2.3, **17–20**, are all more acidic than fluoradene (pK_a by AF method = 14.0),[284] which, as we have seen, is significantly more acidic than triphenylmethane; **20** is only slightly less acidic than fluoradene. These results are all consistent with the traditional view that increasing conjugation increases delocalization of charge, which, in turn, leads to stabilization of the carbanion.

Is there a limit to the degree of stabilization that can be imparted by conjugation? The answer appears to be yes. Tolbert and Ogle approached the problem by examining the ^{13}C NMR spectra of a series of 1,*n*-diphenylalkapolyenyl anions (where the number of intervening carbons ranged from 1 in diphenylmethane to 17 and where the anions were examined in DMSO-d_6 in the presence of 18-crown-6 so as to avoid ion pairing effects on the spectra).[285,286] From the chemical shifts of the carbons in the chain and the use of the empirical Spiesecke–Schneider equation[289] (in the form of eq. 2.39)[288] that relates π-charge density (ρ_C) to the ^{13}C NMR chemical shift (δ_C) for a given carbon, Tolbert and Ogle could calculate the charges and their sign at each carbon along the alka-polyene chain, and in the phenyl end groups.

$$\delta_C = \kappa\rho_C + \delta_o \qquad (2.39)$$

In equation 2.39, as used by Tolbert and Ogle, δ_o represents the chemical shift of the relevant carbon in a neutral alkene; κ is a constant meant to reflect the sensitivity of the chemical shifts to changes in the π-charge density, and while it is usually found to have a value of about 160 ppm, other values have also been found.[289,290] In this case, κ is 187.3 ppm/charge and δ_o is 132.7 ppm and after rearrangement equation 2.39 becomes equation 2.40:

$$\rho_C = (\delta_C - 132.7)/187.3 \qquad (2.40)$$

Although empirical in nature, equation 2.40 has proven useful in analyzing the ^{13}C NMR spectra of a variety of carbanions[289,290] and has the virtue of simplicity,

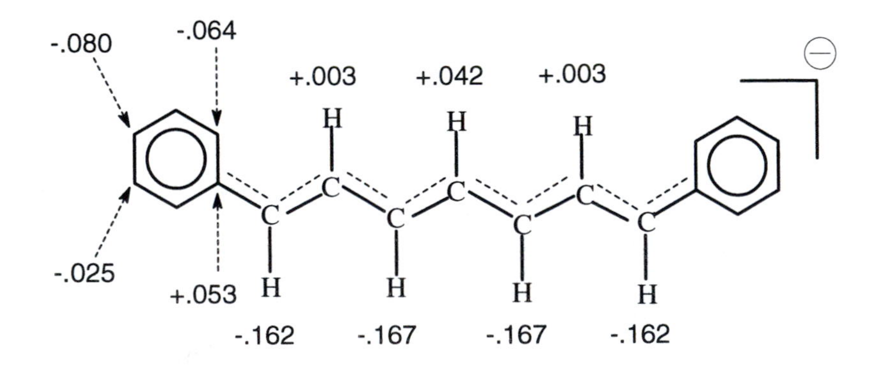

CHART 2.4

if not theoretical rigor.[291,292] It must be stipulated, of course, that equation 2.39 assumes that all charge density is located on the carbons of the molecule.

Nonetheless, within the limitations that apply to these equations, the charge density at each carbon in the 1,*n*-diphenylalkapolyene anions could be calculated from the ^{13}C NMR spectra. For example, for the carbanion generated from 1,7-diphenyl-1,3,5-heptatriene (or a double-bond isomer) by reaction with potassium dimsyl in DMSO-d_6 the π-charge distribution (equation 2.40) is given in Chart 2.4. The salient conclusions obtained from examination of this anion, which also apply to the other members of the series studied by Tolbert and Ogle, are as follows: (1) charges alternate down the polyene chain; (2) there is a buildup of charge at the center of the odd-alternant anion, as previously found in the work of Kloosterziel,[293,294] and; (3) by examining the full series of conjugated anions and comparing chemical shifts to those for the neutral even-alternant hydrocarbons, an estimate can be made of the "distance" a carbanionic charge can be delocalized. The experimental distance, as determined by the comparison in chemical shift outlined, was determined to be between 26 and 31 carbons in the polyenyl chain.[285,286] In other words, if the terminal phenyl groups can be thought of as charge density sinks and the ^{13}C NMR shifts of the aromatic carbons in the anions, as compared with the neutral polyenes, taken as monitors of the delocalization of this density into the phenyl rings, then the point at which the NMR shifts of the phenyl groups for anions and neutral species coincide marks the point beyond which classical resonance delocalization is practically ineffective. By these criteria, a carbanionic charge cannot be delocalized beyond about 30 conjugated carbons.

Importantly, this "delocalization distance" is in reasonable agreement with the "soliton width", where a soliton is defined as a mobile charge carrier in a conductive polymer: reductively/oxidatively doped polyacetylene.[295,296] The polyenes of the Tolbert and Ogle study are oligomeric models for conductive polyacetylene. The conductivity is ascribed to these solitons. Solitons, in this case, are considered resonance-stabilized carbanions/carbocations with a fixed width whose charge density wave travels down the unsaturated polymer chain. Conversely, in order for the soliton to propagate down the chain, it follows that it must have a finite width.[297] (Conduction may, however, also involve interaction of solitons between chains and not just up or down a given conjugated polymer chain.[297]) Regardless of the implications for the further development of conductive polymers, these results prod us to recognize the limits of static resonance delocalization as a mode of stabilization of conjugated carbanions.

Interest in the reactivity and potential applications of the reactions and physical properties of the recently discovered allotropes of carbon, C_{60} and C_{70}, the buckminsterfullerenes (more familiarly, fullerenes and, jocularly, "buckyballs"), has literally exploded, spurred partly by improved methods of isolation[298–303] that have placed larger amounts of these allotropes in the hands of chemists. The general chemistry of fullerenes is the subject of a number of recent reviews and books[304–306] and we will concentrate on the more limited information concerning the carbanions of these allotropes.

Buckminsterfullerene (C_{60}) itself is electrophilic and is easily reduced electrochemically to yield C_{60}^{n-} anions with charges, $n-$, of 1– to 6–,[307] as had earlier been predicted on theoretical grounds.[308,309] Significantly, the reduction potentials ($E_{1/2}$ relative to the ferrocene/ferrocenium couple) at −0.98, −1.37, −1.87, −2.35, −2.85, and −3.26 V were all found to be reversible at slow cyclic voltammetry scan rates.[310] The electrophilicity of fullerenes also accounts for the relative ease of addition of carbanions to the fullerene skeleton.[311,312] Monofullerenide ion (C_{60}^{-}) is not formed as readily as a distinct species; further reduction to more highly charged ions occurs easily, the fullerenide ion is air sensitive and may also disproportionate to give neutral buckminsterfullerene and the C_{60}^{2-} ion.[312] This relatively low persistence explains the lack of accounts concerning the pK_a of the conjugate acid of fullerenide, namely, C_{60}-**H**.

However, the ready addition of other carbanions to the fullerene skeleton provided another route to preparation of fullerenide ions. Thus, reaction of C_{60} with *tert*-butyllithium led to formation of the mono-*t*-butylfullerenide ion (R = *t*-butyl).[313] Based on titrations of *t*-BuC_{60}^{-} against saccharin, 2,4-dinitrophenol, and dichloroacetic acid, the pK_a of *t*-BuC_{60}-**H** was estimated as 5.7 (±0.1). Comparison with the pK_a values of highly conjugated carbon acids (Table 2.3) shows that *t*-BuC_{60}-**H** is a very strong carbon acid *that consists of only carbon and hydrogen.* Since the β-*tert*-butyl group would not be expected to have a large effect on the acidity of *t*-butylfullerene-**H** (although it likely enhances the persistence of the *t*-butylfullerenide ion by hindering approach to the anion), the acidity of C_{60}-**H** is also likely to be high for a hydrocarbon and probably close to the value found for *t*-butylfullerene-**H**.

Importantly, semi-empirical calculations (AM1) reveal that the charge in *t*-butylfullerenide ion is largely distributed on carbons C-2, C-4, and C-11 (cf. structure **21**).[314] Thus, while resonance delocalization stabilizes the anion, extensive resonance delocalization does not appear to be critical, suggesting similarities between the results of Tolbert and Ogle on their linear conjugated systems and the "spherical conjugation" in the *t*-butylfullerenide ion.

Advantage has been taken of the high stability of the *t*-butylfullerenide ion (and the stability of the 1,4-dicyclopropyltropylium cation that is stabilized by

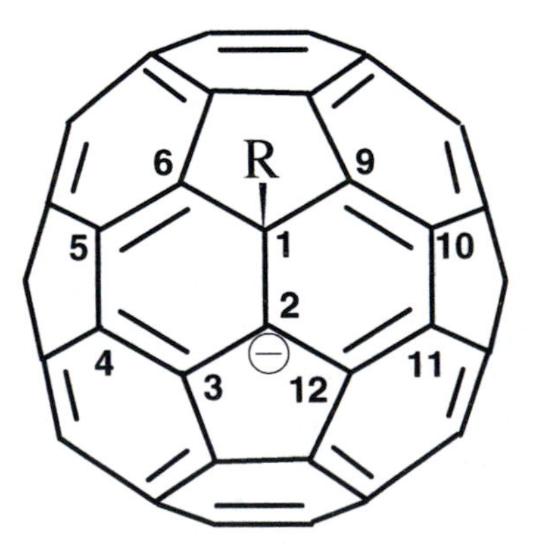

21

Hückel aromaticity) to prepare the covalently bonded hydrocarbon, 1-*tert*-butyl-4-(3,6-dicyclopropyl-2,4,6-cycloheptatrienyl)buckminsterfullerene.[315] This hydrocarbon readily dissociates in high-polarity solvents (e.g., DMSO, DMSO-CS_2) to give equilibrium amounts of the original stabilized carbanion and carbocation; ΔG of the heterolysis was calculated from the dissociation equilibrium constant and found to be 7.7 kcal, a relatively low value that reflects the steric congestion about the relevant C–C bond.

It should be noted, in conclusion, that both *t*-butylfullerenide ion[313] and the longer carbanions formed by the 1,*n*-diphenylalka-polyenes[286] undergo facile one-electron oxidation to yield radicals. As a possible competitive process to carbanion formation (and reaction), it deserves further consideration in the following section.

2.4 COMPETITIVE OXIDATION TO RADICAL SPECIES

Carbanions are susceptible to oxidation to radical species. Examples of single electron transfer are enumerated in a review,[316] as well as a chapter[317] in a volume concerning carbanion chemistry. Although in-depth discussion of the $S_{RN}1$ mechanism[318–320] is beyond the scope of this book, it should be noted that carbanions can act as initiating electron donors (D^-) in the $S_{RN}1$ mechanism:[321,322]

$$R\text{–}X + D^- \longrightarrow R\text{–}X^{\bullet -} + D^{\bullet} \tag{2.41}$$

$$R\text{–}X^{\bullet -} \longrightarrow R^{\bullet} + X^- \tag{2.42}$$

$$R^{\bullet} + D^- \longrightarrow R\text{–}D^{\bullet -} \tag{2.43}$$

$$R\text{–}D^{\bullet -} + R\text{–}X \longrightarrow R\text{–}D + R\text{–}X^{\bullet -} \tag{2.44}$$

Single electron transfer (SET) from a carbanion to an acceptor substrate that bears a leaving group results in formation of a substrate radical anion ($R\text{–}X^{\bullet -}$) in step 2.41. The radical anion can dissociate in the usual rate-determining unimolecular step 2.42. The resultant radical can combine with a donor, a carbanion in our example, in step 2.43 and, finally, in step 2.44 back electron transfer from the new radical anion to the substrate regenerates the substrate radical anion and propagates the radical chain. In the electron transfer reactions we will examine, elements of the $S_{RN}1$ mechanism will be apparent.

Among the studies that have appeared concerning SET processes in carbanion chemistry,[323–334] the work of Ashby and his co-workers is particularly pertinent.[326,327] Thus, the Ashby group, among others, has examined the reactions of haloalkanes and ketones with Grignard reagents (RMgX),[328,329] oganolithium reagents (RLi),[330,331] organolithiocuprates [Corey–House reagents, $LiCu(R)_2$], and enolates, all carbanion or carbanion-like reagents.[332] Specifically, the aldol condensation of enolates with ketones,[333] the Claisen–Schmidt condensation of enolates with aromatic esters,[334] and the Cannizzaro reaction of hydroxide with aromatic aldehydes[316] were analyzed using product and deuterium incorporation studies and cyclizable "radical clocks".[335,336]

Although electron transfer mechanisms may prevail in carbanion reactions that have traditionally been considered in terms of the two-electron "curvy arrow formalism" of the Lewis acid–base description, the choice of pathway (polar vs SET) is linked to reaction conditions, as well as to the nature of the reactants. Thus, Ashby has pointed out the preference for SET mechanisms for enolates with aromatic ketones, because of the more favorable redox potential that aromatic ketones have compared with alkanones, and that use of alkyl iodides in alkylation reactions should favor SET pathways as compared with alkyl chlorides or fluorides, on the same grounds of redox potential.[326]

In this regard, Arnett and co-workers have investigated the reaction of lithium pinacolonate [lithium salt of 3,3-dimethylbutanone; $Li^{+-}CH_2C(O)–C(CH_3)_3$] with *o*- and *p*-methylbenzaldehyde (i.e., the aldol reaction) and with ethyl 4-nitrobenzoate (i.e., the Claisen–Schmidt condensation).[337] The aldol reaction was studied by low-temperature (−80 °C in methylcyclohexane-d_{14}) rapid injection NMR spectroscopy. According to this technique, the kinetic behavior of the aldol reaction was found to conform to a simple second-order rate law and radical intermediates could not be detected; that is, no chemically induced dynamic nuclear polarization (CIDNP) effects were noted.[338–340] Cyclizable probes were used in both the aldol and Claisen–Schmidt reaction systems, but no cyclized products were obtained from either reaction. Finally, theoretical rate constants for single electron transfer between lithium pinacolonate and *p*- and *o*-methylbenzaldehyde were calculated as 2.1×10^{-26} and 2.6×10^{-30} s^{-1}, respectively, based on the reduction potentials of the benzaldehydes;[337,341] these rate constants yield first-order half-lives that Arnett notes "are greater than present estimates for the age of the universe!"[337] The actual second-order rate constants for the *p*- and *o*-methylbenzaldehydes were found to be 0.407 and 0.346 $M^{-1}{\cdot}s^{-1}$, respectively, with corresponding half-lives of 12.7 and 43.2 s. Although not definitive, the sum of this evidence argues against SET as the mechanism in these particular cases.

Clearly, a range of mechanisms from classical spin-paired two-electron processes[342] to SET[326] may obtain, depending on the nature of the system under examination. Continued investigation and controversy over the degree to which standard carbanion reactions can be described as SET systems may be expected.

In theory, however, it should be possible to predict whether an electron transfer reaction with carbanions will occur on the basis of the reduction potential of the acceptor compound and the oxidation potential of the donor carbanion.[343,344] This predictive ability has been hampered by the paucity of reliable oxidation data; electrochemical studies generally have required the use of high-dielectric-constant (polar) solvents that will dissolve the supporting electrolyte, and these results may not be transferable to less polar solvents. Many redox potentials have been obtained from irreversible systems and only reversible half-wave potentials have thermodynamic significance.

The criterion for the feasibility of an electron transfer process has been developed more formally by Eberson from Marcus theory.[325,344] The free energy of electron transfer is determined from electrochemical measurements and then used to estimate the rate constant for the electron transfer step. This is the approach used in the aforementioned study by the Arnett group of the aldol and Claisen–Schmidt reactions of lithium pinacolonate.[337] Generally, the calculated electron transfer rate constant is compared with the actual rate constant for the reaction; large discrepancies between the two rate constants is taken as evidence against electron transfer as the mechanism for the reaction.

Choice of solvent has been found to be useful in biasing the pathway for reaction in some carbanionic systems. For example, in their studies of electron transfer processes involving triphenylmethide ion and a variety of acceptors, including nitrobenzene, the R. D. Guthrie group found that SET could be suppressed through choice of solvent. Single electron transfer from triphenylmethide to nitrobenzene would yield the nitrophenyl radical anion and the triphenylmethyl radical, which would subsequently dimerize. The triphenylmethyl anion could also react with *tert*-butyl alcohol-d_1 to give triphenylmethane deuterated at the benzylic position. The ratio of the rates of the two processes was ca. 400 in favor of SET (i.e., formation of the dimer). Addition of an aprotic dipolar solvent, HMPA in this case, or the addition of crown ether, reduced the size of the ratio in favor of SET.[345–347] On the other hand, Ashby has concluded that in the reaction of a series of endo-5-(2-haloethyl)-2-norbornenes with *tert*-butyllithium, complexing agents enhanced the carbanionic character of the organolithium reagent but did not eliminate electron transfer pathways.[328]

Thus far, we have considered SET processes and standard carbanion mechanisms as an "either/or" proposition. Of course, oxidation of carbanions to radicals may compete with the deprotonation that yields the carbanions in the first place. In the gas phase, Han and Brauman have found competitive carbon acid ionization and oxidation via SET with benzylic carbanions.[348] It is reasonable to think that if the carbanion were stabilized, say by remote electron-withdrawing groups, deprotonation of the carbon acid would become the preferred path. However, when 4-nitrotoluene was treated with a range of bases (KH, dimyl potassium, triphenylmethyl potassium, fluorenyl potassium, and potassium 4-nitroanilide) in a variety of solvents (DMSO, THF, and DME, where the latter two systems contained 18-crown-6 polyether), the nitrobenzyl radical and radical anion were formed competitively with the nitrobenzyl carbanion; that is, SET processes vie with deprotonation of the carbon acid.[349] The assignment of the spectroscopic data for the 4-nitrobenzyl anion was later confirmed in a separate study of the 4-, 2,4-di-, and 2,4,6-trinitrobenzyl carbanions formed through decarboxylation of the corresponding arylacetate anions.[350]

Since stability of the carbanion relative to the carbon acid is governed by the pK_a of the carbon acid, and ease of formation of a radical or radical ion from the carbanion is dependent on the reduction–oxidation (redox) potentials, we might anticipate a mathematical relationship between the two. In fact, the Edwards[351] equation:

$$\log K = \alpha E_n + \beta H \tag{2.45}$$

relates the stability of a given Lewis complex ($\log K = -pK$) to the redox potential, E_n, and Brønsted acidity, $H (= pK_a + 1.74$ in water at 25 °C); α and β are proportionality constants that reflect the importance of each term to the process of Lewis adduct formation.

The situation as it relates to the differential effect of substituents on Brønsted acidity as compared with the effect on the tendency to singly transfer electrons might eventually lead to control of the favored pathway. In this context, Kern and Federlin[352] devised a scale of half-wave oxidation potentials for carbanions. The oxidation potentials correlate with the pK_a values of the parent hydrocarbons.[151,251,352–355] Extensive work has been reported by the Bordwell group, who have used and expanded their comprehensive compilation of pK_a values of carbon acids in DMSO[151] in thermochemical cycles to estimate bond dissociation enthalpies (BDE) for several classes of substituted aryl carbon acids.[251,271,354,355] The following empirical equation (equation 2.46) was derived:

$$\mathrm{BDE} = 1.37\mathrm{pK}_{\mathrm{a(DMSO)}} + 23.1E_{\mathrm{OX}}(\text{conjugate base anion}) + C \qquad (2.46)$$

The BDE values determined from equation 2.46 are in good agreement with directly measured gas-phase values (±2 kcal/mol correspondence), regardless of the use of irreversible oxidation potentials in some of the cases (evaluated from cyclic voltammetry), the fact that the BDE values are gas-phase ones, and, the fact that the pK_a values are measured in dimethyl sulfoxide. The constant, C, is an empirical one, although originally it was designed to convert the E_{OX} values determined in DMSO to the standard hydrogen electrode (SHE). Based on comprehensive thermochemical cycles devised by Nicholas and Arnold,[356] C was assigned a value of 56 kcal/mol. Later, this value was revised; relative to the ferrocene/ferrocenium couple (875 mV), C was assigned as 93.3 kcal/mol. If BDE, pK_a, and E_{OX} terms in equation 2.46 are replaced by ΔBDE, ΔpK_a, and ΔE_{OX} terms and the constant is dropped from the resultant equation, the radical stabilization enthalpy (RSE) can be determined (= ΔBDE) from the modified equation.

This approach by the Bordwell group[180,251,272,354,355] holds the possibility of ultimately controlling the dichotomy between the polar and SET pathways in carbanion reactions. Although most remote *para*-substituents on aromatic ring systems that stabilize anions appear to stabilize radicals (with the possible exception of fluoro and fluorinated substituents that seem to destabilize benzylic radicals),[357–360] comparison of the effect of substituents on pK_a values in a series of carbon acids with that on RSE should reveal differences large enough in energy to favor one path (carbanionic or SET) over the other. It should also be noted here that electron transfer is frequently a rapid but reversible process (cf. the $S_{RN}1$ mechanism; equation 2.44). If the reaction of a carbanion is reasonably fast so that it completes with electron transfer *and is irreversible*, the equilibrium governing electron transfer may be shifted from the radical (or radical anion) to the carbanion as its concentration becomes depleted in solution.

Clearly, the boundary between SET and two-electron reactions of carbanions will remain an area of active and fruitful investigation. We can anticipate further progress in understanding the processes of proton transfer and the reactions of carbanions that will arise from the combination of kinetic isotope effects, including those of heavy atoms, calculational studies, and interpretations of the underlying causes of Non-Perfect Synchronization in proton transfers, as highlighted by the work of Saunders and collaborators.[361–365]

Notes and References

1. Buncel, E.; Saunders, W.H. Jr., Eds.; *Isotopes in Organic Chemistry. Vol. 8. Heavy Atom Isotope Effects*; Elsevier: Amsterdam, 1992.
2. Melander, L. *Isotope Effects on Reaction Rates*; The Ronald Press: New York, 1960.
3. Melander, L.; Saunders, W.H. Jr., *Reaction Rates of Isotope Molecules*; Wiley: New York, 1980.
4. Shiner, V.H. Jr.; Wilgis, F.P. *Isotopes in Organic Chemistry*; Elsevier: Amsterdam, 1992.
5. Buncel, E.; Lee, C.C., Eds.; *Isotopes in Organic Chemistry. Vol. 1. Isotopes in Molecular Rearrangements*; Elsevier: Amsterdam, 1975.
6. Buncel, E.; Lee, C.C., Eds.; *Isotopes in Organic Chemistry. Vol. 2. Isotopes in Hydrogen Transfer Processes*; Elsevier: Amsterdam, 1976.
7. Buncel, E.; Lee, C.C., Eds.; *Isotopes in Organic Chemistry. Vol. 3. Carbon-13 in Organic Chemistry*; Elsevier: Amsterdam, 1977.
8. Buncel, E.; Lee, C.C., Eds.; *Isotopes in Organic Chemistry. Vol. 4. Tritium in Organic Chemistry*; Elsevier: Amsterdam, 1978.
9. Buncel, E.; Lee, C.C., Eds.; *Isotopes in Organic Chemistry. Vol. 5. Isotopes in Cationic Reactions*; Elsevier: Amsterdam, 1980.
10. Buncel, E.; Lee, C.C., Eds.; *Isotopes in Organic Chemistry. Vol. 6. Recent Developments in Theory and Experiment*; Elsevier: Amsterdam, 1984.
11. Buncel, E.; Lee, C.C., Eds.; *Isotopes in Organic Chemistry. Vol. 7. Secondary and Solvent Isotope Effects*; Elsevier: Amsterdam, 1987.
12. Streitwieser, A.; Jagow, R.H.; Fahey, R.C.; Suzuki, S. *J. Am. Chem. Soc.* **1958**, *80*, 2326.
13. Melander, L. *Isotope Effects on Reaction Rates*; The Ronald Press: New York, 1960; p 9.
14. Lowry, T.H.; Richardson, K.S. *Mechanism and Theory in Organic Chemistry*, 3rd ed.; Harper & Row: New York, 1987; p 234.
15. Buncel, E. *Carbanions: Mechanistic and Isotopic Aspects*; Elsevier: Amsterdam, 1975; p 25–27.
16. Swain, C.G.; Stivers, E.C.; Reuwer, J.F., Jr.; Schaad, L.J. *J. Am. Chem. Soc.* **1958**, *80*, 5885.
17. Kresge, A.J.; *J. Am. Chem. Soc.* **1992**, *114*, 3981.
18. Thornton, E.R. *Annu. Rev. Phys. Chem.* **1966**, *17*, 349.
19. Halevi, E.A. *Prog. Phys. Org. Chem.* **1963**, *1*, 109.
20. Caldin, E.F. *Chem. Rev.* **1969**, *69*, 135.
21. Harmony, M.D. *Chem. Soc. Rev.* **1972**, *1*, 211.
22. Lewis, E.S.; Funderburk, L.H. *J. Am. Chem. Soc.* **1967**, *89*, 2322.
23. Lewis, E.S.; Robinson, J.K. *J. Am. Chem. Soc.* **1968**, *90*, 4337.
24. Wilson, H.; Caldwell, J.D.; Lewis, E.S. *J. Org. Chem.* **1973**, *38*, 564.
25. Kwart, H. *Acc. Chem. Res.* **1982**, *15*, 401.
26. Saunders, W.H. Jr., *J. Am. Chem. Soc.* **1984**, *106*, 2223.
27. Westheimer, F.H. *Chem. Rev.* **1961**, *61*, 265.
28. More O'Ferrall, R.A. In *Proton-Transfer Reactions*; Caldin, E.; Gold, V., Eds.; Chapman & Hall, London, 1975; pp 201–263.
29. Wiberg, K.B. *Physical Organic Chemistry*; Wiley: New York, 1970; pp 332–361.

30. Bell, R.P.; Sachs, W.H.; Tranter, R.L. *Trans. Faraday Soc.* **1971**, *67*, 1995.
31. Bell, R.P.; Cox, B.G. *J. Chem. Soc. B* **1971**, 783.
32. Melander, L.; Bergman, N.A. *Acta. Chem. Scand.* **1971**, *25*, 2264.
33. Kresge, A.J.; Chiang, Y. *J. Am. Chem. Soc.* **1969**, *91*, 1025.
34. Bell, R.P. *Chem. Soc. Rev.* **1974**, *3*, 513.
35. Koch, H. F. *Acc. Chem. Res.* **1984**, *17*, 137.
36. Cram, D.J.; Kingsbury, C.A.; Rickborn, B. *J. Am. Chem. Soc.* **1961**, *83*, 3688.
37. Hofman, J.E.; Schriesheim, A.; Nickols, R.E. *Tetrahedron Lett.* **1965**, *22*, 1725.
38. Streitweiser, A.; Hollyhead, W.; Sinnichsen, G.; Pudjaatmaka, A.; Chang, C.J.; Kruger, T. *J. Am. Chem. Soc.* **1971**, *93*, 5096.
39. More O'Ferrall, R.A. *J. Chem. Soc. B* **1970**, 785.
40. Vitale, A.A.; San Filippo, J. Jr., *J. Am. Chem. Soc.* **1982**, *104*, 7341.
41. Kwart, H.; Wilk, K.A.; Chatellier, D. *J. Org. Chem.* **1983**, *48*, 756.
42. Anhede, B.; Bergman, N.A. *J. Am. Chem. Soc.* **1984**, *106*, 7634.
43. Stewart, R. *The Proton: Applications to Organic Chemistry*; Academic Press: Orlando, FL, 1985; p 148–159.
44. Crooks, J.E.; Robinson, B.H. *Faraday. Symp. Chem. Soc.* **1975**, *10*, 29.
45. Johnson, S.L.; Rumon, K.A. *J. Phys. Chem.* **1965**, *69*, 74.
46. Leclerq, J.M.; Dupuis, P.; Sandorfy, C. *Croat. Chem. Acta* **1982**, *55*, 105.
47. Brown, T.L.; Butler, L.G.; Curtin, D.Y.; Hiyama, T.; Paul, I.C.; Wilson, R.B. *J. Am. Chem. Soc.* **1982**, *104*, 1172.
48. Bell, R.P.; Timiwi, B.A. *J. Chem. Soc., Perkin Trans. 2* **1973**, 1518.
49. Coward, J. K.; Bruice, T.C. *J. Am. Chem. Soc.* **1969**, *91*, 5339.
50. Kreevoy, M.M. In *Isotopes in Organic Chemistry. Vol. 2. Isotopes in Hydrogen Transfer Processes*; Buncel, E.; Lee, C.C., Eds.; Elsevier: Amsterdam, 1976; Chapter 1.
51. Dixon, R.E.; Streitwieser, A. *J. Org. Chem.* **1992**, *57*, 6125.
52. Koch, H.F. In *Comprehensive Carbanion Chemistry. Part C. Ground and Excited State Reactivity*; Buncel, E.; Durst, T., Eds.; Elsevier: Amsterdam, 1987; pp. 321–360.
53. Koch, H.F.; McLennan, D.J.; Koch, J.G.; Tumas, W.; Dobson, B.; Koch, N.H. *J. Am. Chem. Soc.* **1983**, *105*, 1930.
54. Koch, H.F.; Lodder, G.; Koch, J.G.; Bogdan, D.J.; Brown, G.H.; Carlson, C.A.; Dean, A.B.; Hage, R.; Han, P.; Hopman, J.C.P.; James, L.A.; Knape, P.M.; Roos, E.C.; Sardina, M.L.; Sawyer, R.A.; Scott, B.O.; Testa, C.A., III; Wickham, S.D. *J. Am. Chem. Soc.* **1997**, *119*, 9965.
55. Koch, H.F.; Dahlberg, D.B.; Toczko, A.G.; Solsky, R.L. *J. Am. Chem. Soc.* **1973**, *93*, 2029.
56. Koch, H.F.; Dahlberg, D.B.; McEntee, M.M.F.; Klecha, C.J. *J. Am. Chem. Soc.* **1976**, *98*, 1060.
57. Zuilhof, H.; Lodder, G.; Koch, H.F. *J. Org. Chem.* **1997**, *62*, 7457.
58. Langner, R.; Zundel, G. *Can. J. Chem.* **2001**, *79*, 1376; and references therein.
59. Brønsted, J.N.; Pedersen, K.J. *Z. Phys. Chem.* **1924**, *108*, 185.
60. Bordwell, F.G. *Acc. Chem. Res.* **1970**, *3*, 456.
61. Blandamer, M.J.; Robertson, R.E.; Scott, J.M.W. *J. Am. Chem. Soc.* **1982**, *104*, 1136.
62. Lin, A.C.; Chiang, Y.; Dahlberg, D.B.; Kresge, A.J. *J. Am. Chem. Soc.* **1983**, *105*, 5380.
63. Bordwell, F.G.; Hughes, D.L. *J. Am. Chem. Soc.* **1985**, *107*, 4737.
64. Bordwell, F.G.; Hughes, D.L. *J. Org. Chem.* **1982**, *47*, 3224.
65. Leffler, J.E.; Grunwald, E. *Rates and Equilibria of Organic Reactions as Treated by Statistical, Thermodynamic and Extrathermodynamic Methods*; Dover: New York, 1989; pp 156–159.

66. Leffler, J.E. *Science* **1953**, *117*, 340.
67. Eyring, H. *J. Chem. Phys.* **1935**, *3*, 107.
68. Bell, R.P. *Proc. R. Soc. London, Ser. A* **1936**, *154*, 414.
69. Evans, M.G.; Polyani, M. *Trans. Faraday Soc.* **1938**, *34*, 11.
70. Marcus, R.A. *J. Phys. Chem.* **1968**, *72*, 891.
71. Hammond, G.S. *J. Am. Chem. Soc.* **1955**, *77*, 334.
72. Thornton, E.R. *J. Am. Chem. Soc.* **1967**, *89*, 2915.
73. Jencks, W.P. *Chem. Rev.* **1985**, *85*, 511.
74. Johnson, C.D. *Chem. Soc. Rev.* **1975**, *75*, 755.
75. Eigen, M. *Angew. Chem., Int. Ed. Engl.* **1964**, *3*, 1.
76. Buncel, E.; Wilson, H. *J. Chem. Educ.* **1987**, *64*, 475.
77. More O'Ferrall, R.A. *J. Chem. Soc. B* **1970**, 274.
78. Jencks, W.P. *Chem. Rev.* **1972**, *72*, 705.
79. Pross, A. *J. Org. Chem.* **1984**, *49*, 1811.
80. Pross, A.; Shaik, S. *J. Am. Chem. Soc.* **1982**, *104*, 1129.
81. Kresge, A.J. *Can. J. Chem.* **1974**, *52*, 1896.
82. Bordwell, F.G.; Boyle, W.J. Jr., *J. Am. Chem. Soc.* **1972**, *94*, 3907.
83. Bordwell, F.G.; Bartmess, J.E.; Hautala, J.A. *J. Org. Chem.* **1978**, *43*, 3107.
84. Bernasconi, C.F. *Adv. Phys. Org. Chem.* **1992**, *27*, 119.
85. Bernasconi, C.F. *Acc. Chem. Res.* **1987**, *20*, 301.
86. Bernasconi, C.F.; Fairchild, D.E. *J. Phys. Org. Chem.* **1992**, *5*, 409.
87. Gandler, J.R.; Bernasconi, C.F. *J. Am. Chem. Soc.* **1992**, *114*, 631.
88. Bunnett, J.F. *Angew. Chem., Int. Ed. Engl.* **1962**, *1*, 225.
89. Jencks, D.A.; Jencks, W.P. *J. Am. Chem. Soc.* **1977**, *99*, 7948.
90. Harris, J.C.; Kurz, J.L. *J. Am. Chem. Soc.* **1970**, *92*, 349.
91. Harris, J.M.; Shafer, S.G.; Moffatt, J.R.; Becker, A.R. *J. Am. Chem. Soc.* **1979**, *101*, 3295.
92. Kreevoy, M.M.; Lee, I.S.H. *J. Am. Chem. Soc.* **1984**, *106*, 2550.
93. Yamataka, H.; Mustanir, M.; Mishima, M. *J. Am. Chem. Soc.* **1999**, *121*, 10223.
94. Keeffe, J.R.; Morey, J.; Palmer, C.A.; Lee, J.C. *J. Am. Chem. Soc.* **1979**, *101*, 1295.
95. Terrier, F.; Lelièvre, J.; Chatrousse, A.P.; Schaal, R.; Farrell, P.G. *Can. J. Chem.* **1987**, *65*, 1980.
96. Terrier, F.; Chatrousse, A.P.; Kizillian, E.; Ignatev, N.V.; Yagupolskii, L.M. *Bull. Soc. Chim. Fr.* **1989**, 627.
97. Terrier, F.; Xie, H.Q.; Farrell, P.G. *J. Org. Chem.* **1990**, *55*, 2610.
98. Terrier, F.; Croisat, D.; Chatrousse, A.P.; Pouet, M.J.; Hallé, J.C.; Jacob, G. *J. Org. Chem.* **1992**, *57*, 3684.
99. Moutiers, G.; Peng, Y.X.; Peignieux, A.; Pouet, M.J.; Terrier, F. *J. Chem. Soc., Perkin Trans. 2* **1999**, 1287.
100. Bernasconi, C.F. *J. Am. Chem. Soc.* **1993**, *115*, 5060.
101. Murdoch, J.R.; Bryson, J.A.; McMillen, D.F.; Brauman, J.I. *J. Am. Chem. Soc.* **1982**, *104*, 600.
102. Bernasconi, C.F.; Wenzel, P.J. *J. Org. Chem.* **2001,** *66*, 968.
103. Gilbert, H.F. *J. Am. Chem. Soc.* **1980**, *102*, 7059.
104. Buncel, E.; Crampton, M.R.; Strauss, M.J.; Terrier, F. *Electron Deficient Aromatic- and Heteroaromatic-Base Interactions. The Chemistry of Anionic Sigma Complexes*; Elsevier: Amsterdam, 1984.
105. Terrier, F. *Nucleophilic Aromatic Displacement. Influence of the Nitro Group*; VCH: New York, 1991.
106. Buncel, E.; Dust, J.M.; Terrier, F. *Chem. Rev.* **1995**, *95*, 2261.
107. Dust, J.M.; Manderville, R.A. *Can. J. Chem.* **1998**, *76*, 622.

108. Vichard, D.; Boubaker, T.; Terrier, F.; Pouet, M.J.; Dust, J.M.; Buncel, E. *Can. J. Chem.* **2001**, *79*, 1617.
109. Makosza, M.; Winiarski, J. *Acc. Chem. Res.* **1987**, *20*, 282.
110. Makosza, M. *Pol. J. Chem.* **1992**, *66*, 33.
111. Makosza, M.; Sienkiewicz, K. *J. Org. Chem.* **1990**, *55*, 4979.
112. Bunnett, J.F.; Zahler, R.E. *Chem. Rev.* **1951**, *49*, 272.
113. Buncel, E.; Tarkka, R.M.; Dust, J.M. *Can. J. Chem.* **1994**, *72*, 1709.
114. Buncel, E.; Eggimann, W. *J. Am. Chem. Soc.* **1977**, *99*, 5958; and references therein.
115. Buncel, E.; Manderville, R.A. *J. Phys. Org. Chem.* **1993**, *6*, 71.
116. Buncel, E.; Dust, J.M.; Jonczyk, A.; Manderville, R.A.; Onyido, I. *J. Am. Chem. Soc.* **1992**, *114*, 5610.
117. Manderville, R.A.; Dust, J.M.; Buncel, E. *J. Phys. Org. Chem.* **1996**, *9*, 515.
118. Simonnin, M.P.; Hallé, J.C.; Terrier, F.; Pouet, M.J. *Can. J. Chem.* **1985**, *53*, 866.
119. Hallé, J.C.; Pouet, M.J.; Simonnin, M.P.; Debleds, F.; Terrier, F. *Can. J. Chem.* **1982**, *60*, 1988.
120. Cox, J.P.L.; Crampton, M.R.; Wight, P. *J. Chem. Soc., Perkin Trans. 2.* **1988**, 25.
121. Crampton, M.R.; Stevens, J.A. *J. Chem. Soc., Perkin Trans. 2* **1991**, 1715.
122. Crampton, M.R.; Kee, T.P.; Wilcock, J.R. *Can. J. Chem.* **1986**, *64*, 1714.
123. Terrier, F.; Goumont, R.; Pouet, M.J.; Hallé, J.C. *J. Chem. Soc., Perkin Trans. 2* **1995**, 1629.
124. Terrier, F. *Nucleophilic Aromatic Displacement. Influence of the Nitro Group*; VCH: New York, 1991; pp 18, 138.
125. Terrier, F.; Kizilian, E.; Hallé, J.C.; Buncel, E. *J. Am. Chem. Soc.* **1992**, *114*, 1740.
126. Bordwell, F.G.; Branca, J.C.; Cripe, T.A. *Isr. J. Chem.* **1985**, *26*, 357.
127. Moutiers, G.; El Fahid, B.; Goumont, R.; Chatrousse, A.P.; Terrier, F. *J. Org. Chem.* **1996**, *61*, 1978.
128. Buncel, E.; Um, I.H.; Hoz, S. *J. Am. Chem. Soc.* **1989**, *111*, 971.
129. Tarkka, R.M.; Park, W.K.C.; Liu, P.; Buncel, E.; Hoz, S. *J. Chem. Soc., Perkin Trans. 2* **1994**, 2439.
130. Tarkka, R.M.; Buncel, E. *J. Am. Chem. Soc.* **1995**, *117*, 1503.
131. Terrier, F.; Moutiers, G.; Xiao, L.; Le Guével, E.; Guir, F. *J. Org. Chem.* **1995**, *60*, 1748.
132. Hoz, S.; Liu, P.; Buncel, E. *J. Chem. Soc., Chem. Commun.* **1996**, 995.
133. Arnaut, L.G. *J. Phys. Org. Chem.* **1991**, *4*, 726.
134. Pross, A.; Shaik, S.S. *J. Am. Chem. Soc.* **1982**, *104*, 1129.
135. Arnaut, L.G.; Formosinho, S.J. *J. Phys. Org. Chem.* **1990**, *3*, 95.
136. Formosinho, S.J. *J. Chem. Soc., Perkin Trans. 2* **1987**, 61.
137. Gross, Z.; Hoz, S. *J. Am. Chem. Soc.* **1988**, *110*, 7489.
138. Pross, A. *Adv. Phys. Org. Chem.* **1985**, *21*, 99.
139. Hoz, S. *Acc. Chem. Res.* **1993**, *26*, 69.
140. Marcus, R.A. *J. Chem. Phys.* **1956**, *24*, 966.
141. Marcus, R.A. *J. Chem. Phys.* **1957**, *26*, 867.
142. Cohen, A.O.; Marcus, R.A. *J. Phys. Chem.* **1968**, *72*, 4249.
143. Keeffe, J.R.; Kresge, A.J. In *Investigations of Rates and Mechanisms of Reactions. Part I.*; Bernasconi, C.F., Ed.; Wiley, New York, 1986; p 747.
144. Bunting, J.W.; Stefanidas, D. *J. Am. Chem. Soc.* **1988**, *110*, 4008.
145. Albery, W.J.; Kreevoy, M.M. *Adv. Phys. Org. Chem.* **1978**, *16*, 87.
146. Lewis, E.S.; Hu, D.D. *J. Am. Chem. Soc.* **1984**, *106*, 3292.
147. Hine, J. *J. Am. Chem. Soc.* **1971**, *93*, 3701.
148. Guthrie, J.P. *J. Am. Chem. Soc.* **1997**, *119*, 1151.
149. Guthrie, J.P. *J. Am. Chem. Soc.* **1996**, *118*, 12878.

150. Terrier, F.; Moutiers, G.; Pelet, S.; Buncel, E. *Eur. J. Org. Chem.* **1999**, 1771.
151. Bordwell, F.G. *Acc. Chem. Res.* **1988**, *21*, 456.
152. Cumming, J.B.; Kebarle, P. *J. Am. Chem. Soc.* **1978**, *100*, 1835.
153. Griller, D.; Ingold, K.U. *Acc. Chem. Res.* **1976**, *9*, 13.
154. McMahon, T.B.; Kebarle, P. *J. Am. Chem. Soc.* **1976**, *98*, 3399.
155. Bohme, D.K.; Lee-Ruff, E.; Young, L.B. *J. Am. Chem. Soc.* **1972**, *94*, 5153.
156. Wheland, G.W.; Farr, J. *J. Am. Chem. Soc.* **1943**, *65*, 1433.
157. Streitwieser, A.; Nebenzahl, L.L. *J. Am. Chem. Soc.* **1976**, *98*, 2188.
158. Brinkman, E.A.; Berger, S.; Brauman, J.I. *J. Am. Chem. Soc.* **1994**, *116*, 8304.
159. DePuy, C.H.; Gronert, S.; Barlow, S.E.; Bierbaum, V.M.; Damrauer, R. *J. Am. Chem. Soc.* **1989**, *111*, 1968.
160. Zhang, S.Z.; Zhang, X.M.; Bordwell, F.G. *J. Am. Chem. Soc.* **1995**, *117*, 602.
161. Buncel, E. *Carbanions: Mechanistic and Isotopic Aspects*; Elsevier: Amsterdam, 1975; p 6.
162. Morrison, R.T.; Boyd, R.N. *Organic Chemistry*, 3rd ed.; Allyn & Bacon: Boston, 1973; pp 256–259.
163. Pellerite, M.J.; Brauman, J.I. In *Comprehensive Carbanion Chemistry. Part A: Structure and Reactivity*; Buncel, E.; Durst, T., Eds.; Elsevier: Amsterdam, 1980; p 88.
164. The most current critical (and searchable) compendium of gas-phase acidities, proton affinities, and other data is found at http://webbook.nist.gov/chemistry.
165. Ervin, K.M.; Gronert, S.; Barlow, S.E.; Gilles, M.K.; Harrison, A.G.; Bierbaum, V.M.; DePuy, C.H.; Lin, W.C. *J. Am. Chem. Soc.* **1990**, *112*, 5750.
166. Bent, H.A. *Chem. Rev.* **1961**, *61*, 275.
167. Streitwieser, A.; Caldwell, R.A.; Young, W.R. *J. Am. Chem. Soc.* **1969**, *91*, 529.
168. So, S.P.; Wong, M.H.; Luh, T.Y. *J. Org. Chem.* **1985**, *50*, 2632.
169. Streitwieser, A.; Ziegler, G.R. *J. Am. Chem. Soc.* **1969**, *91*, 5081; and references therein.
170. Dixon, R.E.; Williams, P.G.; Saljoughian, M.; Long, M.A.; Streitwieser, A. *Magn. Reson. Chem.* **1991**, *29*, 509.
171. Streitwieser, A.; Xie, F.; Speers, P.; Williams, P.G. *Magn. Reson. Chem.* **1998**, *36*, S209.
172. Boerth, D.W.; Streitwieser, A. *J. Am. Chem. Soc.* **1981**, *103*, 6443.
173. Jaun, B.; Schwarz, J.; Breslow, R. *J. Am. Chem. Soc.* **1980**, *102*, 5741.
174. Breslow, R. *Pure Appl. Chem.* **1974**, *40*, 493.
175. Hückel, E. *Z. Physik.* **1931**, *70*, 204.
176. Hückel, E. *Z. Physik.* **1932**, *76*, 628.
177. Breslow, R. *Chem. Eng. News.* **1965**, 90.
178. Dewar, M.J.S. *Tetrahedron* **1966** (Suppl. 8), 75.
179. Dewar, M.J.S.; Dougherty, R.C. *The PMO Theory of Organic Chemistry*; Plenum: New York, 1975.
180. Bordwell, F.G.; Zhang, X.M. *Acc. Chem. Res.* **1993**, *26*, 510.
181. Streitwieser, A.; Juaristi, E.; Nebenzahl, L.L. In *Comprehensive Carbanion Chemistry. Part A. Structure and Reactivity*; Buncel, E.; Durst, T., Eds.; Elsevier: Amsterdam, 1980; pp 367–369.
182. Mathews, W.S.; Bares, J.E.; Bartmess, J.E.; Bordwell, F.G.; Cornforth, F.J.; Drucker, G.E.; Margolin, Z.; McCallum, R.J.; McCollum, G.J.; Vanier, N.R. *J. Am. Chem. Soc.* **1975**, *97*, 7006.
183. Hoffmann, R.; Bissell, R.; Farnum, D. *J. Phys. Chem.* **1969**, *73*, 1789.
184. Brooks, J.J.; Stucky, G.D. *J. Am. Chem. Soc.* **1972**, *94*, 7333.
185. Buncel, E.; Menon, B.C. *J. Am. Chem. Soc.* **1977**, *99*, 4457.

186. Buncel, E.; Menon, B.C. In *Comprehensive Carbanion Chemistry. Part. A. Structure and Reactivity*; Buncel, E.; Durst, T., Eds.; Elsevier: Amsterdam, 1980; p 77.
187. Edlund, U.; Buncel, E. *Prog. Phys. Org. Chem.* **1993**, *19*, 255.
188. Bonhoeffer, K.F.; Gieb, K.H.; Reitz, O. *J. Chem. Phys.* **1939**, *7*, 664.
189. Richard, J.P.; Williams, G.; Gao, J. *J. Am. Chem. Soc.* **1999**, *121*, 715.
190. Streitwieser, A.; Nebenzahl, L.L. *J. Org. Chem.* **1978**, *43*, 598.
191. Cockerill, A.F.; Lamper, J.E. *J. Chem. Soc. B* **1971**, 503.
192. Streitwieser, A.; Murdoch, J.R.; Hafelinger, G.; Chang, C.J. *J. Am. Chem. Soc.* **1973**, *95*, 4248.
193. Bordwell, F.G.; Bares, J.E.; Bartmess, J.E.; McCollum, G.J.; Van Der Puy, M.; Vanier, N.R.; Mathews, W.S. *J. Org. Chem.* **1977**, *42*, 321.
194. Rabideau, P.W.; Wind, B.K.; Sygula, A. *Tetrahedron Lett.* **1991**, *32*, 5659.
195. Bordwell, F.G.; Van Der Puy, M.; Vanier, N.R. *J. Org. Chem.* **1976**, *41*, 1883.
196. Bordwell, F.G.; Zhao, Y. *J. Org. Chem.* **1995**, *60*, 6348.
197. Andreades, S. *J. Am. Chem. Soc.* **1964**, *86*, 2003.
198. Adolph, H.G.; Kamlet, M. J. *J. Am. Chem. Soc.* **1966**, *88*, 4761.
199. Wiberg, K.B.; Castejon, H. *J. Org. Chem.* **1995**, *60*, 6327.
200. Chambers, R.D.; Bryce, M.R. In *Comprehensive Carbanion Chemistry. Part C. Ground and Excited State Reactivity*; Buncel, E.; Durst, T., Eds.; Elsevier: Amsterdam, 1987; p 276.
201. Streitwieser, A.; Mares, F. *J. Am. Chem. Soc.* **1968**, *90*, 2444.
202. Holtz, D. *Prog. Phys. Org. Chem.* **1971**, *8*, 1.
203. Klabunde, K.J.; Burton, D.J. *J. Am. Chem. Soc.* **1972**, *94*, 5985.
204. Hine, J.; Mahone, L.G.; Liotta, C.L. *J. Am. Chem. Soc.* **1967**, *89*, 5991.
205. Hine, J.; Dalsin, P.D. *J. Am. Chem. Soc.* **1972**, *94*, 6998.
206. Jernigan, M.T.; Eliel, E.L. *J. Am. Chem. Soc.* **1995**, *117*, 9638.
207. Roberts, J.D.; Webb, R.L.; McElhill, E.A. *J. Am. Chem. Soc.* **1950**, *72*, 408.
208. Hine, J. *J. Am. Chem. Soc.* **1963**, *85*, 3239.
209. Dixon, D.A.; Fukunaga, T.; Smart, B.E. *J. Am. Chem. Soc.* **1986**, *108*, 4027.
210. Farnham, W.B.; Smart, B.E.; Middleton, W.J.; Calabrese, J.C.; Dixon, D.A. *J. Am. Chem. Soc.* **1985**, *107*, 4565.
211. Hoffmann, R.; Radom, L.; Pople, J.A.; Schleyer, P. von R.; Hehre, W.J.; Salem, L. *J. Am. Chem. Soc.* **1972**, *94*, 6221.
212. Schleyer, P. von R.; Kos, A. *Tetrahedron* **1983**, *24*, 1141.
213. Salzner, U.; Schleyer, P. von R. *Chem. Phys. Lett.* **1992**, *190*, 401.
214. Streitwieser, A.; Holtz, D.; Ziegler, G.R.; Stroffer, G.R.; Brokaw, M.L.; Guibé, F. *J. Am. Chem. Soc.* **1976**, *98*, 5229.
215. Andreades, S. *J. Am. Chem. Soc.* **1964**, *86*, 2006.
216. Streitwieser, A.; Holtz, D. *J. Am. Chem. Soc.* **1967**, *89*, 692.
217. Koppel, I.A.; Pihl, V.; Koppel, J.; Anvia, F.; Taft, R.W. *J. Am. Chem. Soc.* **1994**, *116*, 8654.
218. Koppel, I.A.; Taft, R.W.; Anvia, F.; Zhu, S.Z.; Hu, L.Q.; Sung, K.S.; DesMarteau, D.D.; Yagupolskii, L.M.; Yagupolskii, Y.L.; Ignatev, N.V.; Kondratenko, N.V.; Volkonskii, A.Y.; Vlasov, V.M.; Notario, R.; Maria, P.C. *J. Am. Chem. Soc.* **1994**, *116*, 3047.
219. Fujio, M.; McIver, R. T., Jr.; Taft, R.W. *J. Am. Chem. Soc.* **1981**, *103*, 4017.
220. Sleigh, J.H.; Stephens, R.; Tatlow, J.C. *J. Fluorine Chem.* **1980**, *15*, 411.
221. Ingold, C.K. *Structure and Mechanism in Organic Chemistry*; Cornell University Press: Ithaca, NY, 1953; Chapter 2.
222. Taft, R.W. *Prog. Phys. Org. Chem.* **1983**, *14*, 247.

223. Pellerite, M.J.; Brauman, J.I. In *Comprehensive Carbanion Chemistry. Part. A. Structure and Reactivity*; Buncel, E.; Durst, T., Eds., Elsevier: Amsterdam, 1980; pp 84–88.
224. Pross, A.; Radom, L. *J. Am. Chem. Soc.* **1978**, *100*, 6572.
225. Bordwell, F.G.; Bares, J.E.; Bartmess, J.E.; Drucker, G.E.; Gerhold, J.; McCollum, G.J.; Van Der Puy, M.; Vanier, N.R.; Mathews, W.S. *J. Org. Chem.* **1977**, *42*, 326.
226. Bordwell, F.G.; Van Der Puy, M.; Vanier, N.R. *J. Org. Chem.* **1976**, *41*, 1885.
227. Doering, W. von E.; Hoffmann, A.K. *J. Am. Chem. Soc.* **1955**, *77*, 521.
228. Hochberg, J.; Bonhoeffer, K.F. *Z. Physik. Chem.* **1939**, *A184*, 419.
229. Buncel, E.; Symons, E.A.; Zabel, A.W. *J. Chem. Soc., Chem. Commun.* **1965**, 173.
230. Hartmann, A.A.; Eliel, E.L. *J. Am. Chem. Soc.* **1971**, *93*, 2572.
231. King, J.F.; Luinstra, E.A.; Harding, D.R.K. *J. Chem. Soc., Chem. Commun.* **1972**, 1313.
232. Paquette, L.A.; Freeman, J.P.; Wyvratt, M.J. *J. Am. Chem. Soc.* **1971**, *93*, 3216.
233. Kluger, R.; Wasserstein, P. *J. Am. Chem. Soc.* **1973**, *95*, 1071.
234. Zhang, X.M.; Bordwell, F.G. *J. Am. Chem. Soc.* **1994**, *116*, 968.
235. Petrov, E.S.; Terekhova, M.I.; Shatenshtein, A.I. *Zh. Obshch. Khim.* **1974**, *44*, 1118. (Engl. transl., p 1075).
236. Petrov, E.S.; Terekhova, M.I.; Shatenshtein, A.I.; Tromfirov, B.A.; Mirskov, R.G.; Voronkov, M.G. *Dokl. Akad. Nauk SSSR* **1973**, 211 (Engl. transl., p 692).
237. Miah, M.A.; Fraser, R.R. *J. Bangladesh Chem. Soc.* **1989**, *2*, 77.
238. Miah, M.A.; Fraser, R.R. *Indian J. Chem.* **1990**, *29A*, 588.
239. Fraser, R.R. *Tetrahedron Lett.* **1982**, *23*, 4195.
240. Fraser, R.R.; Savard, S.S.; Mansour, T.S. *Can. J. Chem.* **1985**, *63*, 3505.
241. Streitwieser, A.; Xie, L.; Wang, P.; Bachrach, S.M. *J. Org. Chem.* **1993**, *58*, 1778.
242. Eaborn, C.; Eidenschink, R.; Jackson, P.M.; Walton, D.R.M. *J. Organomet. Chem.* **1975**, *101*, C40.
243. Wetzel, D.M.; Brauman, J.I. *J. Am. Chem. Soc.* **1988**, *110*, 8333.
244. Zhang, S.; Bordwell, F.G. *J. Org. Chem.* **1996**, *61*, 51.
245. Hierl, P.M.; Henchman, M.J.; Paulson, J.F. *Int. J. Mass. Spectrom. Ion. Processes* **1992**, *117*, 475.
246. Ingemann, S.; Nibbering, N.M.M. *J. Chem. Soc., Perkin Trans. 2* **1985**, 837.
247. Grabowski, J.J.; Roy, P.D.; Leone, R. *J. Chem. Soc., Perkin Trans. 2* **1988**, 1627.
248. Moran, S.; Ellison, G.B. *J. Phys. Chem.* **1988**, *92*, 1794.
249. Gunion, R.F.; Gilles, M.K.; Polak, M.L.; Lineberger, W.C. *Int. J. Mass Spectrom. Ion Processes* **1992**, *117*, 601.
250. Bartmess, J.E.; Scott, J.A.; McIver, R.T., Jr., *J. Am. Chem. Soc.* **1979**, *101*, 6047.
251. Taft, R.W.; Bordwell, F.G. *Acc. Chem. Res.* **1988**, *21*, 463.
252. Cutress, N.C.; Katritzky, A.R.; Eaborn, C.; Walton, D.R.M.; Topson, R.D. *J. Organomet. Chem.* **1972**, *43*, 131.
253. Streitwieser, A.; Ewing, S.P. *J. Am. Chem. Soc.* **1975**, *97*, 190.
254. King, J.F.; Rathmore, R.A. *J. Am. Chem. Soc.* **1990**, *112*, 2001.
255. Epiotis, N.D.; Yakes, R.L.; Bernardi, F.; Wolfe, S. *J. Am. Chem. Soc.* **1976**, *99*, 5435.
256. Larson, J.R.; Epiotis, N.D. *J. Am. Chem. Soc.* **1981**, *103*, 410.
257. Wolfe, S.; LaJohn, L.A.; Weaver, D.F. *Tetrahedron Lett.* **1984**, *25*, 2863.
258. Bernardi, F.; Bottoni, A.; Valli, G.S.; Venturini, A. *Gazz. Chim. Ital.* **1990**, *120*, 301.
259. Gutsev, G.L.; Ziegler, T. *Can. J. Chem.* **1991**, *69*, 993.
260. Rauk, A.; Wolfe, S.; Csizmadia, I.G. *Can. J. Chem.* **1969**, *47*, 113.
261. Bernardi, F.; Csizmadia, I.G.; Mangini, A.; Schlegel, H.B.; Whangbo, M.H.; Wolfe, S. *J. Am. Chem. Soc.* **1975**, *97*, 2209.
262. Wolfe, S. *Acc. Chem. Res.* **1972**, *5*, 102.

263. Schleyer, P. von R.; Clark, T.; Kos, A.J.; Spitznagel, G.W.; Rohde, C.; Arad, D.; Houk, K.N.; Rondan, N.G. *J. Am. Chem. Soc.* **1984**, *106*, 6467.
264. Hopkinson, A.C.; Lien, M.H. *J. Org. Chem.* **1981**, *46*, 998.
265. Wolfe, S. In *Organic Sulfur Chemistry. Theoretical and Experimental Advances*; Bernardi, F.; Csizmadia, I.G.; Mangini, A., Eds.; Elsevier: Amsterdam, 1985; pp 133–190; and references therein.
266. Wolfe, S.; Stolow, A.; LaJohn, L.A. *Tetrahedron Lett.* **1983**, *24*, 4071.
267. Bors, D.A.; Streitwieser, A. *J. Am. Chem. Soc.* **1986**, *108*, 1397.
268. Koch, R.; Anders, E. *J. Org. Chem.* **1994**, *59*, 4529.
269. Raabe, G.; Gais, H.J.; Fleischhauer, J. *J. Am. Chem. Soc.* **1996**, *118*, 4622.
270. Bordwell, F.G.; Bausch, M.J. *J. Am. Chem. Soc.* **1986**, *108*, 1979.
271. Bordwell, F.G.; Bausch, M.J.; Branca, J.C.; Harrelson, J.A., Jr., *J. Phys. Org. Chem.* **1988**, *1*, 225.
272. Bordwell, F.G.; Cheng, J.P.; Bausch, M.J.; Bares, J.E. *J. Phys. Org. Chem.* **1988**, *1*, 209.
273. Marriott, S.; Reynolds, W.F.; Taft, R.W.; Topsom, R.D. *J. Org. Chem.* **1984**, *49*, 959.
274. Exner, O.; Krygowski, T.M. *Chem. Soc. Rev.* **1996**, *25*, 71.
275. Ehrenson, S.; Brownlee, R.T.C.; Taft, R.W. *Prog. Phys. Org. Chem.* **1973**, *10*, 1.
276. Kolthoff, I.M.; Chantooni, J.K., Jr.; Bhowmik, S. *J. Am. Chem. Soc.* **1968**, *90*, 23.
277. Lelièvre, J.; Farrell, P.G.; Terrier, F. *J. Chem. Soc., Perkin Trans. 2* **1986**, 333.
278. Terrier, F.; Lelièvre, J.; Chatrousse, A.P.; Farrell, P.G. *J. Chem. Soc., Perkin Trans. 2* **1985**, 1479.
279. Simonnin, M.P.; Xie, H.Q.; Terrier, F.; Lelièvre, J.; Farrell, P.G. *J. Chem. Soc., Perkin Trans. 2* **1989**, 1553.
280. Buncel, E.; Murarka, S.K.; Onyido, I. *Gazz. Chim. Ital.* **1988**, *118*, 25.
281. Keeffe, J.R.; Kresge, A.J.; Lin, Y. *J. Am. Chem. Soc.* **1988**, *110*, 1982.
282. Ross, A.M.; Whalen, D.L.; Eldin, S.; Pollack, R.M. *J. Am. Chem. Soc.* **1988**, *110*, 1981.
283. Keeffe, J.R.; Kresge, A.J.; Yin, Y. *J. Am. Chem. Soc.* **1988**, *110*, 8201.
284. Kuhn, R.; Rewicki, D. *Justus Liebigs Ann. Chem.* **1969**, *706*, 250.
285. Tolbert, L.M.; Ogle, M.E. *J. Am. Chem. Soc.* **1989**, *111*, 5958.
286. Tolbert, L.M.; Ogle, M.E. *J. Am. Chem. Soc.* **1990**, *112*, 9519.
287. Spiesecke, H.; Schneider, W.G. *Tetrahedron Lett.* **1961**, 468.
288. Eliasson, B.; Edlund, U.; Müllen, K. *J. Chem. Soc., Perkin Trans. 2* **1986**, 937.
289. Müllen, K. *Chem. Rev.* **1984**, *84*, 603.
290. O'Brien, D.H. In *Comprehensive Carbanion Chemistry. Part A. Structure and Reactivity*; Buncel, E.; Durst, T., Eds.; Elsevier: Amsterdam, 1980; pp 298–304.
291. Tokuhiro, T.; Fraenkel, G. *J. Am. Chem. Soc.* **1969**, *91*, 5005.
292. Strub, H.; Beeler, A.J.; Grant, D.M.; Michl, J.; Cutts, P.W.; Zilm, K.W. *J. Am. Chem. Soc.* **1983**, *105*, 3333.
293. Kloosterziel, H.; Werner, M.A. *Recl. Trav. Chim. Pays-Bas* **1975**, *94*, 124.
294. Kloosterziel, H. *Recl. Trav. Chim. Pays-Bas* **1974**, *93*, 215.
295. Heeger, A.J.; Kivelson, S.; Schrieffer, J.R.; Su, W.P. *Rev. Mod. Phys.* **1988**, *60*, 781.
296. Roth, S. *Mater. Sci. Forum* **1989**, *42*, 1.
297. Su, W.P. In *Handbook of Conducting Polymers*; Skotheim, T.J., Ed.; Marcel Dekker: New York, 1986; pp 757–794.
298. Krätschmer,W.; Lamb, L.D.; Fostiropouolos, K.; Huffman, D.R. *Nature* **1990**, *347*, 354.
299. Bhyrappa, P.; Pénicaud, A.; Kawamoto, M.; Reed, C.A. *J. Chem. Soc., Chem. Commun.* **1992**, 936.
300. Khemani, K.C.; Prato, M.; Wudl, F. *J. Org. Chem.* **1992**, *57*, 3254.

301. Gügel, A.; Becker, M.; Hammel, D.; Mindach, L.; Räder, J.; Simon, T.; Wagner, M.; Müllen, K. *Angew. Chem., Int. Ed. Engl.* **1992**, *31*, 644.
302. Scrivens, W.A.; Bedworth, P.V.; Tour, J.M. *J. Am. Chem. Soc.* **1992**, *114*, 7917.
303. Hirsh, A. *The Chemistry of the Fullerenes*; Georg Thieme Verlag: Stuttgart, Germany, 1994.
304. Diederich, F.; Isaacs, I.; Philip, D. *Chem. Soc. Rev.* **1994**, *23*, 1186.
305. Hammond, G.S.; Kuck, V.J., Eds.; *Fullerenes*; ACS Symposium Series 481; American Chemical Society: Washington, DC, 1992.
306. An issue of *Accounts of Chemical Research* (**1992**, *25*) has been published that reviews several aspects of fullerene chemistry.
307. Xie, Q.; Pérez-Cordero, E.; Echegoyen, L. *J. Am. Chem. Soc.* **1992**, *114*, 3978.
308. Haddon, R.C.; Brus, L.E.; Raghavachari, K. *Chem. Phys. Lett.* **1986**, *125*, 459.
309. Scuseria, G.E. *Chem. Phys. Lett.* **1991**, *176*, 423.
310. Benito, A.M.; Darwish, A.D.; Kroto, H.W.; Meidine, M.F.; Taylor, R.; Walton, D.R.M. *Tetrahedron Lett.* **1996**, *37*, 1085.
311. Hirsch, A.; Lamparth, I.; Karfunkel, H.R. *Angew. Chem., Int. Ed. Engl.* **1994**, *33*, 437.
312. Stinchcombe, J.; Pénicaud, A.; Bhyrappa, P.; Boyd, P.D.W.; Reed, C.A. *J. Am. Chem. Soc.* **1993**, *115*, 5212.
313. Fagan, P.; Krusic, P.J.; Evans, D.H.; Lerke, S.A.; Johnston, E. *J. Am. Chem. Soc.* **1992**, *114*, 9697.
314. Hirsch, A.; Grösser, T.; Skiebe, A.; Soi, A. *Chem. Ber.* **1993**, *126*, 1061.
315. Toshikazu, K.; Tanaka, T.; Takata, Y.; Takeuchi, K.; Komatsu, K. *J. Org. Chem.* **1995**, *60*, 1490.
316. Eisch, J.J. *Res. Chem. Intermed.* **1996**, *22*, 145.
317. Guthrie, R.D. In *Comprehensive Carbanion Chemistry. Part A. Structure and Reactivity*; Buncel, E.; Durst, T., Eds.; Elsevier: Amsterdam, 1980; pp 197–271.
318. Bunnett, J.F. *Acc. Chem. Res.* **1978**, *11*, 413.
319. Rossi, R.A.; de Rossi, R.H. *Aromatic Substitution by the $S_{RN}1$ Mechanism*; ACS Monograph Series 178; American Chemical Society: Washington, DC, 1983.
320. Rossi, R.A.; Pierini, A.B.; Penenory, A.B. In *The Chemistry of Functional Groups, Suppl. D., The Chemistry of Halides, Pseudo-halides and Azides*; Patai, S.; Rappoport, Z.; Eds.; Wiley: New York, 1995; Chapter 24.
321. Santiago, A.N.; Lassaga, G.; Rappoport, Z.; Rossi, R.A. *J. Org. Chem.* **1996**, *61*, 1125.
322. Nazareno, M.A.; Rossi, R.A. *J. Org. Chem.* **1996**, *61*, 1645.
323. Guthrie, R.D.; Hartmann, C.; Neill, R.; Nutter, D.E. *J. Org. Chem.* **1987**, *52*, 736.
324. Galli, C.; Gentili, P.; Rappoport, Z. *J. Org. Chem.* **1994**, *59*, 6786.
325. Eberson, L. *Adv. Phys. Org. Chem.* **1982**, *18*, 79.
326. Ashby, E.C. *Acc. Chem. Res.* **1988**, *21*, 212.
327. Ashby, E.C.; Mehdizadeh, A.; Deshpande, A.K. *J. Org. Chem.* **1996**, *61*, 1322.
328. Ashby, E.C. *Pure Appl. Chem.* **1980**, *52*, 545.
329. Holm, T.; Crossland, I. *Acta Chem. Scand.* **1971**, *25*, 59.
330. Ashby, E.C.; Tung, N.P. *J. Org. Chem.* **1987**, *53*, 1291.
331. Bailey, W.F.; Patricia, J.J.; Del Gobbo, V.C.; Jarent, R.M.; Okarma, P.J. *J. Org. Chem.* **1985**, *50*, 1999.
332. Ashby, E.C.; Argyropoulos, J.N. *J. Org. Chem.* **1986**, *51*, 472.
333. Ashby, E.C.; Park, W.S. *Tetrahedron Lett.* **1983**, *24*, 1667.
334. Ashby, E.C.; Argyropoulos, J.N. *J. Org. Chem.* **1987**, *52*, 4079.
335. Ingold, K.U.; Griller, D. *Acc. Chem. Res.* **1980**, *13*, 317.
336. Mattalia, J.M.; Chanon, M.; Stirling, C.J.M. *J. Org. Chem.* **1996**, *61*, 1153.

337. Palmer, C.A.; Ogle, C.A.; Arnett, E.M. *J. Am. Chem. Soc.* **1992**, *114*, 5619.
338. Lowry, T.H.; Richardson, K.S. *Mechanism and Theory in Organic Chemistry, third ed.*; Harper & Row: New York, 1987; pp 827–839.
339. Barclay, L.R.C.; Dust, J.M. *Can. J. Chem.* **1982**, *60*, 607.
340. Pine, S.H. *J. Chem. Educ.* **1972**, *49*, 664.
341. Arnett, E.M.; Fisher, F.J.; Nichols, M.A.; Ribiero, A.A. *J. Am. Chem. Soc.* **1989**, *111*, 748.
342. House, H.O. *Acc. Chem. Res.* **1976**, *9*, 596.
343. Martens, F.M.; Verhoeven, J.W.; Case, R.Z.; Pandit, U.K.; De Boer, T.J. *Tetrahedron* **1978**, *34*, 443.
344. Eberson, L. *Electron Transfer Reactions in Organic Chemistry*; Springer Verlag: New York, 1987.
345. Guthrie, R.D.; Weisman, G.R.; Burdon, L.G. *J. Am. Chem. Soc.* **1974**, *96*, 6955.
346. Guthrie, R.D.; Weisman, G.R.; Burdon, L.G. *J. Am. Chem. Soc.* **1974**, *96*, 6962.
347. Guthrie, R.D.; Pendygraft, G.W.; Young, A.T. *J. Am. Chem. Soc.* **1976**, *98*, 5877.
348. Han, C.C.; Brauman, J.I. *J. Am. Chem. Soc.* **1988**, *110*, 4048.
349. Buncel, E.; Menon, B.C. *J. Am. Chem. Soc.* **1980**, *102*, 3499.
350. Buncel, E.; Venkatachalam, T.K.; Menon, B.C. *J. Org. Chem.* **1984**, *49*, 413.
351. Edwards, J.O. *J. Am. Chem. Soc.* **1954**, *76*, 1540.
352. Kern, J.M.; Federlin, P. *Tetrahedron Lett.* **1977**, 827.
353. Bank, S.; Gernon, M. *J. Org. Chem.* **1987**, *52*, 5105.
354. Bordwell, F.G.; Harrelson, J.A., Jr.; *Can. J. Chem.* **1990**, *68*, 1714.
355. Bordwell, F.G.; Cheng, J.P.; Satish, A.V.; Twyman, C.L. *J. Org. Chem.* **1992**, *57*, 6542.
356. Nicholas, A. de P.M.; Arnold, D.R. *Can. J. Chem.* **1982**, *60*, 2165.
357. Dust, J.M.; Arnold, D.R. *J. Am. Chem. Soc.* **1983**, *105*, 1221.
358. Wayner, D.D.M.; Arnold, D.R. *Can. J. Chem.* **1984**, *62*, 1164.
359. Neumann, W.P.; Penenory, A.; Stewen, U.; Lehnig, M. *J. Am. Chem. Soc.* **1989**, *111*, 5945.
360. Jackson, R.A.; Sharifi, M. *J. Chem. Soc., Perkin Trans. 2* **1996**, 775.
361. Saunders, W.H., Jr. *Croatica Chem. Acta* **2001**, *74*, 575.
362. Harris, N.; Wei, W.; Saunders, W.H., Jr.; Shaik, S. *J. Am. Chem. Soc.* **2000**, *122*, 6754.
363. Van Verth, J.E.; Saunders, W.H., Jr. *Can. J. Chem.* **1999**, *77*, 810.
364. Harris, N.; Wei, W.; Saunders, W.H., Jr.; Shaik, S. *J. Phys. Org. Chem.* **1999**, *12*, 259.
365. Van Verth, J.E.; Saunders, W.H., Jr. *J. Org. Chem.* **1997**, *62*, 5743.

3

Stereochemistry of Carbanions

3.1 INTRODUCTION: PLANAR VERSUS PYRAMIDAL CONFIGURATIONS

In the previous two chapters, we have connected hydrocarbon Brønsted acidity, as demonstrated by the relevant equilibrium acidity constants, K_a, to the stability of the resultant carbanions, and thence to their intrinsic reactivity and their structures. We have seen, for example, that the acidity of carbon acids α-substituted with fluorine is dependent upon whether the conjugate carbanide base is anticipated to be planar or pyramidal. A further important consequence of this structural dichotomy is that carbanionic reactions will have stereochemical outcomes.[1-3] Stereochemistry is sufficiently broad in scope to merit study in its own right.[4] Consequently, we will restrict our discussion in this chapter to those aspects of stereochemistry most important in the reactions of carbanions, particularly with regard to carbanions that are stabilized by α-sulfur or phosphorus substituents. (The rich reaction chemistry of enolate ions, including the stereochemistry involved, will be discussed in detail in Chapter 6.)

First, let us consider the geometry about the central carbon in the prototypical carbanion, methide anion, *in the gas phase*. As a species that is isoelectronic with ammonia[5,6] and, in accord with the prediction made of VSEPR (valence shell electron-pair repulsion) theory,[7] we would anticipate that methide would take up a trigonal pyramidal structure,[8,9] consistent with formal sp^3 hybridization. As illustrated in equation 3.1, the methide anion may take up either of two equivalent pyramidal forms that could interconvert via a process of inversion.

The energetics of inversion are intimately linked to the stereochemistry found for carbanions.

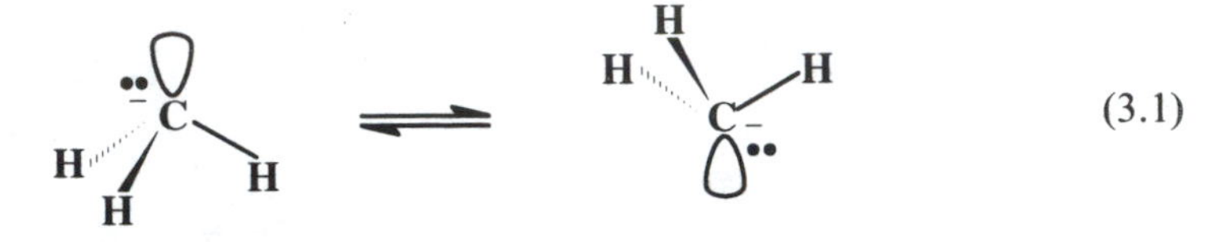

Nonetheless, until 1978 there had been no definitive observation of the methide anion in the gas phase nor experimental determination of its geometry. The overwhelming majority[10–13] of calculational studies[14,15] before and since[16] have confirmed the prediction that methide should preferentially take up a pyramidal structure. Experimental evidence for the pyramidal geometry of the methide ion was provided by examination of the vibrational structure of its photoelectron spectrum.[17] Further, the experimental photoelectron spectrum was shown to correspond well with a spectrum derived from a calculated potential energy profile for methide and methyl radical.[18] The inversion barrier determined in this way is about 2 kcal/mol.[18]

Prior to the mid-1980s, it was generally accepted that the process of pyramidal inversion could occur in two ways.[19,20] In a classical sense, inversion was expected to proceed via a transition state that would correspond in our example to planar methide ion and, thence, to the other pyramidal *invertomer*. Thus, for a general trisubstituted center [e.g., X(a,b,c) as shown in Figure 3.1], inversion proceeds along a normal bending vibrational mode that is parallel to the axis of the pyramid. A qualitative energy profile that corresponds to this understanding is shown in Figure 3.1. The process of inversion is associated with an activation barrier, labeled E_i in the figure.

An alternative mechanism for pyramidal inversion of X(a,b,c) involves quantum mechanical tunneling. We have already discussed tunneling in connection with kinetic isotopic effects (Chapter 2). Suffice it to say that the rate of tunneling is dependent upon a number of factors that include mass of the species, the height and shape of the inversion barrier, and the energy of the species.[21] Generally, tunneling becomes important when the inversion barrier is relatively low (i.e., E_i of about 5 kcal/mol or less) and the atoms substituting the center (X) are low in mass, such as protons [i.e., X(a,b,c) = XH_3].[19] Ammonia is such an XH_3 species; microwave spectroscopy displays the characteristic splitting of rotational–vibrational levels indicative of quantum mechanical tunneling. The inversion barrier (E_i) in this case was determined to be 5.3–5.8 kcal/mol,[22,23] in reasonable agreement with the calculated value for the barrier.[24,25]

Another process of inversion has been advanced[26–29] in which the transition state has a T-shape (C_{2v} symmetry).[27] This finding is of particular importance for the phosphorus-centered X(a,b,c) molecules.[30] In brief, PH_3, NCl_3, and NF_3, invert according to the pyramidal–planar–pyramidal profile shown in Figure 3.1, whereas PF_3, PCl_3, and AsF_3 follow the pyramidal–T-shaped–pyramidal pathway for inversion.

Regardless, for the systems of most interest to us, the classical process of inversion (Figure 3.1) will be presumed to be predominant. Which mode of inver-

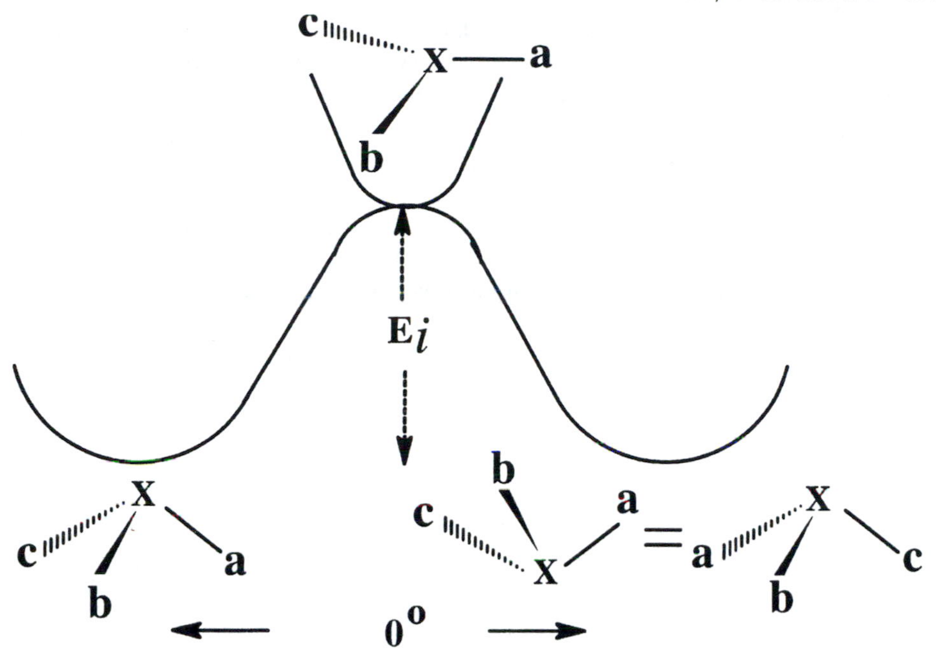

Figure 3.1. Qualitative energy profile for inversion of a general trisubstituted center. The energy maximum corresponds to a planar tricoordinate species (0° from planar), while either energy minimum corresponds to a pyramidal species (sp^3 hybridized at 19.5° from planar, i.e., the a–X–b bond angle is 109.5°).

sion, in fact, is predominant is not relevant to the discussion of the relative magnitudes of inversion barriers, but it is essential to the calculation of a given inversion barrier *from a rate of inversion.*

As we have seen, the experimental evidence,[17] as well as calculational results,[10–13,16,18] support the view that the simplest carbanion, methide, has a pyramidal structure. Calculated inversion barriers have values ranging over ca. 0–24 kcal/mol (Table 3.1), with the best values (1.7–2.2 kcal/mol)[16] in reasonable agreement with experiment;[17,31] H–C–H bond angles vary, but are generally less than 109.5° (e.g., 106.8° according to one calculation),[25] again as expected from VSEPR considerations.[7,8] Further examination of Figure 3.1 shows that inversion of the left-hand pyramidal structure results in formation of the right-hand pyramidal structure that is the corresponding enantiomer.

Several points should be emphasized at this juncture. Clearly, the *configurational integrity* of an idealized tricoordinate pyramidal species, X(a,b,c), including a suitably substituted carbanion, depends directly on the inversion energy (E_i). (Note, however, that a *planar* tricoordinate species may also be chiral if one of the groups attached to the center X is not conically symmetrical and if the barrier to rotation about this bond is high.[10]) On the other hand, the ability to *separately detect* the two stereoisomeric pyramidal forms, and, for example, to study the kinetics of their interconversion, depends on the technique used in the investigation. Thus, a barrier of 20–40 kcal/mol is required for *direct* determination of inversion kinetics by standard laboratory methods. Barriers of 5–35 kcal/mol and 10–20 kcal/mol are typically studied, respectively, by vibrational (infrared and Raman) spectroscopy and by dynamic nuclear magnetic resonance (DNMR) spectroscopy. Relatively small barriers associated with E_i of 0–5 kcal/mol may be measured by microwave spectroscopy.[19] As we have seen, thus far, various molecular orbital calculational methods may be applied to the analysis of invertomers. Alone, these calculational studies are of value in considering very unstable species or when applied to situations where the inversion barrier might be expected to be either very small or very large. Such calculations also complement experimental evidence, as we shall see below. Table 3.1 lists a selection of pyramidal X(a,b,c) species with their calculated and, where available, experimentally

Table 3.1 Inversion barriers of selected pyramidal species [X(a,b,c)].

	Barriers (kcal/mol)		
Species	Experimental	Calculated	Method (basis set)[a]
Methide	2[d]	23.9	ab initio (STO-3G)
		20.2	MINDO[b]
		12.9	ab initio (6-31G*)
		5.6	ab initio[c] (various)
		2.6	ab initio (6-31G**)
		2.2	ab initio[d] (6-31 + G*)
		0	AM1
		0	MNDO
Ethyl anion		3.3	ab initio[e] (MP2/6-31 + G*)
Ethenyl (vinyl) anion		31.5	ab initio
		31.1	MINDO[b]
Cyanomethyl anion		0.15	ab initio[e] (MP2/6-31 + G*)
		0	ab initio (4-31G)
Cyclopropyl anion		36.6	MINDO[b]
		19.4	ab initio (various)
		15.1	ab initio[e] (MP2/6-31 + G*)
Ammonia	5.8[h]	6.5	ab initio[f] (HF/6-31G)
	5.3[i]	5.5	ab initio (6-31 + G*)
		3.7	MINDO[b]
Nitrogen trifluoride	50[j]	62.6	INDO[g]
Phosphine		39.6	ab initio[g] (various)

SOURCES: [a]Data are taken from reference 16 unless otherwise specified; [b]Reference 35; [c]Reference 13; [d]Reference 18; [e]Reference 36; [f]Reference 25; [g]Reference 37; [h]Reference 22; [i]Reference 23; [j]Reference 19.
AM1, Austin Model 1; MNDO, modified neglect of diffential overlap; MINDO, modified intermediate neglect of differential overlap; LNDO, local neglect of diffential overlap.

determined inversion barriers. These X(a,b,c) species are believed to invert by the classical profile (Figure 3.1) involving the planar transition state.

Up to this point, we have also tacitly linked the pyramidal structure of the carbanide with sp^3 hybridization of the central carbon orbitals used for bonding. Examination of the X-ray structure database suggests that sp^3 hybridization of carbon that would lead to perfect tetrahedral (109.5°) interatomic bond angles is rare, even in relatively simple hydrocarbons such as propane.[32] Instead, the atomic orbitals of the central carbon, whether in chloromethane or methide anion, may be thought of as having variable or fractional hybridization.[33] Where otherwise not expressed, carbons that are approximately tetrahedral will be considered to have "formal sp^3" hybridization. Recall, however, that the degree of s-character involved in hybridization is an important factor in the stabilization of carbanionic centers and in the acidity of the conjugate carbon acids.

From our consideration of methide ion, we could conclude that carbanions show a preference for the pyramidal structure, and studies of bridgehead systems suggest that carbanions may readily take up the pyramidal form.[34] However, as we have seen, the barrier to inversion (i.e., the energy difference between pyramidal and planar forms) is not great in methide, and we can expect that either pyramidal or planar carbanions will exist, depending on other factors that stabilize the carbanionic center.[35] A recent ab initio calculational study of a series of substituted methide ions has shown that the cyanomethide, isocyanomethide, and ethanide ions are all pyramidal at the carbanionic center, but inversion barriers are relatively low (0.15 kcal/mol in the case of the anion derived from acetonitrile and up to 3.3 kcal/mol in the case of ethyl anion).[36] On the other hand, anions such as nitromethide and the anion of acetaldehyde (ethanal), which we have previously inferred to be planar *as an explanation* for the enhanced acidity of the corresponding carbon acids, are predicted to be planar according to current calculations. (We recognize, however, that increasingly sophisticated calculational methods may show that even the nitromethide anion is not completely planar.) Furthermore, cyclopropyl anions were shown to be strongly pyramidalized and to have relatively high barriers to inversion (e.g., 20.3 kcal/mol for the isocyano-substituted cyclopropyl anion, but falling to 0.7 kcal/mol for the anion of cyclopropylcarboxaldehyde).[36] Of course, our discussion thus far has been limited to these structurally simple carbanions *in the gas phase*. What is the situation *in solution* and *with more complex* (and, arguably, more challenging) carbanions?

We have previously seen the important role played by solvent in determining carbon acid acidity in solution. The presence or absence of ion pairing, the nature of the counterion and structure of the ion pair(s), or the degree and nature of aggregation, are all factors intrinsic to carbanion chemistry in solution. These factors also play a major role in determining the stereochemical outcome of carbanion reactions and, thus, the structure of carbanions in solution.[37] In fact, many species of interest to us are carbanion-like rather than carbanions in the full sense of the word.

In the following section, we will consider the methods used to examine the stereochemistry of carbanions and the results of these studies.

3.2 PROBES OF CARBANION STEREOCHEMISTRY

The established method of studying the stereochemistry of carbanions involves trapping of the carbanion with electrophiles (CO_2, benzaldehyde, etc.) or by abstraction of an isotopic label from the solvent. The latter approach was developed by the groups of Cram,[1–3,38–40] Corey,[41–43] and Doering,[44,45] among others.[46] In this method, the ratio of the rate constants for isotopic exchange (k_{ex}) and for base-catalyzed racemization (k_α) for an optically active carbon acid in an alkaline medium that contains labile deuterium is determined from the product ratios for the two processes. As shown in equation 3.2, the horizontal series of steps that proceeds from carbon acid to deuterated carbon acid represents the isotopic exchange, while the vertical portion indicates racemization via inversion of the intermediate carbanion. In the first equilibrium, protonation of the carbanion by B–H (or BH_2^+) represents internal return (i.e., the return of the H^+ to the $R_1R_2R_3C^-$ before deuteration can occur).

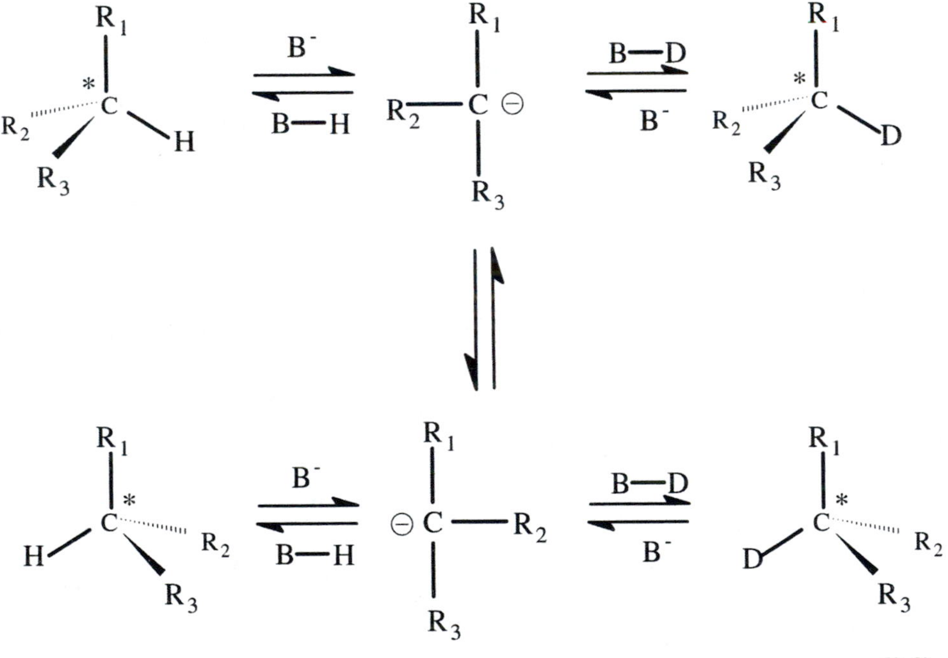

(3.2)

In this single-point kinetic treatment, the ratio, k_{ex}/k_α, can be classified as follows.[1,2] First, if the carbanion is pyramidal and the process of inversion (that leads to racemization) is much slower than isotopic exchange, k_{ex}/k_α will be a large number approaching infinity in the limit. This corresponds to complete retention of configuration; the carbon acid will retain its optical activity. (As we shall see, retention is also possible with a planar carbanion, under suitable conditions.) In practice, any k_{ex}/k_α ratio greater than unity is taken to indicate appreciable retention of configuration and, therefore, configurational stability of the carbanion. If the rates are approximately the same, k_{ex}/k_{α} will approach unity. Here, the inversion barrier is low and the pyramidal carbanion can convert to and from its invertomer at a rate similar to that at which it may be trapped as its deuterium-labeled carbon acid. Alternatively, the carbanion may, in fact, be planar *and* relatively free (i.e., not contact ion paired). Such a carbanion could exchange deuterium from either face and net racemization would result. The k_{ex}/k_{α} ratio would be approximately equal to 0.5 for complete inversion of configuration. A planar carbanion that is strongly ion paired may exchange from the opposite face, leading to inversion. Finally, if the ratio approaches zero then *isoracemization*, as it is termed by Cram,[47] is indicated. Isoracemization may involve transfer of deuterium (or hydrogen) by the base from the front to the back of the carbon acid (i.e., an intramolecular proton transfer), or, in the absence of an acid strong enough to deuterate the carbanion, the original hydrogen of the carbon acid may return to either face of the carbanion, resulting in formation of racemic carbon acid *without isotopic incorporation*. This can be considered to be a form of internal return (equation. 3.2);[48–50] Koch has reviewed the effect of internal return on primary kinetic isotopic effects in reactions that involve hydrogen-bonded carbanions.[51,52] (In Chapter 2, we considered internal return[53,54] and the tests for it with regard to the kinetic isotopic effect. Note that internal return does not appear to be a problem when water or deuterium oxide is the solvent.[55])

We shall now examine some H/D exchange results and the carbanion geometries that may be inferred from these results. First, we will briefly examine systems in which the carbanion would be expected to be pyramidal, but may be configurationally mobile, namely, the simple sp^3 systems (Section 3.2.1.1). Then, systems will be considered that would be expected to show greater configurational stability, as a result of an inherent factor built into their structure, that is, the alkenyl (vinyl) and cyclopropyl anions. Finally, planar carbanions will be examined through the lens of Cram's kinetic criteria (Section 3.2.1.3).

Following these kinetic results, we will consider the structure at the carbanionic center as deduced from spectroscopic data for selected systems (Section 3.2.2). Finally, we will examine in detail the geometry at the carbanionic center for carbanions that are adjacent to phosphorus- or sulfur-centered groups (Section 3.3).

3.2.1 Kinetic Methods

3.2.1.1 sp^3-Hybridized Systems

It is understandable that few simple alkyl carbanion systems have been studied by kinetic methods. In line with our understanding of the gas-phase inversion barrier for methide, for example (Table 3.1), we would expect such simple carbanions to invert much more rapidly than proton (deuteron) transfer could occur. Hence, a simple carbanion, if chiral, would racemize.

The simplest carbanion that retains its chirality is the bromochlorofluoromethide ion (^-CBrClF).[56,57] The bromochlorofluoromethide ion is formed in a haloform reaction, a reaction known to proceed via a carbanionic mechanism (equation 3.3).

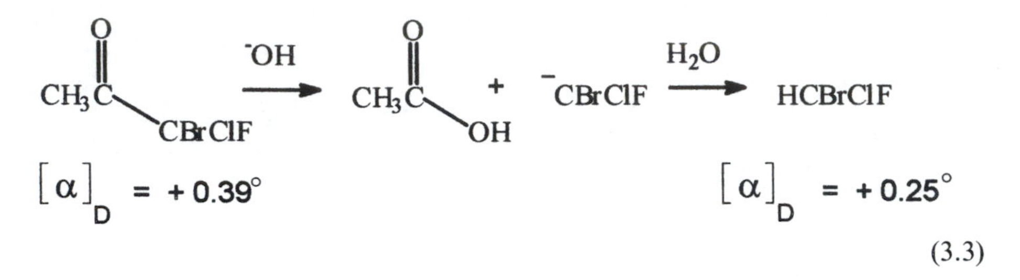

(3.3)

Protonation of the pyramidal and optically active anion leads to the optically active bromochlorofluoromethane (HCBrClF) and demonstrates that the barrier to inversion of the carbanion is high enough to permit the anion to be trapped with retention of chirality. The effect of α-substitution of the carbanionic center with inductively electron-withdrawing halogens is to increase the inversion barrier.[19] [See also the discussion of the effect of α-fluoro substitution on carbon acid acidity in Section 2.3.3.2.(a)]. This effect is also expressed for NF_3, which is isoelectronic with $^-CF_3$; the barrier to inversion here is about 50 kcal/mol (Table 3.1). The increase in the magnitude of the inversion barrier for these X(a,b,c) species may be linked to destabilization of the HOMO of the planar transition state.[58,59]

However, without this α-substitution by inductively electron-withdrawing halogens, simple "free" carbanions would be expected to invert rapidly. An example of the complexity found even in simple systems is offered by the case of the *exo*- and *endo*-2-norbornyl anions (equation 3.4).

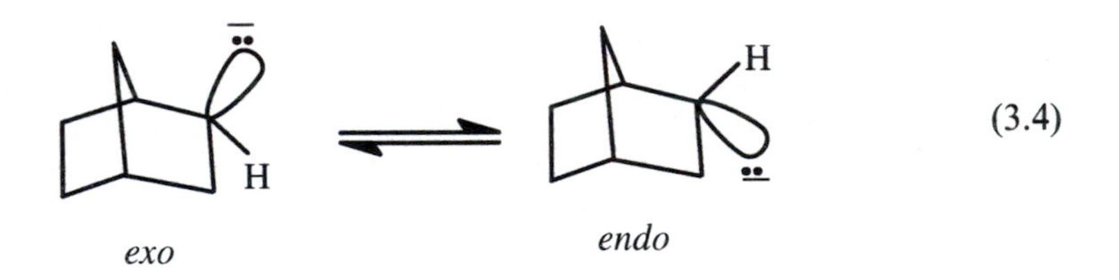

(3.4)

These were formed by oxidation of the respective *exo*- and *endo*-2-norbornyl hydrazines followed by treatment with NaOD, leading to loss of nitrogen from the intermediate norbornyldiimides in deuterium oxide. Regardless of the configuration of the starting hydrazine, *exo* deuteration predominates. This was taken to indicate rapid equilibration of the two free carbanions, favoring the *exo* carbanion. However, the intervention of asymmetrically hydrogen-bonded carbanions was also presumed to be important.[60] Therefore, for the reaction of the *exo*-hydrazine, ca. 87% *exo*-deuterated norbornane was formed along with 13% *endo* product; the latter product was argued to arise from the hydrogen-bonded carbanion pathway where deuterium capture occurs with inversion. Reaction of the *endo* hydrazine gave exclusively the *exo*-deuterated bicycloalkane, again reflecting the equilibrium between the two pyramidal anions that favors the *exo* carbanion, as well as delivery of the deuteron from the deuterium oxide medium to the *exo* face of the *endo* anion, accompanied by inversion.[60] This study illustrates the difficulties involved in extricating information in configurationally mobile systems.

3.2.1.2 Alkenyl Carbanions

Alkenyl carbanions, of which the vinyl anions are prototypical, would be expected to be formally sp^2 hybridized at the carbanionic center. The greater s-character involved in the C–H bond in the carbon acid and in the conjugate carbanion base might be expected not only to result in enhanced acidity for the carbon acid (e.g., ethene C–H acidity compared with the C–H acidity of ethane) but also to enhance configurational stability for the anion.

Deuteroxide-ion-catalyzed H/D exchange with *Z*- and *E*-1,2-dichloroethene has been shown to proceed without isomerization; that is, the intermediate vinyl carbanions, in which the carbanionic center is formally sp^2 hybridized, retain their geometry (e.g., equation 3.5 for the *E*- or *trans* stereoisomer).[61]

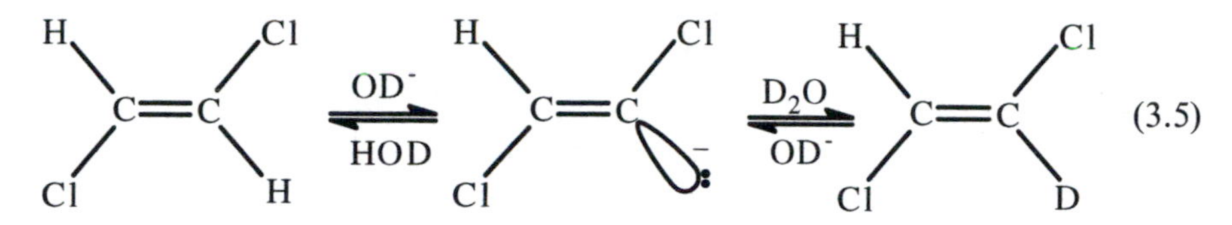

The barrier to isomerization of the 1,2-dichloroethenes was estimated to be 25–35 kcal/mol, while rotation about the double bond, via π-bond cleavage, would require 40–60 kcal/mol; isomerization through C–C bond rotation could be ruled out. Isomerization could conceivably proceed through an addition–elimination process in which deuteroxide would first add to the double bond to yield an alkyl carbanion that would undergo deuterium exchange and from which HOD would eliminate to give the deuterated alkene. Here, the carbanion is alkyl rather than vinyl in structure. However, it would be unlikely that such an isomerization process could proceed with the retention of configuration observed in the H/D exchange. It may be concluded that inversion of vinylic anions occurs via an in-plane wagging vibrational mode that involves a linear transition state.[61] These results are also consistent with trapping experiments that involve formation of the *cis* or *trans* 1-lithiopropene from 1-bromo-1-propene; the lithium carbanide has been shown to react with benzaldehyde with retention of configuration.[62]

Cram's kinetic criterion has been applied to analysis of the configurational stability of the vinyl anion derived from the chiral compound 2,2,4,6,6-pentamethylcyclohexylidenacetonitrile (Scheme 3.1). In this system the k_{ex}/k_{α} ratio was found to be 140 for isotopic exchange of the vinyl hydrogen as compared with the racemization rate for the carbanion (90 °C, $NaOCH_3/CH_3OH$).[63] Since exchange (i.e., trapping of the vinyl anion) occurs 140 times as rapidly as racemization, it can be concluded that the vinyl anion is formed with almost complete retention (99.3%). By way of comparison, the benzoyl vinyl analogue shows only limited retention of configuration (30%).[64] These observations can be rationalized if the second anion is essentially planar, as a result of efficient conjugative delocalization of the carbanionic charge into the carbonyl group, whereas the cyano group (as we have seen in our consideration of the acidity of carbon acids) is less

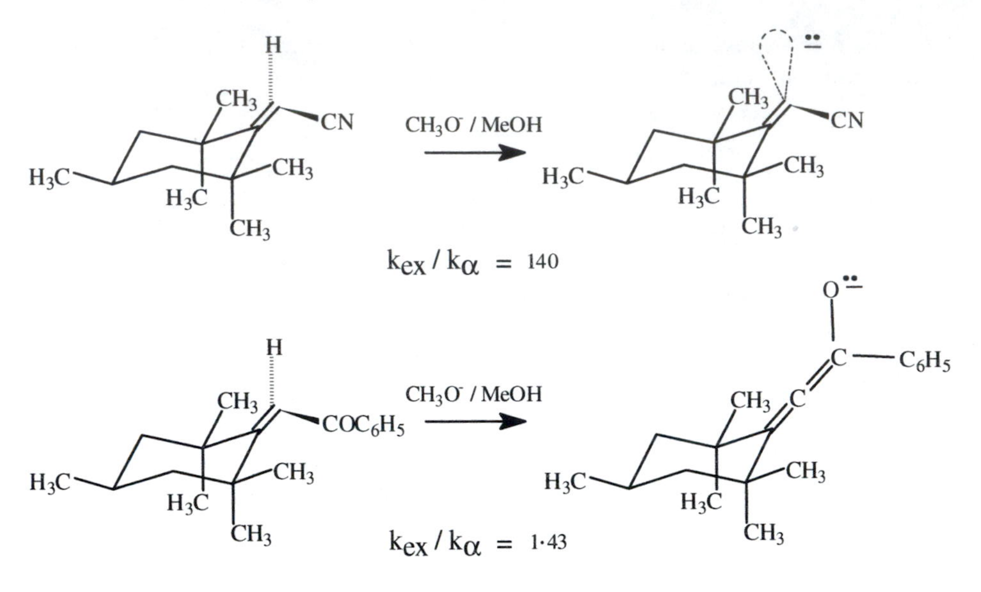

SCHEME 3.1

capable of delocalizing charge by conjugation and, so, of enforcing planarity on the carbanion.

Substituents on the α-carbon that stabilize the vinylic carbanion by an inductive mechanism will enhance configurational stability. A short list of vinyl carbanions (usually formed via equilibration with alkyllithium reagents) that retain configuration include those substituted at the carbanionic center by alkyl and arylthio (RS)[65–67] and arylseleno (RSe)[68] groups, as well as by alkyl groups.[69,70] Isocyanide substitution at the carbanionic center also induces configurational stability.[71] Conversely, α-substituents that delocalize the negative charge through pπ–pπ conjugation enhance stereoisomerism; carbonyl-containing substituents dominate this group.[64,72–74] These results are in qualitative agreement with calculated stabilization energies of the sp^2-type bent structure as compared with a linear form for vinyl carbanions α-substituted with H, CN, CO_2R, and CHO groups; the first two substituents strongly favor the bent form, the next favors the bent form (albeit less so), while only with the last is the linear form energetically preferred over the bent structure.[75]

Alkenyllithiums appear to show a greater tendency to retain configuration in solvents that encourage tight ion pairing. For example, *Z*-1,2-diphenylethene (*cis*-stilbene) reacts with the strong base lithium diisopropylamide (LDA) in benzene–pentane (1:1) to give a carbanion (or likely a tight ion-paired species or structured aggregate) that is quenched with benzophenone or carbon dioxide with complete retention of configuration. In a somewhat more ionizing medium (benzene–diethyl ether, 1:1), complete racemization is observed.[76] These results are supported in other alkenyl systems where more powerful ionizing solvents[77] or crown ether complexing agents were used (e.g. dibenzo[12]-crown-4 in diethyl ether–hexane, 4:1, at −75°C, where the stereochemistry was determined by trapping the lithium carbenide).[78] Clearly, the variable extent of aggregation and its role in controlling the stereochemistry found in these lithium alkenide systems has not been probed fully at this time.

Notwithstanding the importance of aggregation in determining configurational stability, electronegative substituents (which generally contain an atom with available lone pairs, i.e., R′X:) attached to the carbon β to the anionic center have been shown to stabilize a vinyl carbanion via internal complexation involving the lithium counterion, the anion, and the substituent, as shown by structure **22**. The carbanionic center in the cases studied was also stabilized by a suitable α-bonded electron-withdrawing group (EWG). Now, the *E*-stereoisomer is preferentially stabilized; this results in enhanced configurational stability for the *E*-carbanion and diminished configurational stability for the *Z*-stereoisomer.[79–81] An alternative mechanism of stabilization of alkenyl carbanions (or carbanion-like species) involves a stereoelectronic interaction of the carbanionic lone pair with the bond of the β-substituent (i.e., an n → σ* interaction).[82]

Substituents at the β-position of other types of carbanions, as in carbanions that are β-substituted by formamidine or amide groups,[83–85] also show significant configurational stability.[86] This is attributed[83–86] to the same type of internal ion pairing (i.e., : N $\cdots$ Li^+ $\cdots$ C^-) as proposed for the β-substituted alkenyl anions. Also, the term "dipole-stabilized anion" has been applied to these types of internally complexed carbanions.[85,87] These two factors may act synergistically to stabilize the carbanion, as illustrated for a general amide-substituted lithium carbanide, **23**.

It should be noted that nucleophilic displacement reactions of vinylic halides may proceed by either a concerted mechanism or one that involves carbanionic intermediates. In the case of vinyl fluoride displacements, the observation that carbanions are trapped by protonation by the medium[88] is coupled with the observation that *E*- and *Z*-alkenyl fluorides yield the same product; that is, racemization of the vinylic carbanions occurs.[89] In this context, retention of configuration with a number of vinyl chlorides and bromides has been taken to support a concerted mechanism.[90] These reactions have been reviewed by Rappoport[90] and, while significant in carbanion chemistry, they are beyond the scope of the current discussion of geometry at the carbanionic center.

3.2.1.3 Cyclopropyl Anions

In considering the configurational stability of cyclopropyl anions, it is tempting to posit that the cyclopropanides will be configurationally less stable than the vinylic

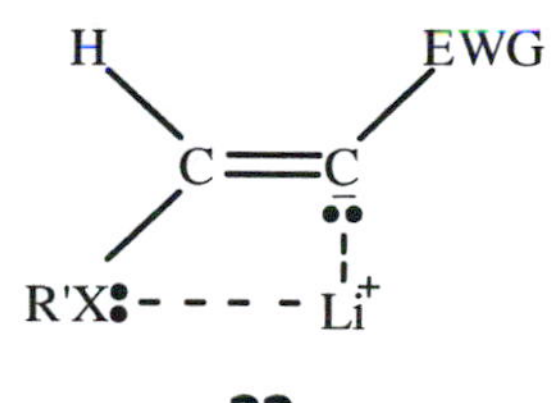

22

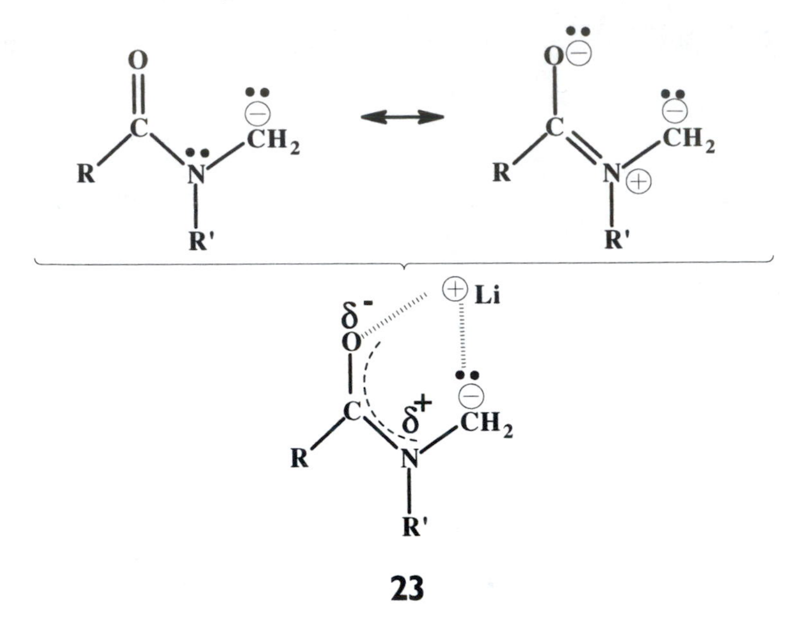

23

carbanions. This is expected because of the decreased s-character of the carbon orbital of the cyclopropyl carbanion as compared with vinyl carbanions (i.e., $sp^{2.28}$ vs sp^2, respectively).[28,63] The s-character of the C–H bond in a carbon acid is a significant factor in determining the acidity of hydrocarbons. Note, for example, that cubane, which utilizes an "s-rich" orbital in bonding to hydrogen, is estimated to be about 60,000-fold stronger as a carbon acid than cyclohexane,[91] and 1,3,5,7-tetranitrocubane has been reported to have a pK_a between 20.5 and 22.5 in THF.[92,93] These cubanides are, of course, constrained in their geometry (C–C–C angle ca. 90°) by the sigma-bonded framework.

Regardless of the hybridization factor, cyclopropyl anions are expected to have high barriers to inversion. Recall that in the energy profile for interconversion of pyramidal species, the transition state is planar (Figure 3.1); the bond angles, therefore, must *open* in going from the pyramidal (a–X–b angle approximately 109°) to the planar (a–X–b angle ca. 120°) form. In the cyclopropyl anion, however, the bond angle of the ring is fixed nominally at 60°. The consequence of this angle fixation is that although the carbanion contains significant angle strain (i.e., on the order of $109° - 60° = 49°$ deviation from the typical pyramidal carbanion bond angle), the planar transition state for inversion is associated with even higher angle strain (i.e., $120° - 60° = 60°$ deviation). We can anticipate a significantly higher barrier to inversion for cyclopropyl carbanions as compared with the prototypical methide ion, and recent calculations are in agreement with this view.[36] Moreoever, empirical inversion barriers for a variety of aziridines, which are isoelectronic with cyclopropyl anions, are significantly higher than those determined for methide anion; the barrier for aziridine itself is estimated to have a lower limit of 11.6 kcal/mol,[94] while even higher barriers have been measured for a number of *N*-alkyl-substituted aziridines.[94–96]

Experimentally, these systems have elicited a great deal of attention. The extensive research of the Walborsky group is particularly notable and it has been reviewed in a monograph.[97] A few examples suffice to illustrate the factors that affect the configurational stability of substituted cyclopropyl anions. In this context, 2,2-diphenylcyclopropyl cyanide reacts with methoxide ion in MeO**D** to give a large k_{ex}/k_{α} ratio (i.e., 913 at 90 °C) indicative of net retention of configuration under these conditions.[98] 1-Lithio- and 1-sodio-1-methyl-2,2-diphenylcyclopropanides have been shown to retain their configurations in aprotic solvents,[99,100] as do the lithium and sodium salts of 1-fluoro-, 1-chloro-, and 1-methoxyl-2,2-diphenylcyclopanes.[101] In contrast, 1-lithio-1-cyano-2,2-diphenylcyclopropanide[102] largely racemizes in aprotic solvents. Although the nature of the solvent, the counterion of the carbanide, and the presence or absence of cation-complexing agents can influence the degree of racemization or retention of configuration, particularly in systems where the carbanion may be planar, the nature of the substituent at the α-position is also a major factor in configurational stability. Therefore, in aprotic solvents, where a solvent-separated carbanion is favored, substituents that stabilize a carbanion by resonance would be expected to enhance planarity and, therefore, racemization. Interestingly, the chiral carbanions derived from (*S*)-phenyl-*N*-benzyl-, and [(*S*)-(1-phenylethyl)]-*N*-aziridin-2-carbothioates displayed complete retention of configuration upon deuteration (MeO**D**) or methylation (MeI/dimethylpropyleneurea). Presumably, here the tendency of the PhSCO substituent to delocalize charge by resonance (and, so, encourage planarity and racemization) is offset by the greater ring strain involved in aziridinyl carbanion inversion.[103]

Cyclopropyl carbanions may also be stabilized inductively and such stabilization does not come at the expense of the pyramidal geometry. As we have seen with the halogen-substituted methide ion (Section 3.2.1.1), α-substitution by halogens increases the barrier to inversion. While substituents like the halogens stabilize the carbanion, partly through inductive electron-withdrawal, substituents such as cyano and isocyano stabilize the carbanionic center primarily via a polarization mechanism; classical resonance delocalization makes a lesser contribution. Hence, the result is found that the 1-lithio-1-isocyano-2,2-diphenylcyclopropanide retains its configuration at low temperature (−72 °C) in the presence of cation-complexing agents [15-crown-5]; it was concluded that the isocyano group does not delocalize the carbanionic charge significantly via resonance.[104]

In considering cyclopropyl systems, however, it should be recognized that isotopic exchange and racemization are not the only possible pathways of reaction for some of these carbanions. When cyclopropanes that are more highly substituted with delocalizing groups (i.e., three phenyl groups, two phenyls and two carbomethoxyls, etc.) are treated with stronger bases [e.g., NaH in DMF, or lithium diisopropylamide (LDA) in THF] the initially formed cyclopropyl carbanions may undergo ring-opening rearrangement via the respective allylic carbanions.[105–107] Such carbanion rearrangements will be considered in Chapter 7.

In summary, vinyl and cyclopropyl carbanions *tend* to maintain their configurations once formed. These anions generally have high barriers to rotation or pyramidal inversion that would lead to racemization, although there are examples where substituents α to the carbanionic center enforce planarity or otherwise lower the inversion barrier. In the next part of this section, we shall consider carbanions that are generally considered planar, in light of Cram's kinetic criteria.

3.2.1.4 Fluorenyl Anion Systems

The 2-dimethylamide of 9-deuterio(or protio)-9-methylfluorene is an optically active carbon acid that would be expected to yield, upon abstraction of the deuteron (or proton) at the 9-position, a planar cyclopentadienide-type carbanion. In his classic study of the kinetics of exchange versus racemization of this planar carbanion, Cram highlighted the role of ion pairing, which is dependent upon the nature of the solvent.[108,109] Naively, with this substituted 9-methylfluorene (pK_a ca. 21), racemization may be expected to dominate because trapping of the planar carbanion by proton abstraction may occur with equal probability from either face (Scheme 3.2). In fact, racemization is observed when isotopic exchange is carried out with potassium *t*-butoxide in the parent alkanol or with tetramethylammonium phenoxide in a benzene:phenol (90:10) medium; here, k_{ex}/k_{α} is equal to unity.

With relatively low concentrations of amine bases (i.e., ammonia, propyl-, and dipropylamine) in THF or *t*-butyl alchol, however, the same substituted 9-methylfluorene system undergoes exchange with predominant retention of configuration ($k_{ex}/k_{\alpha} = 9$–150).[108,109] The salient point is that, in these low-dielectric-constant solvents, the initially formed carbanion exists as an ion pair and protonation by the counterion (e.g., ammonium ion) of the planar carbanion occurs primarily on a single face. This requires a rotation of the counterion to occur in the ion pair, that is, to rotate the previously abstracted deuteron away from the carbanionic center (so that internal return[51,52] does not occur) and simultaneously rotate a proton toward the carbanionic center. This pathway is illustrated in Scheme 3.3.

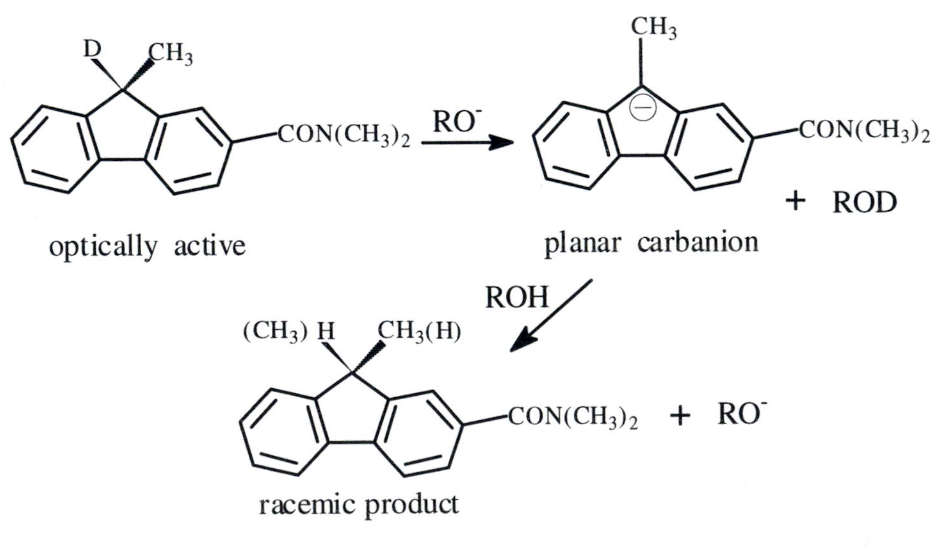

SCHEME 3.2

Evidence in partial support of the mechanism comes from an NMR study in which the hydrogen-bonded carbanion derived from 4-(9-fluorenyl)-2,3,4-trimethyl-2-pentanol was directly observed.[110] This hydrogen-bonded carbanion can, of course, be viewed as a model for the hydrogen-bonded ammonium ion–fluorenide ion pair postulated in Cram's scheme. (Note also the extensive work of H. Koch's group on the role of hydrogen-bonded carbanions versus free carbanions in dehydrohalogenation reactions.[51–54])

Because the process (Scheme 3.3) requires an ion pair (e.g., substituted 9-methylfluorenyl anion···NH_3D^+) racemization should again be observed if the solvent is changed from a low-dielectric-constant solvent like THF to a cation-

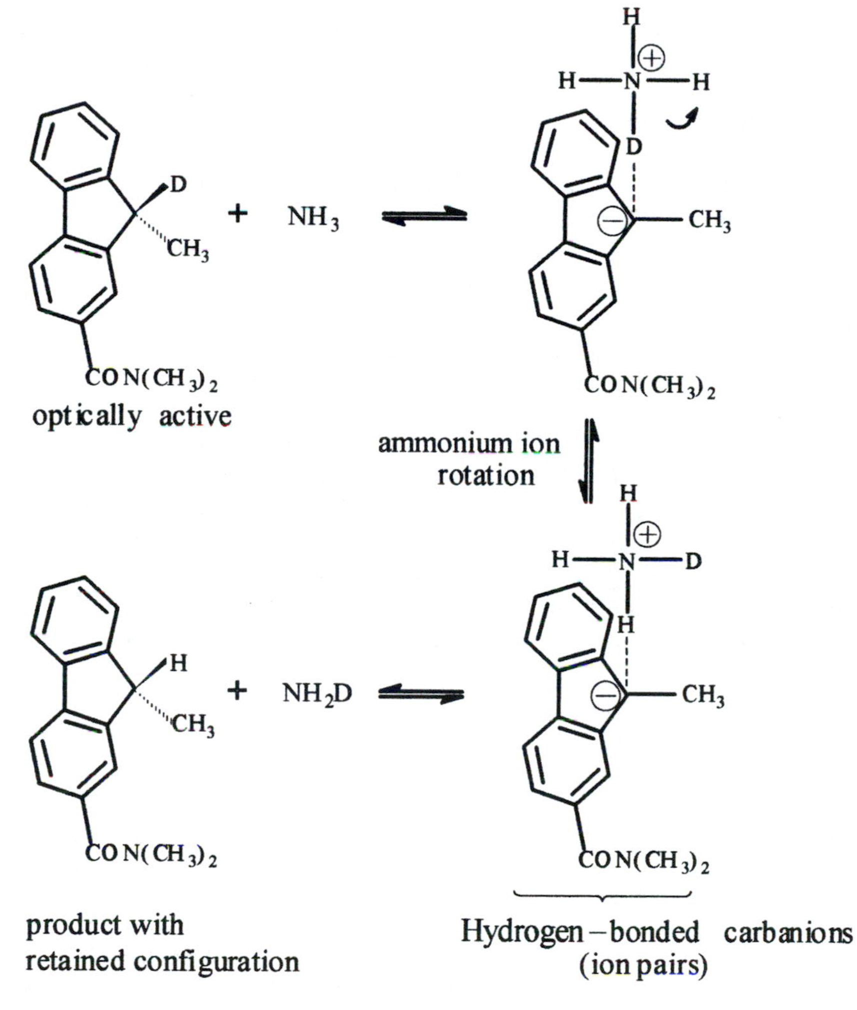

SCHEME 3.3

solvating medium like dimethyl sulfoxide. As anticipated, then, ammonia in DMSO effects exchange with the 2-dimethylamide of 9-deutero-9-methylfluorene to give racemized product ($k_{ex}/k_{\alpha} = 1.0$). In essence, here the ion pairs dissociate faster than the process of rotation of the deuteroammonium ion can occur and the free ion may be trapped from either face to yield the racemic product.

Racemization is linked to the rate of proton transfer in these systems and, as a first approximation, the rate of proton transfer parallels the difference in acidity (ΔpK_a) between carbanion and proton donor acid. (However, see Section 2.2, which focuses on the intrinsic reactivity of carbanions.) In this regard, the 2-dimethylamido-9-deutero-9-methylfluorene has been shown to exchange in the THF–propylamine system with majority retention of configuration ($k_{ex}/k_{\alpha} > 56$), while the corresponding 2-dimethylamido-9-deutero-9-methyl-6-nitrofluorene, *in the same alkaline medium*, exchanges with complete racemization ($k_{ex}/k_a = 1$).[111] The difference in pK_a between the latter fluorene derivative and the former is about 3; the ΔpK_a for the propylammonium ion and the 2-dimethylamido-9-methylfluorenyl carbanion is ca. 11, while the corresponding pK_a difference drops to 8 with the 6-nitro-substituted fluorenyl anion. On the basis of this and comparison with other systems Cram concluded that a pK_a difference of around 10 is required in these systems for the carbanion to capture a proton fast enough to retain configuration.[111]

In general, the observation of ratios of rate constants for isotopic exchange to racemization with a value of 0.5 indicates that complete inversion has occurred. However, it is more typical to find values for k_{ex}/k_{α} that fall between 0.5 and 1. These values suggest that there is a degree of inversion in competition with racemization, or a degree of isoracemization (of which internal return is one mechanism) in competition with retention or racemization. Clearly, these intermediate rate constant ratios require careful interpretation to have any utility in our understanding of the stereochemistry involved.

In the case of the 2-dimethylamido fluorene derivative that we have been considering, a value for k_{ex}/k_{α} of 0.65 is found when the isotopic exchange is carried out at 75 °C using tri-*n*-propylamine in methanol. This value actually rises to 0.78 when aqueous methanol (25% H_2O) is used as the medium. Both results, however, may be understood in terms of a carbanion that is *asymmetrically solvated.* Therefore, proton donation to give the product occurs primarily from the face opposite to the one where proton abstraction (and preferential solvation) happens.[109,111] Competitive racemization raises the ratio in each case and it likely proceeds, as before, through a symmetrically hydrogen-bonded ion pair.

As we have seen, isoracemization is a process wherein racemization occurs faster than isotopic exchange, where the isotopic donor solvent pool is present in excess, and it is identified by a k_{ex}/k_{α} ratio of less than 0.5 (and, in the extreme, zero). Again, various mixtures of competing processes may exist, as indicated by values that fall between zero and 0.5. Interesting examples of isoracemization have been reported by Cram and co-workers for 2-deutero-2-phenylbutanenitrile and, more convincingly, for 2-dimethylamido-9-deutero-9-methyl-6-nitrofluorene;[108,109] these involve the "conducted-tour mechanism".[112] Thus, $k_{ex}/k_{\alpha} = 0.05$–0.2 for 2-deutero-2-phenylbutanenitrile (in solvent media ranging from THF–1.5 M *t*-butyl alcohol–0.1 M tetra(*n*-butyl)ammonium iodide to *t*-butyl alcohol) and 0.1 for the fluorene derivative (THF–

1.5 M *t*-butyl alcohol–0.1 mM tetrapropylammonium iodide), where tri-*n*-propylamine acted as the base in all reaction systems examined.[108,109] Scheme 3.4 shows the conducted-tour mechanism for the 9-nitrofluorene derivative.

In this scheme, the nitro group acts as the "conductor of the tour" in that, after the deuterium is abstracted by the tri-*n*-propylamine, the initially formed hydrogen-bonded carbanion partly dissociates and then re-forms its hydrogen-bond to the nitro substituent of the ring (shown as a canonical form that bears extra charge delocalized from the carbanion). A new hydrogen-bond is then established via the other face of the planar carbanion and the deuterium is transferred back to the carbanion (Scheme 3.4), generating the enantiomeric fluorene derivative. Since dissociation of any of the ion pairs could occur and, consequently, the carbanion could become trapped by proton abstraction, racemization is a minor competitive pathway.

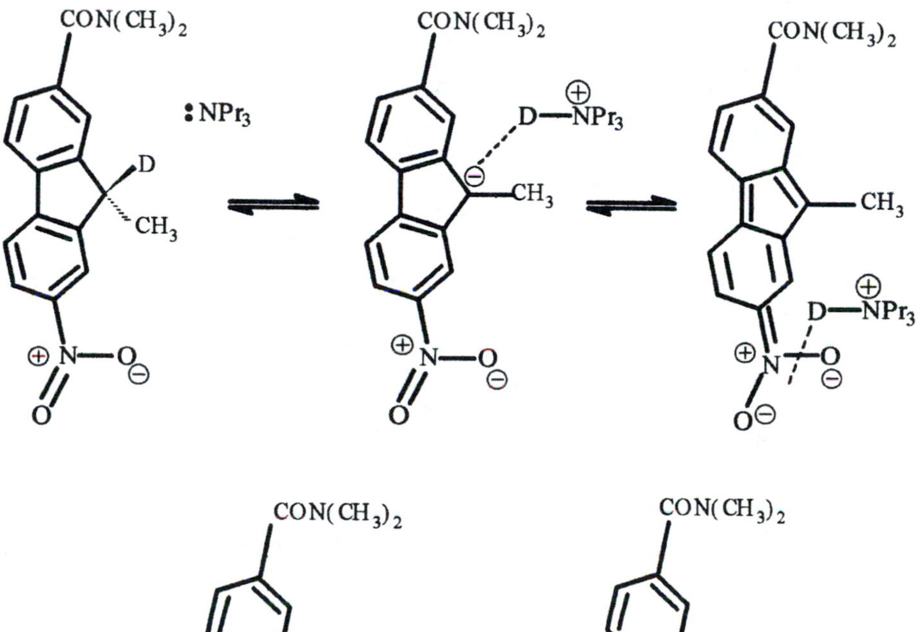

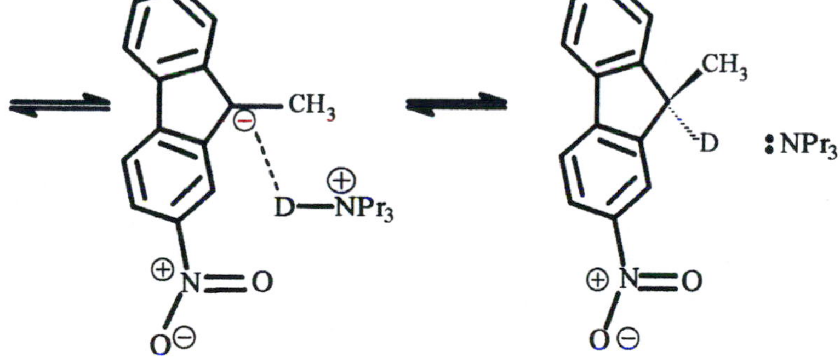

SCHEME 3.4

These kinetic approaches have been successfully applied in a number of systems.[113–115] It is apparent that care must be taken in assessing the various factors that account for the k_{ex}/k_{α} value in a given reaction system. Solvent, ion pairing, and the nature of the expected carbanion (planar or pyramidal) are all important factors in determining the stereochemical outcome even of an apparently simple process such as isotopic exchange. Planar carbanions are not restricted to racemization as a stereochemical consequence of their planarity if asymmetrical solvation or ion pairing operates. On the other hand, rapidly inverting pyramidal carbanions may show complete racemization.

The next section explores the spectroscopic methods that may be employed to complement kinetic approaches to assessing the stereochemistry at the carbanionic center.

3.2.2 Spectroscopic Methods

Infrared spectrophotometry[116] as well as nuclear magnetic resonance spectroscopy[117] have been harnessed to the study of the problem of the geometry of carbanionic centers. In this section, we will only consider the application of these spectroscopic techniques to the stereochemistry at the carbanionic site. More general discussion of the spectroscopic characteristics of these species (using a consistent set of resonance-stabilized carbanides), including analyses of ion pairing and/or aggregation, will be presented in Chapter 4.

Electronic spectra of carbanions have proven useful in assessing the nature and degree of ion pairing, particularly with delocalized carbanions.[118] To the extent that comparison of UV–vis spectra of localized carbanions and structurally related anions that are delocalized through conjugation serves to identify the conjugated ions, it is a useful technique to probe the geometry at the carbanionic center. In other words, UV–vis spectroscopy can probe the planarity of the carbanionic center. On the other hand, the method *is* limited to examination of conjugated carbanionic systems where a significant degree of coplanarity of the α-substituents with the carbanionic center is already expected. As will become apparent in Chapter 4, electronic absorption spectroscopy is of more value in assessing the degree and nature of ion pairing of carbanions in solution. Consequently, the remainder of this section will focus on IR and NMR spectroscopic investigations of carbanion stereochemistry.

3.2.2.1 Infrared Spectroscopic Studies

In principle, infrared spectroscopy can offer a full description of the vibrations of the bonds in a carbanion, as well as the force constants for all bonds, including that for the metal-carbanionic center. In fact, assignment of the various bands in IR is nontrivial and, for carbanions and carbanion-like compounds, may require isotopic labeling, comparison with precursors and model compounds, and, importantly, comparison with calculated vibrational frequencies.[116] To quote from one comprehensive review: "The structure of the molecular or ionic species is both the starting and end point of the spectral assignment. The assumed geometrical structure is first used to calculate the number of the vibrational modes and to derive the selection rules to be compared with observed spectra. When a complete assignment of the spectra has been achieved, the geometrical structure is used to... calculate the force field of the entity ... which give information on the electronic structure of the molecule and on the bond lengths and hybridization angles."[119]

An example of the difficulties involved in the assignment of a structure is that of the salts of nitromethane and other nitroalkanes. Initial analyses by Feuer[120] and Jonathan[121] of the IR spectra of sodium nitrocarbanides favored a planar structure at the carbanionic center with the negative charge dispersed over the α-nitro group; bands in the regions 1605–1587, 1316–1225, 1175–1149, and 734–700 cm^{-1} were ascribed to the C=N vibrational mode, the antisymmetric NO_2 stretch, the symmetric NO_2 stretch, and an NO_2 bending mode, respectively. This assignment was challenged[122] after consideration of the CH_2 vibrational modes in sodium nitromethide; a structure in which the carbanionic center was pyramidal was favored here. Raman spectroscopy of the related sodium 2-nitropropanide anion in water reaffirms the planar geometry at the carbanionic center of carbanions that are α-substituted by nitro groups.[123] It should be noted that these difficulties in making full and unambiguous assignments of the various vibrational modes in these and other carbanions likely account for the relatively few IR studies of carbanions extant.

Infrared spectroscopic studies of the carbanions of simple alkanenitriles has been hampered by dimerization, as in the case of ethanenitrile (acetonitrile).[124] However, an IR spectroscopic investigation of the ion-paired carbanions of ethanenitrile, propanenitrile, 2-methylpropanenitrile (dimethylacetonitrile), cyanocyclopropane, cyanocyclobutane, cyanocyclopentane, and cyanocyclohexane (lithium, sodium, and potassium salts) has been reported; here, hexamethylphosphoramide (HMPA), a powerful cation-solvating solvent, was used[125] and, consequently, the interference due to dimerization is apparently suppressed. The carbanide spectra were assigned partly by comparison of the CN vibrational mode observed with the aid of ab initio (3-21G basis set) force field calculations for the *free* carbanions. Since the agreement between the experimental vibrational bands and the calculated ones was good, it was concluded that the calculated gas-phase structure for each carbanion would also be valid in solution. For these carbanions, with the exception of the cyanocyclopropanide ion, the carbanionic center was found to be planar.[125] In accord with recent ab initio calculations, the cyanocyclopropyl carbanion is pyramidal at the carbanionic center, although the calculations show a lower barrier to inversion for the cyanocyclopropyl anion than for the unsubstituted cyclopropanide.[36] Similarly, in a companion IR and calculational study, it was found that the carbanides of 2-, 3-, and 4-cyanophenylethanenitrile (Li, Na, and K salts in DMSO) were planar.[126] However, this should not be taken to indicate that the α-CN group is wholly responsible for imposing planarity via conjugation with the carbanionic center. Examination of the IR spectra, together with ^{13}C NMR studies, of the potassium and lithium carbanides of phenylethanenitrile and 3-pyridylethanenitrile (in THF with 2.2.2 cryptand, THF–HMPA, or DMSO media) support the view that the carbanionic center in these salts is planar, but that this planarity is largely the result of con-

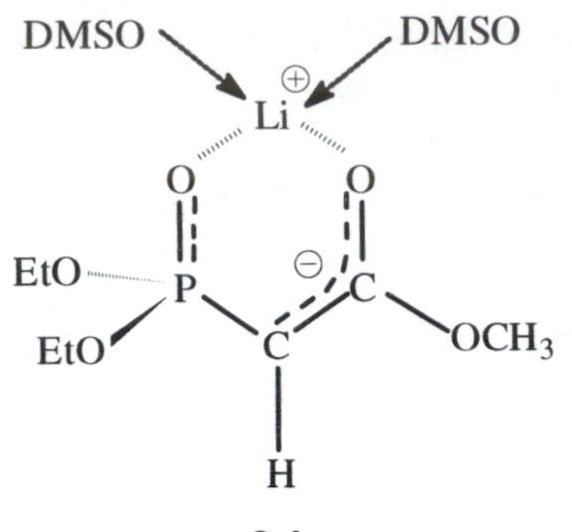

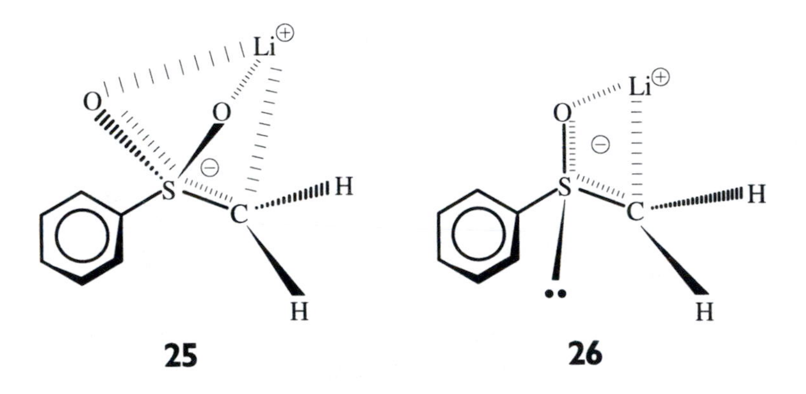

jugative delocalization into the aromatic/heteroaromatic ring rather than into the α-CN function.[127]

As might be expected, the IR spectra (solid state) of enolate salts formed from 2,4-pentanedione (acetylacetone) are in accord with a planar structure, where the counterion is chelated to the two carbonyl oxygens.[128] As far as structure of the carbanion is concerned, this situation likely also exists in solution.[129]

Formation of the sodium salts of diethoxyphosphonylacetylmethane [$(EtO)_2$ $POCH_2COCH_3$] and dimethoxyphosphonylacetylmethane [$(MeO)_2POCH_2$ $COCH_3$] leads to loss of the C=O and P=O stretching vibrations and the appearance of new bands at 1530, 1415, and 1170 cm^{-1} in the IR spectrum recorded in the solid state.[130] The new bands for the carbanides were assigned to P···O, and coupled C···O and C···C stretching vibrations, respectively; this assignment is consistent with a planar geometry at the carbanion center and largely reflects chelation of the counterion. Similarly, the lithium carbanide formed from diethyl[(carbomethoxy)methyl]phosphonate [$(EtO)_2POCH_2COOCH_3$] has been shown to be planar (DMSO solvent), reflecting the chelation of the lithium ion (**24**), as well as stabilization by complexation with the solvent.[131,132] Further IR spectroscopic data concerning the geometry at the carbanionic center in α-phosphonyl-substituted carbanions will be presented below along with relevant NMR spectroscopic information.

The structure of carbanions substituted at the α-position by thio, sulfinyl, and sulfonyl groups has been an area of intense investigation, utilizing computational, kinetic, and spectroscopic tools.[133,134] As with other carbanion systems, few IR spectroscopic studies have been reported. However, IR spectroscopic examination of the lithium and potassium carbanide salts of phenyl methyl sulfone (**25**), sulfoxide (**26**, complexing solvent molecules omitted for simplicity), and sulfide (in THF) led to the conclusion that the first two are planar about the carbanionic center; chelated structures with lithium ion were proposed.[135] On the other hand, the phenylthiomethide ion was found to be pyramidal. These conclusions will be considered briefly in Section 3.2.2.2, with respect to the NMR spectroscopic evidence, and in detail in Section 3.3.

3.2.2.2 Nuclear Magnetic Resonance Spectroscopic Studies

A variety of techniques may be used in NMR spectroscopic studies to probe the stereochemistry at the carbanionic center. In theory, in proton NMR the stereochemistry at the carbanionic center may be determined by analysis of vicinal H–H couplings. For a system in which inversion is rapid on the NMR timescale, the *cis* and *trans* vicinal coupling constants for model compounds may be compared with the observed coupling constants for the carbanion and, therefore, the relative conformer populations estimated. Inversion may be slow or may be slowed on the NMR timescale by lowering the temperature to the point where the individual peaks of the two invertomers can be observed and the two concentrations determined directly by integration of the peaks.[117] In the case of 2-methylbutyllithium, the low-temperature spectrum displays α-proton signals as AB of an ABX pattern, but as the temperature was increased this spin system collapsed to A_2 of the corresponding A_2X pattern [in solvents such as pentane, diethyl ether, and toluene, or in pentane with added tetramethylenediamine (TMEDA)].[136] In short, at the higher temperature the two α-protons become equivalent and this could be interpreted as being due to rapid inversion of the two pyramidal lithium carbanides (equation 3.6, where the relevant protons are labeled A, B, and X). Similar results were found for neohexyllithium under the same conditions.[137]

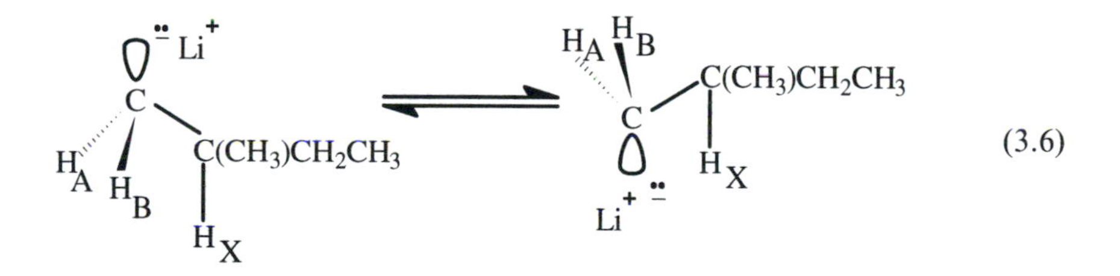

(3.6)

On the other hand, the polymeric/aggregate nature of such simple alkyllithiums makes definitive assignment of the microscopic processes that are associated with the spectroscopic parameters difficult. Thus, from the temperature dependence of the ^{1}H NMR spectrum of 2-methylbutyllithium, rates of inversion were determined;[138,139] inversion barriers ($\Delta G^{\ddagger}$) ranged from 14.9 kcal/mol in diethyl ether to 16.5 kcal/mol in toluene.[137] The increased barrier in toluene may reflect the "tighter" structure of the ion pair or aggregate in this solvent as compared with ether. However, in a more complete lineshape analysis[140] of the same system, the averaging of the α-H signals was attributed to a combination of a fast process associated with interaggregate exchange and a slower process ascribed to pyramidal inversion; at 8 °C, the free energy of activation for inversion for (2*R*)-2-methylbutyllithium in pentane was estimated to be 15.8 kcal/mol.[141] Clearly, dissection of the factors influencing lineshape, resolution, and assignment of signals is fraught with difficulties; the attribution of derived energetics is equally challenging.

The temperature dependence of the ^{1}H and ^{13}C NMR spectra of arylmethyl anions may permit estimation of the barriers for rotation about the carbanionic–aryl C–C bond with barriers of 14–18 kcal/mol having been reported from various benzylic anions (in THF[142] or isooctane-TMEDA[143]) and 18.7–22.7 kcal/mol for pyridyl-methyl and -ethyl anions (Li salts in THF[144] or the Li salt in tetraglyme, the Na salt in THF and the Na salt in THF/tetraglyme[145]). One interpretation is that such barriers to rotation arise exclusively from conjugative delocalization of carbanionic charge into the aryl ring with concomitant increase in double-bond character of the C–C bond; a high barrier to rotation would imply an approach to planarity at the carbanionic center, in this case. In line with this interpretation, the barrier to rotation about the carbanionic–phenyl C–C bond in the lithium and potassium salts of diphenylmethane in THF or DME has been similarly determined to be ca. 11 kcal/mol,[146] which is similar to the values for the activation barrier to carbanionic–phenyl C–C rotation in the 2-, 3-, and 4-pyridylbenzyl anions (Li, Na, and K counterions in THF).[147] However, the barrier to rotation about the carbanionic–pyridyl C–C bond of the lithium salts was found to decrease in the order 4-pyridyl, 2-pyridyl, 3-pyridyl for these lithium carbanides in THF (19, 14.8, 14.5 kcal/mol, respectively). It may be concluded that charge is preferentially delocalized into the pyridyl rings relative to the phenyl rings and the carbanion is essentially planar.[147] On the other hand, it must be emphasized that we are not considering free ions in any of the foregoing examples and that C–C bond rotation must also involve rearrangement of the ion pair or aggregate structure. This is particularly true of the pyridyl systems where specific complexation of the cation to the pyridine ring nitrogen no doubt also occurs. Thus, the magnitudes of the barriers measured as a function of changing solvent and counterion also functions as a probe of the medium.

^{13}C NMR spectroscopy offers advantages in the examination of carbanions. First, in ^{13}C NMR spectroscopy the carbanionic center can be observed directly. Secondly, the scale of carbon chemical shifts is about a factor of 20 larger than the comparable scale of proton chemical shifts and, therefore, small changes in electron density at a carbon site may exhibit larger changes in carbon chemical shift than in proton shift.

It may be argued that from the magnitude of the one-bond ^{13}C–^{1}H coupling constants ($^{1}J_{C\text{–}H}$), the hybridization of the carbanionic center could be determined. In the work of Karplus, Grant, and Lichtman, $^{1}J_{C\text{–}H}$ at a carbon was shown to be related to a term that includes the square of the s-character (s^2) of the

C–H bond and the cube of the effective nuclear charge (Z^3), which reflects C–H covalency.[148,149] If a comparable dependency on hybridization is assumed for the carbanion (regardless of degree of ion pairing), then the difference in C–H coupling constant for the carbanion, as compared with the neutral carbon acid, should be associated with the change in s-character at the position. Since the starting carbon acids are usually sp^3 hybridized and a pyramidal carbanion is formally sp^3 hybridized, only a small change in the C–H coupling constant would be expected. Conversely, a large change would be expected if the carbanion is approximately planar and the carbanionic center is nominally sp^2 hybridized. Partly on the basis of such an analysis, Waack and co-workers found the carbanionic carbon of benzyllithium to be essentially sp^3 hybridized.[150] The hybridization was suggested to change in the direction of greater s-character as the solvent was modified from benzene to THF. Similar studies on the simple lithium carbanides, methyllithium, *tert*-butyllithium, and *n*-butyllithium, in ethereal solvents, showed all three to be approximately sp^3 hybridized.[151] Importantly, an examination of the small ^{13}C–^{6}Li coupling constant in benzyllithium and its analogues (ca. 2.5–4 Hz) suggests that the hybridization of benzyllithium is intermediate between sp^3 and sp^2 with partial delocalization into the benzene ring and with the lithium counterion above the benzylic carbon.[152] Interestingly, this view of the structure of benzyllithium in ethereal solution is in agreement with its solid-state structure, as determined from X-ray crystallographic data.[153,154]

Bradamante and Pagani and their co-workers[155] have applied the following equation[156] to the analysis of $^1J_{C\text{–}H}$:

$$(0.2)\Delta^1 J_{C-H} = \Delta(\%s) \tag{3.7}$$

where the $\Delta^1 J_{C\text{–}H}$ represents the change in coupling constant at the carbanionic center for the carbanion as compared with the same carbon in the neutral carbon acid and %s is the percentage s-character. From the equation, it is apparent that complete rehybridization from a pure sp^3 to a trigonal planar carbanion that is sp^2 hybridized (i.e., $\%\ s = 33 - 25 = 8$) should give rise to a change in the coupling constant of 40 Hz. (This assumes that the initial carbon acid is perfectly sp^3 hybridized, whereas only approximate sp^3 hybridization is more likely to obtain.[32,33]) The values found for a series of symmetrically (X_2CH^-) and asymmetrically ($XYCH^-$) substituted carbanions (sodium salts; X = Y = Ph, $CONMe_2$, CO_2Me, COMe, COPh, CN, SO_2Me, SOPh, $PO(OEt)_2$; X = CO_2Me, Y = COMe, SOPh, SO_2Ph, $PO(OEt)_2$ in DMSO where the carbanions would be expected to be approximately "free") all fall within a range of 13–30 Hz (Table 3.2).[155] The less-than-theoretical value in $\Delta^1 J_{C\text{–}H}$ was attributed to an excess of π negative charge that typically lowers the $^1J_{C\text{–}H}$ at planar carbons.[157] Importantly, the ^{13}C chemical shifts and C–H coupling constants were found to be virtually the same in methanol or DMSO. It was concluded that all of the carbanions examined were planar.[155] Comparable values of $\Delta^1 J_{C\text{–}H}$ were also taken to indicate significant planarity at the benzylic carbon in sodium benzyl carbanides α-substituted at the benzylic position by groups including Ph, COPh, NO_2, CN, SOPh, SO_2Ph, and $PO(OEt)_2$ (i.e., X = Ph, Y = substituent; Table 3.2).[158] In the case of strong conjugative (pπ–pπ) electron-withdrawing groups such as nitro, it is reasonable that the carbanion would be planar. The conclusion

Table 3.2. Change in $^{13}C-^{1}H$ coupling constants (DMSO solvent) for selected sodium carbanides, $XYCH^{-a}$

X	Y	$\Delta^{1}J_{C-H}{}^{a}$
Ph	Ph	20.3
$CONMe_2$	$CONMe_2$	20.2
CO_2Me	CO_2Me	25.3
COPh	COPh	16.2
SO_2Me	SO_2Me	30.7
SOPh	SOPh	18.2
CO_2Me	COMe	23.7
CO_2Et	SOPh	27.0
Ph	$CONMe_2$	21.3
Ph	NO_2	29.2
Ph	CN	12.6
Ph	SOMe	31.9
Ph	SO_2Me	23
Ph	SOPh	26
Ph	SO_2Ph	20.7
Ph	$PO(OEt)_s$	20

SOURCE: Data taken from references 155 and 158.
^{a}Absolute value of the difference in $^{1}J_{C-H}$ for the carbanion as compared with the neutral carbon acid.

of carbanion planarity will be examined further below for the α-phosphorus and -sulfur group substituted carbanions.

In summary, on the basis of coupling constant analysis, benzyllithium has a carbanionic center of intermediate hybridization,[152,140,150] while addition of further conjugatively delocalizing substituents (e.g.,NO_2, COR, COOR, etc.) to the carbanionic center tends to make the center more nearly planar.[158]

The ambiguities associated with coupling constant analysis for carbanions have led some authors to suggest that only a thorough examination of the ^{13}C chemical shifts can reveal the structure of the carbanion,[159] or that a combination of analysis of coupling constants and chemical shift are required in order to assign the geometry of a carbanionic center with confidence.[160-162] In this context, the downfield shift of the benzylic carbon that is observed upon formation of the lithium carbanides of triphenylmethane, diphenylmethane, and toluene may be taken to indicate a degree of delocalization of charge into the phenyl rings.[163] In fact, two major opposing effects can be envisaged: (1) shielding, caused by the negative charge on the carbon in the carbanion, leads to an upfield shift and (2) deshielding of the carbon, caused by rehybridization, leads to a downfield shift. On the other hand, the chemical shift may also be influenced by the type and degree of ion pairing and degree of covalency at the carbanionic center,[164,165] as will be explored further in the next chapter, using the arylmethyllithium family as an appropriate general model. Thus, assessment of the geometry at the carbanionic center requires examination of the influence of all of these factors on the spectroscopic data, combined with kinetic and calculation data. Typically, a series of structurally related compounds must be studied in order to answer the simple question: Is the carbanion center planar or pyramidal?

3.3 CHIRALITY OF CARBANIONS STABILIZED BY PHOSPHORUS- AND SULFUR-CENTERED GROUPS

3.3.1 Kinetic Isotopic Exchange Studies

Both pK_a measurements[166] and relative rate comparisons[55,166–168] support the view that sulfur- and phosphorus-containing functionalities attached to the acidic carbon site enhance the acidity of the carbon acid and may stabilize the resultant carbanion (cf. Section 2.3.3.3, but see also reference 169). As but one further indication of the ease of formation of carbanions from such carbon acids, it has been reported that when methyl phenyl sulfone is treated with butyllithium in THF (0°C) it reacts with a range of bromoalkanes and bromoalkenes to give dialkylated product as well as the expected monoalkylated sulfone.[170] Generalized structures for some of the sulfur- and phosphorus-substituted systems that have been studied are shown (structures **27–37**). Cram's criteria form the starting point for our discussion of the structure and potential chirality of the carbanions formed from structures **27–37**. In this regard, predominant retention of configuration has been found in H/D exchange (usually using NaOD-D_2O, *t*-BuOK-*t*-BuOD, or Me_4NOD-*t*-BuOD) for sulfonium salts (ylides, **27**),[171–173] sulfoxides (**29**),[174–179] and sulfones (**30**)[49,180–183] and for sulfonates (**31**), sulfonamides (**32**), and phosphonates (**36**), with the degree of retention varying with the base–solvent system employed, for reasons outlined earlier.[46,182] However, as we have found, it is often difficult to connect clearly the results of isotopic exchange studies to the geometry at the carbanionic center. A carbanion that would reasonably be expected to be planar may yet show retention of configuration as a result of ion-pairing effects or asymmetric solvation. On the other hand, a pyramidal chiral carbanion may not show retention as a result of rapid inversion. Further, isotopic exchange results may be complicated by internal return, as we have discussed previously, and this is illustrated by the work of the H. Koch group.[51–54] Finally, it is apparent that the substituents on the carbanionic center (other than the P- or S-groups under consideration, i.e., Y or Z) may impose either planarity or pyramidality on the carbanion.

3.3.2 Bonding in Sulfonium and Phosphonium Ylides: Structural Consequences at the Carbanion Center

What geometry would we expect for α-P(P^+, PO, PO_2, etc.)- and α-S(S^+, SO, SO_2, etc.)-carbanions in the absence of ion pairing or structural modification associated with the nature of Y and Z (structures **27–37**)? More to the point, what is the nature of bonding between the α-substituent and the carbanion? To answer these

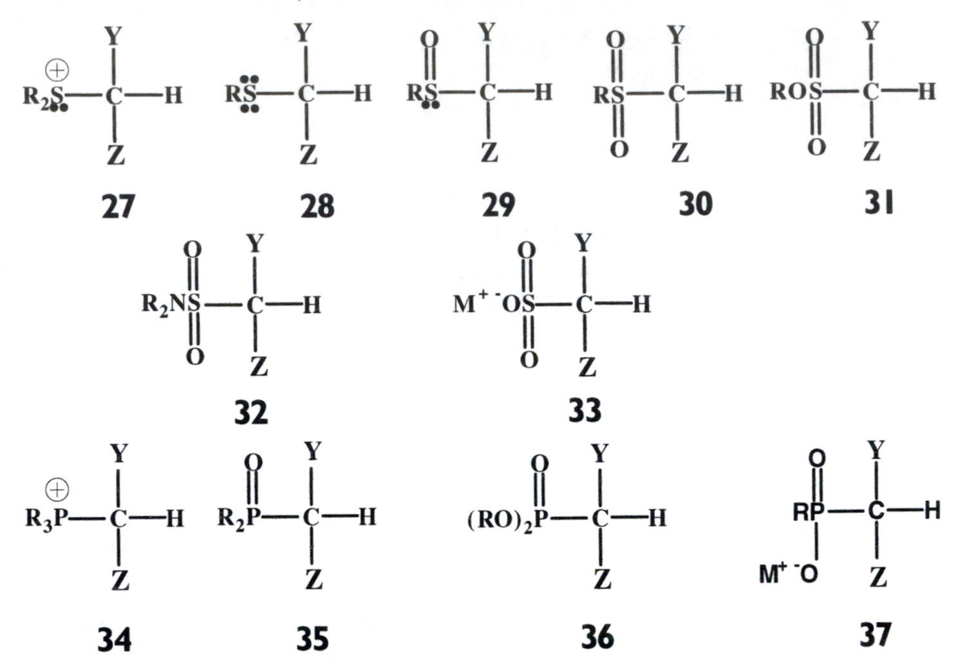

questions we will first consider the sulfonium and phosphonium ylides that arise from hydrogen-abstraction by base from structures **27** and **34**, respectively. In examining these ylides, it should be remembered that carbanions formed by phosphonates (**36**) react in the Horner[184,185] (or Horner–Emmons or Wadsworth–Emmons[186,187]) modification of the Wittig[188] reaction with carbonyl compounds to give the same alkenes that would result from the Wittig reaction with the corresponding phosphoranes.[189] Similarly, a sulfur ylide may be formed from either structure **27** or **29** and, although there are slight differences in their chemistry, both act as CR_2 transfer agents.[190]

In the past, ylides derived from structures **27** and **34** have been described as the resonance hybrids of two canonical forms (**27a,b**) and (**34a,b**). However, the resonance forms **27b** and **34b** have been taken to represent pπ–dπ overlap, where p-orbital electron density is donated into low-lying virtual d-orbitals of the sulfur or phosphorus heteroatoms. Such structures would further seem to indicate that the carbanion-like centers of ylides should be planar. A number of calculational studies have probed the question of whether such an interaction is significant or whether it even exists. The best current evidence is that resonance forms such as **27b** and **34b** do not contribute significantly to the stability of these ylides (and related S- and P-group substituted carbanions) and that the d-orbitals used in current calculations are "polarization functions" rather than actual valence orbitals; inclusion of these d-orbitals as polarization functions, however, *is* required to obtain correct structures.[30,133,134] The enhanced stability of the carbanions and the enhanced acidity of the parent carbon acids arises from a combination of the polarizability of S and P (as compared with N, for example) and negative hyperconjugation in the anions.[36,191–200]

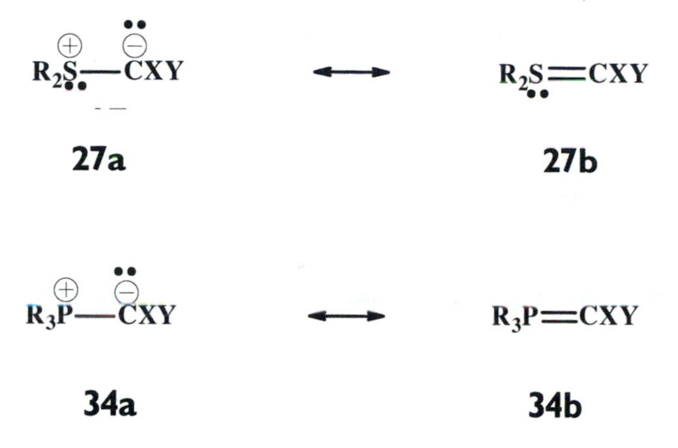

The thorough review by Wolfe considers the historical development of calculational methods, and the interplay between calculation and experimental data, and provides a qualitative understanding of the preferred structures of various α-S-substituent carbanions, based on perturbation molecular orbital (PMO) theory.[134] In the gas phase and in the absence of a counterion, calculations showed that the archetypal *unstabilized* (i.e., groups Y, Z do not further stabilize the carbanion) sulfonium ylide ($^{-}CH_2SH_2^{+}$), α-thiocarbanion ($^{-}CH_2SH$), α-sulfinyl carbanion ($^{-}CH_2SOH$), and α-sulfonyl carbanion ($^{-}CH_2SO_2H$) have structures that are, respectively, planar, pyramidal, pyramidal, and planar.

As shown in the sulfonium ylide (structure **38**), the planar carbanionic center places its lone-pair roughly perpendicular to the sulfur lone-pair axis; this arrangement permits a stabilizing interaction between the nonbonding carbon orbital and the unoccupied π-antibonding orbital of the sulfur of suitable symmetry (i.e., π_y*-SH_2^{+}). In comparison, Mitchell, Wolfe, and Schlegel[201] found two structures (**39, 40**) in which stabilizing two-electron donation of electron density into antibonding orbitals could operate. Hence, for the prototypical phosphonium ylide, $^{-}CH_2PH_3^{+}$, the structure (**39**) corresponding to that of the simple sulfonium ylide (i.e., **38**) permits carbon lone-pair donation into the π_y* orbital of

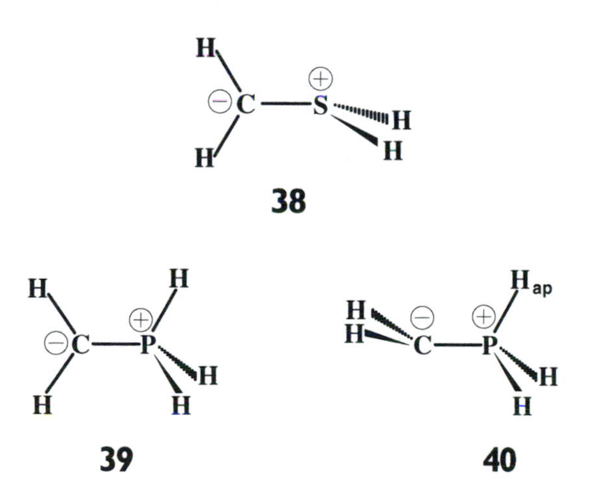

the PH_3 fragment. Note that this n → π_y* donation more than offsets a destabilizing four-electron interaction involving the carbanionic lone pair and the π-bonding orbital of the PH_3 moiety of the same symmetry (i.e., π_y); this is a function of the presumed energy levels of the interacting orbitals. However, rotation of the CH_2 portion by 90° to give structure **40** now permits a large stabilizing n → π_x* interaction that more than makes up for the destabilizing interaction of the lone pair (n) orbital and the filled π_x-bonding orbital.[201] Since similar stabilization is available whether the approximately planar $^-CH_2$ center is in the parallel or perpendicular orientation, the barrier to rotation about the P–CH_2 bond is low, as both calculation[201] and experiment[202] confirm.

A secondary n → σ* interaction (carbanion lone-pair donation into the approximately antiperiplanar P–H σ-antibonding orbital) favors the second perpendicular structure, **40**. In this context, it is notable that in the solid state triphenylphosphonium methide (Ph_3P^+–CH_2^-) apparently takes up a structure analogous to **40** and the carbanion center is approximately pyramidal,[203] rather than the approximately planar structure determined by ab initio calculations.[201,204] On the other hand, in other systems where the carbanion was shown to be pyramidal, the carbanionic center was found to undergo facile inversion at room temperature, as found in NMR studies.[205]

In structure **40**, one of the H's has been subscripted with an "ap". This subscript is a reminder that in this structure this hydrogen is antiperiplanar to the lone-pair orbital of the carbanionic center and, therefore, is favorably aligned for an n → σ* interaction, which may also be recast as anionic hyperconjugation. Two structural features would be expected to arise from this anionic hyperconjugation: a shortening of the P–C bond length and a lengthening of the P–H_{ap} bond length. In fact, in the calculational study of Mitchell et al., the P–H_{ap} bond length (with all basis sets studied—STO-3G, STO-3G*, 4-31G, and 4-31G*) was found to be longer than the bond length of P to either of the other attached hydrogens in the second structure shown above.[201] Strong experimental confirmation *in the solid state* comes from the studies of Grützmacher and Pritzkow[206] who obtained the X-ray structure of bis(*tert*-butyl)chlorophosphonium diphenylmethide [(*t*-$C_4H_9)_2ClP^+$–$C(C_6H_5)_2^-$] and, indeed, found that the P–Cl bond was significantly lengthened relative to standard P–Cl bonds. Note that σ* for the Cl fragment would be lower in energy than the corresponding antibonding orbital for the *tert*-butyl moieties and, consequently, the n → σ* orbital interaction would be most favorable for P–Cl. The unit cell in this X-ray study included three separate ylide molecules in which the $^-C(C_6H_5)_2$ moieties were oriented somewhat differently. Importantly, the trend to longer P–Cl bonds correlated with the degree of perpendicularity of the $^-C(C_6H_5)_2$ group (and, hence, the approach of the carbanionic lone pair to the anti-orientation with respect to the P–Cl bond).

Undoubtedly, our perception of the structure of all of the phosphorus- and sulfur-group substituted carbanions is informed by our understanding of the nature of bonding and stabilization in these species. [An alternative to the view of bonding in these systems as involving anionic hyperconjugation[196,197,207,208] is presented by a localized MO approach or by general valence bond (GVB) theory; both envisage banana bonds between the C and heteroatom centers.[209–211] These alternatives will not be discussed further here.]

Bernasconi and Kittredge have probed the nature of stabilization of α-thiocarbanions (cf. structure **28**).[212] This study capitalizes on the observation that delocalization of carbanionic charge into α-attached pπ–pπ acceptor groups (e.g., nitro) lags behind the proton transfer that yields such carbanions. This *imbalanced transition state* (cf. Section 2.2.1.5) gives rise to lower intrinsic rate constants (k_o) for the deprotonation of carbon acids that would give resonance-stabilized carbanions as compared with k_o for structurally similar carbon acids whose carbanionic center would not be stabilized by pπ–pπ conjugation. Reasoning then in the opposite direction, if anionic hyperconjugation is a major factor in stabilization of α-thio-substituted carbanions, a low intrinsic rate constant should be found. (Similarly, dπ–pπ bonding, albeit discredited in current thinking, would also result in a low intrinsic reactivity for the α-thio carbanion.) However, Bernasconi and Kittredge also argued that polarizability effects that could operate readily on an essentially pyramidal carbanion would lead to an enhanced intrinsic reactivity. The result of this comparison of the deprotonation kinetics of (phenylthio)nitromethane and 1-nitro-1-phenylethane (using primary and secondary alicyclic amines in water and 90% DMSO–10% water) shows that k_o is larger for (phenylthio)nitromethane than it is for 1-nitro-1-phenylethane or, in fact, than it is for nitromethane itself (statistically corrected).[212] In accord with the reasoning outlined, Bernasconi and Kittredge concluded that a polarizability effect was the major factor, providing between half and two-thirds of the stabilization of the (phenylthio)nitromethide anion, with inductive effects providing most of the remainder of the stabilization.

Negative hyperconjugation is a useful concept in describing the preferred geometry at the carbanionic center in carbanions that are α-substituted by sulfur- and phosphorus-centered functions, but it is not necessarily the dominant mechanism of stabilization of these carbanides.

3.3.3 Intrinsically Asymmetric Carbanions: The Carbanion of Benzyl Methyl Sulfoxide

Up to this point, we have only considered the carbanion structure for these systems. Significantly, the sulfur centers in the parent sulfonium salts ($RR'R''S^+X^-$), sulfoxides (RSOR′), sulfinates (RSO–OR′), and sulfites (RO–SO–OR′) are pyramidal and, when chiral (i.e., $R \neq R' \neq R''$), are configurationally stable;[213,214] a range of methods exist to resolve enantiomers of these compounds.[215-217] Similarly, compounds with configurationally stable, tetrahedral phosphorus have also been resolved into the homochiral compounds.[218] These chiral carbon acids are, consequently, precursors for carbanions that are inherently asymmetric and may yield diastereomers.

It should be acknowledged that there has been a scientific explosion in the field of asymmetric synthesis in recent years[217,219–225] and the use of carbanions generated from chiral sulfur- and phosphorus-containing substrates[226–232] is only one of the current synthetic strategies, albeit an important one. In this regard, one text delineates four generations of methods and ranks reactions with nucleophiles that bear a chiral auxiliary (e.g., carbanions formed from chiral sulfoxides) among the second-generation methods.[233]

We focus now on a chiral sulfoxide system that has been considered in detail, both experimentally and by means of high-level calculations. Hydrogen/deuterium exchange from benzyl methyl sulfoxide is illustrative both of the potential for chiral α-sulfinyl carbanions in stereoselective synthesis and of some of the pitfalls innate to the approach, as will be shown.

Continuing on the H/D exchange studies with dimethyl sulfoxide[55] to benzyl methyl sulfoxide (NaOH–D_2O) revealed three sites for exchange.[175] Thus, the diastereotopic protons (underlined in structure **41**) adjacent to the (approximately) tetrahedral sulfur exchange at different rates.[175–180,234,235] (Note, that the bond angles around sulfur, i.e., C–S–C and C–S–O angles, differ from tetrahedral, and that structure **41** is not strictly analogous to an ethane derivative.[134]) In an interesting extension of this work, Wolfe and Rauk demonstrated that chirality could be transferred from sulfur to carbon, allowing optically active benzyl methyl sulfoxide to undergo stereoselective H/D exchange, followed by oxidation of the sulfoxide to yield the optically active sulfone.[234]

Later work provided evidence that for the *S*-sulfoxide depicted, it is the pro-*R* hydrogen (**41**) that exchanges the most readily.[236] In D_2O, the selectivity in exchange was found to be 17:1.[175] However, this selectivity depends markedly on the base–solvent system employed. Thus, in NaOD–acetonitrile–D_2O the selectivity is 2:1, while in NaOD–pyridine–D_2O selectivity has disappeared (i.e., 1:1)[134] and, finally, has inverted in *t*-BuONa–*t*-BuOD.[237]

The explanation for this behavior is intimately tied to the following assumptions: (1) the selectivity found in the aqueous system (NaOD–D_2O) reflects the actual propensity for removal of one prochiral proton as compared with the other, free of ion-pairing effects or differential solvation effects; (2) the preference corresponds to formation of the carbanion of preferred (i.e., most stable) configuration; and (3) the most stable structure for the carbanion is in accord with the geometrical preferences found in the gas-phase (ab initio) calculations for the unstabilized α-sulfinyl carbanion ($^{-}CH_2SOH$).[134] Further, our discussion will focus on the chiral *S*-sulfur benzyl methyl sulfoxide (structure **41**).

With these provisos in mind, we recall that the model unstabilized α-sulfinyl carbanion is calculated to be pyramidal at the carbanionic center. Moreover, in this geometry-optimized structure the carbanionic lone pair is approximately anti-

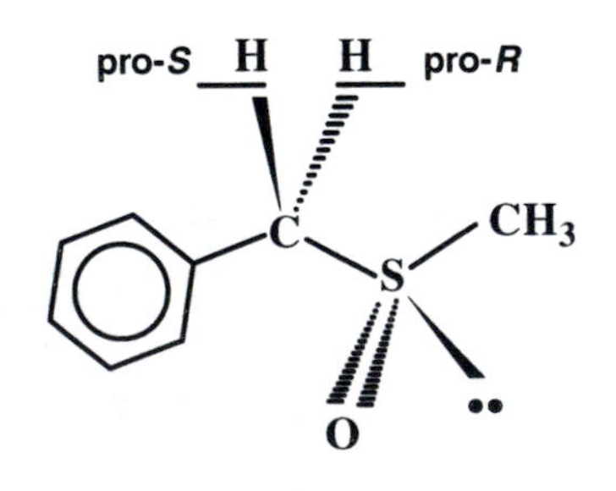

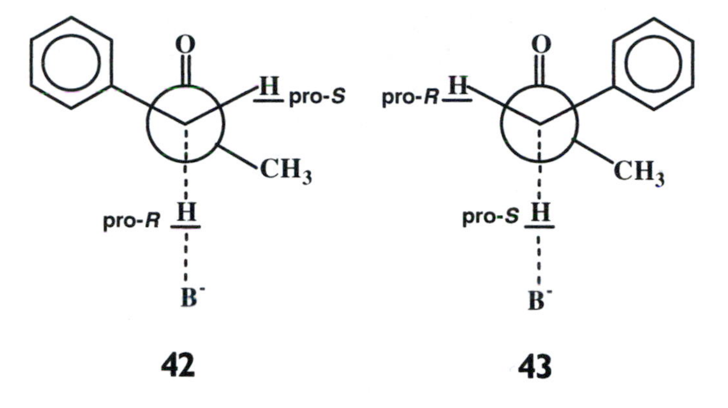

periplanar to the sulfoxide sulfur–oxygen bond.[134] Therefore, the enhanced kinetic acidity of the pro-*R* hydrogen in the *S*-enantiomer of the sulfoxide arises from the fact that this proton is already approximately antiperiplanar to the acceptor S–O bond. In accord with the Curtin–Hammett principle, the two transition states (i.e., that involving pro-*R* hydrogen-abstraction and that involving pro-*S* hydrogen-abstraction) must differ in energy. As pointed out by Wolfe,[134] a further factor is that the transition state for removal of the pro-*R* hydrogen (**42**) involves less strain than the transition state for removal of the pro-*S* hydrogen (**43**), in that the former has a methyl group–hydrogen gauche interaction while the latter has a methyl group–phenyl group gauche interaction. This is illustrated for two generalized transition states (structures **42**, **43**).

Similar predictions can be made for sulfonium ylides, α-thio- and α-sulfonyl-substituted carbanions, using similar assumptions.[134] In each case, the optimum structure for the carbanion is presumed to be that of the simple *unstabilized* prototypes. The nature of the base used, the solvent (and whether asymmetric solvation of the carbanion is likely), and the nature and degree of ion pairing (or more complex aggregation) all impinge on the utility of these carbanions as intermediates in asymmetric synthesis.

3.3.4 Comparison of H/D Exchange and Alkylation for Chiral Carbanions

It is often easiest to use carbon acids that would give rise to carbanions that are stabilized—for example, by conjugation with an aromatic moiety or with a powerful electron-withdrawing group like NO_2—and which therefore may be deprotonated readily. Such stabilized carbanions, even if also α-substituted by an alkylsulfinyl group, may be planar rather than pyramidal. Recall, however, that if the barrier to rotation about the carbanionic C–S bond is high enough, the asymmetry of the carbanion may also be retained.

It must also be conceded that in actual synthetic systems, the stereoselectivity determined by isotopic exchange may not hold. In this regard, it has been found that the lithium carbanide of racemic benzyl methyl sulfoxide reacted with various sources of deuterium [e.g., D_2O, *t*-BuOD, etc.] in THF to give predominant retention, but that reaction with methylating agents [e.g., CH_3I, $(CH_3O)_2SO_2$, etc.] leads to inversion of configuration.[237,238] Quenching the lithium carbanide of benzyl methyl sulfoxide with CO_2 or acetone is also highly stereoselective and gives retention, though degree of retention does vary with solvent and abstracting base.[237,239,240] Significantly, the diastereomer favored in the NaOD–D_2O H/D exchange experiments is not the preferred diastereomer in the deuterium quenching experiments where the solvent is THF and the carbanion counterion is lithium.

Similar differences in stereochemical outcome for protonation/deuteration as compared with alkylation have been found in a number of related lithiated sulfoxide systems.[241,242] These results, as well as calculational studies,[243] support the view that these lithium carbanides are chelated structures (**44**, charges omitted for simplicity) in which the lithium cation is *cis* to the sulfinyl oxygen. Association of D_2O (ROD) with the chelated lithium could direct deuterium to the top face of the carbanion, and so, give retention, while the bulkier iodomethane electrophile, for example, would have to attack via the bottom face of structure **44**, resulting in inversion at the carbanionic center. Note also that chelation has the effect of raising the barrier to rotation about the C–S bond, which is the mechanism of racemization for planar, chiral carbanions.

In general, it would appear that substrate structural features, solvent, and counterion are all important in determining whether these α-S- and α-P-group substituted carbanions are pyramidal or planar and, more importantly, whether the carbanionic center is configurationally stable. Another factor that can be important is the potential for π–π stacking in the putative transition state of any carbanion reaction. An interesting example of this is provided by the reaction of (*S*)-lithiomethyl 1-naphthyl sulfoxide (formed by treatment of the sulfoxide with Et_2NLi in THF), which reacts with various alkyl phenyl ketones and, after reductive workup, leads to formation of only the *S*-enantiomeric alkanols.[244] This high stereoselectivity of addition to the carbonyl face appears to reflect not only chelation of the lithium counterion by both the sulfoxide oxygen and the carbonyl oxygen but also π–π stacking of the naphthyl and phenyl rings in the transition state, as shown in Scheme 3.5. As might be expected, the stereoselectivity decreases inversely with the bulk of the alkyl group, R. Presumably, advantage may be taken of such structural features in cases where the carbanionic center would also be chiral.

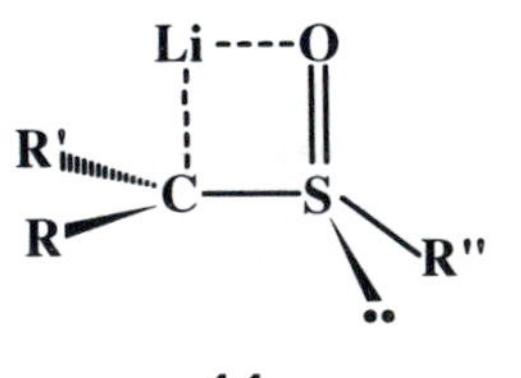

44

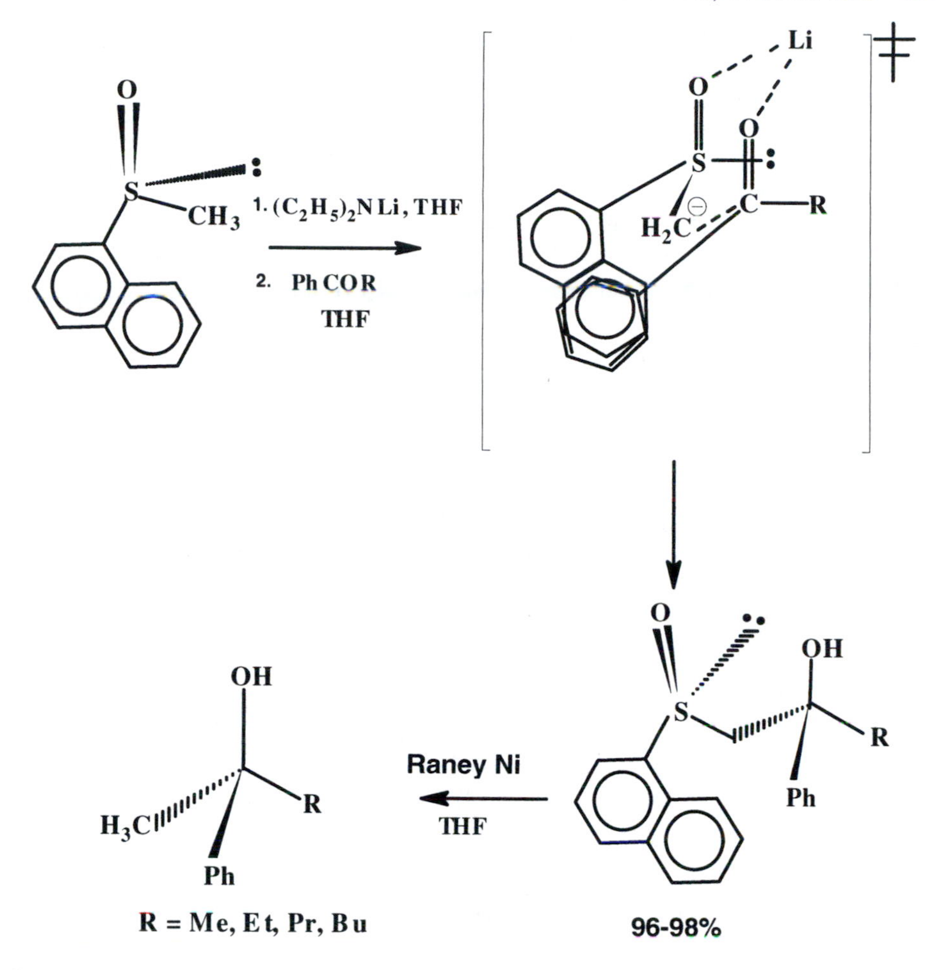

SCHEME 3.5

Ion pairing with lithium is also important in the stereoselective reactions of carbanions derived from sulfones, where the lithium usually sits in between the two oxygens of the sulfone and the carbanionic lone pair is gauche to both oxygens. Interestingly, the lithium does not appear to be coordinated to the carbanionic center.[207,245–247] As has been summarized,[207] the lithium carbanides of sulfones appear to be either solvated monomeric[248] or dimeric[249–251] contact ion pairs in the solid state according to X-ray crystallographic studies, but they exist primarily as monomeric contact ion pairs in THF;[246,250] if the carbanionic center is substituted with at least one aryl group it is usually found to be planar,[246,249,250] whereas substitution of the carbanionic center exclusively with alkyl groups leads to a pyramidal carbanion.[251,252] Importantly, it was shown by Raabe, Gais, and Fleischhauer[207] that greater configurational stability is conferred on an α-sulfonyl carbanide by also substituting the sulfur center with fluorine or with a trifluoromethyl group, as in fluorosulfonyl methide or trifluoromethylsulfonyl methide; calculations (HF/6-31+G*//HF/6-31+G* level) showed that such fluorine substitution acts to enhance the $n_C \rightarrow \sigma^*_{SO\text{-}R}$ interaction. In turn, this anionic hyperconjugation raises the carbanionic C–S rotation barrier.[207] Again, fluorine substitution of this sort may be profitably used to increase configurational stability in chiral α-sulfonyl carbanions.

Another approach to control of configurational stability in these and related systems involves the inclusion of the carbanionic center in a ring system or as a substituent attached to a conformationally locked ring system. Chiral cyclopropyl carbanions often retain their configuration in H/D exchange although, as usual, this is dependent also on other ring substituents and solvent, counterion, and so on. In the crystal, the carbanionic center of 2,2-diphenyl-1-(phenylsulfonyl)cyclopropyllithium (with dimethoxyethane in a 2:3 ratio) is pyramidal and is apparently stable in solution (THF/DMSO).[252] It was shown by Padwa and Wannamaker[253] that deprotonation of the *trans*-1-methoxy-2,2-dimethyl-3-phenylsulfonylcyclopropane with lithium diisopropylamide (LDA in THF) at −78 °C, followed by quenching with a variety of electrophiles (E-LG = D_2O, CH_3I, $CH_2{=}CHCH_2Br$, and $ClCOOCH_3$) led to formation of a single stereoisomer, as shown in equation 3.8.

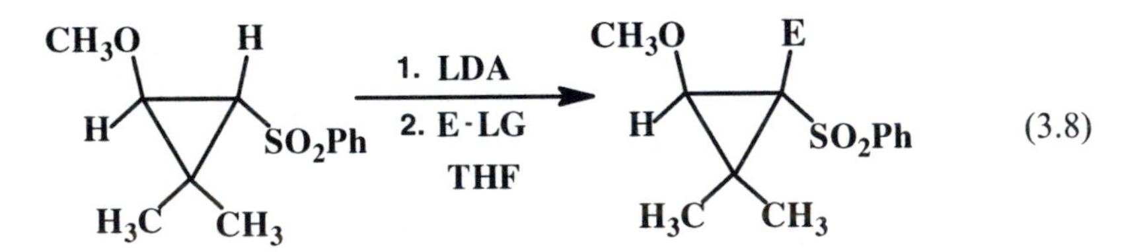

(3.8)

However, treatment of the *cis* isomer of the sulfone in the same manner also gave the same single isomeric product, indicating that the intermediate carbanions could interconvert but that the final product arises from the preferred carbanion (i.e., the carbanion that could be formed directly from the *trans* sulfone). It was suggested that the preference reflected both chelation of the *trans* anion by lithium, using the methoxyl group, and $n \rightarrow \sigma^*$ stabilization of the carbanion.[253]

Considerable stereoselectivity in deprotonation and reaction of the resultant carbanion with electrophiles has been found for constrained ring systems in which the carbanion is α-substituted by a sulfonium center,[174] a sulfoxide[177,253] or a sulfone.[178,179] Carbanions derived from 1,3-dithianes (e.g., structure **45**) also exhibit high stereoselectivity in their reactions with electrophiles in ethereal solvents and, as found for cyclic lithiated sulfoxides,[254] the role of the lithium ion in coordinating to the carbanionic center (**45**), as well as the ring sulfurs, was initially advanced as a significant factor in determining the course of the reactions.[255] Later, the importance of stereoelectronic factors was emphasized.[256] Reaction of the lithium carbanide of the 1,3-dithiane (**45**) with electrophiles, including DCl, iodomethane, and formaldehyde in ethereal media, was shown to give almost entirely products of equatorial substitution.

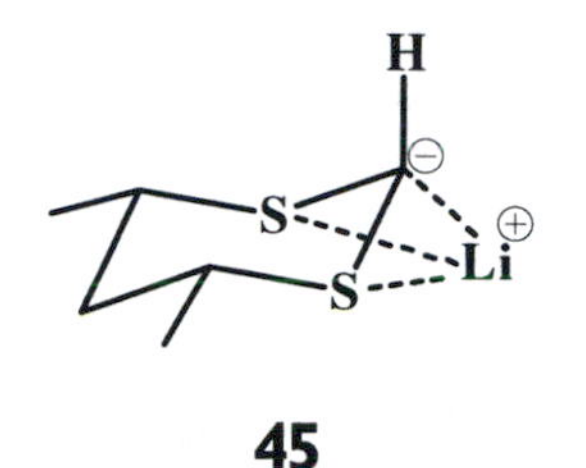

45

3.3.4.1 Disastereoselectivity from Carbanions Adjacent to Six-Membered P-Heterocycles

A vigorous program of research into carbanions stabilized by α-substitution by various PO–X and PS–X groups has been engaged in by the Denmark group.[257–265] As but one illustrative example, Denmark and Dorow[257] examined the ^{1}H, ^{13}C, and ^{31}P NMR spectra (THF-d_8, 20 to −90 °C) of the following two carbon acids, 2-benzyl-5,5-dimethyl-2-oxo-1,3,2-dioxaphosphorinane (**46**) and 2-benzyl-1,3-dimethyl-2-oxo-1,3,2-diazaphosphorinane (**47**), and their lithium carbanides. While structure **46** proved to be configurationally mobile, the conformer shown is the preferred chair form at 20 °C. At lower temperatures (−76 °C), the other chair form is favored. No such mobility was observed for the phosphonamide (**47**).

The lithium carbanide of structure **46** had the following salient features. First, the *ortho* H's of the carbanion are anisochronous in the ^{1}H NMR spectrum at −86 °C, which is indicative of slow rotation about the carbanion α-C to the phenyl ring bond and suggests that the carbanion is planar. Secondly, increases in the coupling constants ($^{1}J_{C\text{–}H}$ and $^{1}J_{P\text{–}C}$) in going from the parent carbon acids to the lithium carbanides are consistent with rehybridization (sp^3 to sp^2). Finally, the ^{31}P downfield shift that occurs upon carbanion formation supports the view that the phosphonyl group stabilizes the carbanion via a polarization mechanism.

The lithium salt derived from structure **47** displays even larger upfield shifts of the signals for the *para*-C and *para*-H of the phenyl ring. This arises from an even greater delocalization of negative charge into the phenyl ring in the carbanion of **47**. Further, observation of a single ^{31}P-coupled methyl resonance at all temperatures investigated (i.e., down to −76 °C) is consistent with a low carbanion C-to-P rotation barrier, in common with the low rotation barrier found in phosphonium ylides.[30,201,202]

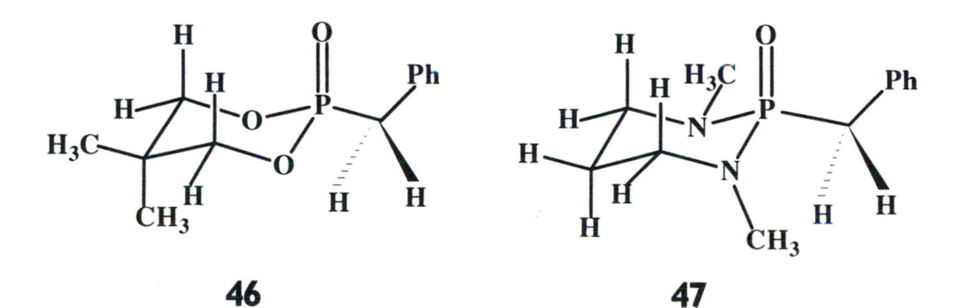

46 **47**

In general, higher P–C rotation barriers have been found for the thiophosphoryl (i.e., P=S) analogues of compounds like **46** and **47**; these have been attributed to the nature of anionic hyperconjugative interactions in the ground-state structures of the carbanions.[261] However, these increased rotation barriers afford the chiral thiophosphoryl carbanions greater configurational stability and lead to greater diastereoselectivity in reaction with electrophiles, as shown in Table 3.3 for the reaction of **48** to give structures **49** and **50** that are illustrated in Scheme 3.6.

Before considering diastereoselectivity in these systems generally, comment is warranted on the effect of the addition of HMPA solvent. The solvent HMPA is a powerful cation-solvating agent and would be expected to convert contact ion pairs into solvent-separated ion pairs (SSIP). It might equally be expected that solvent-separated ion pairs would undergo racemization faster than contact ion pairs, and yet diastereoselectivity in this case is enhanced by addition of HMPA. The explanation is that carbanion C–P bond rotation is the rate-limiting step in racemization for these carbanions, rather than ion separation, and that in the "looser" SSIP there is more probability that the carbanion can take up suitable orientations that would maximize $n \rightarrow \sigma^*$ stereoelectronic stabilization.[248] Similar behavior has been reported for lithium carbanides α-substituted by both phenylthio and dimethylphenylsilyl groups[266] and by the Denmark group for a related lithiated cyclic thiophosphonamide.[260]

The diastereoselectivity found in Table 3.3 clearly depends upon many factors that have been enumerated in a paper by Kranz and Denmark:[263] "(1) location of the lithium ion, (2) the conformation of the heterocycle, (3) the orientation of the carbanion substituents in the reactive conformation, and (4) the relative rates of P–C bond rotation and of the reactions."

X-ray crystallographic analysis of a number of comparable systems in the solid state (i.e., lithiated oxo and thiooxophosphoryl diamides) shows that the lithium counterion is always bonded to the group-16 atom on phosphorus (i.e., O or S). Importantly, and in contrast to many of the α-sulfur group substituted carbanions, the lithium *is not bonded to the carbanionic center*. The carbanion itself is essentially planar in the systems studied thus far and the carbanion is oriented

Table 3.3. Diastereoselectivity in methylation of **48** (See Scheme 3.6)

Y	X	Additive	Temperature (°C)	Ratio 49:50
Ph	O	—	−7.0	34:1
Ph	S	—	−7.0	> 1000:1[a]
MeO	O	—	−7.0	2.25:1
MeO	S	—	−7.0	1:1.9
MeO	S	HMPA	−95	16:1

SOURCE: Data taken from reference 262.
[a]Estimate based on NMR; the ratio represents the limit of detection.[259]

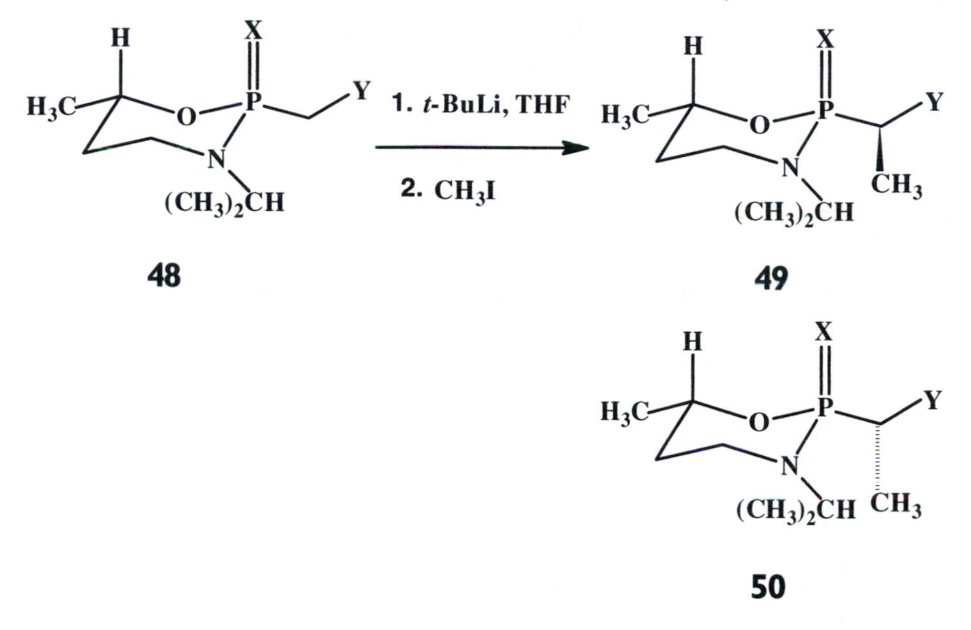

SCHEME 3.6

parallel to the P=X bond, as expected for a significant n→ σ^* (P=X) stabilizing interaction.

In conclusion, carbanions may take up either planar or pyramidal geometries at the carbanionic center. Either geometry may lead to a chiral configurationally stable structure, depending on such factors as the height of the energy barrier to inversion, in the case of pyramidal carbanions, and the height of the barrier to bond rotation, in the case of planar carbanions. The nature of the counterion, whether it is chelated to the carbanion and an adjacent group (C=O, S=O, etc.), and the solvent and its influence on the structure of ion pairs or higher-order aggregates are all important considerations in determination the preferred structure of a carbanion.

That investigations into the geometry at the carbanionic center are proceeding at what may seem a frantic pace reflects the need to understand these important potentially chiral reagents. In this regard, the work of the Ahlberg group[267–271] and that of the Saunders group[272–275] merit closer attention. Given the importance of ion pairing, the next chapter will examine the nature of ion pairing/aggregation/polymerization using spectroscopic methods for a set of arylmethyllithiums.

References

1. Cram, D.J. *Fundamentals of Carbanion Chemistry*; Academic Press: New York, 1966; Chapters 2–4.
2. Cram, D.J.; Allinger, J.; Langemann, A. *Chem. Ind. (London)*, **1955**, 919.

3. Cram, D.J.; Neilsen, W.D.; Rickborn, B. *J. Am. Chem. Soc.* **1960**, *82*, 6415.
4. Eliel, E.L.; Wilen, S.H.; Mander, L.N. *Stereochemistry of Organic Compounds*; Wiley: New York, 1994.
5. Walsh, A.D. *J. Chem. Soc.* **1953**, 2260.
6. Walsh, A.D. *J. Chem. Soc.* **1953**, 2296.
7. Gillespie, R.J.; Robinson, E.A. *Angew. Chem., Int. Ed. Engl.* **1996**, *35*, 495; but also see reference 9.
8. Gillespie, R.J.; Hargittai, I. *The VSEPR Model of Molecular Geometry*, Allyn & Bacon: Boston, 1991; particularly p 76.
9. Wolfe, S.; Tel, L.M.; Haines, W.J.; Robb, M.A.; Csizmadia, I.G. *J. Am. Chem. Soc.* **1973**, *95*, 4863.
10. Driessler, F.; Ahlrichs, R.; Staemmler, V.; Kutzelnigg, W., *Theor. Chim. Acta (Berlin)*, **1973**, 315.
11. Williams, J.E.; Streitwieser, A. *J. Am. Chem. Soc.* **1975**, *97*, 2634.
12. Surratt, G.T.; Goddard, W.A., III *Chem. Phys.* **1977**, *23*, 39.
13. Kari, R.E.; Csizmadia, I.G. *J. Chem. Phys.* **1969**, *50*, 1443; but see references 14 and 15.
14. Lewis, T.P. *Tetrahedron* **1969**, *25*, 4117.
15. Kalcher, J.; Janoschek, R. *Chem. Phys.* **1986**, *104*, 251.
16. Li, W.K.; Nobes, R.H.; Poppinger, D.; Radom, L. In *Comprehensive Carbanion Chemistry. Part C. Ground and Excited State Reactivity*; Buncel, E.; Durst, T., Eds.; Elsevier: Amsterdam, 1987; pp 16–20.
17. Ellison, G.B.; Engelking, P.C.; Lineberger, W.C. *J. Am. Chem. Soc.* **1978**, *100*, 2556.
18. Marynick, D.S.; Dixon, D.A. *Proc. Natl. Acad. Sci. U.S.A.* **1977**, *71*, 410.
19. Rauk, A.; Allen, L.C.; Mislow, K. *Angew. Chem., Int. Ed. Engl.* **1970**, *9*, 400.
20. Henderson, J.W. *Chem. Soc. Rev.* **1973**, *2*, 397.
21. Bell, R.P. *The Tunnel Effect in Chemistry*; Chapman & Hall: New York, 1980.
22. Swalen, J.D.; Ibers, J.A. *J. Chem. Phys.* **1962**, *36*, 1914; and references therein.
23. Papoušek, D. *J. Mol. Struct.* **1983**, *100*, 179.
24. Rodwell, W.R.; Radom, L. *J. Am. Chem. Soc.* **1981**, *103*, 2865.
25. Schlegel, H.B.; Skancke, A. *J. Am. Chem. Soc.* **1993**, *115*, 7465.
26. Arduengo, A.J.; Dixon, D.A.; Roe, D.C. *J. Am. Chem. Soc.* **1986**, *108*, 6821.
27. Dixon, D.A.; Arduengo, A.J.; Fukanaga, T. *J. Am. Chem. Soc.* **1986**, *108*, 2321.
28. Dixon, D.A.; Arduengo, A.J. *J. Am. Chem. Soc.* **1987**, *109*, 338.
29. Clotet, A.; Rubio, J.; Illas, F. *J. Mol. Struct. THEOCHEM* **1988**, *164*, 351.
30. Gilheany, D.G. *Chem. Rev.* **1994**, *94*, 1339.
31. Millie, P.; Berthier, G. *Int. J. Quantum Chem.* **1968**, *2S*, 67.
32. Boese, R.; Bläser, D.; Niederprüm, N.; Nüsse, M.; Brett, W.A.; Schleyer, P. von R.; Bühl, M.; Hommes, N.J.R. van E. *Angew. Chem., Int. Ed. Engl.* **1992**, *31*, 314.
33. Carroll, F.A. *Perspectives on Structure and Mechanism in Organic Chemistry*; Brooks/Cole: Pacific Grove, CA, 1998; pp 33–45; and references therein.
34. Applequist, D.E.; Roberts, J.D. *Chem. Rev.* **1954**, *54*, 1065.
35. Dewar, M.J.S.; Shanshal, M. *J. Am. Chem. Soc.* **1969**, *91*, 3654.
36. Wiberg, K.B.; Castejon, H. *J. Org. Chem.* **1995**, *60*, 6327.
37. Lambert, J.B. *Top. Stereochem.* **1971**, *6*, 19.
38. Cram, D.J.; Whitney, T.A. *J. Am. Chem. Soc.* **1967**, *89*, 4651.
39. Ratajczak, A.; Anet, F.A.L.; Cram, D.J. *J. Am. Chem. Soc.* **1967**, *89*, 2072.
40. Cram, D.J.; Trepka, R.D.; St. Janiak, P. *J. Am. Chem. Soc.* **1964**, *86*, 2731.
41. Corey, E.J.; Lowry, T.H. *Tetrahedron Lett.* **1965**, 793.
42. Corey, E.J.; Lowry, T.H. *Tetrahedron Lett.* **1965**, 803.
43. Corey, E.J.; König, H.; Lowry, T.H. *Tetrahedron Lett.* **1962**, 515.

44. Doering, W. von E.; Levy, K.K. *J. Am. Chem. Soc.* **1955**, *77*, 509.
45. Doering, W. von E.; Gaspar, P.P. *J. Am. Chem. Soc.* **1963**, *85*, 3043.
46. Zimmerman, H.E.; Thyagarajan, B.S. *J. Am. Chem. Soc.* **1960**, *82*, 2505.
47. Ford, W.T.; Graham, E.W.; Cram, D.J. *J. Am. Chem. Soc.* **1967**, *89*, 4551.
48. Streitwieser, A.; Owens, P.H.; Sonnichsen, G.; Smith, W.K.; Ziler, G.R.; Neimeyer, H.M.; Kruger, T.L. *J. Am. Chem. Soc.* **1973**, *95*, 4254.
49. Hunter, D.H.; Lin, Y.T.; McIntyre, A.L.; Shearing, D.L.; Zvagulis, M. *J. Am. Chem. Soc.* **1973**, *95*, 8327.
50. Fraser, R.R.; Ng, L.K. *J. Am. Chem. Soc.* **1976**, *98*, 4334.
51. Koch, H.F. In *Comprehensive Carbanion Chemistry, Part C. Ground and Excited State Reactivity*; Buncel, E.; Durst, T., Eds.; Elsevier: Amsterdam, 1987; pp 321–360.
52. Koch, H.F. *Acc. Chem. Res.* **1984**, *17*, 137.
53. Koch, H.F.; Lodder, G.; Koch, J.G.; Bogdan, D.J.; Brown, G.H.; Carlson, C.A.; Dean, A.B.; Hage, R.; Han, R.; Hopman, J.C.P.; James, L.A.; Knape, P.M.; Roos, E.C.; Sardina, M.L.; Sawyer, R.A.; Scott, B.O.; Testa, C.A., III; Wickham, S.D. *J. Am. Chem. Soc.* **1997**, *119*, 9965.
54. Zuilhof, H.; Lodder, G.; Koch, H.F. *J. Org. Chem.* **1997**, *62*, 7457.
55. Buncel, E.; Symons, E.A.; Zabel, A.W. *J. Chem. Soc., Chem. Commun.* **1965**, 173.
56. Hargreaves, M.K.; Modarai, B. *J. Chem. Soc., Chem. Commun.* **1969**, 16.
57. Hargreaves, M.K.; Modarai, B. *J. Chem. Soc. C* **1971**, 1013.
58. Jolly, C.A.; Chan, F.; Marynick, D.S. *Chem. Phys. Lett.* **1990**, *174*, 320.
59. Cherry, W.; Epiotis, N. *J. Am. Chem. Soc.* **1976**, *98*, 1135.
60. Stille, J.K.; Feld, W.A.; Freeburger, M.E. *J. Am. Chem. Soc.* **1972**, *94*, 8485.
61. Miller, S.I.; Lee, W.G. *J. Am. Chem. Soc.* **1959**, *81*, 6313.
62. Seyferth, D.; Vaughan, L.G. *J. Am. Chem. Soc.* **1964**, *86*, 883.
63. Walborsky, H.M.; Turner, L.M. *J. Am. Chem. Soc.* **1972**, *94*, 2273.
64. Arnett, J.F.; Walborsky, H.M. *J. Org. Chem.* **1972**, *37*, 3678.
65. Schmidt, R.R.; Schmid, B. *Tetrahedron Lett.* **1977**, *18*, 3583.
66. Schmidt, R.R.; Speer, H.; Schmid, B. *Tetrahedron Lett.* **1979**, *20*, 4277.
67. Muthukrishnan, R.; Schlosser, M. *Helv. Chim. Acta* **1976**, *59*, 13.
68. Raucher, R.; Koople, G. *J. Org. Chem.* **1978**, *43*, 3794.
69. Millon, J.; Lorne, R.; Linustumella, G. *Synthesis* **1975**, 434.
70. Dreiding, A.S.; Pratt, R.J. *J. Am. Chem. Soc.* **1954**, *76*, 1902.
71. Schöllkopf, U.; Staforst, D.; Jeutsch, R. *Justus Liebigs Ann. Chem.* **1977**, *714*, 1167.
72. House, H.O.; Weeks, P.D. *J. Am. Chem. Soc.* **1975**, *97*, 2785.
73. Shabtai, J.; Inger, E.N.; Pines, H. *J. Org. Chem.* **1981**, *46*, 3795.
74. Feit, B.A.; Melamed, U.; Speer, H.; Schmidt, R.R. *J. Chem. Soc., Perkin Trans. 1* **1984**, 775.
75. Caramella, P.; Houk, K.N. *Tetrahedron Lett.* **1981**, *22*, 819.
76. Curtin, D.Y.; Koehl, W.J., Jr. *J. Am. Chem. Soc.* **1962**, *84*, 1967.
77. Panek, E.J.; Neff, B.L.; Cha, H.; Panek, M.J. *J. Am. Chem. Soc.* **1975**, *97*, 3996.
78. Feit, B.A.; Melamed, U.; Schmidt, R.R.; Speer, H. *Tetrahedron* **1981**, *37*, 2143.
79. Schmidt, R.R.; Speer, H.; Schmid, B. *Tetrahedron Lett.* **1979**, 4277.
80. Schmidt, R.R.; Talbiersky, J.; Russegger, P. *Tetrahedron Lett.* **1979**, 4273.
81. Vlattas, I.; Vecchina, L.D.; Lee, A.O. *J. Am. Chem. Soc.* **1976**, *98*, 2008.
82. Feit, B.A.; Haag, B.; Kast, J.; Schmidt, R.R. *J. Chem. Soc., Perkin Trans. 1* **1986**, 2027.
83. Beak, P.; Zajdel, W.J.; Reitz, D.B. *Chem. Rev.* **1984**, *84*, 471.
84. Gawley, R.E.; Rein, K.; Chemburkar, S. *J. Org. Chem.* **1989**, *54*, 3002; and references therein.

85. Chiang, Y.; Griesbeck, A.G.; Heckroth, H.; Hellrung, B.; Kresge, A.J.; Meng, Q.; O'Donoghue, A.C.; Richard, J.P.; Wirz, J. *J. Am. Chem. Soc.* **2001**, *123*, 8979.
86. Pearson, W.H.; Lindbeck, A.C. *J. Am. Chem. Soc.* **1991**, *113*, 8546.
87. Beak, P.; Reitz, D.B. *Chem. Rev.* **1978**, *78*, 275.
88. Marchese, G.; Naso, F. *La. Chim. E. Ind. (Milan)* **1971**, *53*, 760.
89. Marchese, G.; Naso, F.; Modena, G. *J. Chem. Soc. B* **1969**, 290.
90. Rappoport, Z. *Acc. Chem. Res.* **1992**, *25*, 474.
91. Streitwieser, A.; Xie, L.; Speers, P.; Williams, P.G. *Magn. Reson. Chem.* **1998**, *36*, S209.
92. Dixon, R.; Streitwieser, A.; Williams, P.G.; Eaton, P.E. *J. Am. Chem. Soc.* **1991**, *113*, 357.
93. Lukin, K.; Li, J.; Gilardi, R.; Eaton, P.E. *Angew. Chem., Int. Ed. Engl.* **1996**, *35*, 864.
94. Gutowsky, H.S. *Ann. N.Y. Acad. Sci.* **1958**, *70*, 786.
95. Lehn, J.M.; Wagner, J. *J. Chem. Soc., Chem. Commun.* **1968**, 148.
96. Jautelat, M.; Roberts, J.D. *J. Am. Chem. Soc.* **1969**, *91*, 642.
97. Boche, G.; Walborsky, H.M. *Cyclopropane Derived Reactive Intermediates*; Wiley: Chichester, U.K., 1990.
98. Walborsky, H.M.; Motes, J.M. *J. Am. Chem. Soc.* **1970**, *90*, 2445.
99. Walborsky, H.M.; Impastato, F.J.; Young, A.E. *J. Am. Chem. Soc.* **1964**, *86*, 3283.
100. Pierce, J.B.; Walborsky, H.M. *J. Org. Chem.* **1968**, *33*, 1962.
101. Walborsky, H.M.; Allen, L.E.; Traenker, H.J.; Powers, J.E. *J. Org. Chem.* **1971**, *36*, 2937.
102. Walborsky, H.M.; Hornyak, F.M. *J. Am. Chem. Soc.* **1955**, *77*, 6026.
103. Häner, R.; Olano, B.; Seebach, D. *Helv. Chim. Acta* **1987**, *70*, 1676.
104. Periasamy, M.P.; Walborsky, H.M. *J. Am. Chem. Soc.* **1977**, *99*, 2631.
105. Huisgen, R.; Eberhard, P. *J. Am. Chem. Soc.* **1972**, *94*, 1346.
106. Boche, G.; Martins, D. *Angew. Chem., Int. Ed. Engl.* **1972**, *11*, 724.
107. Newcomb; M.; Ford, W.T. *J. Am. Chem. Soc.* **1973**, *95*, 7186.
108. Cram, D.J.; Gosser, L. *J. Am. Chem. Soc.* **1963**, *85*, 3890.
109. Cram, D.J.; Gosser, L. *J. Am. Chem. Soc.* **1964**, *86*, 5445.
110. Ahlberg, P.; Davidsson, O.; Johnson, B.; McEwen, I.; Rönnqvist, M. *Bull. Soc. Chim. Fr.* **1988**, 177.
111. Cram, D.J.; Gosser, L. *J. Am. Chem. Soc.* **1964**, *86*, 2950.
112. Cram, D.J.; Willey, F.; Fischer, H.P.; Scott, D.A. *J. Am. Chem. Soc.* **1964**, *86*, 5510.
113. Ford, W.T.; Graham, E.W.; Cram, D.J. *J. Am. Chem. Soc.* **1967**, *89*, 4661.
114. Chu, K.C.; Cram, D.J. *J. Am. Chem. Soc.* **1972**, *94*, 3521.
115. Almy, J.; Hoffman, D.H.; Chu, K.C.; Cram, D.J. *J. Am. Chem. Soc.* **1973**, *95*, 1185.
116. Corset, J. In *Comprehensive Carbanion Chemistry. Part A. Structure and Reactivity*; Buncel, E.; Durst, T., Eds.; Elsevier: Amsterdam, 1980; pp 125–197.
117. O'Brien, D.H. In *Comprehensive Carbanion Chemistry. Part A. Structure and Reactivity*; Buncel, E.; Durst, T., Eds.; Elsevier: Amsterdam, 1980; pp 271–323.
118. Buncel, E.; Menon, B. In *Comprehensive Carbanion Chemistry. Part A. Structure and Reactivity*; Buncel, E.; Durst, T., Eds.; Elsevier: Amsterdam, 1980; pp 97–114.
119. Corset, J. In *Comprehensive Carbanion Chemistry. Part A. Structure and Reactivity*; Buncel, E.; Durst, T., Eds.; Elsevier: Amsterdam, 1980; p 152.
120. Feuer, H.; Savides, C.; Rao, C.N.R. *Spectrochim. Acta* **1963**, *19*, 431.
121. Jonathan, N. *J. Mol. Spectrosc.* **1961**, *7*, 105.
122. Yarwood, J.; Orville-Thomas, W.J. *J. Chem. Soc.* **1963**, 5991.
123. Brookes, M.J.; Jonathan, N. *J. Chem. Soc. A* **1968**, 1529.
124. Juchnovski, I.N.; Binev, I.G. *J. Organomet. Chem.* **1975**, *99*, 1.
125. Juchnovski, I.N.; Tsenov, J.A.; Binev, I.G. *Spectrochim. Acta, Part A* **1996**, *52*, 1145.

126. Binev, I.G.; Tsenov, J.A.; Velcheva, E.A.; Radomirska, V.B.; Juchnovski, I.N. *J. Mol. Struct.* **1996**, *14*, 133.
127. Croisat, D.; Seyden-Penne, J.; Strzalko, T.; Wartski, L.; Corset, J.; Froment, F. *J. Org. Chem.* **1992**, *57*, 6435.
128. Lawson, K.E. *Spectrochim. Acta* **1961**, *17*, 248.
129. Cambillau, C.; Bram, G.; Corset, J.; Riche, C. *Nouv. J. Chim.* **1979**, *3*, 9.
130. Cotton, F.A.; Schunn, R.A. *J. Am. Chem. Soc.* **1963**, *85*, 2394.
131. Bottin-Strzalko, T.; Seyden-Penne, J.; Pouet, M.J.; Simonnin, M.P. *J. Org. Chem.* **1978**, *43*, 4346.
132. Bottin-Strzalko, T.; Corset, J.; Froment, F.; Pouet, M.J.; Seyden-Penne, J.; Simonnin, M.P. *J. Org. Chem.* **1980**, *45*, 1270.
133. Oae, S.; Uchida, Y. In *The Chemistry of Sulphones and Sulphoxides*; Patai, S.; Rappoport, Z.; Stirling, C., Eds.; Wiley: Chichester, U.K., 1988; pp 583–665.
134. Wolfe, S. In *Organic Sulfur Chemistry: Theoretical and Experimental Advances*; Bernardi, F.; Csizmadia, I.G.; Mangini, A., Eds.; Elsevier: Amsterdam, 1985; pp 133–190.
135. Chassaing, G.; Marquet, A.; Corset, J.; Froment, F. *J. Organomet. Chem.* **1982**, *232*, 293.
136. Fraenkel, F.; Dix, D.T.; Marlson, M. *Tetrahedron Lett.* **1968**, 579.
137. Fraenkel, G.; Subramanian, S.; Chow, A. *J. Am. Chem. Soc.* **1995**, *117*, 6300.
138. Whitesides, G.M.; Kaplan, F.; Roberts, J.D. *J. Am. Chem. Soc.* **1963**, *85*, 2167.
139. Fraenkel, G.; Dix, D.T. *J. Am. Chem. Soc.* **1966**, *88*, 979.
140. Fraenkel, G. In *Techniques in Chemistry: Investigations of Rates and Mechanisms of Reactions. Part 2*, 4th ed.; Bernasconi, C.F., Ed.; Wiley-Interscience: New York, 1986; pp 357–604.
141. Seebach, D.; Siegel, H.; Gabriel, J.; Hässig, R. *Helv. Chim. Acta* **1980**, *63*, 2046.
142. Bywater, S.; LaChance, P.; Worsfold, D.J. *J. Phys. Chem.* **1975**, *79*, 2148.
143. Fraenkel, G.; Russel, J.G.; Chen, V.H. *J. Am. Chem. Soc.* **1973**, *95*, 3208.
144. Konishi, K.; Matsumoto, H.; Saito, K.; Takahashi, K. *Bull. Chem. Soc. Jpn.* **1985**, *58*, 2294.
145. Hogen-Esch, T.E.; Jenkins, W.L. *J. Am. Chem. Soc.* **1981**, *103*, 3666.
146. Bank, S.; Marcantonio, R.P.; Bushweller, C.H. *J. Org. Chem.* **1984**, *49*, 5091.
147. Bank, S.; Dorr, R. *J. Org. Chem.* **1987**, *52*, 501.
148. Karplus, M.; Grant, D.M. *Proc. Natl. Acad. Sci. U.S.A.* **1959**, *45*, 1269.
149. Lichtman, W.M.; Grant, D.M. *J. Am. Chem. Soc.* **1967**, *89*, 2228.
150. McKeever, L.D.; Waack, R. *J. Chem. Soc., Chem. Commun.* **1969**, 750.
151. McKeever, L.D.; Waack, R.; Doran, M.A.; Baker, E.B. *J. Am. Chem. Soc.* **1969**, *91*, 1057.
152. Fraenkel, G.; Martin, K.V. *J. Am. Chem. Soc.* **1995**, *117*, 10336; and references therein.
153. Zarges, W.; Marsch, M.; Harms, K.; Boche, G. *Chem. Ber.* **1989**, *122*, 2304.
154. Zarges, W.; Marsch, M.; Harms, K.; Koch, W.; Frenking, G.; Boche, G. *Chem. Ber.* **1991**, *124*, 543.
155. Barchiesi, E.; Bradamante, S.; Ferracciola, R.; Pagani, G.A. *J. Chem. Soc., Perkin Trans. 2* **1990**, 375.
156. Jackman, L.M.; Sternhall, S. *Applications of Nuclear Magnetic Resonance Spectroscopy in Organic Chemistry*, 2nd ed.; Pergamon Press: Oxford, U.K., 1969; p 345.
157. McNab, H. *J. Chem. Soc., Perkin Trans. 2* **1981**, 1287.
158. Bradamante, S.; Pagani, G.A. *J. Chem. Soc., Perkin Trans. 2* **1986**, 1035.

159. Browne, S.E.; Asher, S.E.; Cornwall, E.H.; Frisoli, J.K.; Harris, L.J.; Salot, E.A.; Sauter, E.A.; Trecoske, M.A.; Veale; P.S., Jr. *J. Am. Chem. Soc.* **1984**, *106*, 1432.
160. Bradamante, S.; Pagani, G.A. *Pure Appl. Chem.* **1989**, *61*, 709.
161. Abbotto, A.; Bradamante, S.; Pagani, G.A. *J. Org. Chem.* **1993**, *58*, 444.
162. Abbotto, A.; Bradamante, S.; Pagani, G.A. *J. Org. Chem.* **1993**, *58*, 449.
163. Waack, R.; Doran, M.R.; Baker, E.B.; Olah, G. *J. Am. Chem. Soc.* **1966**, *88*, 1272.
164. Peoples, P.R.; Grutzner, J.B. *J. Am. Chem. Soc.* **1980**, *102*, 4709.
165. Fraenkel, G.; Qui, F. *J. Am. Chem. Soc.* **2000**, *122*, 12806; and references therein.
166. Bordwell, F.G. *Acc. Chem. Res.* **1988**, *21*, 456.
167. Doering, W. von E.; Hoffmann, A.K. *J. Am. Chem. Soc.* **1955**, *77*, 521.
168. Oae, S.; Tagaki, W.; Ohno, A. *Tetrahedron* **1964**, *20*, 417.
169. Oae, S.; Tagaki, W.; Ohno, A. *Tetrahedron* **1964**, *20*, 427.
170. Speers, P.; Laidig, K.E.; Streitwieser, A. *J. Am. Chem. Soc.* **1994**, *116*, 9257.
171. Pine, S.H.; Shen, G.; Bautista, J.; Sutton, C., Jr.; Yamada, W.; Apodaca, L. *J. Org. Chem.* **1990**, *55*, 2234.
172. Hofer, O.; Eliel, E.L. *J. Am. Chem. Soc.* **1973**, *95*, 8045.
173. Barbarella, G.; Garbesi, A.; Fava, A. *J. Am. Chem. Soc.* **1975**, *97*, 5883.
174. Barbarella, G.; Dembech, P.; Garbesi, A.; Bernardi, F.; Bottoni, A.; Fava, A. *J. Am. Chem. Soc.* **1978**, *100*, 200.
175. Rauk, A.; Buncel, E.; Moir, R.Y.; Wolfe, S. *J. Am. Chem. Soc.* **1965**, *87*, 5498.
176. Bullock, E.; Scott, J.M.W.; Golding, P.D. *J. Chem. Soc., Chem. Commun.* **1967**, 168.
177. Hutchinson, B.J.; Andersen, K.K.; Katritzky, A.R. *J. Am. Chem. Soc.* **1969**, *91*, 3839.
178. Fraser, R.R.; Schuber, F.J. *Can. J. Chem.* **1970**, *48*, 633.
179. Fraser, R.R.; Schuber, F.J.; Wigfield, Y.Y. *J. Am. Chem. Soc.* **1972**, *94*, 8795.
180. Nishihata, K.; Nishio, M. *Tetrahedron Lett.* **1972**, *4839*.
181. Cram, D.J.; Scott, D.A.; Neilsen, W.D. *J. Am. Chem. Soc.* **1961**, *83*, 3696.
182. Corey, E.J.; Kaiser, E.T. *J. Am. Chem. Soc.* **1961**, *83*, 490.
183. Cram, D.J.; Trepka, R.D.; St. Janiak, P. *J. Am. Chem. Soc.* **1966**, *88*, 2749.
184. Brown, M.D.; Cook, M.J.; Hutchinson, B.J.; Katritzky, A.R. *Tetrahedron* **1971**, *27*, 593.
185. Horner, L.; Hoffmann, H.; Wippel, H.G. *Chem. Ber.* **1958**, *91*, 61.
186. Stec, W.J. *Acc. Chem. Res.* **1983**, *16*, 411.
187. Wadsworth, W.S. Jr., Emmons, W.D. *J. Am. Chem. Soc.* **1961**, *83*, 1733.
188. Wittig, G. *Pure Appl. Chem.* **1964**, *9*, 245.
189. March, J. *Advanced Organic Chemistry*: *Reactions, Mechanisms and Structure;* 4th ed.; Wiley: New York, 1992; pp 956–963.
190. March, J. *Advanced Organic Chemistry*: *Reactions, Mechanisms and Structure;* 4th ed.; Wiley: New York, 1992; pp 974–976.
191. Streitwieser, A.; Williams, J.E. *J. Am. Chem. Soc.* **1975**, *97*, 191.
192. Lehn, J.M.; Wipff, G. *J. Am. Chem. Soc.* **1976**, *98*, 7498.
193. Bernardi, F.; Csizmadia, I.G.; Mangini, A.; Schlegel, H.B.; Wangbo, M.H.; Wolfe, S. *J. Am. Chem. Soc.* **1975**, *97*, 2209.
194. Hopkinson, A.C.; Lien, M.H. *J. Org. Chem.* **1981**, *46*, 998.
195. Schleyer, P. von R.; Kos, A.J. *Tetrahedron* **1983**, *39*, 1141.
196. Schleyer, P. von R.; Clark, T.; Kos, A.J.; Spitznagel, G.W.; Rohde, C.; Arad, D.; Houk, K.N.; Rondan, N.G. *J. Am. Chem. Soc.* **1984**, *106*, 6467.
197. Cuevas, G.; Juaristi, E. *J. Am. Chem. Soc.* **1997**, *119*, 7545.
198. Bors, D.A.; Streitwieser, A. *J. Am. Chem. Soc.* **1986**, *108*, 1397.
199. Van Verth, J.E.; Saunders; W.H., Jr. *J. Org. Chem.* **1997**, *62*, 5743.
200. Saunders; W.H., Jr.; Van Verth, J.E. *J. Org. Chem.* **1995**, *60*, 3452.
201. Mitchell, D.J.; Wolfe, S.; Schlegel, H.B. *Can. J. Chem.* **1981**, *59*, 3280.

202. Liu, Z.P.; Schlosser, M. *Tetrahedron Lett.* **1990**, *31*, 5753.
203. Schmidbaur, H.; Jeong, J.; Schier, A.; Graf, W.; Wilkinson, D.L.; Müller, G. *New J. Chem.* **1989**, *13*, 341.
204. Francl, M.M.; Pellow, R.C.; Allen, L.C. *J. Am. Chem. Soc.* **1988**, *110*, 3723.
205. Schmidbaur, H.; Schier, A.; Milewski-Mahrla, B.; Schubert, U. *Chem. Ber.* **1982**, *115*, 722.
206. Grützmacher, H.; Pritzkow, H. *Angew. Chem., Int. Ed. Engl.* **1992**, *31*, 99.
207. Raabe, G.; Gais, H.J.; Fleischhauer, J. *J. Am. Chem. Soc.* **1996**, *118*, 4622.
208. Wiberg, K.B.; Castejon, H. *J. Am. Chem. Soc.* **1994**, *116*, 10489.
209. Lischka, H. *J. Am. Chem. Soc.* **1977**, *99*, 353.
210. Dixon, D.A.; Dunning, T.H.; Eades, R.A.; Gassman, P.G. *J. Am. Chem. Soc.* **1983**, *105*, 7011.
211. Patterson, C.H.; Messmer, R.P. *J. Am. Chem. Soc.* **1989**, *111*, 8059.
212. Bernasconi, C.F.; Kittredge, K.W. *J. Org. Chem.* **1998**, *63*, 1944.
213. Mislow, K.; Siegel, J. *J. Am. Chem. Soc.* **1984**, *106*, 3319.
214. Solladié, G. *Synthesis* **1981**, 185.
215. Andersen, K.K. In *The Chemistry of Sulphones and Sulphoxides;* Patai, S.; Rappoport, Z.; Stirling, C.J.M., Eds.; Wiley: Chichester, U.K., 1988; Chapter 3, pp 55–94.
216. Solladié, G.; Carreño, M.C. In *Organosulphur Chemistry: Synthetic Aspects;* Page, P.C.B., Ed.; Academic Press: New York, 1995; Chapter 1, pp 1–47.
217. Barbachyn, M.R.; Johnson, C.R. In *Asymmetric Synthesis: The Chiral Carbon Pool and Chiral Sulfur, Nitrogen, Phosphorus and Silicon Centers*; Morrison, J.D.; Scott, J.W., Eds.; Academic Press: New York, 1984; Vol. 4, Chapter 2, pp 227–263.
218. Valentine, D., Jr. In *Asymmetric Synthesis: The Chiral Carbon Pool and Chiral Sulfur, Nitrogen, Phosphorus and Silicon Centers*; Morrison, J.D.; Scott, J.W., Eds.; Academic Press: New York, 1984; Vol. 4, Chapter 3, pp 263-313.
219. Lowe, G. *Acc. Chem. Res.* **1983**, *16*, 244.
220. Posner, G.H. In *The Chemistry of Sulphones and Sulphoxides*; Patai, S.; Rappoport, Z.; Stirling, C.J.M., Eds.; Wiley: Chichester, U.K., 1988; Chapter 16, pp 823–850.
221. Kukhar, V.P.; Svistunova, N.Yu.; Solodenko, V.A.; Soloshonok, V.A. *Russ. Chem. Rev.* (Engl. Transl.) **1993**, *62*, 261.
222. Vedejs, E.; Peterson, M.J. In *Advances in Carbanion Chemistry*; Snieckus, V., Ed.; JAI Press: Greenwich, CT, 1996; Vol. 2, pp 1–86.
223. Tagliavini, E.; Trombini, C.; Umani-Ronchi, A. In *Advances in Carbanion Chemistry*; Snieckus, V., Ed.; JAI Press: Greenwich, CT, 1996; Vol. 2, pp 111–146.
224. Braun, M. In *Advances in Carbanion Chemistry*; Snieckus, V., Ed.; JAI Press: Greenwich, CT, 1992; Vol. 1, pp 177–248.
225. Hua, D.H. In *Advances in Carbanion Chemistry*; Snieckus, V., Ed.; JAI Press: Greenwich, CT, 1992; Vol. 1, pp 249–282.
226. Carreño, M.C. *Chem. Rev.* **1995**, *95*, 1717.
227. Hanessian, S.; Gomtsyan, A.; Payne, A.; Hervé, Y.; Beaudoin, S. *J. Org. Chem.* **1993**, *62*, 261.
228. Pietrusiewicz, K.M.; Zablocka, M. *Tetrahedron Lett.* **1989**, *30*, 477.
229. Sawamura, N.; Ito, Y.; Hayashi, T. *Tetrahedron Lett.* **1989**, *30*, 2247.
230. Hanessian, S.; Bennani, Y.L.; Delorme, D. *Tetrahedron Lett.* **1990**, *31*, 6461.
231. Hanessian, S.; Bennani, Y.L.; Delorme, D. *Tetrahedron Lett.* **1990**, *31*, 6465.
232. Sting, M.; Steglich, W. *Synthesis* **1990**, 132.
233. Aitken, R.A.; Kilényi, S.N., Eds.; *Asymmetric Synthesis*; Blackie Academic & Professional: Glasgow, U.K., 1992.
234. Wolfe,S.; Rauk, A. *J. Chem. Soc., Chem. Commun.* **1966**, 778.

235. Baldwin, J.E.; Hacker, R.E.; Scott, R.M. *J. Chem. Soc., Chem. Commun.* **1969**, 1415.
236. Durst, T.; Fraser, R.R.; McClory, M.R.; Swingle, R.B.; Viau, R.; Wigfield, Y.Y. *Can. J. Chem.* **1970**, *48*, 2148.
237. Beillmann, J.F.; Vicens, J.J. *Tetrahedron Lett.* **1974**, 2915.
238. Beillmann, J.F.; Vicens, J.J. *Tetrahedron Lett.* **1978**, 467.
239. Durst, T. *J. Am. Chem. Soc.* **1969**, *91*, 1034.
240. Durst, T.; Molin, M. *Tetrahedron Lett.* **1975**, 63.
241. Tanikaga, R.; Hamamura, K.; Hosoya, K.; Kaji, A. *J. Chem. Soc., Chem. Commun.* **1988**, 817.
242. Tanikaga, R.; Murashima, T. *J. Chem. Soc., Perkin Trans. 1* **1989**, 2142.
243. Wolfe, S.; LaJohn, L.A.; Weaver, D.F. *Tetrahedron Lett.* **1984**, *25*, 2863.
244. Sakuraba, H.; Ushiki, S. *Tetrahedron Lett.* **1990**, *31*, 5349.
245. Gais, H. J.; Hellmann, G. *J. Am. Chem. Soc.* **1992**, 114, 4439.
246. Gais, H.J.; Hellmann, G.; Günther, H.; Lopez, F.; Lindner, H.J.; Braun, S. *Angew. Chem., Int. Ed. Engl.* **1989**, *28*, 1025.
247. Aggarwal, V.K. *Angew. Chem., Int. Ed. Engl.* **1994**, *33*, 175.
248. Gais, H.J.; Vollhardt, J.; Krüger, C. *Angew. Chem., Int. Ed. Engl.* **1988**, *27*, 1092.
249. Boche, G. *Angew. Chem., Int. Ed. Engl.* **1989**, *28*, 277
250. Gais, H.J.; Hellmann, G.; Lindner, H.J. *Angew. Chem., Int. Ed. Engl.* **1990**, *29*, 100.
251. Gais, H.J.; Hellmann, G. *J. Am. Chem. Soc.* **1991**, *113*, 4002.
252. Hollstein, W.; Harms, K.; Marsch, M.; Boche, G. *Angew. Chem., Int. Ed. Engl.* **1988**, *27*, 846.
253. Padwa, A.; Wannamaker, M.W. *Tetrahedron Lett.* **1986**, *27*, 2555.
254. Chassaing, G.; Lett, R.; Marquet, A. *Tetrahedron Lett.* **1978**, 471.
255. Eliel, E.L.; Hartmann, A.A.; Abatjoglou, A.G. *J. Am. Chem. Soc.* **1974**, *96*, 1087.
256. Abatjoglou, A.G.; Eliel, E.L.; Kuyper, L.F. *J. Am. Chem. Soc.* **1977**, *99*, 8262.
257. Denmark, S.E.; Dorow, R.L. *J. Am. Chem. Soc.* **1990**, *112*, 864.
258. Denmark, S.E.; Stadler, H.; Dorow, R.L.; Kim, J.H. *J. Org. Chem.* **1991**, *56*, 5063.
259. Denmark, S.E.; Swiss, K.A.; Wilson, S.R. *J. Am. Chem. Soc.* **1993**, *115*, 3826.
260. Denmark, S.E.; Swiss, K.A. *J. Am. Chem. Soc.* **1993**, *115*, 12195.
261. Denmark, S.E.; Chen, C.T. *J. Org. Chem.* **1994**, *59*, 2922.
262. Denmark, S.E.; Chen, C.T. *J. Am. Chem. Soc.* **1995**, *117*, 11879.
263. Kranz, M.; Denmark, S.E. *J. Org. Chem.* **1995**, *60*, 5867.
264. Denmark, S.E.; Swiss, K.A.; Wilson, S.R. *Angew. Chem., Int. Ed. Engl.* **1996**, *35*, 2515.
265. Kranz, M.; Denmark, S.E.; Swiss, K.A.; Wilson, S.R. *J. Org. Chem.* **1996**, *61*, 8551.
266. Reich, H.J.; Dykstra, R.R. *Angew. Chem., Int. Ed. Engl.* **1993**, *32*, 1469.
267. Arvidsson, P.I.; Ahlberg, P.; Hilmersson, G. *Eur. J. Chem.* **1999**, *5*, 1348.
268. Arvidsson, P.I.; Hilmersson, G.; Ahlberg, P. *J. Am. Chem. Soc.* **1999**, *121*, 1883.
269. Ahlberg, P.; Karlsson, A.; Davidsson, O.; Hilmersson, G.; Loewendahl, M. *J. Am. Chem. Soc.* **1997**, *119*, 1751.
270. Ahlberg, P.; Davidsson, O.; Loewendahl, M.; Hilmersson, G.; Karlsson, A.; Haakansson, M. *J. Am. Chem. Soc.* **1997**, *119*, 1745.
271. Karlsson, A.; Hilmersson, G.; Davidsson, O.; Loewendahl, M. *Acta Chem. Scand.* **1999**, *53*, 693.
272. Van Verth, J.E.; Saunders, W.H., Jr.; Kermis, T.W. *Can. J. Chem.* **1998**, *76*, 821.
273. Saunders, W.H., Jr. *J. Org. Chem.* **1999**, *64*, 861.
274. Van Verth, J.E.; Saunders, W.H., Jr. *Can. J. Chem.* **1999**, *77*, 810.
275. Saunders, W.H., Jr. *J. Org. Chem.* **1997**, *62*, 244.

4

Spectroscopy of Carbanions

4.1 INTRODUCTION AND SCOPE

The presentation in this chapter will concentrate on some selected aspects of carbanion structure from the viewpoint of how spectroscopy has shed light on structure; for more general treatments, including other spectroscopic studies of diverse carbanions and organometallics, the reader is referred to different reviews and texts.[1–12]

It is now generally recognized that, in solution in nonpolar solvents, such as tetrahydrofuran, in which they are normally studied, carbanions are associated to some degree with the metal counterions; free carbanion species exist only under exceptional circumstances. The bonding between the carbon that bears negative charge and the metal atom (generally alkali-metal) can be considered as having appreciable, or predominant, ionic character, depending on the nature of the metal, the medium, and the molecular environment—the moieties attached to the carbanionic center. In a formal sense, one can describe this bond as a resonance hybrid of covalent and ionic canonical forms:

$$\mathrm{R}\dot{-}\mathrm{M} \quad = \quad \mathrm{R}^{\delta-}-\mathrm{M}^{\delta+} \quad \leftrightarrow \quad \mathrm{R}^{-}-\mathrm{M}^{+} \tag{4.1}$$

One parameter by which bond polarity in the R-M species may be estimated is the electronegativity difference between the two elements forming the C–M bond; thus, bond polarity, $C^{\delta-}-M^{\delta+}$, will increase as the magnitude of the difference in electronegativity for the two elements increases. As an approximation, we can use Pauling's relationship[13] between electronegativity differences, and this leads to the following estimates for the percentage ionic character: C–Li, 42%; C–Cs,

55%. There is some controversy as to the validity of this approach[14] and recent theoretical calculations point to much greater degrees of ionic character.[15–18] There has been vigorous controversy concerning the precise nature of the carbon–lithium bond, with strictly covalent or strictly ionic bonding at the extremes. A complete historical account of this controversy has been published and the reader is referred to this interesting review.[3] Spectroscopic methods have provided insight into this controversy.

Another important feature that concerns carbanions in solution in nonpolar solvents is that, commonly (i.e., the non-delocalized carbanides such as butyllithium, etc.), they do not exist as monomeric species but rather as dimers, tetramers, and higher aggregates as indicated by equation 4.2.

$$(\mathrm{R-M})_n \leftrightarrows (\mathrm{R-M})_{n-x} \leftrightarrows (\mathrm{R-M})_{n-y} \leftrightarrows \mathrm{R-M} \qquad (4.2)$$

While measurements of colligative properties, such as measurements of vapor–pressure barometry, boiling point elevation, freezing point depression, and so on, were the first to provide evidence on aggregation,[19–23] an increased understanding of these aggregation phenomena has accrued through spectroscopic investigations that have afforded detailed structural information of the aggregates.

A facet where spectroscopy has been uniquely suitable in shedding light is the phenomenon of different types of ion pairing that are now recognized as existing for the R–M molecules in solution in nonpolar solvents. The various stages of ionization from dissociation of the solvated R–M, namely, $(\mathrm{R{-}M})_{\mathrm{solv}}$, to the "free ions" (FI), are shown in equation 4.3:

$$(\mathrm{R-M})_{\mathrm{solv}} \leftrightarrows \underset{\mathrm{CIP}}{(\mathrm{R^-M^+})_{\mathrm{solv}}} \leftrightarrows \underset{\mathrm{SSIP}}{(\mathrm{R^-\|M^+})_{\mathrm{solv}}} \leftrightarrows \underset{\mathrm{FI}}{(\mathrm{R^-})_{\mathrm{solv}} + (\mathrm{M^+})_{\mathrm{solv}}} \qquad (4.3)$$

Through the pioneering work of Winstein,[24] Grunwald,[25] Hogen-Esch and Smid,[26] and Szwarc especially,[27,28] ion pairs can be defined as pairs of oppositely charged ions with a common solvation shell whose lifetimes are sufficiently long for the pairs to be recognizable kinetic entities in solution, with the two ions being held together by electrostatic binding forces.[29] Two distinct forms of ion pairs can be discerned: contact (tight) ion pairs or CIPs, and solvent-separated (loose) ion pairs or SSIPs. In the former, the distance between the cation and anion is small, solvation is peripheral or external to the RM pair, and no solvent molecules are present between the two ions, while in the latter case the solvated cations and anions are separated by solvent molecules. Again, while classical physical methods such as conductivity[30] and electric permittivity[31] have provided valuable information, definitive evidence has come from spectroscopic studies. The type of ion pairing in a given system is known to have important influence on stereoselectivity of reactions, and reactivity in general.[1–12,32–34]

Coverage in this chapter will be limited in the main to arylmethyl anions, with occasional reference to related species. For example, comparison will be made to fluorenyl anions, which are constrained planar analogues of diphenylmethide ion. Also, we will consider briefly aryl-silyl, -germyl, -stannyl, and -plumbyl anions, that is, anions where the carbanionic center is replaced by other group 14 elements. Arylmethyl carbanions such as the benzyl, diphenylmethyl, and triphenylmethyl anions, have played a central role in the development of the concepts of delocalization in carbanion systems,[35–37] and have been focal in structure–reactivity relationships.[12,35–38] Together with other delocalized anions, these archetypal species have formed the basis for the determination of solution acidities of weak carbon acids,[39,40] including that of even the simplest binary hydride molecule, H_2.[41,42] The experimental methods for measurements of solution acidities of these species have generally been spectroscopic,[38,39] though other methods described in Chapter 1 have given complementary information. Spectroscopy of the arylmethyl anions is therefore of some interest.

Importantly, the arylmethyl carbanions have been studied by an array of spectroscopic methods: electronic absorption (UV–vis) spectroscopy, vibrational (IR, Raman) spectroscopy, nuclear magnetic resonance (NMR) spectroscopy, and by X-ray diffraction studies. The following presentation will bring together, for the first time, the different spectroscopies as they have been directed towards elucidation of structure of these carbanions. Structure of, and substituent effects on, benzylic carbanions functionalized by electron-withdrawing groups (EWG), including their spectroscopic properties, have been addressed in reference 35 (but also see Chapter 3).

The common charge-localized alkyl (methyl, propyl, butyl) anions, which are typically studied as the lithium derivatives, will not be discussed here. They are known to exist in solution or in the solid state as dimers, tetramers, hexamers, and higher aggregates with fascinating, diverse structures, as revealed by ^{13}C NMR, and especially through measurement of ^{13}C–$^{6(7)}$Li couplings.[43-58] Phenyl and allyl anions, also studied as the lithium compounds, will not be discussed here and the interested reader is referred to the original literature.[43–55]

In the following presentation, the main emphasis will be given to UV–vis and NMR spectroscopy because these, while highly informative techniques, are available in most laboratories. The section on UV–vis spectroscopy will precede that on NMR, partly for historical reasons, since many of the characteristics of carbanions, such as ion-pairing equilibria, were first observed by electron absorption spectroscopy before the huge developments in NMR methodology took place. That will be followed by the section on IR and Raman spectroscopy; however, first in our presentation is a brief account of results obtained by X-ray studies. Though based on principles that differ from the various other spectroscopic methods considered here (i.e., transitions between different quantum energy levels with absorption or emission of radiation), X-ray diffraction studies provide definitive information on carbanion structure, including position of the counterion, in the solid state versus solution.

4.2 SPECTROSCOPIC STUDIES OF ARYLMETHYL CARBANIONS: X-RAY, UV–VIS, NMR

4.2.1 X-ray Studies

X-ray diffraction studies provide the ultimate proof of structure and, accordingly, a number of organolithium compounds have been examined by this method. It is evident also that X-ray studies are complementary to solution studies, with the limitation of availability of suitable crystals. Often, amines such as quinuclidine or tetramethylethylenediamine (TMEDA) are used as ligands to complex the cation. However, although the preferred cation position and state of ion pairing or aggregation can be determined unambiguously in this way, one can always dispute whether the structure prevails in solution. These and other facets are discussed in reviews on X-ray structures of organolithium compounds.[56–58]

Benzyllithium[59] and triphenylmethyllithium[60] and -sodium[61] were the first arylmethyl anions to be studied, as the TMEDA complexes. The latter are contact ion pairs between the triphenylmethyl anion, $[Ph_3C]^-$, and the alkali-metal cation coordinated to the bidentate ligand $[TMEDA{\cdot}Li(Na)]^+$. On the other hand, a polymeric structure prevails in the case of benzyllithium tetramethylethylenediamine. The repeating unit consists of a benzyl carbanion and two half TMEDA cages coordinated to a lithium cation. The Li atom is within bonding distance to the α-carbon, and also to the *ipso* and *ortho* carbons of the phenyl ring.[59] This type of structural environment for Li is also found in the polymeric species $[PhCH_2Li{\cdot}DABCO]_n$, **51**,[60,62] and $[PhCH_2Li{\cdot}Et_2O]_n$, **52**,[63] as shown, with the charges omitted for clarity. In triphenylmethyllithium, the lithium cation is asymmetrically positioned and bridges the two phenyl rings in such a way that charge delocalization is equally important for both rings; the third ring is twisted out of conjugation.

Arylmethyl anions with the counterion complexed by crown ether have enabled the observation of the carbanions in the absence of coordinating amine ligands.[64] The diphenylmethyl-crown-ether-complexed lithium $[4(12\text{-crown-}4)_2][CHPh_2]$ (**54**) and the triphenylmethyllithium analogue $[Li(12\text{-crown-}4)_2][CPh_3]$ THF (**53**) were isolated as orange and red needles, respectively, stable in the absence of air or moisture. Both anions are found to have planar geometry at the central carbon atom in accord with extensive delocalization as illustrated in Figures 4.1 and 4.2. Also noteworthy is the coplanarity of the two phenyl rings in the diphenylmethyl system, whereas in the triphenylmethyl case the rings assume a propeller arrange-

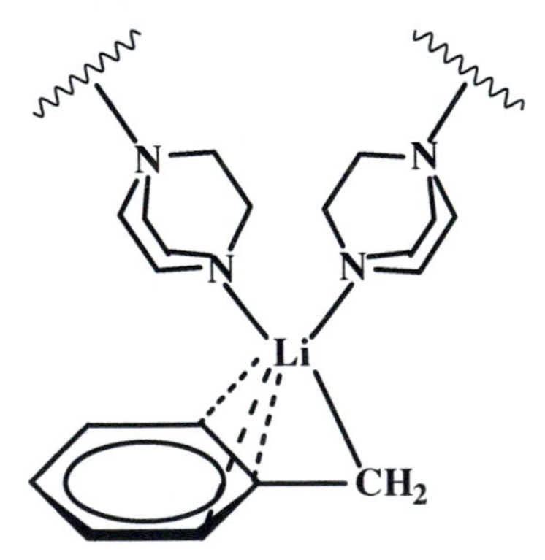

51

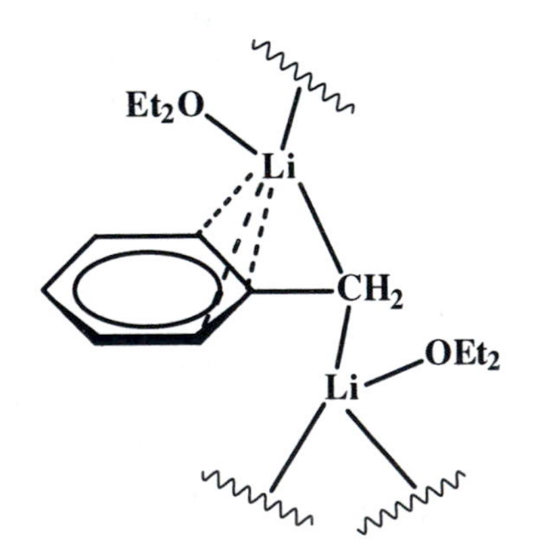

52

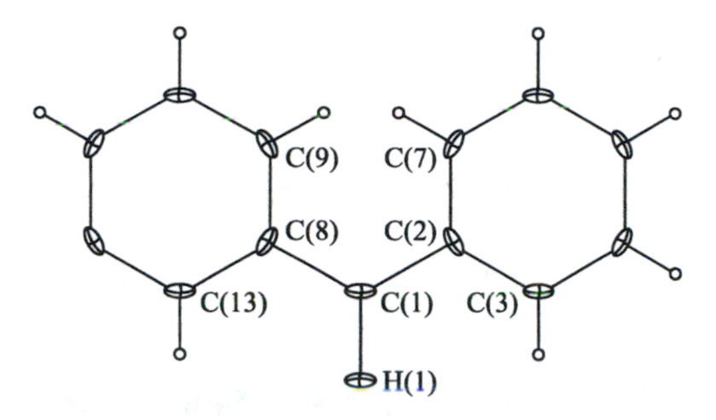

Figure 4.1. Bond distances (Å) and angles (deg) for **54**: C(1)–C(2) = 1.435 (6); C(1)–C(8) = 1.404(6); C(2)"C(3) = 1.433(7); C(3)–C(4) = 1.399 (7); C(4)–C(5) = 1.429 (8); C(5)–C(6) = 1.367 (7); C(6)–C(7) = 1.376 (7); C(7)–C(2) = 1.439 (6); C(1)C(2)C(7) = 119.4 (4); C(2)C(3)C(4) = 123.0 (4); C(3)C(4)C(5) = 120.9 (5); C(4)C(5)C(6) = 116.7 (5); C(5)C(6)C(7) = 123.0 (5); C(2)C(7)C(6) = 123.5 (4). Distances and angles in the C(8) phenyl ring are similar. (Reproduced from reference 64. Copyright 1985 American Chemical Society.)

ment as a result of steric constraints. Unfortunately, the authors were unable to isolate crystals in an analogous preparation of [Li(12-crown-4)$_2$][CH$_2$Ph].

An interesting X-ray study in the context of the classical ion-pair concept is an investigation of the dilithium bisfluorenylethane compounds.[65] These THF-complexed salts were found to exist as SSIPs, with the lithium atoms coordinated to four THF molecules.

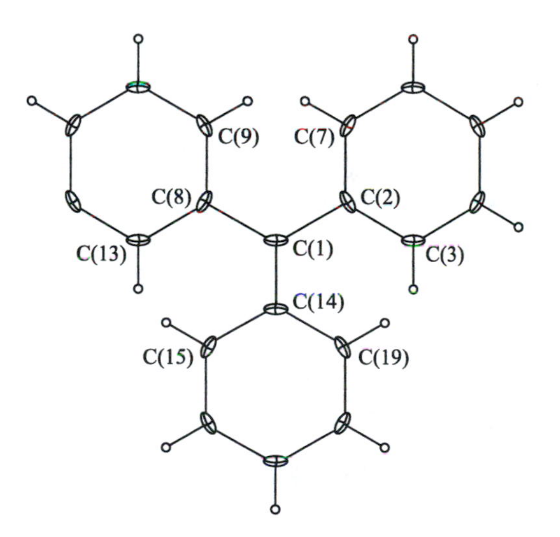

Figure 4.2. Bond distances (Å) and angles (deg) for **53**: C(1)–C(2) = 1.451 (5); C(1)–C(8) = 1.459 (5); C(1)–C(14) = 1.450 (4); C(2)C(1)C(8) = 119.4 (3); C(8)C(1)C(14) = 122.3 (3). Distances and angles in phenyl rings are similar to those in **54**. (Reproduced from reference 64. Copyright 1985 American Chemical Society.)

4.2.2 Electron Absorption Spectroscopic (UV–vis) Studies

4.2.2.1 General Considerations

Continuing in the spirit of Chapter 1, where carbanions are considered in relation to their conjugate acids, one can note the pronounced bathochromic shift on deprotonation of most carbon acids that give rise to delocalized anions. Taking toluene as an example, the ionization $PhCH_3 \rightarrow PhCH_2^-$ (Na^+,THF) is accompanied by a shift in λ_{max} from 250 nm to 355 nm,[66] while for triphenylmethane, $Ph_3CH \rightarrow Ph_3C^-$ (K^+, THF), the shift is from 254 nm to 486 nm.[67] Also noted here is the bathochromic shift accompanying increased resonance delocalization in the anions. Thus, resonance stabilization of the anions must be greater than that of their carbon acid precursors.

The electronic absorption (UV–vis) spectra of carbanion alkali-metal salts exhibit a marked dependence on the nature of the solvent, cation, temperature, and the presence of cation-complexing agents such as the macrocyclic crown ethers and cryptands. The theoretical basis that has been advanced to account for these spectral shifts by a number of workers[68–71] need only briefly be indicated here.

As would be expected, UV–vis spectra of SSIPs are indistinguishable from the spectra of the free (or crown/cryptand complexed) anions. In the case of CIPs, however, ion association leads to considerable perturbation of molecular orbital energy levels, the result of which is greater stabilization of the ground state than of the excited state, which induces a hypsochromic shift in λ_{max} that increases with decreasing cation radius. Figure 4.3 illustrates these ideas for both absorption and emission.[71]

Thus, the bathochromic shift in UV–vis absorption spectra associated with an increase in cationic radius in nonpolar solvents where contact ion pairs predominate is attributed to a greater destabilization of the ground state than of the excited state with the larger cation. In the ground state, the dipole is minimized

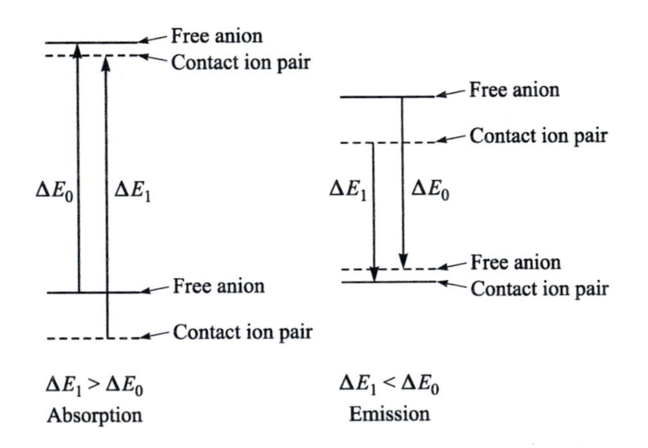

Figure 4.3. Effect of ion pairing on transition energies. (Reproduced with permission from reference 71. Copyright 1987 Elsevier.)

by locating the cation close to the center of negative charge, but upon excitation there is a redistribution of charge density, and hence an increased dipole, since the position of the cation is unaffected according to the Franck–Condon principle. This leads to a lower energy transition with the larger cation. A linear plot between ν_{max} and $1/r_C + x)$, where r_C is the cationic radius, x is an empirically determined constant, and ν_{max} is the frequency of the absorption maximum, is generally observed. This is consistent with an expected reciprocal relationship between the electrostatic interaction for the two ions forming the ion pair, which will depend on the interionic separation and the transition energy due to the perturbation of the molecular energy levels of the anion by the cationic field.

An increase in polarity of the solvent, leading to formation of solvent-separated ion pairs, will also decrease the interaction between anion and cation and is hence also expected to cause a bathochromic shift according to the preceding discussion. This is, indeed, found to be the case for arylmethyl as well as other carbanions.[72–75] A similar result is consequent to complexation of the cation on addition of a crown ether or cryptand, or on lowering the temperature, which favors the formation of SSIPs.

4.2.2.2 UV–vis characterization of CIPs and SSIPs

The systems consisting of triphenylmethyllithium (TPM^-Li^+) and triphenylmethylpotassium (TPM^-K^+), and structurally similar carbanions, in ethereal solvents, provide good examples for illustrating the spectroscopic characteristics of different types of ion pairs. Solvents examined here are 1,2-dimethoxyethane, tetrahydrofuran, and diethyl ether, as these form a gradation of solvating power (donicity),[76] DME > THF > Et_2O. It can be assumed that the extent of ionization into fully dissociated ions is small under these conditions.

The UV–vis spectra of TPM^-Li^+ in THF, DME, and Et_2O at ambient temperature are shown in Figure 4.4, while the wavelengths of the absorption maxima are given in Table 4.1.[67] The change in λ_{max} in the different solvents is striking and is indicative of a change in the type of ion pair present. Comparable results have been reported for fluorenyl carbanions.[73] One can conclude that TPM^-Li^+ is present predominantly as CIP in Et_2O and as SSIP in THF and DME, at ambient temperature, on the basis of the following considerations:[74–76] (1) absorption maxima for SSIP occur at longer wavelengths than for CIP; (2) for a given type of ion pair, λ_{max} is only slightly dependent on the solvent medium over wide ranges of solvent type and polarity; and (3) the tendency for SSIP formation increases in the order of the solvating powers of the solvents, that is, Et_2O < THF < DME.

For TPM^-K^+, the corresponding results are included in Table 4.1. In this case, there is predominant formation of CIP in Et_2O and SSIP in DME, but in THF both types of ion pairs are present, based on the following considerations: (1) the tendency for SSIP formation is expected to be greater for Li^+-anion pairs than for K^+-anion pairs; and (2) for CIP species there is a bathochromic shift upon changing the cation from Li^+ to K^+, but for SSIP species λ_{max} varies only slightly with change in counterion.

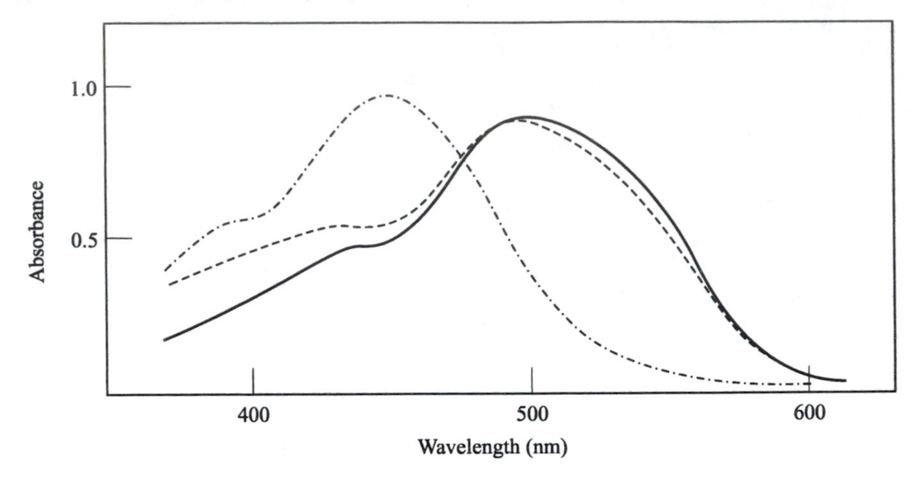

Figure 4.4 Visible absorption spectra of triphenylmethyllithium in ether (—·—·—), tetrahydrofuran (——), and dimethoxyethane (- - - -) at room temperature. (Reproduced from reference 67. Copyright 1979 American Chemical Society.)

Table 4.1 also shows results obtained following addition of the cation complexing agent 18-crown-6 ether (X). It is seen that this converts CIP species to crown-ether-complexed species (R^-, X, M^+); the latter has λ_{max} closely similar to SSIP species. (See reference 77 for complexation of crowns with fluorenyl carbanions and evidence for formation of crown-DME-separated ion pairs, as well as external coordination of contact ion pairs by these agents.)

Further evidence on CIP ⇆ SSIP equilibria is obtained from their temperature dependence. Generally, such equilibria are characterized by negative ΔH values

Table 4.1. Contact (R^-M^+), solvent-separated ($R^-\|M^+$), and 18-crown-6 ether complexed (R^-,X,M^+) ion pairs of triphenylmethyl carbanions in ethereal solvents at room temperature: effect of solvent and counterion

R^-M^+	Solvent	$\lambda_{max}(\in)^b$ R^-,M^+	$R^-\|M^+$	Fraction of $R^-\|M^+$ at 25 °C	λ_{max} R^-X,M^+
TPM^-Li^+	Et_2O	446,390 sh(21,900)		0.15	494,430 sh
	THF		500,435 sh(28,300)	0.95	500,435 sh
	DME		496,432 sh(30,000)	1.00	496,432 sh
TPM^-K^+	Et_2O	476,414 sh(24,800)		0.00	492,430 sh
	THF		486,420 sh(21,300)	0.65	495,430 sh
	DME		494,430 sh(30,000)	0.85	494,430 sh

SOURCE: Reproduced from reference 67. Copyright 1979 American Chemical Society.
[a]Extinction coefficient at absorption maximum [L/(mol·cm)]; ∈(sh) is approximately half the absorption maximum value.

owing to the gain in solvation energy on formation of SSIP. Also, the SSIP formation process is associated with negative ΔS values as a result of immobilization of solvent molecules around the cation.[78]

Illustrative spectral changes of the temperature dependence are shown in Figure 4.5 for the case of $TPM^{-}Li^{+}$ in Et_2O. The results permit evaluation of equilibrium constants for this CIP $\leftrightarrows$ SSIP process and, hence, the respective ΔH and ΔS values can be calculated. Results for $TPM^{-}Li^{+}$ and $TPM^{-}K^{+}$ in the ethereal solvents are given in Table 4.2 and these confirm the anticipated negative ΔH and ΔS values.

It is noteworthy that rather similar results have been reported in a ^{1}H NMR study of $TPM^{-}Li^{+}$ and $DPM^{-}Li^{+}$ (*vide infra*). The importance of using different experimental methods to glean information on ion-pair equilibria has been emphasized.[26,74]

A UV–vis study of diphenylmethyllithium and diphenylmethylpotassium in ethereal solvents provided evidence of another type of ion pair that has spectroscopic properties intermediate between CIP and SSIP species.[79] This may be termed as having peripheral or external solvation, as had previously been proposed in a study of 1,3-diphenylpropene anion in a number of ethereal solvents.[80] Alternatively, one could visualize the intermediate state as one in which the distance between the ions is larger than that in the contact ion pair, but not yet so

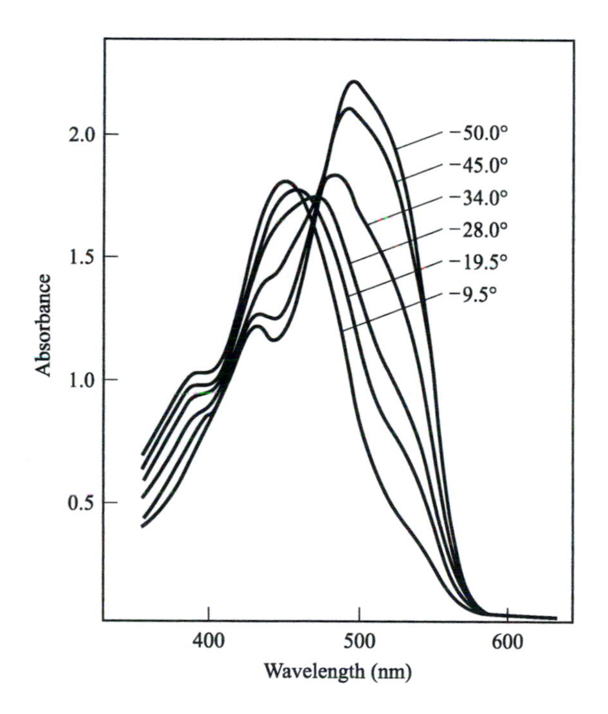

Figure 4.5. Temperature dependence on absorption spectra of triphenylmethyllithium (7.51×10^{-5} M) in ether. (Reproduced from reference 67. Copyright 1979 American Chemical Society.)

Table 4.2. Thermodynamic parameters for contact ⇆ solvated ion-pair equilibria of triphenymethyllithium and potassium in ethereal solvents

R^-M^+	Solvent	Temperature Range (°C)	K^a	ΔH^b(kcal/mol)	ΔS^beu
TPM^-Li^+	Et_2O	−54° to +10°	0.1	−11.9	−46
TPM^-Li^+	THF	+ 2° to +45°	39	−9.2	−24
TPM^-K^+	THF	−53° to +23°	2.0	−6.7	−21
TPM^-K^+	DME	+ 1° to +34°	5.7	−11.8	−36

SOURCE: Reproduced from reference 67.

[a] $K = [R \parallel M^+]/[R^-,M^+]$ at 25 °C.

[b] Uncertainty in ΔH, ΔS is ± 10% to ± 20%.

large that a solvent molecule can be entirely placed between the two ions. The perturbation of the energy levels of the carbanion by the electric field of the cation will be influenced by this incipient solvation, with consequent effect on the spectral absorption. If this interpretation is correct, then the interesting point is that such intermediate states can have sufficient stability so as to be observable as distinct species on the timescale of electronic absorption spectroscopy. Ion-pair equilibria in these systems can be denoted by equations 4.4 and 4.5 (and similarly extended to DPM^-K^+):

$$DPM^-Li^+pS \leftrightarrows DPM^-, Li^+, Sp \qquad (4.4)$$

$$DPM^-, Li^+, S_p + qS \leftrightarrows DPM^- \parallel Li^+ \qquad (4.5)$$

4.2.3 Nuclear Magnetic Resonance (NMR) Studies of Arylmethyl Anion

4.2.3.1 General Considerations

Nuclear magnetic resonance studies, particularly proton and ^{13}C chemical shifts, have been highly informative toward gaining an in-depth description of carbanions. Qualitatively, hydrogens attached to carbons bearing high electron density will be subject to upfield chemical shift changes relative to an appropriate neutral precursor. However, the induced changes in ^1H chemical shifts are also influenced by factors such as magnetic anisotropy effects, ring currents, and electric field effects, as well as by solvent effects and the nature of the metal counterion. On the other hand, ^{13}C shift data show greatly increased sensitivity to electronic charge and offer a more reliable probe of electron density changes, being less influenced by ring currents.[81]

Empirically, it has been found that ^{13}C shifts are approximately 15 times more sensitive to π-charge compared with proton sensitivity.[82,83] Equation 4.6 gives the relationship between the chemical shift δ and the π-electron density ρ, where κ represents the sensitivity of the chemical shift to π-electron density:

$$\delta = a - \kappa\rho \tag{4.6}$$

$$\text{for } ^1\text{H}: \quad a = 18.2, \kappa = 10.7$$

$$\text{for } ^{13}\text{C}: \quad a = 290, \kappa = 160$$

Thus, ^{13}C chemical shifts allow direct monitoring of π-electron density changes at various carbon positions. If the ion-pair structure changes and the anion–cation interaction increases, then the carbons closest to the preferred cation position will be shielded while the more remote conjugated positions of the carbanion will be deshielded.[84–86]

Many of the problems encountered with ^1H NMR will also be present when $^{6(7)}Li$ NMR is used to gain information about ion pairing.[50–52,87–90] Lithium chemical shifts of these anion systems reflect the structure of the first solvation shell. Hence, the chemical shift should be dependent on the nature of the anion, solvent, and temperature. Shift changes will arise from ring currents of the phenyl rings, electrostatic polarization of the lithium nucleus, and the orbital overlap between lithium and solvent anion.[87–90] The latter two effects will cause downfield shifts and all shifts will be modulated by the internuclear distance. Ring-current effects on the cation resonance will in addition be determined by the actual position of the cation relative to the anion.

Scalar spin–spin coupling between lithium and carbon provides direct evidence of the cation position (e.g., indicative of the nature of bonding) and state of aggregation. The method, pioneered by Fraenkel[91] and by Seebach,[92] exploits the multiplicity of the $^{13}C_\alpha$ signal under the conditions of slow interaggregate exchange and slow relaxation of the lithium nucleus. The quadrupolar relaxation of the most abundant isotope (92.6%) ^7Li or Li-7 (spin $I = 3/2$) is normally too rapid to resolve coupling, but for the rare isotope (7.4%) ^6Li (spin $I = 1$) nucleus this relaxation is slow. The multiplicity (N) of resonances of the ^{13}C nucleus bonded to lithium will then depend on the number of equivalently coupled ^6Li nuclei, $N = 2nI+1$. Carbon–lithium coupling has been observed for a number of alkyl, alkenyl, and alkynyl, as well as aryllithium compounds,[42–54,93–95] but was only realized recently with any phenylmethyllithiums (*vide infra*). On the other hand, phenylsilyllithiums also exhibit ^{29}Si–^6Li coupling.[96,97] Such measurements allow differentiation between a monomeric structure versus a static aggregate and a fluxional one involving rapid intramolecular exchange.

The NMR studies have yielded insight into carbanion stereochemistry via analysis of vicinal (*cis*/*trans*) proton–proton couplings. A variety of acyclic (allylic anions, enolate ions, etc.) and cyclic delocalized (pentadienyl, homoaromatic ions, etc.) have been elucidated in this manner.[98–102] Charge delocalization in conjugated systems leads to development of partial π-bonding, and isomer distributions as well as barriers to rotation can be determined through dynamic nuclear magnetic resonance (DNMR) studies. This applies also to arylmethyl anions as can be seen in Tables 4.3 and 4.4 (where the spectra were measured in ethereal solvents, generally THF or DME).[81,103–114]

4.2.3.2 Electronic Effects in Arylmethyl and Other Group 14 Anions

The NMR studies of arylmethyl anions have been widely reported, with respect to variation of solvent, temperature, and counterion, through the nuclei ^1H, ^{13}C, and $^{6(7)}Li$. Also of interest has been the comparison of structural and electronic aspects as one proceeds down the group 14 elements, changing the anionic center from carbon to silicon, germanium, tin, and lead. One aspect that has received special scrutiny is whether the delocalization known to be important in arylmethyl anions is maintained as one proceeds down the family of group 14 elements.

The ^1H NMR shifts for arylmethyl anions in THF solvent at ambient temperature with Li^+ and K^+ as counterions are given in Table 4.3.[81] Upfield shifts with respect to benzene ($\delta = 7.27$ ppm) are observed, in the order *para*-H > *ortho*-H ≈ *meta*-H, and the α-hydrogen signal is also found upfield (with respect to toluene), in accord with development of negative charge at these centers. The shifts become attenuated on going from the benzyl system to diphenylmethyl and to triphenylmethyl anion, as the charge becomes spread over more than one ring. In the diphenylmethyl and triphenylmethyl anions, the *ortho*-hydrogens are less shielded than *meta*-hydrogens because of ring current arising from the adjacent phenyl group(s).

Carbon-13 chemical shifts of the arylmethyl anions, listed in Table 4.4,[81] show much larger upfield changes than ^1H shifts (relative to benzene, $\delta = 128.6$ ppm),

Table 4.3. Proton chemical shifts (ppm) for benzyl, diphenylmethyl, and triphenylmethyl anions in THF (unless otherwise specified)

H_α	H_2	H_3	H_4	M^+
Benzyl				
1.62	6.09	6.30	5.50	Li[a]
0.88	6.41	6.77	6.48	Li (PhH)[b]
2.24	5.59	6/12	4.79	K[c]
Diphenylmethyl				
4.22	6.51	6.54	5.65	Li[d]
	7.28	6.80	6.31	Li (Et_2O)[e]
Triphenylmethyl				
	7.28	6.48	5.93	Li[f]
	7.30	6.58	6.03	Na[e]
	7.30	6.60	6.05	K[fe]
	7.28	6.63	6.07	Rb[e]
	7.23	6.64	6.09	Ca[e]

SOURCE: Data taken from reference 81.
ORIGINAL SOURCES: [a]References 103–105; [b]Reference 106; [c]Reference 107; [d]References 103 and 104; [e]Reference 108; [f]References 103, 108, and 109.

Table 4.4. Carbon-13 shifts (ppm) and couplings (Hz) for benzyl, diphenylmethyl, and triphenylmethyl anions in THF (unless otherwise specified)

C_a	C_1	C_2	C_3	C_4	$J(C_{\alpha}–H_{\alpha})$	M^+
Benzyl						
35.7	161.0	116.7	128.3	104.4	132	Li[a]
30.1	157.8	120.7	130.8	113.6	116	Li (PhH)[b]
52.8	153.0	110.9	130.7	95.7	153	K[c]
Diphenylmethyl						
78.5	147.4	117.5	128.1	107.1	143	Li[d]
68.7	147.9	118.9	130.2	112.0		Li (PhH)[e]
78.8	145.7	116.9	129.4	108.2		K[f]
Triphenylmethyl						
90.5	149.9	124.0	128.0	113.1		Li[g]
87.1	150.4	125.1	129.5	115.3		Li (PhH)[e]
88.3	148.8	123.7	128.9	114.3		K[h]

SOURCE: Data taken from reference 81. [b]ORIGINAL SOURCES: [a]References 104, 105, and 110; [b]References 104 and 111; [c]References 111 and 113; [d]References 104, 108, and 111; [e]Reference 111; [f]References 110–114; [g]References 104, 110, and 114; [h]References 111 and 113.

in the order *para*-C > *ortho*-C, while *meta*-carbons remain almost unchanged (see above on ring-current effect on *meta*-H chemical shift). On the contrary, the *ipso*-carbon is appreciably deshielded relative to toluene ($\delta = 137.8$). These shifts are in general accord with expectations for a resonance effect (R_π) being operative, as opposed to a π-polarization mechanism (F_π) found for the other group 14 anions (M = Si, Ge, Sn, Ph).[115–117] The two mechanisms are illustrated in Figure 4.6.

As was found with ^{1}H chemical shifts, there is an attenuation of the *para*-carbon chemical shifts as one goes from benzyl to diphenylmethyl to triphenyl-

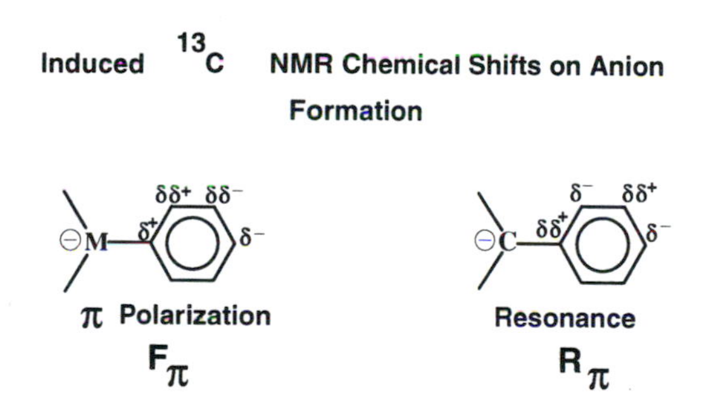

Figure 4.6. Illustration of two mechanisms that induce chemical shift changes upon anion formation. The π-polarization (F_π) mechanism found in group 14 anions (M = Si, Ge, Sn, and Pb) is shown by the structure of the left-hand side, while the resonance effect (R_π) is shown on the right-hand side and applies to carbon-centered benzylic anions.

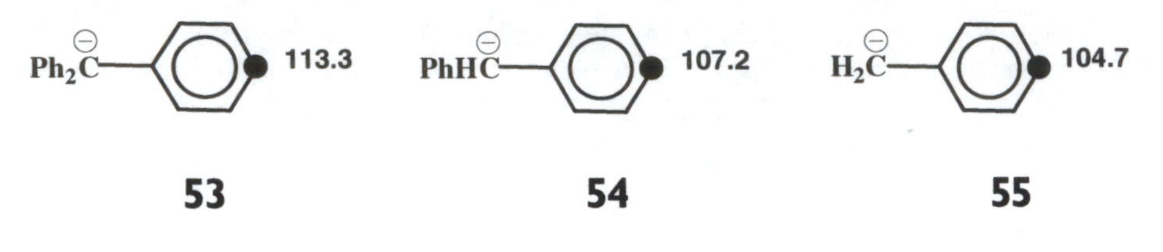

methyl anions, in accord with increasing distribution of charge along the series, as shown in structures **53** to **55** (*para*-C shifts listed).

π-Electron densities calculated from ^{1}H and ^{13}C chemical shifts via equation 4.6 are given in Table 4.5 and compared with values obtained from self-consistent field (SCF) calculations.[81] Considering the aforementioned approximations involved in derivation of equation 4.6 in the case of ^{1}H shifts, the general agreement in π-densities apparent in Table 4.5 could be fortuitous [see references 118 and 119 for charge density calculations in carbanions via intermediate neglect of differential overlap (INDO), complete neglect of differential overlap II (CNDO II), and Slater-type orbital, three Gaussian (STO-3G) methods]. In fact, the data quoted in Table 4.5 have to be further qualified. The so-called "π-electron densities" (first introduced by Mulliken) are really "electron populations" and the SCF-calculated net charges are heavily dependent on the basis set function selected. On the other hand, atomic charges calculated by the "atoms-in-molecules" (AIM) methodology[120] have gained favor in later years. For critical assessments of population analysis and atomic charges, the reader is referred to appropriate reviews.[121,122]

Table 4.5. Empirical and calculated (SCF) π-electron densities (ρ) for arylmethyl anions

Carbanion		α	1	2	3	4	Σρ
$C_6H_5CH_2^-$ (1, 2, 3, 4, α)	ρ_H	1.55		1.12	1.10	1.18	
	ρ_C	1.51	0.88	1.14	1.02	1.24	7.9
	ρ_{SCF}	1.50	0.87	1.14	1.07	1.23	8.0
$(C_6H_5)_2CH^-$ (1, 2, 3, 4, α)	ρ_H	1.31		1.09	1.09	1.17	
	ρ_C	1.35	0.92	1.11	1.02	1.16	14.0
	ρ_{SCF}	1.32	0.86	1.07	1.06	1.23	14.0
$(C_6H_5)_3C^-$ (1, 2, 3, 4)	ρ_H			1.02	1.09	1.15	
	ρ_C	1.29	0.90	1.06	1.03	1.12	19.9
	ρ_{SCF}	1.25	0.85	1.05	1.06	1.19	20.0

SOURCE: Data taken from reference 81.

4.2.3.3 Changes in Ion Pairing

(a) Solvent Effects The most comprehensive ion-pair study of phenylmethyllithiums was reported by O'Brien.[114] By studying the α-carbon shift as a function of counterion and temperature in several ethereal solvents, detailed information was obtained concerning charge distribution and entropies and enthalpies of solvation. Three ethereal solvents were studied: 2-methyltetrahydrofuran (MTHF), tetrahydrofuran, and 1,2-dimethoxyethane. The TPM^-Li^+ compound was found to be a solvent-separated ion pair in all solvents at room temperature. Only in MTHF at higher temperature could a contribution from contact ion pairs be detected. The DPM^-Li^+ structure was best described as an equilibrium between SSIP and CIP in all solvents examined over the available temperature range, and the SSIP/CIP ratio increased depending on the solvent in the order MTHF < THF < DME. For benzyllithium, it was found that the α-carbon shift was practically invariant with solvent change, which indicated that the benzyl anions are contact ion pairs at room temperature. For this relatively small anion, the energy gain on solvation is comparable to the energy of cation separation, which makes solvation much less favorable thermodynamically. Thermodynamic parameters (ΔH and ΔS) in various solvents[81] give further credence to this argument.

The deshielding of the α-carbon on going from a CIP to a SSIP could be calculated in MTHF and THF, and for TPM^-Li^+ this value was about 7.4 ppm in MTHF and 6.0 ppm in THF. The DPM^-Li^+ salt showed even larger deshielding effects by this ion-pair change, 12.8 ppm in MTHF and 10.6 ppm in THF. The smaller value obtained in THF can generally be ascribed to better external solvation of the CIP in this solvent. The corresponding shielding observed in the *para*-carbon position is of smaller magnitude, but, as a guideline, one could mention that for DPM^-Li^+ or TPM^-Li^+ a 10-ppm decrease in the α-position gives rise to approximately 2 × 4.4-ppm increase in shifts of the *para* positions.

The magnitude of the α-carbon shift change depends on the size of the π-system and the change in the interionic distance on solvation. For the 14π-electron systems, DPM^- undergoes a larger change than does fluorenide because of the larger interionic distance required in the solvent-separated ion. The solvated cation may get closer to the planar fluorenide system than to the DPM^- anion in which rotation about the C_α-C_1 renders the anion bulkier. For the 20π-electron system, TPM^-, a smaller change of the α-carbon might be expected because of the larger π-system, but this effect is compensated for by the larger interionic distance required in the SSIP state. This is caused by the steric influence of the three phenyl groups. O'Brien also reported that the α-carbon shifts of these systems were cation independent when the systems were described as SSIP.[114]

Systematic study of temperature dependence for the spectra of the carbanides (above) demonstrates that the α-carbon signals shift upfield as a function of temperature increase. If a limiting value is obtained, this can be taken as corresponding to a CIP, where the cation resides close to the most electron-rich position, C_α. Of course, the π-electron density is also strongly polarized toward this carbon. In contrast, at low temperature, an SSIP state prevails, where the cation is wrapped with solvent molecules and the carbanion responds by spreading the π-electron density more evenly over the system to give a deshielded chemical shift at C_α. The size of that deshielding is, as mentioned above, dependent on the interionic distance, and the observation that only lithium and sodium salts undergo appreciable temperature-dependent carbon chemical shift changes is therefore expected.

From the sigmoid curves obtained for the C_α carbon peaks of the lithium and sodium salts, it has been possible to calculate the ion-pair equilibrium constants and obtain the thermodynamics of solvation, provided that one could measure or reasonably estimate the limiting shifts for the CIP and SSIP states.[114]

(b) Cation Effects Besides the general observation that the degree of CIP formation increases as the size of the cation gets larger, one can note several important characteristics by studying the temperature and solvent effects of various alkali salts of phenylmethyllithiums.[81,110,113,123] The most apparent observation is that the CIP shift is dependent on the cation size. One could also note that the chemical shift change experienced by the carbons decreases as the size of the cation increases. The α-carbon shift is thus proportional to the reciprocal of the interionic distance in the contact ion pair. As the distance between the carbanion and the cation increases, the contact ion-pair shift of C_α becomes more deshielded because the positive fields of the larger cations are less effective in polarizing the electron density toward that position. It is important to keep this information in mind when discussing ion-pair-induced chemical shifts of alkali-metal associated anions. In this regard, solvent-induced chemical shifts of the larger alkali salts are strongly suppressed; in fact, Rb and Cs salts do not give any information of significance concerning solvation. To illustrate this condition, note that the contact ion-pair shift difference of C_α for DPM^-Li^+ on going from MTHF to DME is about 4 ppm, while for the corresponding potassium salt the shift change is about 1 ppm.[114] The importance of external solvation of smaller cations is thus evident. The shift change on going from CIP to a SSIP for DPM^-Li^+/THF is about 12 ppm for C_α, while for the potassium salt the deshielding of the α-position amounts to only about 1.6 ppm.

4.2.3.4 The Nature of the Carbon–Lithium Bond in Arylmethyl Anions

In earlier work on phenylmethyllithiums, no $^{13}C{-}(d)^{6(7)}Li$ scalar coupling was observed even at low temperatures.[42] This would be consistent with a fast lithium exchange process, and/or an essentially ionic nature of the carbon–lithium bonding. However, recent studies of benzyllithium compounds have revealed new evidence on this problem.

In one study, the benzyllithium was enriched in both ^{13}C and ^{6}Li by deprotonation of toluene–$^{13}C_\alpha$ with *n*-butyllithium-^{6}Li in pentane containing TMEDA.[124] Whereas with ^{7}Li, the most abundant isotope of lithium, quadruple relaxation is fast and the $^{13}C{-}^{7}Li$ coupling constant becomes averaged, the ^{6}Li relaxation is sufficiently slow for observation of $^{1}J^{13}C{-}^{6}Li$). A solution of the benzyllithium-$^{13}C_\alpha{-}^{6}Li$ and TMEDA (5×10^{-3}M) in THF-d_8 at 180 K gave

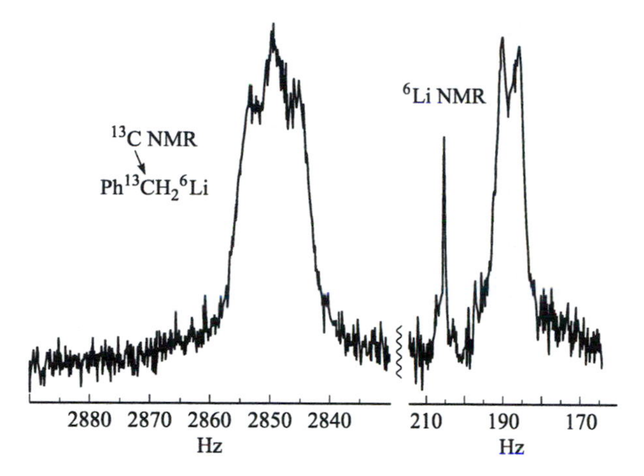

Figure 4.7. Left: NMR of benzyllithium-$^{13}C_{\alpha}$ ^{6}Li, 0.005 M, with 0.005 M TMEDA in THF-d_8; right: Same sample ^{6}Li NMR at 180 K. (Reproduced from reference 124. Copyright 1995 American Chemical Society.)

rise to the ^{13}C and ^{6}Li NMR spectra shown in Figure 4.7. The spectrum reveals a triplet for $^{13}C_{\alpha}$ NMR and a doublet for ^{7}Li NMR, with a splitting of 3.8 Hz in both. Note that under these conditions of dilute solution, low temperature, and ^{13}C enrichment, the bimolecular intermolecular C–Li exchange (equation 4.7), which would normally average the ^{13}C–^{6}Li coupling, is greatly slowed down, thus accounting for the observed splitting.

$$^{13}C^{*6}Li^{*} + ^{13}C^{6}Li \leftrightarrows ^{13}C^{*6}Li + ^{13}C^{6}Li^{*} \qquad (4.7)$$

The observation of ^{13}C–^{6}Li coupling in the benzyllithium of 3–4 Hz is indicative of a small but detectable degree of covalent character for the C_{α}–Li bond. Previous work showed that for a number of monomeric organolithium compounds, $^{1}J(^{13}C\text{–}^{6}Li) = 16 \pm 1$Hz. The low $^{1}J(^{13}C\text{–}^{6}Li)$ value observed for benzyllithium was, therefore, taken as the "missing link", in a continuum of C–Li covalency, between the "common pattern" monomers and separated ion pairs. The geometric arrangement that would fit the low degree of covalency was considered to be one where the lithium is situated above C_{α} and normal to the benzyl π-plane. This proposal is corroborated by X-ray crystallographic structural studies of benzyllithium complexed to donors such as TMEDA (*vide supra*). The lithium position in benzyllithium is also deduced on the basis of FMO[125] and electrostatic[126] models. Note, however, that even a purely ionic C–Li bond is known to make the carbanide C-pyramidal by electrostatic attraction alone.

Two structurally modified benzyllithiums in which the lithium is encapsulated in crown-ether-type arrangements, **56** and **57** (charges omitted for simplicity), exhibit $^{1}J(^{13}C\text{–}^{7}Li)$ values of 7 and 9 Hz, respectively.[127] Clearly, C–Li exchange in these molecules must also be slow.

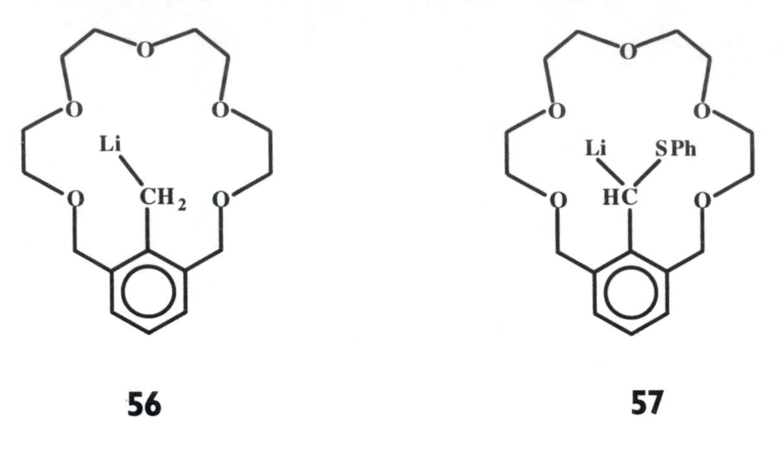

4.2.3.5 Restricted Rotation in Diphenylmethyl Anions

In arylmethyl anions, the delocalization of charge into the phenyl rings with the accompanying development of π-bonding across the benzylic carbon–phenyl bond results in restricted phenyl rotation. This can be monitored by DNMR methods.

An ingenious demonstration of this type of hindered rotation is the case of a diphenylmethyl anion where a rather moderate structural change has led to *different* rates of rotation of the phenyl rings. The study concerns the lithium salts of anions **58** and **59**, with evidence obtained by ^{1}H and $^{13}C[^{1}H]$ DNMR.[128,129] At 195 K in THF, two separate sets of resonances for each ring were seen and a complete line shape analysis allowed unambiguous phenyl ring assignment to the differential aryl barriers to rotation. A ready explanation of these observations is that the electron-releasing methyl group leads to reduced π-bonding of the methylated ring to the α-carbon and thus to a reduced barrier to rotation of the *p*-tolyl moiety compared with the deutero-phenyl moiety; that is, the former rotates faster.

Table 4.6 presents the barriers to aryl rotation for anion **58** in THF and DME.[129] Since it is reasonable to expect ion pairing in the more polar DME

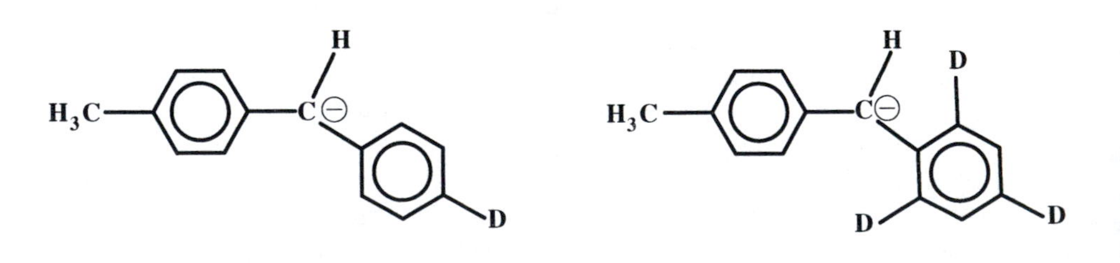

Table 4.6. Barriers to aryl rotation for [4-deuteriophenyl-(4′-methylphenyl)]-lithium in various solvents

Solvent	Ring	$\Delta G^{\neq}$ (kcal/mol)[a]
THF	4′-methyl	10.7
	deuteriophenyl	11.6
DME	4′-methyl	11.3
	deuteriophenyl	12.9

SOURCE: Reproduced from reference 129. Copyright 1984 American Chemical Society.
[a]At 223 K (50 °C).

solvent to shift to the more solvent-separated type, with consequent greater delocalization of charge, this will lead to the observed overall increases in the barriers to aryl rotation in DME. Moreover, the differential charge between the two rings induced by the methyl becomes magnified in the more polar solvent, in accord with observations. The data in Table 4.6 may be compared with the barrier to phenyl rotation in the parent diphenylmethide, which has been measured as 11.2 kcal/mol.[130]

In a related study[131] concerned with the effect of the Me_3Si substituent on carbanion stability, it was shown by DNMR that in anion **60** the rotation of the two phenyl rings occurs at different rates. The 4-methylphenyl ring was found to rotate faster than the 4′-(trimethylsilyl)phenyl by a factor of 200 at 255 K in THF. This is consistent with preferential delocalization of charge into the 4′-(trimethylsilyl)-substituted ring that raises the rotation barrier of the silicon-bearing phenyl moiety. While the mechanism of this delocalization has not been elucidated, the result is consistent with some degree of p_{π}–d_{π} overlap between the ring π-system and a vacant d-orbital on silicon. Rotational barriers in diphenylmethyl anions stabilized by $Cr(CO)_3$ moieties have also been determined.[132]

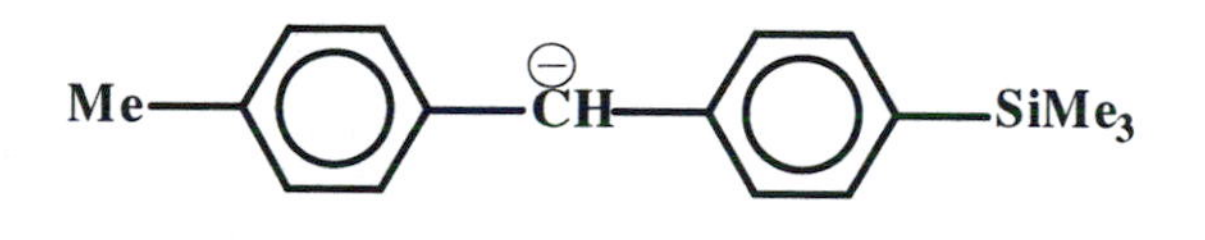

60

4.2.3.6 Solid-State NMR Studies

Complementary to X-ray crystal structure studies (Section 4.2.1) are the solid-state NMR studies, where recent advances have been highly informative in relating X-ray and NMR data to each other and to solution results. Figure 4.8[133] shows the ^{13}C cross-polarization magic angle spin (CP/MAS) NMR spectrum of TMEDA-complexed ^{7}Li triphenylmethide, $TPM^{-}Li^{+}$. The *ipso*- and *para*-carbon signals from the phenyl ring twisted out of conjugation are starred (cf. X-ray structure discussed previously). Noteworthy also is the deshielding of 10 ppm of the *para*-carbon on decreasing conjugation.

A related solid-state NMR study of TMEDA and quinuclidine complexes of fluorenyllithium involved measurement of ^{7}Li chemical shifts and quadrupolar coupling constants and has provided direct information on location of the lithium cation relative to the carbanion framework.[134] Structural information was also gathered through a solid-state ^{6}Li–^{13}C rotational echo double resonance (REDOR) NMR study of the TMEDA complex of fluorenyllithium performed on ^{6}Li-enriched samples.[135]

Clearly, these novel solid-state NMR techniques hold great promise in being uniquely able to provide information on carbanion structure, and especially on the interaction with the metal counterion.

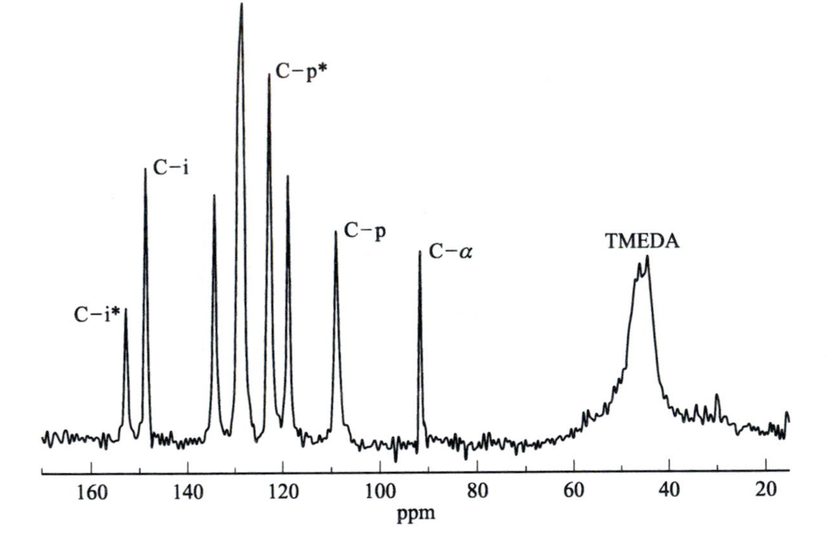

Figure 4.8. Carbon-13 CP/MAS NMR spectrum of TMEDA-complexed ^{7}Li triphenylmethide, $TPM^{-}Li^{+}$. [Crystals obtained from a hexane solution. A pulse repetition time of 2.5 s and a contact time of 1 ms was used at a frequency of 25.178 MHz (Bruker MSL-100). The *ipso*- and *para*-carbon signals from the phenyl ring twisted out to conjugation are starred. Note the deshielding of 10 ppm of the *para*-carbon by decreasing conjugation.] (Reproduced from reference 115. Copyright 1993, Wiley.)

4.3 VIBRATIONAL SPECTROSCOPY OF ARYLMETHYL CARBANIONS

4.3.1 Overview of Present Coverage

The study of carbanions by vibrational spectroscopy[136-139] can potentially provide information on structure and bonding, including charge distribution, geometry, and the nature of the interaction between the metal and the carbande residues, particularly in the case of solid-phase or matrix studies. In the case of solution studies, there are additional factors relating to the interaction with solvent molecules, ion-pairing phenomena, and the effect thereon of complexing agents such as crown ethers and cryptands.

Past studies by vibrational spectroscopy of organometallic compounds derived from nitrile[140–142] nitro,[143,144] carbonyl,[145–152] and phosphonate[153] compounds have been most informative, as have been studies of carbanions stabilized by adjacent sulfur functionalities.[154] The structural complexity of Grignard reagents has likewise received clarification through IR and Raman spectroscopy.[155] However, simple alkali-metal alkyl derivatives have so far received limited investigation as a result of their high reactivity with typical solvents. Nevertheless, even these limited studies have provided important structural information, including information on aggregation: (RLi) $\leftrightarrows$ $(RLi)_2$ $\leftrightarrows$ $(RLi)_4$, and so on.[156,157] Allylic alkali-metal compounds have proven more amenable to such study and, for allylsodium, the IR and Raman spectral assignments are in complete accord with charge delocalization in this carbanion.[158–160] Simple organolithiums such as CH_3Li have been examined as the monomer trapped in an argon matrix.[161] In this detailed study, isotopic substitution, $^1H/^2H$, $^6Li/^7Li$, and $^{12}C/^{13}C$, facilitated complete vibrational analysis. Solution spectra have also been determined, for example, for *t*-butyllithium in benzene[162] and in methylcyclohexane,[163] and these indicated a tetrameric structure in both solvents.

4.3.2 Diphenylmethyllithium: Charge Delocalization and Ion Pairing—IR Evidence

Two vibrational spectroscopic studies of arylmethyl carbanions have been reported, in both cases using alkali-metal salts of diphenylmethane. However, the two studies are complementary. The first refers to DPM^-K^+ in liquid ammonia, formed from diphenylmethane (DPM) by the action of KNH_2 (KND_2) in NH_3 (ND_3), and views the C–H stretching vibrations, yielding information on C_α hybridization.[164] In the second study, DPM^-Li^+ was generated by the action of *n*-BuLi on DPM in several ethereal solvents; this enabled C=C vibrational frequencies to be determined and provided information on charge delocalization, as well as on ion-pairing phenomena.[165]

Formation of DPM^-K^+ from DPM and potassium amide in ammonia is accompanied by decrease or disappearance of bands at 2910 and 2840 cm^{-1}, corresponding to the asymmetric and symmetric aliphatic C–H stretch.[164] Concurrently, a new absorption at 3010 cm^{-1} appears; such an absorption is ascribable to an sp^2 C–H stretching mode. These spectroscopic changes are in accord with the rehybridization from sp^3 to sp^2 that accompanies formation of the carbanionic center from the neutral substrate.

Infrared absorptions and band assignments of DPM^-Li^+ in Et_2O, THF, and DME are given in Table 4.7, while Table 4.8 gives corresponding data for diphenylmethane.[165,166] Clearly, the 1597 cm^{-1} absorption in DPM is shifted to 1560–1565 cm^{-1} on formation of DPM^-Li^+. This lowering of $\nu_{C=C}$ by 30–40 cm^{-1} is in accord with charge delocalization in the carbanion brought about by a lowering of the force constants of C=C stretching vibrational modes.

Similarly, the band for DPM at 695–700 cm^{-1} that corresponds to an out-of-plane ring deformation also undergoes a shift to lower frequencies, namely, to 675–684 cm^{-1}, on formation of the carbanion. This, once again, is in accord with charge delocalization in this species.

In contrast to the above, the absorption occurring at 735–739 cm^{-1} in DPM undergoes minimal change in the formation of DPM^-Li^+. It is reasonable that this out-of-plane aromatic C–H vibrational mode would be less sensitive to charge delocalization compared with the C=C vibrational modes.

The differences in the absorption frequencies exhibited by DPM^-Li^+ as the solvent is varied (Table 4.7) still remain to be considered. Such differences have previously been ascribed to the presence of different types of ion pairs and a similar conclusion can be drawn in the present case. The pertinent vibrational modes to be considered in this respect are the skeletal in-plane stretching vibration ($\nu_{C=C}$) and the out-of-plane ring deformation ($\phi_{C=C}$). These bands have been found to undergo characteristic shifts on formation of the carbanion from its precursor and hence they provide a measure of the accompanying changes in electron-density distribution. Thus, these bands can serve as probes of the interaction between the carbanion and its environment, that is, solvent molecules and counterion. On the other hand, the aromatic C–H out-of-plane bending vibration (γ_{C-H}; Tables 4.7, 4.8) experiences minimal shift in frequency on formation of DPM^-Li^+ from DPM, which is in accord with a smaller sensitivity to charge delocalization compared with vibrational modes that involve the carbon skeleton.

Table 4.7. Infrared absorptions (cm^{-1}) of diphenylmethyllithium (DPM^-Li^+) in ethereal solvents

	Solvent		
Vibrational mode	Et_2O	THF	DME
$\nu_{C=C}$[a]	1565, 1600[d]	1560, 1601[d]	1560, 1601[d]
γ_{C-H}[b]	736	735	740
$\phi_{C=C}$[c]	684	678	675

SOURCE: Reproduced from reference 165. Copyright 1982, Elsevier.

[a]Skeletal in-plane stretching vibration. [b]Out-of-plane bending vibration. [c]Out-of-plane ring deformation. [d]The band at 1600 or 1601 cm^{-1} was shown to result from reaction of *n*-BuLi (used in reaction with DPM) with solvent.

Table 4.8. Infrared absorptions (cm^{-1}) of diphenylmethane (DPM) in ethereal solvents

	Solvent		
Vibrational mode	Et_2O	THF	DME
$\nu_{C=C}$[a]	1596	1597	1598
γ_{C-H}[b]	735	736	739
$\phi_{C=C}$[c]	697	695	700

SOURCE: Reproduced from reference 165. Copyright 1982, Elsevier.
[a]Skeletal in-plane stretching vibration. [b]Out-of-plane bending vibration. [c]Out-of-plane ring deformation.

Hence, the $\gamma_{C\text{-}H}$ vibrational frequency may be disregarded in considering ion-pairing effects of DPM^-Li^+ in the three solvents.

Recall that examination of the electronic absorption spectra of DPM^-Li^+ in ethereal solvents[79] indicated the presence of three spectral species that were ascribed to the contact ion pair, DPM^-, Li^+, the externally solvated contact ion pair, DPM^- Li^+, S_p, and the solvent-separated ion pair, $DPM^-\|Li^+$ (see Section 4.2.2.2 and equations 4.4 and 4.5). The spectral evidence was in accord with DPM^- Li^+ being the predominant species present in Et_2O; DPM^- Li^+, S_p the predominant species in THF; and $DPM^-\|Li^+$ the predominant species in DME. This trend is also in accord with the solvating capabilities (donicities) of these ethers.

The IR spectral data are subject to a qualitatively similar interpretation. Since charge delocalization in the carbanion can take place to a greater extent in solvent-separated ion pairs than in contact ion pairs, the former are expected to be characterized by absorption bands occurring at relatively lower frequencies. From the data compiled in Table 4.7, it is seen that the $\phi_{C=C}$ band shows a uniform shift to lower frequencies on going from Et_2O, the medium of least solvating ability, to DME, which has the highest solvating capability in this series. The results indicate that in Et_2O the predominant species present is the contact ion pair, DPM^-Li^+, while in DME the solvent-separated ion pair, $DPM^-\|Li^+$, predominates. The intermediate value of the frequency shift in THF could be interpreted as arising from the externally solvated contact ion pair, DPM^-Li^+, S_p, although a mixture of the contact and solvent-separated ion-pair species could possibly also be accommodated by the results.

The $\nu_{C=C}$ band also undergoes an overall shift to lower frequency on going from Et_2O to the more solvating media THF and DME (Table 4.7). The observation that the frequency shift is identical for THF and DME suggests that in THF the predominant species is also $DPM^-\|Li^+$ and argues against the previously advanced alternative explanation that the DPM^-Li^+, S_p species could be present in THF.

Qualitatively similar conclusions concerning the types of ion pairs that are present follow if the argument is based on individual differences in frequencies between DPM and $DPM^{-}Li^{+}$ for each solvent (i.e., differences in data of Tables 4.7 and 4.8), rather than considering, as above, only trends in frequency data found in Table 4.7 as the solvent is varied. On this new basis, the frequency differences in the case of $\phi_{C=C}$ for Et_2O, THF, and DME are 13, 17, and 25 cm^{-1}, while for $\nu_{C=C}$ the corresponding differences are 31, 37, and 38 cm^{-1} respectively. An advantage of the second method is that it involves a larger span of frequency differences, which should increase the sensitivity of the method. Thus, both methods lead to qualitatively similar conclusions.

References

1. Snieckus, V., Ed. *Advances in Carbanion Chemistry*; JAI Press: Greenwich, CT, 1996; Vol. 2.
2. Snieckus, V., Ed. *Advances in Carbanion Chemistry*; JAI Press: Greenwich, CT, 1992; Vol. 1.
3. Sapse, A.-M.; Schleyer, P. von R., Eds. *Lithium Chemistry: A Theoretical and Experimental Overview*; Wiley: New York, 1995.
4. Buncel, E.; Durst, T., Eds. *Comprehensive Carbanion Chemistry. Part A. Structure and Reactivity*; Elsevier: Amsterdam, 1980.
5. Buncel, E.; Durst, T., Eds. *Comprehensive Carbanion Chemistry. Part B. Selectivity in Carbon–Carbon Bond Forming Reactions*; Elsevier: Amsterdam, 1984.
6. Buncel, E.; Durst, T., Eds. *Comprehensive Carbanion Chemistry. Part C. Ground State and Excited State Reactivity*; Elsevier: Amsterdam, 1987.
7. Bates, R.B.; Ogle, C.A. *Carbanion Chemistry*; Springer Verlag: Berlin, 1984.
8. Buncel, E. *Carbanions, Mechanistic and Isotopic Aspects*; Elsevier: Amsterdam, 1975.
9. Wakefield, B.J. *The Chemistry of Organolithium Compounds*; Pergamon Press: New York, 1974.
10. Cram, D.J. *Fundamentals of Carbanion Chemistry*; Academic Press: New York, 1965.
11. Haiduc, I.; Zuckerman, J.J. *Basic Organometallic Chemistry*; Walter de Gruyter: Berlin, 1985.
12. Wilkinson, G.; Stone, F.G.A.; Abel, E.W. Eds. *Comprehensive Organometallic Chemistry*; Pergamon Press: Oxford, U.K., 1982.
13. Pauling, L. *The Nature of the Chemical Bond*; Cornell University Press: Ithaca, New York, 1960; p 96.
14. Cotton, F.A.; Wilkinson, G. *Advanced Inorganic Chemistry*, 3rd ed.; Wiley-Interscience: New York, 1972; pp 114–116.
15. Clark, T.; Rohde, C.; Schleyer, P. von R. *Organometallics* **1983**, *2*, 1344.
16. Bachrach, S.M.; Streitwieser, A. *J. Am. Chem. Soc.* **1984**, *106*, 2283.
17. Reed, A.E.; Weirstock, R.B.; Weinhold, F. *J. Chem. Phys.* **1985**, *83*, 735.
18. Streitwieser, A.; Bachrach, S.M.; Dorigo, A.; Schleyer, P. von R. In *Lithium Chemistry: A Theoretical and Experimental Overview*; Sapse, A.-M.; Schleyer, P. von R., Eds.; Wiley: New York, 1995; Chapter 1.
19. Brown, T.L. *Adv. Organomet. Chem.* **1966**, *3*, 365.
20. Bauer, V.; Seebach, D. *Helv. Chim. Acta* **1984**, *67*, 1972.
21. West, P.; Waack, R. *J. Am. Chem. Soc.* **1967**, *89*, 4395.
22. Jackman, L.M.; DeBrosse, C.W. *J. Am. Chem. Soc.* **1983,** *105*, 4177.
23. Fraenkel, G.; Beckenbaugh, V.E.; Yang, P.P. *J. Am. Chem. Soc.* **1976**, *98*, 6878.
24. Winstein, S.; Clippenger, E.; Fainberg, A.H.; Robinson, G.C. *J. Am. Chem. Soc.* **1954**, *76*, 2597.

25. Grunwald, E. *Anal. Chem.* **1954**, *26*, 1696.
26. Hogen-Esch, T.E.; Smid, J. *J. Am. Chem. Soc.* **1966**, *88*, 307.
27. Szwarc, M., Ed. *Ions and Ion Pairs in Organic Reactions*; Wiley: New York, 1972; Vol. 1.
28. Szwarc, M., Ed. *Ions and Ion Pairs in Organic Reactions*; Wiley: New York, 1974; Vol. 2.
29. Price, E. In *The Chemistry of Non-Aqueous Solvents*; Lagowski, J.J., Ed.; Academic Press: New York, 1966; Vol. 1, p 86.
30. Nichols, D.; Sutphen, C.; Szwarc, M. *J. Phys. Chem.* **1968**, *72*, 1021.
31. Grunwald, E.; Highsmith, S.; Ting-Po, R. In *Ions and Ion Pairs in Organic Reactions*; Szwarc, M., Ed.; Wiley: New York, 1972; Vol. 1, Chapter 5.
32. Briggs, T.F.; Winemiller, M.D.; Xiang, B.; Collum, D.B. *J. Org. Chem.* **2001**, *66*, 6291.
33. Abbotto, A.; Leung, S.S.-W.; Streitwieser, A.; Kilway, K.V. *J. Am. Chem. Soc.* **1998**, *120*, 10807; and references therein.
34. Jackman, L.M.; Bortiatynski, J. In *Advances in Carbanion Chemistry*; Snieckus, V., Ed.; JAI Press: Greenwich, CT, 1992; Vol. 1, pp 45–88.
35. Bradamante, S.; Pagani, G.A. In *Advances in Carbanion Chemistry*; Snieckus, V., Ed.; JAI Press: Greenwich, CT, 1996; Vol. 2, pp 189–264.
36. Jones, J.R. *Ionization of Carbon Acids*; Academic Press: London, 1973.
37. Stewart, R. *The Proton: Applications to Organic Chemistry*; Academic Press: Orlando, FL, 1985.
38. Buncel, E.; Menon, B.C. In *Comprehensive Carbanion Chemistry. Part A. Structure and Reactivity*; Buncel, E.; Durst, T., Eds.; Elsevier: Amsterdam, 1980; Chapter 3.
39. Streitwieser, A., Jr.; Juaristi, E.; Nebenzahl, L.L. In *Comprehensive Carbanion Chemistry. Part A. Structure and Reactivity*; Buncel, E.; Durst, T., Eds.; Elsevier: Amsterdam, 1980; Chapter 7.
40. Streitwieser, A.; Wang, D.Z.; Stratakis, M.; Facchetti, A.; Gareyev, R.; Abbotto, A.; Krom, J.A.; Kilway, K.V. *Can. J. Chem.* **1998**, *76*, 765.
41. Buncel, E.; Menon, B.C. *J. Am. Chem. Soc.* **1977**, *99*, 4457.
42. Buncel, E.; Menon, B.C. *Can. J. Chem.* **1976**, *54*, 3949.
43. Brown, T.L. *Acc. Chem. Res.* **1968**, 1, 23.
44. Brown, T.L. *Pure Appl. Chem.* **1970**, 23, 447.
45. McKeever, L.D. In *Ions and Ion Pairs in Organic Reactions*; Szwarc, M., Ed.; Wiley: New York, 1972; Vol.1, Chapter 6.
46. McGarrity, J.F.; Ogle, C.A. *J. Am. Chem. Soc.* **1985**, *107*, 1805.
47. Parsons, R.L., Jr.; Fortunak, J.M.; Dorow, R.L.; Harris, G.D.; Kauffman, G.S.; Nugent, W.A.; Winemiller, M.D.; Briggs, T.F.; Xiang, B.; Collum, D.B. *J. Am. Chem. Soc.* **2001**, *123*, 9135.
48. Bauer, W.; Winchester, W.R.; Schleyer, P. von R. *Organometallics* **1987**, *6*, 2371.
49. Günther, H.; Moskau, D.; Bast, P.; Schmalz, D. *Angew. Chem. Int. Ed. Engl.* **1987**, *26*, 1212.
50. Johnels, D.; Edlund, U. *J. Organomet. Chem.* **1990**, *393*, C35.
51. Thomas, R.D. In *Isotopes in the Physical and Biomedical Sciences, Vol. 2: Isotopic Applications in NMR Studies*; Buncel, E.; Jones, J.R., Eds.; Elsevier: Amsterdam, 1991, Chapter 7.
52. Bauer, W.; Schleyer, P. von R. In *Advances in Carbanion Chemistry*; Snieckus, V., Ed.; JAI Press: Greenwich, CT, 1992; Vol. 1 pp 89–176.
53. Bauer, W. In *Lithium Chemistry: A Theoretical and Experimental Overview*; Sapse, A.-M.; Schleyer, P. von R., Eds.; Wiley: New York, 1995.

54. Fraenkel, G. In *Techniques in Chemistry, Investigation of Rates and Mechanisms of Reactions. Part. 2*; 4th ed.; Bernasconi, C.F., Ed.; Wiley-Interscience: New York, 1986; pp 357–604.
55. Fraenkel, G.; Qui, F. *J. Am. Chem. Soc.* **2000**, *122*, 12806.
56. Weiss, E. *Angew. Chem., Int. Ed. Engl.* **1993**, *32*, 1501.
57. Boche, G.; Lohrenz, J.C.W.; Opel, A. In *Lithium Chemistry: A Theoretical and Experimental Overview*; Sapse, A.-M.; Schleyer, P. von R., Eds.; Wiley: New York, 1995; Chapter 7.
58. Pauer, F.; Power, P.P. In *Lithium Chemistry. A Theoretical and Experimental Overview*; Sapse, A.-M.; Schleyer, P. von R., Eds.; Wiley: New York, 1995; Chapter 9.
59. Patterman, S.P.; Karle, I.L.; Stucky, G.D. *J. Am. Chem. Soc.* **1970**, *92*, 1150.
60. Brooks, J.J.; Stucky, G.D. *J. Am. Chem. Soc.* **1972**, *94*, 7333.
61. Köster, H.; Weiss, E. *J. Organomet. Chem.* **1979**, *168*, 273.
62. Zarges, W.; Marsch, M.; Harms, K.; Koch, W.; Frenking, G.; Boche, G. *Chem. Ber.* **1991**, *124*, 543.
63. Beno, M.A.; Hope, H.; Olmstead, M.M.; Power, P.P. *Organometallics* **1985**, *4*, 2117.
64. Olmstead, M.M.; Power, P.P. *J. Am. Chem. Soc.* **1985**, *107*, 2174.
65. Becker, B.; Enkelmann, V.; Mullen, K. *Angew. Chem.* **1989**, 101.
66. Asami, R.; Levy, M.; Szwarc, M. *J. Chem. Soc.* **1961**, 361.
67. Buncel, E.; Menon, B.C. *J. Org. Chem.* **1979**, *44*, 317.
68. Carter, H.V.; McClelland, B.J.; Warhurst, E. *Trans. Faraday Soc.* **1960**, *56*, 455.
69. Bayliss, N.S.; McRae, E.G. *J. Phys. Chem.* **1954**, *58*, 1002.
70. Griffiths, T.R.; Symons, M.C.R. *Mol. Phys.* **1960**, *3*, 90.
71. Tolbert, L.M. In *Comprehensive Carbanion Chemistry. Part C. Ground State and Excited State Reactivity*; Buncel, E.; Durst, T., Eds.; Elsevier: Amsterdam, 1987; Chapter 4.
72. Buncel, E.; Menon, B.C. *J. Chem. Soc., Chem. Commun.* **1978**, 758.
73. Hogen-Esch, T.E.; Smid, J. *J. Am. Chem. Soc.* **1966**, *88*, 307; 318.
74. Smid, J. In *Ions and Ion Pairs in Organic Reactions*, Szwarc, M., Ed.; Wiley: New York, 1972; Vol. 1, Chapter 3.
75. Hogen-Esch, T.E. *Adv. Phys. Org. Chem.* **1977**, *15*, 153.
76. Gutmann, V. *Coordination Chemistry in Non-Aqueous Solutions*; Springer-Verlag: Berlin, 1968.
77. Chan, L.L.; Wong, K.H.; Smid, J. *J. Am. Chem. Soc.* **1970**, *92*, 1955.
78. Parkes, H.M.; Young, R.N. *J. Chem. Soc., Perkin Trans. 2* **1980**, 1137.
79. Buncel, E.; Menon, B.C.; Colpa, J.P. *Can. J. Chem.* **1979**, *57*, 999.
80. Burley, J.W.; Young, R.N. *J. Chem. Soc., Perkin Trans. 2* **1972**, 835.
81. O'Brien, D.H. In *Comprehensive Carbanion Chemistry. Part A. Structure and Reactivity*; Buncel, E.; Durst, T., Eds.; Elsevier: Amsterdam, 1980; Chapter 6.
82. Spiesecke, H.; Schneider, W.G. *Tetrahedron Lett.*, **1961**, 468.
83. Eliasson, B.; Edlund, U.; Mullen, K. *J. Chem. Soc., Perkin Trans. 2* **1986**, 937.
84. O'Brien, D.H.; Russell, C.R.; Hart, A.H. *J. Am. Chem. Soc.* **1979**, *101*, 633.
85. Takahashi, K.; Kondo, Y.; Asami, R.; Inoue, Y. *Org. Magn. Reson.* **1974**, *6*, 580.
86. Edlund, U. *Org. Magn. Reson.* **1974**, *12*, 661.
87. McKeever, L.D.; Waack, R. *J. Organomet. Chem.* **1971**, *28*, 145.
88. Hogan, R.J.; Scherr, P.A.; Weibel, A.T.; Oliver, J.P. *J. Organomet.Chem.* **1975**, *85*, 265.
89. Assadourian, L.; Faure, F.; Gau, G. *J. Organomet. Chem.* **1985**, *280*, 153.
90. Cox, R.H.; Janzen, E.G.; Harrison, W.B. *J. Magn. Reson.* **1972**, *4*, 274.

91. Fraenkel, G.; Henrichs, M.; Hewitt, J.M.; Su, B.M.; Geckle, J.M. *J. Am. Chem. Soc.* **1980**, *102*, 3345.
92. Seebach, D.; Hassig, R.; Gabriel, J. *Helv. Chim. Acta* **1983**, *66*, 308.
93. Setzer, N.; Schleyer, P. von R. *Adv. Organomet. Chem.* **1985**, *24*, 353.
94. Edlund, U.; Johnels, D. *J. Am. Chem. Soc.* **1990**, *112*, 1647.
95. Cabral, J.; Fraenkel, G. *J. Am. Chem. Soc.* **1992**, *114*, 9067.
96. Edlund, U.; Lejon, T.; Venkatachalam, T.K.; Buncel, E. *J. Am. Chem. Soc.* **1985**, *107*, 6408.
97. Buncel, E.; Venkatachalam, T.K.; Edlund, U. *Can. J. Chem.* **1986**, *64*, 1674.
98. Katz, T.J. *J. Am. Chem. Soc.* **1966**, *88*, 4732.
99. Rieke, R.; Ogliaruso, M.; McClung, M.; Winstein, S. *J. Am. Chem. Soc.* **1966**, *88*, 4729.
100. Grutzner, J.B.; Winstein, S. *J. Am. Chem. Soc.* **1972**, *94*, 2200.
101. Jackman, L.M.; Lange, B.C. *Tetrahedron* **1977**, *33*, 2737.
102. Seebach, D. *Angew. Chem., Int. Ed. Engl.* **1988**, *27*, 162.
103. Sandel, V.R.; Freedman, H.H. *J. Am. Chem. Soc.* **1963**, *85*, 2328.
104. Waack, R.; Doran, M.A.; Baker, E.B.; Olah, G.A., *J. Am. Chem. Soc.* **1966**, *88*, 1272.
105. McKeever, L.D.; Waack, R. *J. Organomet. Chem.* **1971**, *28*, 145.
106. Bywater, S.; Worsfold, D.J. *J. Organomet. Chem.* **1971**, *33*, 273.
107. Takahashi, K.; Takaki, M.; Asami, R. *Org. Magn. Reson.* **1971**, *3*, 539.
108. Grutzner, J.B.; Lawlor, J.M.; Jackman, L.M. *J. Am. Chem. Soc.* **1972**, *94*, 2306.
109. Cox, R.H.; Janzen, E.G.; Harrison, W.B. *J. Magn. Reson.* **1971**, *4*, 274.
110. van Dongen, J.P.C.M; van Dijkman, H.W.D.; de Bie, M.J.A. *Recl. Trav. Chim. Pays-Bas* **1974**, *93*, 29.
111. Takahashi, K.; Kondo, Y.; Asami, R.; Inoue, Y. *Org. Magn. Reson.* **1974**, *6*, 580.
112. O'Brien, D.H.; Russell, C.R.; Hart, A.J. *J. Am. Chem. Soc.* **1976**, *98*, 7427.
113. O'Brien, D.H.; Hart, A.J.; Russell, C.R. *J. Am. Chem. Soc.* **1975**, *97*, 4410.
114. O'Brien, D.H.; Russell, C.R.; Hart, A.J. *J. Am. Chem. Soc.* **1979**, *101*, 633.
115. Edlund, U.; Buncel, E. *Prog. Phys. Org. Chem.* **1993**, *19*, 225.
116. Buncel, E.; Venkatachalam, T.K.; Eliasson, B.; Edlund, U. *J. Am. Chem. Soc.* **1985**, *107*, 303.
117. Edlund, U.; Lejon, T.; Pyykkö, P.; Venkatachalam, T.K.; Buncel, E. *J. Am. Chem. Soc.* **1987**, *109*, 5982.
118. Browne, S.E.; Asher, S.E.; Cornwall, E.H.; Frisoli, J.K.; Harris, L.J.; Salot, E.A.; Sauter, E.A.; Trecoske, M.A.; Veale, P.S., Jr. *J. Am. Chem. Soc.* **1984**, *106*, 1432.
119. Bushby, R.J.; Tytko, M.P. *J. Organomet. Chem.* **1984**, *270*, 265.
120. Bader, R.F.W. *Chem. Rev.* **1991**, *91*, 893.
121. Bachrach, S.M. In *Reviews in Computational Chemistry, Vol. 5*; Lipkowitz, K.; Boyd, D.B., Eds.; VCH; New York, 1994; Chapter 3.
122. Williams, D.E. In *Reviews in Computational Chemistry, Vol. 2*; Lipkowitz, K.; Boyd, D.B., Eds.; VCH, New York, 1991; Chapter 6.
123. Edlund, U. *Org. Magn. Reson.* **1979**, *12*, 661.
124. Fraenkel, G.; Martin, K.V. *J. Am. Chem. Soc.* **1995**, *117*, 10366.
125. Bauer, W.; Winchester, W.R.; Schleyer, P. von R. *Organometallics* **1987**, *6*, 2371.
126. Stucky, G. *Adv. Chem. Ser.* **1974**, *130*, 56.
127. Ruhland, T.; Hoffmann, R.W.; Schade, S.; Boche, G. *Chem. Ber.* **1995**, *128*, 551.
128. Bushweller, C.H.; Sturges, J.S.; Cipullo, M.; Hoogasian, S.; Gabriel, M.W.; Bank, S. *Tetrahedron Lett.* **1978**, 1359.
129. Bank, S.; Marcantonio, R.P.; Bushweller, C.H. *J. Org. Chem.* **1984**, *49*, 5091.
130. Bank, S.; Sturges, J.; Bushweller, C.H. *Stereodynamics of Molecular Systems*; Sarma, R.H., Ed.; Pergamon Press: New York, 1979; p 147.

131. Bank, S.; Sturges, J.S.; Heyer, D.; Bushweller, C.H. *J. Am. Chem. Soc.* **1980**, *102*, 3982.
132. Top, S.; Jaouen, G.; Sayer, B.G.; McGlinchey, M.J. *J. Am. Chem. Soc.* **1983**, *105*, 6426.
133. Boman, A.; Johnels, D. *Magn. Reson. Chem.* **2000**, *38*, 853.
134. Johnels, D.; Andersson, A.; Boman, A.; Edlund, U. *Magn. Reson. Chem.* **1996**, *34*, 908.
135. Quist, P.-O.; Förster, H.; Johnels, D. *J. Am. Chem. Soc.* **1997**, *119*, 5390.
136. Lambert, J.B.; Shurvell, H.F.; Lightner, D.A.; Cooks, R.G. *Organic Structural Spectroscopy*; Prentice Hall: Upper Saddle River, NJ, 1998.
137. Corset, J. In *Comprehensive Carbanion Chemistry. Part A. Structure and Reactivity*; Buncel, E.; Durst, T., Eds.; Elsevier: Amsterdam, 1980; Chapter 4.
138. Greenwood, N.N. *Spectroscopic Properties of Inorganic and Organometallic Compounds, A Specialist Periodical Report*. The Chemical Society; London, 1976.
139. Maslowsky, E. Jr., *Vibrational Spectra of Organometallic Compounds*; Wiley: New York, 1977.
140. Juchnovski, I.N.; Dimitrova, J.S.; Binev, I.G.; Kaneti, J. *Tetrahedron* **1978**, *34*, 779.
141. Juchnovski, I.N.; Radomirska, V.B.; Binev, I.G.; Grekova, E.A. *J. Organomet. Chem.* **1977**, *128*, 139.
142. Mayer, E.; Gardiner, D.J.; Hester, R.E. *J. Mol. Struct.* **1974**, *20*, 127.
143. Yarwood, J.; Orville-Thomas, W.J. *J. Chem. Soc.* **1963**, 5991.
144. Brookes, M.J.; Jonathan, N. *J. Chem. Soc. A* **1968**, 1529.
145. Zook, H.D.; Russo, T.J.; Ferrand, E.F.; Stotz, D.S. *J. Org. Chem.* **1968**, *33*, 2222.
146. House, H.O.; Trost, B.M. *J. Org. Chem.* **1965**, *30*, 2502.
147. Meyer, R.; Gorrichon, L.; Maroni, P. *J. Organomet. Chem.* **1977**, 129, C7.
148. Lochmann, L.; De, R.L.; Trekoval, J. *J. Organomet. Chem.* **1978**, *156*, 307.
149. DePalma, V.M.; Arnett, E.M. *J. Am. Chem. Soc.* **1976**, *98*, 7447.
150. DePalma, V.M.; Arnett, E.M. *J. Am. Chem. Soc.* **1978**, *100*, 3514.
151. Cambillau, C.; Bram, G.; Corset, J.; Riche, C.; Pascard-Billy, C. *Tetrahedron* **1978**, *34*, 2675.
152. Cambillau, C.; Bram, G.; Corset, J.; Riche, C. *Nouv. J. Chim.* **1979**, *3*, 9.
153. Bottin-Strzalko, T.; Corset, J.; Froment, F.; Pouet, M.J.; Seyden-Penne, J.; Simonnin, M.R. *J. Org. Chem.*, **1980**, *45*, 1270.
154. Chassaing, G.; Marquet, A. *Tetrahedron* **1978**, *34*, 1399.
155. Kress, J.; Novak. A. *J. Organomet. Chem.* **1975**, *86*, 281; **1976**, *121*, 7.
156. Oliver, J.P. *Adv. Organomet. Chem.* **1977**, *15*, 235.
157. Brown, T.L. *Adv. Organomet. Chem.* **1965**, *3*, 365.
158. Sourrisseau, C.; Pasquier, B.; Hervieu, J. *Spectrochim. Acta, Part A* **1975**, *31*, 287.
159. Kress, J.; Sourrisseau, C.; Novak, A. *Metal Ligand Interactions in Organic Chemistry and Biochemistry*; Pullman, B.; Goldblum, N., Eds. D. Reidel: Dordrecht, The Netherlands, 1977, Part 2, p 299.
160. Sourrisseau, C. *J. Mol. Struct.* **1977**, *40, 167*.
161. Andrews, L. *J. Chem. Phys.* **1967**, *47*, 4834.
162. Weiner, M.; Vogel, C.; West, R. *Inorg. Chem.* **1962**, *1*, 654.
163. Sovell, V.M.; Kimura, B.Y.; Spiro, T.G. *J. Coord. Chem.* **1971**, *1*, 107.
164. Kane, A.A.; Tupitsyn, I.F. *Tr. Gos. Inst. Prikl. Khim.* **1970**, *66*, 132; *Chem. Abstr.*, **1972**, *76*, 133820h.
165. Menon, B.C.; Shurvell, H.F.; Colpa, J.P.; Buncel, E. *J. Mol. Struct.* **1982**, *78*, 29.
166. *Aldrich Library of FT Infrared Spectra*, 2nd ed.; Sigma-Aldrich Co.: Milwaukee, WI, **1997**; Vol. 2, p 1631c.

5

Heterogeneous Carbanion Reactions Catalyzed by Solid Bases

5.1 INTRODUCTION: CATALYSIS BY SOLID BASIC CATALYSTS

In our consideration of the chemistry of carbanions, thus far, we have examined the properties of carbon acids and carbanions in solution and, to a lesser extent, in the gas phase. However, the solid phase or, more properly, the interface between the solid phase and the gas or solution phases, has not been discussed yet. In this chapter, we will explore this area of study, with an emphasis less on industrial uses and gas–solid phase reactions than on the experience in the synthetic laboratory.

Industrially, solid bases are important catalysts for organic reactions, although the range of applications involving base catalysis rather than acid catalysis is smaller.[1–5] For example, the use of hetereogeneous acidic catalysts in petroleum catalytic cracking has led to extensive investigation of solid acidic catalysts (e.g., amorphous silica-alumina and acidic zeolites) where the central intermediates are carbocations.[6,7] On the other hand, the use of solid basic catalysts to effect organic transformations has been known since the classic study of Pines and co-workers into double-bond migration in alkenes, as catalyzed by sodium metal dispersed on alumina.[8] Yet it is generally conceded that less extensive study has been made of basic catalysts, where carbanions are presumably the key intermediates.[2–5,8,9] In part, this reflects the ease of electron transfer from the catalyst to the nascent carbanion, leading to radical formation and unwanted side reaction. Nonetheless, contemplation of the wide span of synthetically important and versatile reactions that involve carbanions suggests that development of solid base catalysts as laboratory reagents, as well as industrial adjuncts, will accelerate in years to come.[2,4,9–11]

In the next section, we will consider the classification of some of these heterogeneous basic materials, then proceed to characterization of the alkaline properties of solid bases and finally to selected classes of reactions promoted by these reagents. Throughout this chapter, we will use the term "catalyst" in a loose sense. Many of the solid basic materials that we will consider readily deactivate under reaction conditions and many are used in stoichiometric rather than catalytic amounts to initiate reactions.

5.2 STRUCTURE AND CLASSIFICATION

While there appears to be no universally accepted classification scheme for solid base catalysts, categorization according to structural and chemical similarity, as well as the types of reactions that may be catalyzed, provides an operational framework for the discussion of these heterogenous systems. Table 5.1 represents one such classification scheme, while references 13 and 14 give other possible lists. Obviously, this classification scheme will expand as the different structural and chemical types of catalysts are developed and, for the present, a number of "miscellaneous" basic catalysts have been omitted from Table 5.1, such as silicon oxynitride, which has been shown to catalyze the Knoevenagel reaction more readily than ammonia.[12] However, a selection of these will be discussed in the section on polymeric basic catalysts (Section 5.2.3). Moreover, it will become apparent that our classification scheme (Table 5.1) involves some overlap, as in the case of single alkaline-earth-metal oxides and the "superbase" systems.

In the following sections, the types of catalysts listed in Table 5.1 will be described and their features briefly highlighted.

In discussing these catalysts, a property of obvious importance must be surface basicity. Although surface basicity will be defined quantitatively in

Table 5.1. Classes of solid base catalysts by structural and chemical similarity

Class	Examples
Single-component metal oxides	Alkali-metal oxides: Li_2O, Na_2O, Rb_2O, etc.
	Alkaline-earth oxides: CaO, MgO, etc.
	Transition-metal oxides: TiO_2, ZrO_2, ZnO, etc.
	Lanthanum oxides: CeO_2, Pr_6O_{11}, etc.
	Solid "superbases": CaO (from calcination of $CaCO_3$ at 900 °C), MgO (heated to 550 °C and then treated with molten NaOH), etc.
Mixed metal oxides	$ZnO–SiO_2$, $ZrO_2–SnO_2$, $MgO–Fe_2O_3$, etc.
Zeolites	Alkali ion-exchanged zeolites
	Alkali ions added to zeolites
Anionic polymers	Amberlite IRA-400 OH type, Dowex-3, polyvinyl pyridine, etc.

Section 5.3, for the present the basicity of these catalysts will be used in a qualitative sense.

Furthermore, the structure of the various types of catalysts, inasmuch as it affects their activity, will be briefly surveyed for each class of solid basic catalyst. Detailed descriptions of the nature of these catalysts can be found in various monographs[4,6] and reviews[9,15] and the reader is referred to these.

5.2.1 Single-Component and Mixed Metal Oxides

Although initial interest in basic surface catalysts was sparked by the observation of the isomerization of alkenes caused by metallic sodium dispersed on alumina,[6] and although study of the alkali-metal oxides might appear to be a reasonable extension, few studies of the basic catalytic properties of alkali-metal oxides have, in fact, been reported. By a number of measures, alkali-metal oxides like Rb_2O could be considered to be "solid superbases". These superbases will be examined separately below. Regardless, alkali-metal oxides (Li_2O, Na_2O, K_2O, Rb_2O, and Cs_2O) have all been found to catalyze the isomerization of 1-butene to *cis*/*trans*-2-butene; the mechanism, which proceeds via allylic anions, appears to be the same for all of the oxides.[16,17]

Of the heterogeneous basic catalysts, the alkaline-earth oxides, which contain highly basic sites, are among the most studied. The base strength has been found to follow the decreasing order BaO > SrO > CaO > MgO.[2,9] Tanabe, Misono, Ono, and Hattori considered MgO to be a benchmark solid base for comparison with other surface alkaline catalysts.[18]

In common with many solid basic catalysts, the degree of activity is dependent on the protocol used in pretreatment of the materials. Generally, heating to high temperatures (> 500 °C) in vacuum is required to activate alkaline-earth-metal oxides.[9] In fact, magnesium oxide shows a range of activities, depending on the temperature of activation. Thus, heating to 527 °C yields MgO that will isomerize 1-butene, whereas ca. 700 °C is needed to activate MgO for methane H/D exchange, and 1100 °C is required to produce MgO that will catalyze the hydrogenation of ethene.[19] The role of such heat treatment appears to be desorption of water and carbon dioxide, as well as decomposition of surface carbonates, hydroxides, and peroxides. Magnesium oxide activated at 600 °C has pore sizes with radii from 25 to 75Å,[20] and as activation temperatures are increased the average radius of the pores increases while the distribution of pore sizes also broadens.[21] This pattern of formation of larger pores, as well as a broader distribution of pore sizes, with higher pretreatment temperatures likely holds for many of the metal oxides. However, pore size and distribution appears less significant for carbanion reactions catalyzed by alkaline-earth metal oxides than for basic zeolite catalysts, for example.

The observation for MgO (and duplicated for other metal oxides) that different treatments lead to different catalytic activity suggests different types of basic sites in the catalyst. Partly on these grounds, but also in agreement with molecular orbital calculations,[22] it has been suggested that magnesium oxide contains Mg^{2+}–O^{2-} ion pairs with different coordination numbers.[9,23] Ion pairs of low coordination number are expected to be found at corners, edges, or other defects in the polycrystalline structure.[24] The strongly basic site is then formed by the three-coordinate Mg^{2+} to three-coordinate O^{2-} surface ion pair. Higher-coordinated ion pairs (including four-fold and five-fold) are less basic and, therefore, less active in catalysis.[9] Magnesium oxide activated at lower temperatures will also have some surface OH groups that may also be important. A total of four distinct basic catalytic sites have been identified for magnesium oxide.[5]

This picture of various basic sites on the magnesium oxide surface is further complicated by the identification of reducing sites, where single electron transfer can occur,[20] and radical sites, where hydrogen atom abstraction can occur.[25] However, for H/D exchange for hydrocarbons at the gas–magnesium oxide interface, Tanabe's group concluded that deprotonation by a basic MgO site was rate determining.[26] (For a review of catalytic exchange in isotopic labeling, the reader is directed to reference 27.) Evidence has also been presented for carbanions as central intermediates in isomerization and hydrogenation of alkenes and dienes, as catalyzed by MgO.[28]

Klabunde and co-workers completed a thorough study of H/D exchange at the MgO surface for a series of very weak carbon acids (e.g., CH_4, C_2H_6, toluene, benzene, etc.)[29] In a typical experiment, toluene (6 Torr) and D_2 (110 Torr) were allowed to recirculate over MgO catalyst (0.075 g; activated at 660 °C under ca. 3×10^{-3} Torr vacuum for 6 h); complete deuteration of the methyl group occurred within 0.5 h at 25 °C. Under these conditions, deuteration was selective and no ring H/D exchange was detected.[30] Significantly, enthalpies of activation for this catalytic H/D exchange for a limited series of hydrocarbons (i.e., $C_2\mathbf{H}_6$, $C_3\mathbf{H}_6$, benzene, CH_3–$C_6H_4C\mathbf{H}_3$, $C\mathbf{H}_3Ph$, $CH_3C\mathbf{H}_2Ph$, and $(CH_3)_2C\mathbf{H}Ph$, where the boldfaced hydrogen is exchanged) correlate well with gas-phase enthalpies for acid dissociation of the carbon acids (see equation 1.36).[31] Rate constants for H/D exchange for a wider range of carbon acids has been shown to correlate with solution (DMSO) pK_a values for the hydrocarbons.[29] Both results provide convincing evidence for carbanions as the key intermediates in isotopic exchange over basic magnesium oxide.

These studies by the Klabunde group demonstrate the range of physical organic tools that may be used to scrutinize the nature of catalytic sites. In summary, for H/D exchange over MgO (activated as described above): (1) entropies of activation were found to be negative, large, and almost constant, regardless of structure of the hydrocarbon acid (i.e., −25.6 eu for propane to −28.4 eu for cumene); (2) the 1° kinetic isotope effect (KIE) for toluene was small, $k_H/k_D = 1.6$ (100 °C); (3) MgO activated at 550, 600, and 660 °C has OH sites (0.644, 0.135, and 0.017 OH/nm^2, respectively), as well as more basic catalytic sites, and rate constants for H/D exchange correlate with the coverage of OH sites (i.e., larger rate constant with higher amount of OH sites); and (4) H/D isotopic exchange occurs with much lower activation energies on magnesium oxide than found in solution (e.g., 0.9 kcal/mol for toluene–MgO as compared with 18 kcal/mol for toluene–DMSO-*t*-butoxide).[29]

The foregoing results, combined with the correlation of exchange rate constants with solution pK_a values, permitted construction of the following picture of the catalytic process. First, the large negative and almost invariant $\Delta S^{\ddagger}$ values found for the range of structurally varied carbon acids was interpreted as favoring a compact, bimolecular complex for all of the carbon acids. The small isotope effect was attributed to a linear *and symmetrical* transition state for proton or deuteron transfer, but one from which a type of facile internal return could operate and so reduce the isotope effect to 1.6 from our ideal zero-point energy value of 4.3 (at 100 °C; see Section 2.1.1 and 2.1.3). Note, however, that the same low KIE could arise either from a nonlinear transition state or from a highly unsymmetrical linear transition state for deuteron/proton transfer. The requirement for OH(or OD) groups on the surface suggests that deuterium is transferred to the carbanion from the OD group, which acts as an acid site in the rate-determining transition state. Thus, while we can characterize MgO as essentially a basic catalyst, it is, in fact, bifunctional by virtue of having a range of sites of different basicity; that is, the OH/OD sites here are acidic relative to the ion-paired oxide sites. Electrostatic attraction between the incipient carbanion and a magnesium ion site on the surface would be expected to lower the activation energy for the process of deprotonation/deuteration. (This "maximum ion pairing" was postulated to account for the difference in activation energy between solution phase and this heterogeneous isotopic exchange). The correlation of rate constants with condensed phase pK_a values confirms the importance of C–H bond breaking in the rate-limiting transition state.[29] Overall, structure **61** was proposed for the transition state for exchange.

In the model above, it was further presumed that the lowest coordination number magnesium ion–oxide anion paired sites (see above) would be taken up in binding D_2 or HD and, consequently, the site of binding of the hydrocarbons was presumed to involve magnesium ions and oxide sites of higher coordination number. The significance of this supposition is that the less basic sites of the magnesium oxide are still basic enough to abstract protons from the weakest carbon acids known (e.g., CH_4 with a $pK_{a(DMSO)}$ estimated to be ca. 56).

Although this model for the surface active sites on magnesium oxide may well also hold for the other alkaline-earth-metal oxides, it does not appear to apply to the lanthanum oxides; La_2O_3, for example, displays none of the temperature-dependent pretreatment behavior of the alkaline-earth-metal oxides, but, rather, achieves maximum activity for all of the test reactions when heated at 650 °C.[9,32] Clearly, the nature of the catalysts, and the number and basicity of the catalytic sites, all depend on the pretreatment of the catalyst. Comparisons of older literature and current results must take such subtleties into account.

Mixed oxide catalysts were originally developed to extend the number of basic catalysts available. These mixed metal oxides (e.g., ZnO–SiO_2,[33] MgO–Al_2O_3,[34] ZrO_2–SnO_2,[35] MgO–Fe_2O_3,[36] etc.) have been reported to show enhanced surface basicity compared with the basicity of the individual component oxides.[37] The

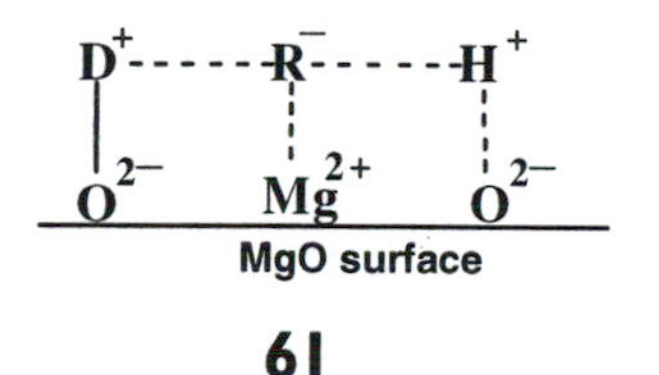

61

structures of these mixed metal oxides are not always well defined and, presumably, vary depending on the oxides mixed, the proportions of each oxide in the mixture, and, importantly, whether mixed phases are formed,[37,38] as well as on the method of pretreatment.[39] Although there is currently no generally recognized model to predict quantitatively the basicity changes that occur upon formation of these binary oxide catalysts,[39] their catalytic properties may be "tuned" empirically. For example, the maximum number of base sites are generated in the ZnO–Al_2O_3 system when the mixed oxide catalyst is composed of 50% alumina,[40] while the TiO_2–MgO system achieves a maximum number of basic sites when it contains 90% magnesia.[41] Our understanding of such systems will also benefit from current advances in the spectroscopy of surfaces (*vide infra*).

5.2.1.1 Superbase Systems

Either high-temperature formation of alkaline-earth-metal oxides from their carbonates or hydroxides[42] or the evaporation of metallic alkali metals such as sodium on an alkaline-earth-metal oxide[43] leads to preparation of what have been termed "superbases".[14,44] By analogy to the superacids,[45] which exhibit very high acidities ($H_0 \leq -12$, i.e., 19 H_0 units from $H_0 = 7$, or neutral), solid superbases have high basicities; any solid basic catalyst with a surface basicity equal to or greater than an H_- value of 26 (i.e., 19 H_- units from $H_- = 7$) is considered a solid superbase.[14]

Standard superacids such as FSO_3H and CF_3SO_3H, (H_0 ca. −15.1 and −14.1, respectively),[46] also find solid counterparts.[47,50,51] Table 5.2 lists a selection of the solid superbases and their reported H_- values.

Comparison can be made briefly to superbase systems in solution. These may be divided into two broad classes, after the suggestion of Caubère:[52] (1) the Lochmann–Schlosser-type superbase systems[53–56] that involve two bases[57] with two different counterions—*n*-butyllithium with potassium *t*-butoxide in THF, for

Table 5.2. Selected solid superbases

Base and method of preparation	H_- values
SrO from calcining $Sr(OH)_2$ at 850 °C	26.5[a]
CaO from calcining $CaCO_3$ at 900 °C	26.5[a]
MgO thermally treated at 550 °C, then treated with molten NaOH	26.5[b]
MgO thermally treated at 650 °C, then Na vapor deposited on the surface	35[c]
Al_2O_3 calcined at 450 °C, then treated with molten NaOH and then molten Na	37[d]

SOURCES: [a]Data taken from reference 43; [b]Data taken from reference 48; [c]Data taken from reference 44; [d]Data taken from reference 49.

example—and are termed "multimetal super bases" by Caubère[52]; and (2) "unimetal super bases" formed from two bases with the same counterion, such as pentylsodium with sodium alkoxides in ethereal solvents.[52,58,59] In either case, the result of these "mixed" basic systems is enhanced basicity. Thus, *n*-butyllithium is ineffective in deprotonating toluene in heptane, but addition of potassium *t*-pentoxide results in rapid proton abstraction to give benzylpotassium.[60] The uses of these two types of solution-phase (though sometimes heterogeneous) superbases in synthesis are the subject of comprehensive reviews[52,61] and their synthetic utility will not be explored further here. Although still controversial,[62] it appears that for at least some of these solution superbases the enhanced deprotonation activity is linked to the formation of mixed aggregates in which the two metals are more strongly coordinated to the alkoxide oxygens and lithium cation (or cations) than to the carbanionic center; this results in an effective increase in the basicity of the carbanide.[63] Support for this view comes partly from X-ray structures of lithium 4,6-dimethyl-2-sodiomethylphenoxide coordinated with TMEDA (tetramethylethylenediamine)[63] and ab initio (MP4SDTQ/6-31 + G^*//6-31 + G^*) calculations.[63,64] The similarity between these solution superbase systems and the solid superbases that we shall discuss in detail in this chapter lies not just in the high basicity of these superbase systems, but also in the importance of structure, whether of the aggregate in solution or of the surface in the solid phase, in determining this basicity.

Note that the solid superbases also fall into two operational categories: those like CaO that are prepared and activated by high temperatures alone and those like MgO that are heated to lower temperatures but are then treated with molten alkali metal or alkali-metal vapor. In the first case, it would seem reasonable to assume that the high temperatures are converting less basic site (e.g., surface $Mg^{2+}OH^-$) groups to more basic sites (i.e., $Mg^{2+}\ O^{2-}$). The model of Kijensky and Malinowski suggests that alkali metals could also perform this function,[65–67] according to equations 5.1–5.3 (where M represents the alkali metal and the subscript "surface" indicates reactive sites on the surface of the catalyst):

$$2OH_{surface} + M \rightarrow OM_{surface} + H_2O_{(g)} \tag{5.1}$$

$$OH_{surface} + M \rightarrow OM_{surface} + 1/2H_{2(g)} \tag{5.2}$$

Furthermore, Kijensky and Malinowski also noted the possibility of holes trapped by oxide anions that could be converted to basic oxide sites, and then act as electron-pair donors, namely, Lewis bases, as shown in equation 5.3:

$$O^-[\text{hole trapped on O anion}] + M \rightarrow O^{2-}_{surface} + M^+ \tag{5.3}$$

On the other hand, alkali-metal atoms could interact with anionic vacancies (e.g., Schottkey defects[68]) in the polycrystalline structure to generate "trapped electrons" or F-centers and alkali-metal cations (equation. 5.4):

$$[\text{anion vacancy}] + M \rightarrow [e^-]\,\text{or F-center} + M^+ \tag{5.4}$$

These F-centers then act as single-electron donors. An alternative view has been put forward, at least for MgO doped with Li [i.e., $MgO/Mg(OH)_2$ aqueous slurry, mixed with aqueous LiOH and, after evaporation of the water, thermally activated]. Partly because of the similar ionic radius of Li^+ and Mg^{2+}, it was suggested that Li might become incorporated in a partial crystallite, $[Li_3Mg_6O_9]^{3-}$. Of course, the overall negative charge of this small surface defect would have to be balanced by anionic vacancies in the bulk crystal. This high negative charge defect would then act as the electron donor.[69] Regardless of the exact nature of the electron donors in superbase systems, it is nonetheless clear than superbases may act as single-electron donors, as well as strong Brønsted bases. Note that solution-phase superbase systems may also act as single-electron donors.[52] Regardless, the structure of these powerful basic solids will be elucidated by the rapidly advancing surface spectroscopic techniques.[70]

Undoubtedly, this demonstrated high basicity suggests that the superbase systems will be most applicable to catalysis of gas-phase carbanionic processes (e.g., alkene isomerization, side-chain alkylation of alkylbenzenes and isotopic exchange, etc.) or in reaction systems where solvent is not required. Otherwise, side reactions involving deprotonation of most common organic solvents could be expected.

5.2.2 Basic Zeolites

Zeolites have become the "workhorses" of industrial catalytic hydrocarbon chemistry in recent years, particularly as acidic catalysts in processes such as the conversion of methanol to gasoline feedstock.[1] It is a rare laboratory that does not have molecular sieves, particularly for drying solvents and gases.[71] However, we shall see that these versatile solids, in the basic form, are also valuable reagents for reactions involving carbanion intermediates.[15] In fact, the field of zeolites is so large and growing at such a pace that only an overview is possible in the context of the current work.

Conventional zeolites are crystalline aluminosilicates. The linked SiO_4 and AlO_4 tetrahedra of these zeolites form a three-dimensional structure that includes regular pores or cages and channels.[15] While modified zeolites are described similarly, the central Si or Al atom of the tetrahedra are partially or wholly replaced by elements such as B, Ga, Fe, Ti, Co, or Ge. These isomorphous substitutions give rise to the AlPONs (mostly Al- and P-centered tetrahedra), SAPO (Si, Al, and P tetrahedra), MAPO (Al, and P tetrahedra with M = Co, Fe, Mg, etc.), and related families.[71–73]

A significant facet of all of these zeolites is that the size of the cages and channels is similar to that of organic molecules. It is this feature, the need for size complementarity between catalyst pores and the carbon acid, that primarily distinguishes basic zeolites from the metal oxide bases we have examined thus far.

Zeolites have been found to have structures similar to those of naturally occurring minerals such as faujasite (Figure 5.1). The faujasite zeolites are further subdivided according to their Si:Al ratios; X zeolites have Si:Al ratios of between 1:1 and 1.5:1, whereas the Y zeolites contain more silicon and have a Si:Al ratio of between 1.5:1 to 3.0:1. Structurally, however, both X and Y zeolites fit the faujasite model. Both could be described as having supercages or cavities linked together via hexagonal prism structures; the maximum pore aperture for either form is 7.4Å.[71,74] Other structural types of zeolites, including types A, L, morde-

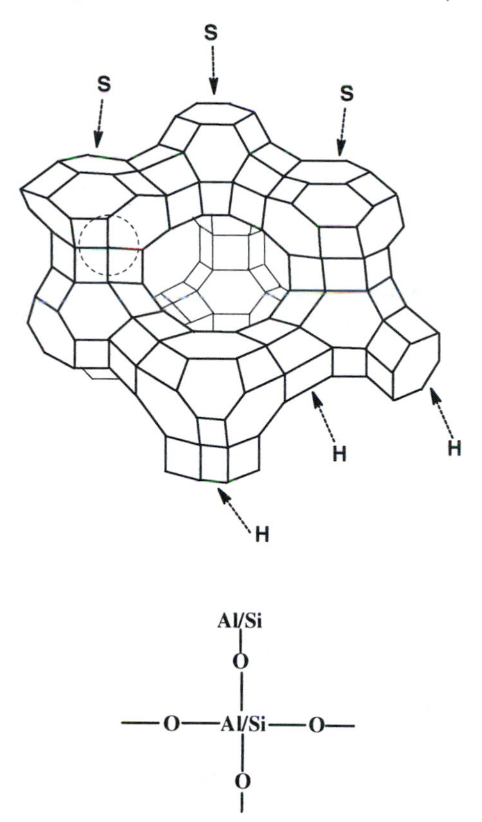

Figure 5.1a. Topology and structure of faujasite-type zeolites. The bold portion of the structure is the outer framework of the supercage. Sections marked H are hexagonal prisms and sections marked S are sodalite prisms. The portion encircled by dashed lines is expanded in Figure 5.1b.

Figure 5.1b. Expanded portion from Figure 5.1a. At each vertex, a silicon or aluminum atom is located. The oxygens are located in between these points. Angles for the SiO_4 and AlO_4 tetrahedra vary from 130° to 170°. The differing angles, as well differing bond lengths between metal and oxygen centers, give rise to differing negative charges on the oxygen atoms.[15]

nite, beta, etc., have been described by Barthomeuf,[15,75] and each has a different, unique three-dimensional topology, which will not be described further here.

Returning to the faujasite-type zeolites (Figure 5.1a), it can be seen that the bold-lined structure represents the outer framework of the supercage, while the thinner-lined structure depicts the inside of this cavity. Each vertex in the structure is occupied by either a silicon or aluminum atom and the oxygen atoms link adjacent tetrahedra as shown in Figure 5.1b. Unlike the alkaline-earth-metal oxides, then, there are no exposed oxides to act as base sites, although the varying tetrahedral angles give rise to different partial negative charges on the linking oxygens and, hence, yield oxygens with different basicities. In the faujasite-type zeolites, four different types of basic oxygens have been identified, but numbers vary; for example, mordenite is reported to have 10 different types of oxygens.[15] The strongest basic oxygens are those involved in AlO_4 tetrahedra.[15,75]

Generally, the basicity of the framework oxygens may be increased by increasing the Al content in the framework Therefore, for maximum intrinsic zeolite basicity, a low Si:Al ratio is required.[73] We may expect X zeolites by nature to be more basic than their Y structural analogues. Clearly, substitution of either Si or Al in the skeleton with other atoms such as Ge for Si or P for Al will also modify the basicity of the zeolite.[76,77]

Although hydroxide ions may be generated in faujasite zeolites, these ions are relatively weak bases and, in general, to be useful as basic catalysts, the zeolite must be modified by treatment with a base or through deposition of alkali metals. Ion-exchanged zeolites are prepared by titrating the naturally acidic zeolite with an alkali-metal hydroxide (MOH), followed by heating to high temperature, under vacuum, to dehydrate the zeolite. Hattori[9] makes a distinction between these alkali ion-exchanged zeolites and "alkali ion-added" zeolites, where concentrations of added MOH exceed the ion-exchange capacity of the zeolite. Such basic zeolites are designated by the alkali-metal cation used for exchange, as well as the type of zeolite, as in NaY or CsX, for example. It has been found that the basicity of ion-exchanged zeolites depends on the alkali-metal hydroxide used. Basicity decreases in the following order: Cs > Rb > Na > Li.[9,15]

Other methods of introducing basicity into these zeolites start with the alkali-metal cation-exchanged zeolite, followed by (1) subjecting the zeolite to alkali-metal vapor;[78–82] or (2) impregnating the zeolite with an alkali-metal azide that is subsequently decomposed thermally;[83–86] or (3) treating the zeolite with aqueous solutions of alkali-metal acetates or methanolic solutions of magnesium methoxide with subsequent complete evaporation of the solvent.[87–89] The first two types of procedures (and related methods)[90,91] generate alkali-metal ion and/or neutral alkali metal clusters.[70,71,92–94] The current view[15] is that the ionic clusters are not involved in base catalysis with these zeolites. On the other hand, the neutral clusters associate with framework oxygens and apparently enhance the basicity of these oxygens, which then act as the actual basic catalytic sites.[91,92] The type of cluster formed, however, is strongly dependent on the method of introduction of the alkali metals to the zeolite and on the activation conditions.[15]

In the case of the magnesium methoxide and related protocols for zeolites, MgO (and, more generally, alkaline-earth-metal oxide, MO) clusters form and are trapped in the pores of the zeolite.[87–89] These clusters act as the basic sites, combining the range of base sites found for polycrystalline alkaline-earth-metal oxides with the structural constraints inherent in zeolites. Two problems have been identified, however, with these MO-cluster zeolites. First, the oxides may be readily converted to the corresponding, and catalytically ineffectual, carbonates by reaction with atmospheric carbon dioxide.[15] Further, some activation procedures have been reported to result in collapse of the zeolite structure.[9]

In Section 5.4, we will highlight the differences in selectivity of some carbanion-centered reactions on basic zeolites as contrasted with the other types of basic catalysts that we have considered, particularly with respect to the requirement for steric complementarity in the zeolites.

5.2.3 Polymeric Organic Resins and Some Miscellaneous Solid Bases

Catalysis of reactions with acidic polymers is overshadowed by those catalyzed by Nafion-H®,[47] polymeric superacids[46] with the following general structure:

$$-(CF_2CF_2)_x - (CF_2CF)_y - (OCF_2-CF)_m - OCF_2CF_2-SO_3H$$

where the subscript $m = 1$ and the $x{:}y$ ratio is between 2:1 and 50:1. In contrast, no such dominant alkaline organic polymer has yet been discovered. Generally, polymeric resins with associated hydroxide ions, such as polymeric quaternary ammonium hydroxide systems, are less basic than common homogeneous bases, namely, aqueous NaOH, and so on. However, the fact that these polymeric resins are not soluble in the (usually aqueous) solvent is an advantage in workup: the resin can simply be filtered off.

(Note, however, that an alternative approach to the use of polymers in the synthetic chemistry of carbanions is provided by solid-phase reactions in which the reactive moiety is bound to a suitable polymer; the carbanion chemistry—often involving carbon–carbon bond formation—is performed and then the moiety is cleaved from the polymeric backbone. This approach has been reviewed by Lorsbach and Kurth, and the interested reader is directed to this review.[95])

Anionic-exchange resins are the polymeric basic systems that have been examined most, although it must be acknowledged that relatively few studies have been undertaken on polymers as basic catalysts (as compared with their use as phase-transfer agents, for example). These ion-exchange resins are essentially polystyrene (PS), cross-linked with a small amount (typically 2%) of divinylbenzene (DVB).[96] The resins are typically of the gel type or the macroreticular type. The latter is formed by polymerization in the presence of an organic solvent that solvates the monomer well but only swells the polymer poorly. The result is a regular three-dimensional structure, shaped by the organic solvent molecules. After copolymerization or polymerization followed by cross-linking, the resins may be functionalized, usually via reaction at the benzene moieties of the polymer. Sulfonation of the polystyrene groups of the PS–DVB copolymer leads to cationic exchange resin (i.e., PS–DVB–$SO_3^-H^+$), such as Amberlite IR-120 gel. Conversely, chloromethylation of the polystyrene residues of the PS–DVB followed by reaction with a trialkyl amine produces anionic-exchange resin (i.e., PS–DVB–$NR_3^+X^-$, where X^- may be OH^-), such as Amberlite IRA-900 (macroreticular) or Amberlite IRA-400 (gel).[96]

These anionic-exchange resins have been shown to catalyze the Knoevenagel and aldol reactions, as well as the cyanoethylation of alkanols. Interestingly, in a study by Astle and Zaslowski, neutral amino-substituted resins (PS–DVB–NR_2) were found to be more effective (i.e., higher yield of condensation product) than the quaternary ammonium type (PS–DVB–$NR_3^+OH^-$).[97] Note, also, that polyvinylpyridine has been reported to be an effective catalyst for the Knoevenagel reaction.[98]

Early reports of catalysis of Michael additions to carbonyl compounds by Amberlite (IRA-400, or -410, in this case) were somewhat disappointing.[99,100] Yields were low and side products of self-condensation or cyclization often

SCHEME 5.1

became the main isolated products.[99] Further, a common problem of these catalysts was identified: the benzylic-type structure of the quaternary ammonium salts (i.e., polymer–C_6H_4–CH_2–$NR_3^+X^-$) implies that nucleophilic displacement at the benzylic position may also occur; obviously, this results in degradation of the polymer catalyst.[99] However, these exploratory studies have led to more successful "fitting" of reactants and catalyst in the Michael reaction.[101] A particularly interesting example is provided by the Michael addition of the enolate of methyl 1-oxoindan-2-carboxylate to 3-butenone (methyl vinyl ketone), which yielded the chiral *S*-addition product (Scheme 5.1) with an optical activity of $\leq 27\%$, when a chiral anionic-exchange resin was used as the catalyst.[102] In this case, the catalyst was prepared by reacting ring chloromethylated PS–DVB cross-linked copolymer with cinchonidine to produce the chiral anionic-exchange resin, **62**. In this system, the counterion of the catalyst (X^-) presumably deprotonates the carbon acid, while the polymeric quaternary ammonium ion forms an ion pair with the newly formed carbanide. Interestingly, in this work a variety of counterions for the ammonium salt were explored ($X^- = Cl^-$, F^-, and OH^-) and all were found to be effective in initiating the addition reaction.[102]

Alkylation of 1-cyano-1-phenylmethane (phenyl acetonitrile) with 1-bromobutane occurs in the presence of various anionic resins (e.g. Dowex 1-X8, –11, 21 K, 44, with Cl^- or OH^- as counterions). Here, the resins may be acting as phase-transfer catalysts as much as bases.[103]

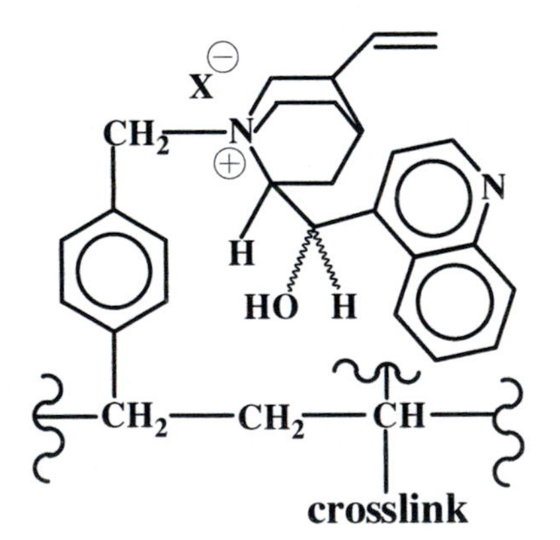

62

At the time of writing, the utility of polymeric catalysts is somewhat hampered by the fragility of these anionic resins.[98–103] Progress may be expected in modification of polymers by attaching strongly basic moieties. For example, the hindered base 4-*N*,*N*-dimethylaminopyridine (DMAP) is an effective homogeneous base. Attachment of its 4-(*N*-benzyl-*N*-methylamino)pyridine analogue to PS–DVB via the *N*-benzyl site yielded a polymeric catalyst that was less active in catalyzing acetylation of alcohols than DMAP, but more effective than 4-(*N*-benzyl-*N*-methylamino)pyridine.[104] Tethering similar bases to more robust polymeric backbones should lead to advances in the development of these solid base catalysts.

As mentioned earlier, Table 5.1 represents but one way of cataloguing the various types of solid base catalysts. A number of miscellaneous catalysts were omitted from the table, but a selection of these should be considered briefly, since they represent the growth occurring in these studies.

Anionic clays show promise as solid bases. The hydrotalcite-type clays have the general composition $M_x^{2+}M_y^{3+}(OH)_{2x+2y}(X^-)_{2y}\cdot mH_2O$.[105] The parent clay, hydrotalcite, has the formula $Mg_6Al_6(OH)_{16}(CO_3^{2-})\cdot 4H_2O$.[93] These clays belong to the class of pyroaurite–sjogrenite minerals and have a layered structure of M^{3+} cation-substituted $M^{2+}(OH)_2$ octahedra.[105,106] The M^{2+}–M^{3+}–OH sheet bears a net positive charge that is balanced by inner anionic layers; water molecules are associated with these anionic inner sheets.[107] As might be expected, activation of the catalyst requires heating at moderately high temperatures (ca. 400 °C)[108] but, once activated, hydrotalcites have been shown to catalyze H/D exchange, alkene isomerization and the aldol condensation,[108,109] and the Claisen–Schmidt condensation of benzaldehyde with acetone,[110] as well as initiating the anionic ring-opening polymerization of propylene oxide.[111] The process of thermal activation removes water, generates oxide basic sites, and introduces a pore structure to the catalyst; activated hydrotalcite has a general formula of $Mg_6Al_2O(OH)_2$.[106–109] The surface basicity has been estimated to be equivalent to an H_- value of ca. 17.[112]

Asbestos-type minerals [e.g., crysotile, $Mg_3Si_2O_5(OH)_4$] were briefly investigated for use as solid bases,[113] but their carcinogenicity[114] has largely halted such activities. In this regard, the work of the Laszlo group with the basic asbestos-replacement mineral xonotlite [$Ca_6Si_6O_{17}(OH)_2$] is noteworthy.[10,115,116] This fibrous calcium silicate has a ladderlike structure formed by parallel chains of silicates that are connected via every fourth silicon. Although xonotlite is apparently moderately basic in its own right, in the work of the Lazslo group the calcium silicate is generally heat treated and allowed to react with potassium *t*-butoxide.[10] The catalyst has been shown to catalyze the Knoevenagel reaction[115] and Michael addition.[116]

Graphite treated with alkali-metal vapor, particularly K, Cs, and Rb, has been investigated as a solid base catalyst. The activity of graphite–alkali-metal intercalation compounds is largely of the single-electron-transfer type and this has been reviewed.[117]

There are a number of reports on "oxynitride" solid bases. These are generally formed by action of ammonia on various calcined oxides, including silica (to give silicon oxynitride)[14,118] and aluminophosphate zeolites (to give AlPONs).[119,120]

These basic[121] materials appear to contain surface $-NH_2$ groups and, in the case of AlPONs, also –NH– groups linking aluminum and phosphorus zeolite skeletal tetrahedra.[120] Such materials all have been shown to catalyze the Knoevenagel condensation, with varying overall yields.

Various basic solids have been shown to be effective in initiating specific carbanion-mediated reactions. Classic solid bases such as KOH have been shown to be effective in initiating alkylation of furfural by the cyanomethide anion,[122] while $Ba(OH)_2$ catalyzes both Claisen–Schmidt[123] and aldol[124] condensations, and even such a mildly alkaline material as *neutral* alumina[125] has been shown to be effective in Knoevenagel condensations.

Clearly, the classification scheme outlined in Table 5.1 is only the starting point in the examination of potential solid bases as reagents and/or catalysts for carbanion reactions. In the next section, we will consider the various methods used to probe and quantify surface basicity.

5.3 CHARACTERIZATION OF SURFACE BASICITY

5.3.1 Acidity Function Measurements

Historically, interest in and examination of solid basic catalysts followed similar investigations of solid acids and, consequently, the methods developed for solid acid catalysts were extended to the basic materials. Following the proposal of Walling,[126] solid acids were first characterized for surface acidity by adding the solid to a benzene solution of a Hammett indicator of the H_0 acidity function scale.[127] Uptake of the indicator by the solid and observation of a color change corresponding to conversion of the indicator from a neutral or basic form to an acidic form would indicate that the solid contained acid sites strong enough to protonate the indicator. Using a series of indicators, the acidity of the surface could be bracketed. In this acidity function approach, the Brønsted acidity of the medium (i.e., the solid surface) is measured.

In a similar fashion, the surface basicity of solid bases can be determined, using the indicator acids that define the H_- acidity function[127] (determined in either DMSO[128–131] or DMF[132]). According to equation 5.5,

$$H_{-(\text{surface})} = \text{p}K_{\text{BH}} + \log [\text{B}^-]_{\text{surface}}/[\text{BH}]_{\text{surface}} \tag{5.5}$$

if the concentration of the undissociated indicator acid, BH, on the surface of the solid base, and the concentration of the conjugate base of the indicator, B^-, also on the surface of the solid base, are known, as well as the pK_a of the indicator (pK_{BH}), then the H_- value of the surface can be determined. In other words, equation 5.5 applies to the following equilibrium between indicator acid, BH, with the basic surface, $\mathbf{B}_{\text{surface}}$, to give the conjugate indicator base, B^-, and protonated surface site, $\mathbf{BH^+}_{\mathbf{surface}}$, as per equation 5.6.

$$\text{BH} + \mathbf{B}_{\text{surface}} \rightleftharpoons \text{B}^- + \mathbf{BH}^+_{\text{surface}} \tag{5.6}$$

The indicators commonly used, their colors, and their H_- values are given in Table 5.3.

Like the original acidity function methods, this approach to surface basicity may be criticized on the theoretical grounds that the assumptions concerning activity coefficient cancellation (cf. equations 1.27 and 1.28) are only approximately valid. However, to date, none of the modified acidity function methods[133–135] have been applied to solid bases.

Let us consider application of this approach to a superbase solid catalyst such as CaO (prepared by calcining $CaCO_3$ at 900 °C; Table 5.2).[43] A small amount of this catalyst is placed in a tube containing an inert solvent, usually benzene, and the first indicator acid is added. If this indicator is observed to have the color of its base form, then the experiment is repeated with the next indicator of higher pK_a. The process continues until the indicator chosen registers only the color of its neutral, acid form. In this example of superbasic CaO, starting with bromothymol blue and continuing up to 4-chloroaniline, each indicator, in turn, changes color from that of its acid form to that of its conjugate base form. However, repetition of the experiment with diphenylmethane shows that the indicator that began colorless, remains so. The conclusion is that the surface basicity of this CaO lies between an H_- value of 26.5 (color change for 4-chloroaniline) and 35.0 (color change for diphenylmethane). To be conservative, the H_- value of this basic catalyst is generally reported as 26.5, but, clearly, it could be more basic.

Various experimental deficiencies must be addressed. First, and unlike the application of H_- in various solvents, the actual ratio of $[B^-]/[BH]$ is not measured. Since the determination is made in the case of surface bases with the naked eye, the color change is typically only noticeable when about 10% conversion of acid to base form has occurred ($[B^-]/[BH] \approx 0.1/0.9$). Increase in the intensity of the color generally registers with the naked eye when 90% conversion has been accomplished ($[B^-]/[BH] \approx 0.9/0.1$). Therefore, the experimenter notes the onset of color when H_- is equal to $pK_{BH} - 1$ and the intensification of color when H_- equals $pK_{BH} + 1$. Observation of the intermediate color is then taken to mean that H_- of the surface is equivalent to pK_a of the indicator acid.[137] The errors involved

Table 5.3. H_- indicator acids used in determination of surface basicity

Indicator acid	Color of BH	Color of B^-	pK_a value
Bromothymol blue	Yellow	Green	7.2
Phenolphthalein	Colorless	Red/magenta	9.3
2,4,6-Trinitroaniline	Yellow	Reddish-orange	12.2
2,4-Dinitroaniline	Yellow	Violet	15.0
4-Chloro-2-nitroaniline	Yellow	Orange	17.2
4-Nitroaniline	Yellow	Orange	18.4
4-Chloroaniline	Colorless	Pink	26.5
Diphenylmethane	Colorless	Yellow-orange	33.0
Cumene	Colorless	Pink	37.0

Source: Data taken from references 129, 130, and 136.

in such a procedure are apparent. Secondly, indicators whose color changes are not intense enough may be masked by the color of the catalyst itself.[138] This limits the choice of indicators for some catalysts. Also, for equations 5.5 and 5.6 to be valid, any color change of the acid indicators must arise from proton transfer to give the conjugate base of the indicator. Binding to the surface alone, single electron transfer or H–atom abstraction may, in some cases, also lead to color changes; indicators susceptible to color changes under these conditions are clearly unsuitable.[9] For highly basic catalysts such as the CaO superbase we have been considering, the experiment must be carried out under inert atmosphere with degassed solvent to avoid potential poisoning of the basic sites by atmospheric carbon dioxide or water. Finally, benzene or another nonpolar solvent is commonly used, following the assumption that such solvents will not preferentially bind to the catalyst surface and so prevent the needed interaction of indicator and surface base sites.[126] This assumption appears reasonable in the case of alkaline-earth metal oxides, for instance, where pore size and distribution are not significant factors in assessing the activity of the basic catalyst, but the assumption is problematic for basic zeolites where size and shape of cavities are important.

Nonetheless, indicator methods have been used for basic zeolites[139–142] and the following correlation of basicity with zeolite type (all faujasites) has been reported:[15,139]

Zeolite :	NaX	<	CsNaX(Cs from CsCl)	<	CsNaX(Cs from CsOH)
pK_a :	7.2	<	8.2	<	18.4

In line with our previous observation that faujasite zeolites with high Al content (i.e., low Si:Al ratio) tend to be more basic, indicator studies also show that NaX has stronger base sites than NaY.[142]

Returning to our example of superbasic CaO and accepting the limitations of the method, we recognize that this catalyst has a surface basicity equivalent to H_- of 26.5. What this means is that there are base sites ranging from much less basic up to (and likely somewhat more basic than) 26.5 on the H_- scale. The question arises of how many base sites are present for a given amount of the base catalyst? This can be determined by an indicator titration. Specifically, a given H_- indicator is adsorbed on the catalyst and changes color, indicating that the indicator is associated with the former base site corresponding to the H_- value of the indicator. The sample is then titrated with benzoic acid. Since benzoic acid is a stronger acid, though otherwise structurally similar to the indicators used (see Table 5.3), it will displace the indicator from the surface and protonate it. The amount of benzoic acid required to displace the indicator and change it back to its acid form (in mmol/g or $mmol/m^2$) is a measure of the number of base sites on the catalyst *at the H_- value of the chosen indicator*.[9] By performing the titration using several different indicators, a distribution of basic sites may be constructed. While valuable in assessing the number and strength of basic sites on the solid base, the method has the disadvantage that equilibration of the benzoic acid with the catalyst may be slow. Dissolution of some of the basic catalyst in the solvent used for the titration may occur in some cases and this results in an overestimation of the number of base sites for the catalyst.[9] This dissolution problem is most apparent in studies where the significantly more acidic 2,2,2-trichloroacetic acid was used to titrate the basic sites.[143,144]

Benzoic acid titration of adsorbed indicator acids with alkali-metal exchange zeolites revealed surprisingly small numbers of base sites.[140] Again, this likely reflects the steric requirements of the zeolites rather than the actual number and distribution of basic sites in these basic zeolites.[15]

Regardless of the various disadvantages of classical titration approaches, these methods do permit an approximate comparison of solid bases with those more commonly encountered in organic chemistry.

5.3.2 Spectroscopic Methods

Diverse types of spectroscopy have been applied to the examination of solid catalysts. While spectroscopic studies of solid acidic catalysts must be acknowledged to be more highly developed than comparable investigation of solid bases,[145] the importance of such investigations is obvious in achieving an understanding of these basic systems.

In general, spectroscopic methods have been used (1) to examine the catalyst surface itself and (2) to examine the binding of "reporter" molecules to the surface or examine the catalyst under simulated reaction conditions. Infrared spectroscopy has been used in both capacities. The use of IR spectroscopy for direct examination of the catalyst surface has been reviewed by Tanabe and the reader is referred to that review.[146]

In the following sections, we will consider the use of spectroscopic techniques, particularly IR, ^{13}C NMR, and X-ray photoelectron (XP) spectroscopies, to examine the catalytic sites of solid bases.

5.3.2.1 Infrared and ^{13}C Nuclear Magnetic Resonance Studies of Molecules Bound to Solid Bases

Infrared spectroscopy has been used as a method of examining reporter molecules, particularly to probe the interactions of carbon dioxide and pyrrole with solid basic surfaces. As we shall see, binding and thermal desorption of carbon dioxide are commonly used to characterize solid base catalysts and, therefore, IR is a valuable adjunct to analyzing these processes. While as many as six different modes of binding of carbon dioxide to a basic (oxide) catalytic surface have been proposed,[147] only three species have been reported to arise from the interaction of CO_2 with the common single-component metal oxides such as CaO, namely, monodentate (**63**) and bidentate carbonates (**64**) and hydrogen carbonates (that may form as a result of interaction of the carbon dioxide with surface hydroxyls rather than oxide sites).[148,149] For MgO, the mode of binding is dependent on the degree of CO_2 coverage: bidentate binding predominates at low CO_2 coverage with monodentate binding becoming dominant at high coverage.[149] On the other hand, with CaO the bidentate mode (**64**) is prevalent regardless of coverage. For ZrO_2, CO_2 binds preferentially by the monodentate mode, again independent of coverage.[150]

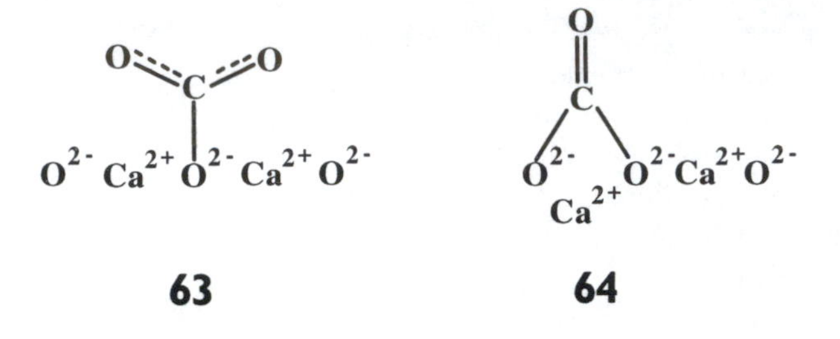

Studies of CO_2 binding with zeolites have also revealed formation of monodentate and bidentate types of carbonates with IR bands in the 1440–1500 cm^{-1} region[151,152] However, the complexity of the spectra, as well as limited stability of the species formed,[153] has led to the conclusion that CO_2 is not an appropriate IR probe molecule for basic zeolites.[15]

Upon interaction of pyrrole with the framework oxygens of basic zeolites, the N–H stretching vibration shifts from 3430 cm^{-1} (pure liquid) to ca. 3200 cm^{-1}.[154,155] More to the point, the basicity of the framework oxygen correlates with this shift of the N–H stretching band: the greater the oxygen basicity, the greater the shift of the band. In the initial study on alkali-metal ion-exchanged zeolites, it was shown that (1) NaY zeolite was less basic than NaX zeolite and (2) NaGeX zeolite was the most basic member of this series.[154] The first result is consistent with indicator methods that confirm the enhanced basicity of X-faujasites compared with their Y-structural analogues.[142] That NaGeX is the most basic of these zeolites reflects the replacement of Si in the framework tetrahedra with Ge. Some studies[156,157] using this pyrrole probe showed that the order of basicity of alkali-metal ion-exchanged zeolites (as determined by the difference in N–H stretching frequency, Table 5.4) decreases in the following order: CsX > NaX > KY > NaY ≥ KL = Na-mordenite = Na-beta. Further, the order

Table 5.4. Change in N–H stretching vibration of adsorbed pyrrole on alkali-metal ion-exchanged zeolites and fractional negative charge on the zeolite oxygens

Zeolite	$\Delta\nu$ (N–H)[a] (cm^{-1})	Fractional charge on oxygen[b]
CsX	240	−0.461
NaX	180	−0.413
KY	70	−0.383
NaY	30–40	−0.352
KL	30	−0.356
Na-mordenite	30	−0.278
Na-beta	30	−0.240

Source: Data taken from references 156 and 157.
[a]Difference between bound N–H stretch and N–H stretch for pyrrole as a pure liquid.
[b]Calculated from Sanderson electronegativities.

correlates with fractional negative charges on the zeolite oxygens (as calculated from Sanderson electronegativities, Table 5.4) for the most basic zeolites (CsX to NaY).[15,156,157] Interestingly, while pyrrole binding, as studied by IR spectroscopy, is firmly entrenched as a tool for evaluating zeolite basicity, it is unlikely to supplant CO_2 for the same use with single-component oxides; pyrrole has been shown to polymerize on alkaline-metal oxides though not on basic zeolites at room temperature.[154,158]

Although it has been much less explored than IR for the characterization of surfaces, ^{13}C NMR (including solid-state magic angle spinning NMR) spectroscopy has proven to be useful in following reactions on basic catalysts and in model studies of the binding of organic carbon acids (e.g., 1- and *cis/trans*-2-butene) to alkali-metal ion-exchanged zeolites[159,160] and mixed metal oxides (SnO–SbO).[160] In the case of 1- and *cis/trans*-2-butene, the molecules were found to form π-type complexes with starting 1-butene and the isomeric 2-butenes;[159,160] preferential *cis* binding was postulated as being important in determining the ratio of the isomerized alkenes.[161] No doubt such NMR studies will extend our understanding of these processes on basic solids. In later sections, results from NMR studies will be introduced as required to clarify reaction mechanisms on solid bases.

5.3.2.2 Spectroscopic Studies of Basic Surfaces

Electronic absorption (UV–vis) and photoluminescence spectroscopies have proved useful in elucidating the surface structures of solid bases. Although MgO, as a single crystal, does not absorb UV light, high-surface-area magnesia does.[162] The absorption bands at 230 and 274 nm were assigned to surface oxide ions with coordination numbers of 4 and 3, corresponding to the two most basic sites identified in the MgO model described in Section 5.2.1, involved in the following electron excitation process:[163]

$$Mg^{2+}O^{2-} + h\nu \rightarrow Mg^{+}O^{-} \tag{5.7}$$

Photoluminescence can be considered to be the reverse process and it supports the assignment of the surface structure of polycrystalline MgO.[164,165] No comparable studies have been reported for zeolites, perhaps reflecting the complexity of composition of these basic catalysts.

Heteronuclear NMR studies have been performed on zeolites and ^{17}O NMR has been used to identify the different types of moieties in various types of zeolites, including the faujasite X and Y zeolites. In this regard, the ^{17}O chemical shift was found to fall between 44 and 52 ppm for Si–O–Si structures, between 31 and 40 ppm for Si–O–Al structures, and at 28 to 29 ppm for Si–O–Ga moieties.[166,167] Correlations of ^{29}Si NMR chemical shifts with Si–O–(Al/Si/,etc.) bond angles and Si–O bond lengths have also been reported,[168–170] but currently no direct correlations of oxygen basicity in basic zeolites and such changes in silicon chemical shift have emerged.[15] No doubt our understanding of surface basicity and structure both for zeolites and for other basic solids will advance, along with improved NMR techniques for studying solids.

X-ray photoelectron spectroscopy (XPS) is particularly valuable in assessing oxygen site basicity spectroscopically. Generally, the binding energy (BE), determined by XPS, is inversely proportional to the base strength of the oxygen.[171] Therefore, for a series of single-component metal oxides the, O_{1s} BE and effective oxygen charge (as a measure of Brønsted basicity) have been shown to correlate.[172] Similar correlations have been found for X- and Y-faujasite alkali-metal ion-exchanged zeolites as well as mordenite and A-type zeolites; BE values of O_{1s} decrease as the Al content increases, in agreement with indicator and other studies.[173,174] Examination of XPS O_{1s} BE for alkali-metal ion-exchanged zeolites, where the alkali metal was varied from Li down the group to Cs, also confirmed previous studies that oxygen basicity increases down this family.[173] Another study has shown that incorporation of P in the framework (i.e., SAPOs) results in a lowering of the O_{1s} binding energy suggesting that introduction of P in the zeolite structure increases oxygen basicity of the zeolite.[175] Interestingly, pyrrole N–H vibration frequencies for bound pyrrole in zeolites determined by IR spectroscopy have also been shown to correlate with the BE for N_{1s} for pyrrole adsorbed in the same zeolites;[176] this cross-correlation emphasizes that both pyrrole adsorption studies by IR and XPS BE studies are measuring the same phenomenon, catalyst basicity.

While XPS studies can be complicated by formation of carbon on zeolite surfaces,[177] and further require a careful choice of accelerating energies when dealing with zeolites that include sodium ions (or ionic clusters),[178] this spectroscopic technique provides valuable direct information concerning the surface basicity of basic zeolites.

Electron spin resonance (ESR) spectroscopy has also been used to examine possible single electron transfer sites on zeolites. This is done through observation of the formation of perylene radical cations (from donation of a single electron from perylene to the zeolite) or observation of the formation of tetracyanoethylene radical anions (from acceptance of a single electron from the zeolite to tetracyanoethylene).[15] Electron spin resonance spectroscopy has also been used to distinguish between neutral sodium and ionic sodium clusters encapsulated in basic zeolites.[93,94] Clearly, ESR will be valuable in differentiating between Brønsted-type basic sites and single electron transfer sites, notably in superbase systems.

5.3.3 Gas Desorption Methods

Gas desorption or, more fully, temperature-programmed desorption of gases (TPD) is another approach to characterization of basic surfaces. Its relative ease of use and low cost often makes it the first choice for rapid characterization of solid bases. Here, the aim is to measure directly the strength of binding of probe molecules to the surface along with the number of basic sites. (Infrared spectroscopic studies of pyrrole binding by comparison provide an indirect or relative measure of binding strength.)

In the TPD technique, a basic oxide, for example, whose surface is covered with acidic probe molecules, most commonly CO_2, is heated gradually (i.e., at a *programmed* rate), while an inert gas is passed over the surface. Molecules that are desorbed are detected by typical gas chromatographic detectors (e.g., thermal conductivity or flame ionization detectors). In a typical TPD trace (detector response vs temperature), the strength of binding is related to the temperatures of onset of desorption and completion of desorption. Under ideal conditions, a sharp peak can be recorded and from it the heat of adsorption may be estimated. The area under the curve of the TPD peak is proportional to the number of molecules desorbed and, hence, proportional to the number of basic sites. (For a detailed description of the supporting theory and applications of TPD, the reader is referred to the reviews by Cventanovic and Amenomiya[179] and Knötzinger.[180])

In principle, the relative base strengths and the number of sites for different catalysts may be obtained from TPD studies of the catalysts run under the same conditions. However, TPD traces do not automatically yield an absolute scale for catalyst basicity (unless heats of adsorption can be obtained for the series of catalysts under study).[9]

The TPD of carbon dioxide from the alkaline-earth metal oxides found the following order of decreasing basicity: $BaO > SrO > CaO > MgO$,[181] in accord with previous classical assessments of relative basicity for these oxides.[2] The number of basic sites or sites for binding of carbon dioxide (per unit mass of catalyst) was found to decrease from CaO to MgO to SrO and, finally, to BaO.[181]

The TPD of hydrogen gas may be used to determine the most basic sites—for example, on superbases.[182]

Somewhat anomalous results have been found from TPD studies of basic zeolites in which carbon dioxide was used as the adsorbed gas. These results may reflect partial compensatory effects in desorption: basicity versus steric complementarity.[183,184]

5.4 MECHANISMS OF REACTIONS CATALYZED BY SOLID BASES

In the subsequent sections, we will examine a variety of reactions that have been found to be catalyzed by solid bases. Where possible, the reactions will be discussed mechanistically. Frequently, however, mechanistic evidence for the solid basic systems may be limited, in which case the mechanism in solution may be discussed.

5.4.1. β-Elimination Reactions on Solid Bases: Dehydration of Alkanols and Dehydrohalogenation of Haloalkanes

Dehydration of alkanols to alkenes falls within the class of elimination reactions. Bunnett was the first to describe a spectrum of elimination pathways[185] from the central concerted (and ideally synchronous) E_2 bimolecular mechanism to the E_1 carbocationic unimolecular and E_{1cb} carbanionic unimolecular mechanisms[186] on the extremes. Various elegant approaches to discern the mechanism of elimination have been developed for solution studies,[187–189] including H/D kinetic isotope effects[190] and examination of the stereochemistry of elimination using suitable compounds.[191,192] Of the possible mechanistic studies that could differentiate between the different mechanisms[189] and that could be carried out on solid bases, such as H/D kinetic isotope effect studies, few have actually been undertaken.

Dehydration of alkanols to give alkenes and ethers is more generally found to be catalyzed by solid acids,[193] where the mechanism may be expected to be more E_1-like (or of the E_2-type, with considerable carbocationic character). On the other hand, dehydration of alkanols on basic solid catalysts might be expected to proceed by a form of E_{1cb} mechanism involving carbanions as intermediates[186,194] or by an E_2-type concerted process where the transition state (TS) contains considerable carbanionic character.

Notwithstanding these mechanistic considerations, alkanol dehydrations have been observed to occur on solid bases. For example, 2-butanol has been shown to dehydrate to the Hofmann-type elimination product, 1-butene, on lanthanide oxides,[195] as well as on ZrO_2.[196,197] Under acidic conditions, the Zaitsev product, 2-butene, would be expected to predominate. Presumably, the initial step in this elimination is abstraction of a proton from the C-1 position to generate the carbanion (i.e., cb = conjugate base of the carbon acid) in an E_{1cb} process.[9] Alternatively, the process is of the E_2-type with significant transition-state carbanionic character. Significantly, in these dehydrations over ZrO_2, the selectivity for Hofmann elimination has been shown to be dependent on the presence or absence of acidic SiO_2 impurities in the zirconia. When these impurities are present, a higher proportion of 2-alkenes (Zaitsev products) are formed; poisoning these acidic sites by treating the ZrO_2 with NaOH leads to higher proportions of the 1-alkenes.[9] Zirconia treated with sodium hydroxide is also the basic catalyst used in the industrial dehydration of 1-cyclohexylethanol to 1-cyclohexylethene (vinylcyclohexane, equation 5.8),[198] which again is the Hofmann elimination product.

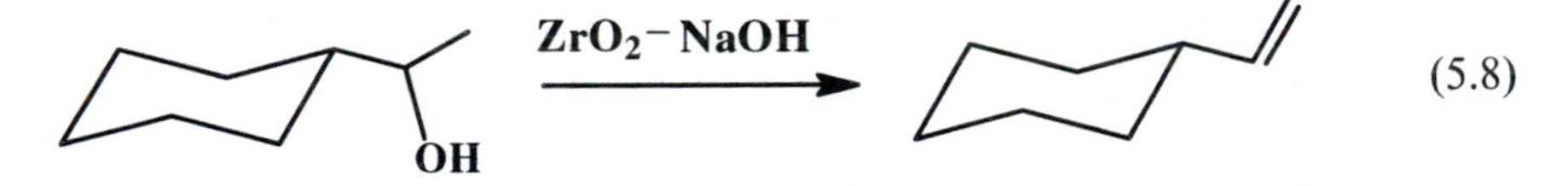

(5.8)

Correlation of yield with catalyst basicity in the dehydration of alkanols,[199] as well as the observation that dehydration of alkanols is not catalyzed by the mildly basic surface hydroxyls of aluminum hydroxide (bayerite),[200] also militate in favor of mechanisms involving carbanions or TS with carbanion character. Stereochemical evidence has also been presented, however, that would favor a concerted process. Thus, the products of solid-base-catalyzed elimination of *cis,-cis*-1-decalol arise from *trans* elimination,[201] and, over alumina, menthol dehydrates to give the *trans* elimination product: 2-menthene.[202]

The proposed mechanism, at least for mildly basic catalysts such as alumina, requires both basic oxide and (relatively) acidic hydroxide sites,[203] as illustrated in Reaction 5.1.

Without doubt, the mechanistic study of such dehydrations will prove a fertile ground for researchers in the future. One further comment, however, should be made concerning solid-base-catalyzed dehydration of alkanols. Generally, these dehydrations are accompanied by dehydrogenation, that is, the formation of

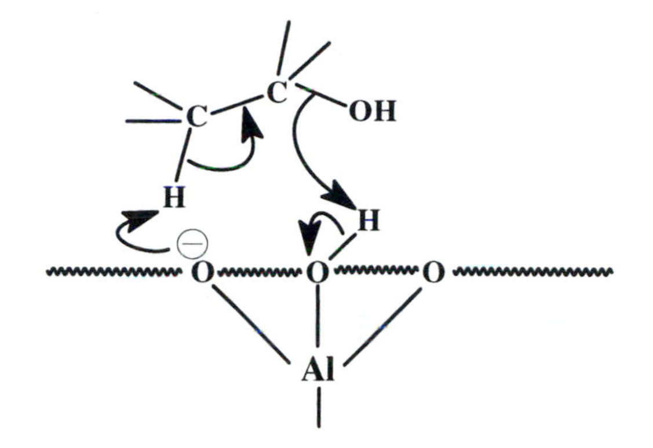

REACTION 5.1

aldehydes and ketones along with alkenes. In this regard, dehydration of 2-propanol to propanone (acetone) proceeds readily on the basic sites of alkali-metal ion-exchange faujasite (X and Y) zeolites.[204] Similarly, methanol is dehydrogenated to methanal (formaldehyde) by basic zeolites and the process has been studied by IR spectroscopy.[205] It has been suggested that, in basic zeolites, dehydrogenation becomes the favored process as the basicity of the zeolite is increased.[15]

β-Elimination of haloalkanes on solid bases has been studied rarely and no significant mechanistic information has been deduced from the limited range of systems examined thus far. However, a number of haloalkanes have been found to give high yields when treated with KF/Al_2O_3, a mildly basic or even acid–base dual-function solid.[206] In the case of 1-bromo-2-phenylethane, reaction for 24 h over KF/Al_2O_3 (CH_3CN solvent, 25 °C) produced a 86% yield of styrene.[206] Again, systematic studies, involving variation of halide-leaving group, H/D kinetic isotope effects, and so on, have not been performed for dehydrohalogenation on basic catalysts.

5.5 ALKENE ISOMERIZATION, DIMERIZATION, POLYMERIZATION

Double-bond migration, notably in the isomerization of 1-butene to 2-butene, is probably the most exhaustively studied class of reactions that are initated by solid bases.[2–4,8,9,207] In fact, observation of the conversion of 1-butene to 2-butene with a high ratio of the *cis* stereoisomer to its *trans* counterpart is taken to be diagnostic of catalytic activity for a solid base, as well as indicative of the intervention of allylic carbanions in the process.[9,208] As shown in the simplified Scheme 5.2, the isomerization commences with formation of the allylic carbanion via abstraction of a proton at C-3 by the basic surface.[209] In principle, either a *trans* or *cis* allylic carbanion may form at this point. Since the *cis* allylic carbanion is more stable than its *trans* isomer,[9,210–213] this accounts for the product distribution favoring

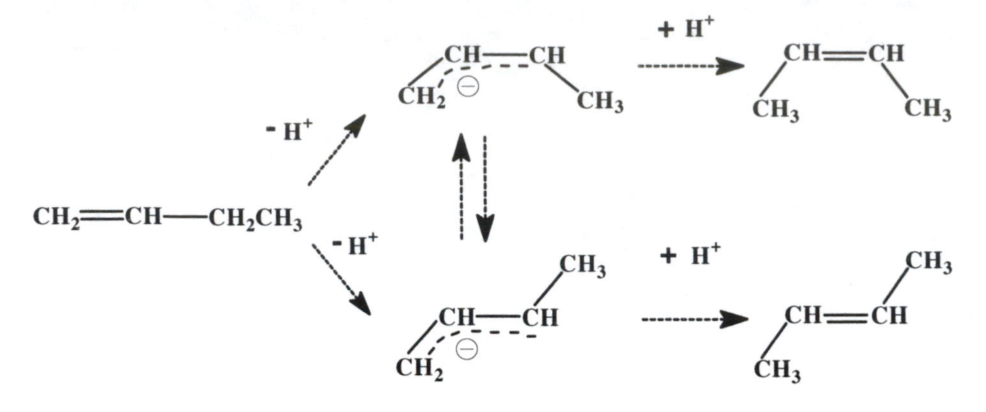

SCHEME 5.2

cis-2-butene. However, kinetic factors are also significant; the *cis* anion is a kinetically preferred intermediate and the final alkene product ratio does not reflect the stability of the two alkenes. Nonetheless, the *cis* anion, bound to the surface as confirmed by ^{13}C NMR studies involving basic zeolites and mixed oxides,[159–161] must be present in higher concentration than the surface-bound *trans* anion.[214] This concentration difference may reflect the high barriers to rotation that would be involved in conversion of one surface-bound anion into the other; such rotational barriers have been shown to be dependent on the nature of the catalyst.[208] Again, it should be emphasized that this interconversion of anions and final loss of the proton to give the product alkenes is dominated by kinetic factors and that the products are not in true equilibrium (Scheme 5.2). Regardless, it is germane to the discussion to note that in acid-catalyzed isomerization of 1-butene the *cis*:-*trans* ratio is close to unity.[215]

Labeling studies have also proven to be useful in elucidating the mechanism of double-bond migration.[216] In this context, when a 1:1 mixture of fully deuterated 1-butene and nondeuterated 1-butene are both subjected to the same active solid base, the only products are the fully deuterated 2-butene and the nondeuterated 2-butene in equal 1:1 ratio. Clearly, all proton/deuteron transfers are occurring within the confines of the catalytic site and proton or deuteron transfer does not involve other molecules of 1-butene, only the molecule bound to the surface and that surface.[217]

From the foregoing discussion, it is apparent that the rates (and energetics) for 1-butene to 2-butene isomerization will be catalyst dependent, as Table 5.5 demonstrates. In some studies, it has been possible to determine the rate constants for the cycle: *trans*-2-butene to *cis*-2-butene, via 1-butene, as shown here in Scheme 5.3 (where $k_{t \to 1}$, for example, represents the rate constant for conversion of *trans*-2-butene into 1-butene; the other rate constants in Scheme 5.3 are defined similarly).

From Table 5.5, it can be seen that the overall apparent equilibrium constant for isomerization of *trans*-2-butene to *cis*-2-butene (Scheme 5.3) favors the *cis* stereoisomer ($K_{eq} = k_{t \to c}/k_{c \to t} = 81.5/48.9 = 1.67$ on MgO). However, this

Table 5.5. Kinetics and energetics of 1-butene isomerization

Base	T (°C)	Rate constants, k (molecules/s·cm^2) × 10^{-9}					
		$k_{1\to c}$	$k_{1\to t}$	$k_{c\to 1}$	$k_{c\to t}$	$k_{t\to 1}$	$k_{t\to c}$
ZnO	28	144	2.4	34.7	48.9	5.9	81.5
[E_a, kcal/mol:		7.9	2.3	11.2	15.3	10.8	7.4]
La_2O_3	30	210,000	23,000	25,000	Low	35	363
[E_a, kcal/mol:		4.1	9.1	7.9	20.0	18.2	13.0]

SOURCE: Data taken from references 208 and 218.

"equilibrium constant" must be taken as an apparent one, including as it does the rates for absorption and desorption from the surface; presumably, if the system achieved true equilibrium, the normal equilibrium mixture of the butenes would be obtained.

Importantly, all of the categories of solid bases demonstrate some capacity to initiate double-bond migration, largely independent of alkene structure.[219] In general, the most basic catalysts are also the most effective in promoting double-bond isomerization. Thus, the superbase systems (e.g., MgO–Na; H_- ca. 35) are more active than single-component metal oxides such as MgO.[14] Zeolites are also effective in initiating the isomerization, notably zeolites containing trapped sodium metal clusters.[220]

Synthetically, double-bond isomerizations over solid bases offer access to interesting and complex compounds that would otherwise be difficult to prepare. Thus, equilibration of (+)-calarene over a K–Al_2O_3 catalyst gave a 9:1 ratio of isomeric (−)-aristolene to the starting (+)-calarene (equation 5.9).[221]

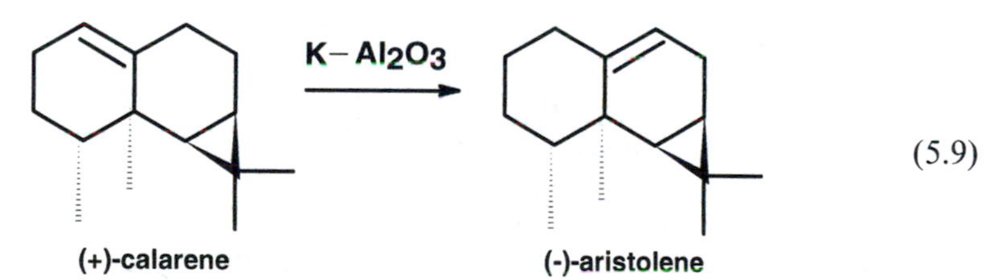

(5.9)

Hattori gives a number of examples of complex natural products that isomerize smoothly over solid bases *at relatively low temperatures*[9]. Importantly, in these examples as in the case of the isomerization of (+)-calarene, strained structures such as cyclopropyl rings are left intact in the double-bond migratory process.[9]

Alkenes that contain heteroatom groups have also been shown to isomerize cleanly. For instance, facile double-bond migration is found for allylamines, such as 1-*N*-pyrrolidinyl-2-propene, which isomerizes to 1-N-pyrrolidinyl-1-propene, catalyzed by MgO, CaO, SrO, or BaO (40 °C).[222] Propenyl ethers are also readily isomerized over alkaline-metal oxides; yields of 72–95% have been reported for the isomerization of ethyl-, phenyl- or isopropyl-2-propenyl ethers under mild conditions (0 °C, 1-min contact time);[223,224] CNDO/2 calculations[224] were used to account for the preference for *cis* stereochemistry in the products. No doubt,

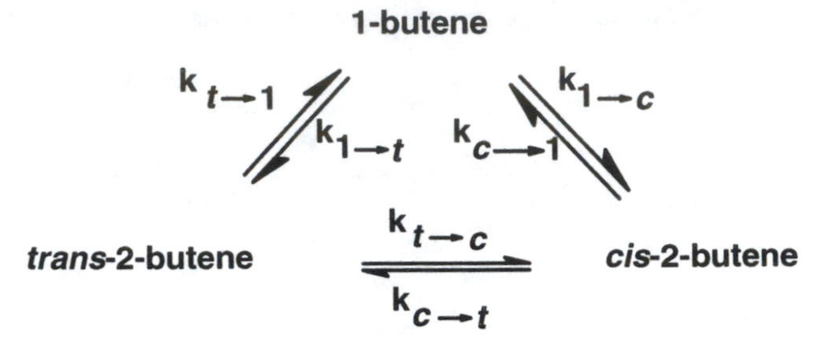

SCHEME 5.3

higher-level calculations such as those performed on the double-bond rearrangement of the lithium salt of 3-hydroxycyclohex-1-ene to the corresponding lithium salt of 4-hydroxycyclohex-1-ene, initiated by $LiNH_2$, will prove insightful here, as well.[225]

In considering the foregoing, it should be remembered that in attempting similar isomerizations under acid conditions, cleavage of the ether and further skeleton rearrangement would compete with the double-bond migration. Therefore, isomerization under solid base conditions offers a clean and usually high-yield alternative to acid-catalyzed procedures.

In a number of systems, dimerization, oligomerization, and, in the extreme, polymerization may be competing processes. Dimerization of propene to a variety of hexenes has been shown to occur over a variety of solid base catalysts.[226–228] Presumably, dimerization along with double-bond migration involves initial formation of an allylic carbanion that adds to another molecule of propene prior to termination via reprotonation from surface sites, likely hydroxyls. Careful choice of catalyst has been shown to yield primarily 4-methyl-1-pentene (a feedstock in separate polymerization processes).[228]

Alkenes that are more bulky than propene may not dimerize, as a result of steric hindrance to addition of the initially formed carbanion to the double bond. However, alkali-metal and alkaline-earth-metal oxides have been shown to initiate polymerization of formaldehyde, ethylene oxide (oxacyclopropane), and propiolactone.[2] In a number of these examples, the solid base is actually acting as a nucleophile (e.g., ring-opening polymerization of epoxides); metal alkoxides act similarly in the formation of poly(lactones).[229] Polymerization is preferred, of course, when addition of the initial carbanion to the double bond of substrate results in formation of a new stabilized carbanion. An example of this is polymerization of acrylamide.[230] Polymerization may be suppressed relative to dimerization, for example, because the growing polymer buries the catalytic base sites.

5.6 BASE-CATALYZED ALKYLATION

Alkylation of carbonyl compounds over various solid bases has been reported by the Malinowski group.[231–235] As but one example, methanal (formaldehyde) is alkylated by nitromethane over silica gel that had been treated with sodium vapor;[231] although this solid catalyst would not necessarily have been very basic, the high acidity of nitromethane would compensate for this factor. In examining a series of such reactions, linear correlations were found between the overall rate constants and the amount of sodium in the catalyst (which itself correlates with the type and number of surface base sites on the catalyst), and between the pK_a of the alkylation agents (e.g., nitromethane, acetonitrile, etc.) and the rate constants for alkylation.[235] On these grounds, the mechanism appears to involve initial deprotonation of the alkylating agent on the basic catalyst to give a stabilized carbanion that then attacks the formaldehyde. Bram and co-workers have reported the alkylation of the anion of 9,10-dihydroacrid-9-one with propargyl bromide and the alkylation of the carbanion of a series of diethyl acylaminomalonates with a range of organohalides.[236] In these cases, a mildly basic KOH–Al_2O_3 catalyst was used.

We have already briefly considered alkylation reactions catalyzed by polymeric basic resins and the ambiguities involved.[103,104] Competitive reactions are also possible on the other solid bases, including possible dehydrohalogenation of any haloalkane reactants, as well as self-condensation involving enolates.

5.6.1 Side-Chain Alkylation of Arenes

The alkylation of toluene by methanol with zeolites, in principle, may give either ring products (i.e., xylenes) or side-chain products (i.e., styrene and ethylbenzene), the former via a carbonium ion mechanism dependent on the acidic sites of the zeolite and the latter according to a carbanion mechanism.[140,237] More basic zeolites have been developed that essentially give only side-chain alkylation.[15,238,239] Although the mechanism is still in some dispute,[240] there is evidence that it involves several surface catalytic steps.[241] In the first, methanol is dehydrogenated to give methanal. There is precedent for this step on its own (cf. Section 5.4.1). The second step involves deprotonation of the toluene to give a benzylic carbanion that adds to the traces of methanal present and is followed by protonation on acidic (or more acidic) zeolite sites (or entrained metal hydroxides). Elimination yields styrene as one product, in an overall process from the addition of the benzylic carbanion to methanal that is reminiscent of the aldol condensation, while hydrogenation of some of the styrene under the reaction conditions provides the other main alkylation product, ethylbenzene.[241] It is important to note that regardless of the basic zeolite type, the temperatures required to initiate reaction and the conditions needed to achieve maximum yield are similar for all of the zeolites examined; this suggests a common rate-determining step in the mechanism. Formation of the benzyl carbanion was taken to be rate limiting.[241]

Computational studies of the alkylation of toluene support the view that both acidic and basic sites are required in the zeolite cage in order to achieve selective side-chain alkylation.[242,243] However, other calculational studies have also highlighted the importance of the steric fit of toluene and methanal in the zeolite cavity in order to effect side-chain alkylation.[244] It appears that toluene and methanal are adsorbed at somewhat different sites,[245] with the interaction of the methyl group of toluene with a zeolite basic center determining the selectivity for side-chain alkylation.[243]

The ^{13}C NMR spectroscopic studies of CsX faujasite have demonstrated that the mobility of the toluene is restricted,[246] in agreement with the results of the calculational studies. The same study identified both methanal, the reagent required for side-chain alkylation, and methyl carbocations, the ring alkylation agent, in the CsX cavity. It has been suggested[15] that, in the CsX zeolite, the methyl carbocation is tightly associated with the surface, as a superficial methoxy group, and this tight association suppresses ring alkylation in the zeolite.

Side-chain alkylation of various aromatic compounds occurs with alkenes as the alkylating agents and solid bases including Na–Al_2O_3, NaH, and graphite treated with potassium metal.[247] Further, K–KOH–Al_2O_3 has been reported to be an efficient base catalyst for side-chain alkylation of 2-phenylpropane (cumene) using alkenes such as ethene and propene at close to ambient temperature (27 °C).[248] Again, the importance of proton abstraction from the benzylic substrate was emphasized as a major factor in deciding the selectivity of the alkylation.

5.7 SELECTED CONDENSATION REACTIONS

Condensation reactions form a class of carbanion reactions whose utility in organic synthesis is so great that many of the reactions are "named".[249] However, regardless of name, there is an underlying similarity in all these reactions, which proceed through carbanion intermediates.[250] Thus, in the base-catalyzed condensations of ketones and aldehydes, an enolate anion, which is resonance stabilized, is formed first, usually in quite low concentration. Reaction of the enolate with another molecule may lead to alkylation if the electrophilic molecule contains a good leaving group, or addition of the enolate to an electrophilic carbonyl (typically of unchanged starting carbonyl compound) may occur, leading to the condensation product. Often, the mechanistic steps are reversible; retroaldol reactions of α,β-unsaturated and β-hydroxy aldehydes in the presence of hydroxide base are also well known.

In solution, observation of either specific or general base catalysis provides evidence upon which to assign the rate-determining step in the mechanism.[250]

We will now consider some examples of condensation reactions that have been found to occur with solid bases. Generally, mechanistic studies remain to be carried out or, alternatively, postulated mechanisms rest largely on product analyses.

5.7.1 Aldol Condensation

The aldol condensation of acetone (self-condensation) has been reported to occur under catalysis by anion-exchange resins (Amberlite IRA-400).[2,26,27,251] Basic alumina has also been shown to successfully induce the self-aldol with acetone.[252] It should be pointed out that, in solution, acetone normally does not undergo aldol condensation to give high yields of diacetone unless special methods or equipment are employed[253,254] The reasonable yields reported with solid basic Al_2O_3, consequently, are all the more remarkable.[252] Zhang, Hattori, and Tanabe examined the acetone self-aldol on the alkaline-earth-metal oxide series and found that yields correlated with the order of surface basicity, that is, the yields declined in the following order: BaO > SrO > CaO > MgO.[255] Basic zeolites that contain neutral sodium metal clusters are also effective in the aldol condensation of acetone.[15]

Ethanal (acetaldehyde) undergoes self-condensation on anion-exchange resins.[2] Propanal (propionaldehyde) has been found to react in the aldol condensation catalyzed by such mild solid bases as lithium phosphate and calcium hydroxide.[256] Barium hydroxide has also been reported to initiate aldol condensations.[124] Several metal oxides have been found to catalyze the crossed-aldol condensation of furfural with ethanal to give 2-furylacrolein. In this case, dehydration follows the aldol; the driving force is reasonably presumed to be extension of conjugation in the final product.[256]

The self-aldol of a number of ketones on solid bases has also been reviewed and the reader's attention is directed to these reviews for further examples of reactions and catalytic conditions.[257,258]

The mechanism of the reaction may be similar to that found in solution. A dependence on the basicity of the catalyst has been found that implicates carbanions in the mechanism.[235] More important, the retroaldol reaction on basic catalysts is well documented,[259] and has even been suggested as an operational measure of catalyst surface basicity.[260,261] This demonstrates the reversibility of the reaction and suggests that both basic and acidic (or less basic) sites (such as surface OH groups) may be important. On a series of alumina and mixed SiO_2–$AlPO_4$ catalysts, dehydration of the diacetone alcohol to mesityl oxide competed with the retroaldol, implicating more acidic sites in the dehydration.[259]

Basic faujasite zeolites, X-[262,263] and Y-types,[264] are effective in catalyzing aldol condensations. In the study of the condensation of butanal to 2-ethyl-2-hexenal, as with other milder solid bases where dehydration follows condensation, the presence of both base and acid sites was deemed important; the dual functionality of the alkali-metal ion-exchanged Y-zeolite was contrasted with the less active, but more basic, MgO catalyst.[265]

5.7.2 Knoevenagel Condensation

The main differences between the aldol condensation and the Knoevenagel reaction, in solution, stem from the use of (1) a diactivated carbon acid, typically a diester (pK_a ca. 9–12, as determined directly in water) and (2) a secondary or tertiary amine as the base to deprotonate the relatively strong carbon acid. With these features in mind, the Knoevenagel reaction is usually performed with carbon acids such as ethylcyanoacetate, malononitrile, and diethyl malonate, which react with benzaldehyde or substituted benzaldehydes. Under the basic reaction conditions in solution, carbanion addition to the carbonyl of the benzaldehyde (or related compound) is generally followed by elimination to give a substituted alkene conjugated with the benzene ring. Thus, as emphasized by Laszlo,[10] the Knoevenagel is but one of a limited repertoire of organic reactions (including the

Wittig and Wittig–Horner reactions) that serve to convert a carbonyl into a carbon–carbon double bond.

The Knoevenagel condensation is a frequently observed and widely studied reaction found to occur on all kinds of solid bases, including the most weakly basic. As such, any listing here of the systems that can catalyze the reaction is bound to be incomplete! With understandable trepidation, then, we note that the Knoevenagel reaction has been initiated by (1) weakly basic anion-exchange resins (Amberlite IR-4B, Dowex-3 and polyvinylpyridine);[97,98] (2) alumina;[125] (3) alumino phosphate oxynitrides (ALPONs),[119-121] and other nitride-type base catalysts such as silicon oxynitride;[14,118] (4) basic clays like xonotlite-potassium *t*-butoxide;[115] (5) hydrotalcite-type minerals;[266] (6) weakly basic alumina;[267] (7) potassium fluoride supported on alumina;[268] (8) alumina-aluminum orthophosphate;[269] (9) MgO and ZnO, which are more highly basic catalysts than the others in this list;[270] (10) $Ba(OH)_2$ (calcined from the octahydrate);[270] (11) the mixed oxide system, MgO–Al_2O_3;[271] and (12) Na_2CO_3 with 4 Å molecular sieves.[272]

Basic alumina, as one pertinent example,[267] was found to catalyze the Knoevenagel reaction of malononitrile, cyanoacetic ester, cyanoacetamide, and cyanophosphonates with a series of carbonyl compounds including benzaldehyde, acetone, and diethyl ketone in yields ranging from quantitative to 53%, with reaction times varying from 3 min to 77 h. The high acidity of the carbon acids accounts for the efficiency of the reaction, even though alumina is not especially basic. Interestingly, yields diminished when solvent was used in the reactions, suggesting that all species, carbanions and substrates, are bound to the surface throughout the course of the reaction. In the Knoevenagel reaction of malononitrile and ethylcyanoacetate with a variety of aldehydes and ketones on alumina-aluminum orthophosphate, again without solvent, the best yields were obtained with aldehydes.[269] A good correlation was found in the Hammett plot for 4-substituted benzaldehydes reacting with either carbon acid. This is consistent with a carbanionic mechanism, while the small Hammett slope parameter, ρ, may indicate that the negative charge in the rate-determining transition state is more delocalized than expected for the substituted benzoic acids that define the Hammett equation. The authors suggested that the reaction mechanism was similar to that in solution.[269]

The Knoevenagel reactions show interesting selectivity on solid bases. For example, α, β-unsaturated carbonyl compounds were found to give exclusively Knoevenagel products without competitive Michael addition to the unsaturated system when alumina-aluminum orthophosphate was used as the catalyst.[269] This suggests, again, that all species are held in suitable positions on the catalyst surface for the Knoevenagel addition but not for the Michael addition. On the mixed magnesia–alumina basic catalyst,[271] or on MgO or ZnO alone,[270] the reaction of benzaldehyde with cyanomethyl diethylphosphonate was found to give both Knoevenagel and Wittig–Horner reaction products (as shown in Scheme 5.4). Importantly, in the MgO–alumina system, addition of salts to increase the number of *acidic* sites on the catalyst was found to favor the Knoevenagel route, whereas use of DMSO or HMPA as solvent encouraged the Wittig–Horner reaction. It was argued that the intermediate product of the

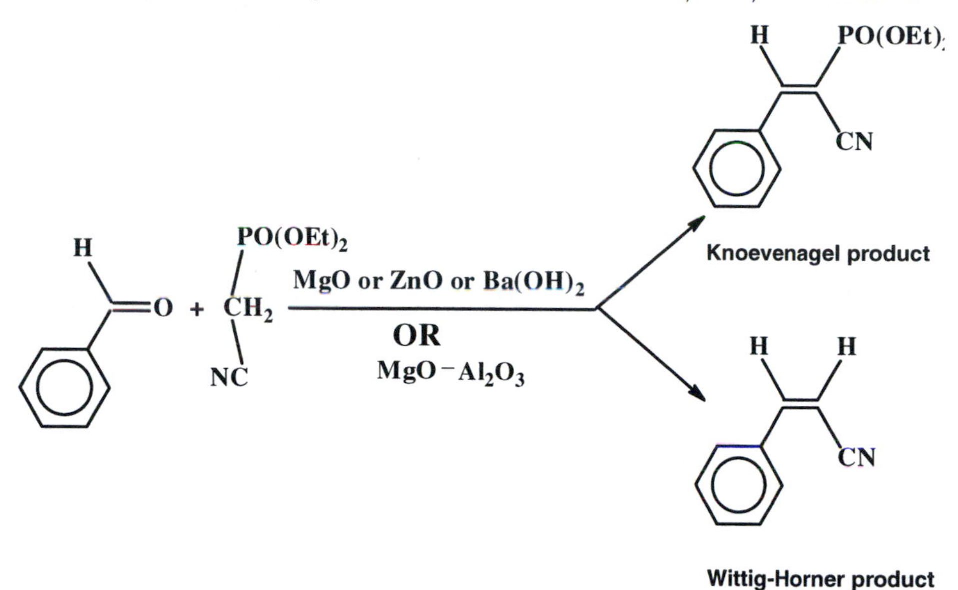

SCHEME 5.4

addition of the carbanion to the carbonyl of benzaldehyde was absorbed on acidic sites of the catalyst through the oxygen of the former carbonyl. Aprotic dipolar solvents such as DMSO would promote the Wittig–Horner route by desorbing this intermediate, which could then react further in solution rather than on the catalyst surface.[271]

Thus far, we have emphasized the necessity of having a basic catalyst in these reactions, but it should also be recalled that, in a number of condensations, protonation of the penultimate intermediate in the sequence is required. Therefore, as in the dehydration reactions considered above, both basic and acidic sites (or *less basic sites* such as surface OH's) may be required, as well.

The list of catalysts that promote the Knoevenagel reaction would be far from complete if we did not consider the alkaline zeolites.[15] Therefore, it should be noted that the reaction of the carbanions of ethylcyanoacetate,[76,273–275] diethyl malonate,[76,273–275] and ethyl acetoacetate[274,275] with substituted benzaldehydes occurs on basic faujasites; CsX and NaGeX types proved to be the most active, in accord with the enhanced basicity of these zeolites relative to Y-type faujasites.[76,273–275] Comparative reactivity in the Knoevenagel reaction catalyzed by the zeolites and by pyridine and piperidine suggested that the zeolites have basic sites with pK_b values of less than 13.3.[274,275] Competitive Michael addition may be suppressed in basic zeolites because of the steric requirements of the cavities, that is, the cavities may not accommodate the bulkier Michael addition products as compared with the Knoevenagel addition products.[276] At the same time, it may be argued that the zeolites are not so basic as to induce competitive aldol or Claisen condensation of the active methylene component.[9]

5.7.3 Claisen–Schmidt Reaction

The condensation of an aromatic aldehyde with an alkyl aryl ketone is termed the Claisen–Schmidt reaction (equation 5.10). The final product, after in situ dehydration, is a chalcone, as shown: in equation 5.10.

$$\mathrm{ArCHO} + \mathrm{RCH_2\overset{\displaystyle O}{\overset{\|}{C}}Ar'} \xrightarrow[-\mathrm{H_2O}]{\mathrm{Base}} \mathrm{ArCH{=\!=}CR{-\!-}\overset{\displaystyle O}{\overset{\|}{C}}Ar'} \tag{5.10}$$

The reaction has been found to be catalyzed by alumina,[266] by barium hydroxide in ethanol,[123] and by hydrotalcites in 95% ethanol.[110] Although current information on these reactions is limited, the evidence that is extant supports the view that much the same carbanionic mechanism applies on the solid catalyst as in solution.[110] Again, basic and acidic (or, at least, much less basic) sites may be involved, although the rate-limiting step appears to be base dependent of the carbanion of the alkyl aryl ketone.[123]

5.7.4 Condensation Reactions: Summary

Beyond the condensation reactions considered briefly above, there are a number of other synthetically useful condensation reactions that have been shown to be catalyzed by solid bases, including the Wittig–Horner reaction, which competes with the Knoevenagel reaction in a number of systems,[270,271] and the Henry reaction.[277] The Tischenko[278] and Cannizzaro reactions[2] also involve C–C bond formation and have been found to occur on solid bases. The Cannizzaro reaction competes with the aldol condensation of propanal with calcium hydroxide as the catalyst.[2]

An interesting study by Climent and co-workers concluded that for the aldol, Wittig–Horner, and Wittig reactions, weak carbon acids (pK_a > 9.0) react on the surface of a variety of $Ba(OH)_2$ catalysts. However, for moderately acidic carbon substrates (pK_a < 9.0), the reaction actually involves barium salt complexes in homogeneous solution.[279] This behavior would appear to be unique to barium hydroxide catalysts and serves to differentiate them from other solid bases that catalyze carbanion reactions.

5.8 MICHAEL ADDITION

Reaction of a carbanion with a conjugated system, such as an α,β-unsaturated carbonyl compound, in a 1,4 manner defines a Michael addition. In its classic form, the Michael addition involves reversible formation of the carbanion from an activated carbon acid, such as diethyl malonate, followed by attack of the anion, as a carbon nucleophile, on the 4-position of the α,β-unsaturated system. The final step is protonation of the carbanion that results from the addition.[280]

An example of a standard Michael addition catalyzed by solid base is the reaction of acetyl ethylmalonate with chalcone in the presence of barium hydroxide. With a catalytic quantity of the hydroxide, the addition occurs in 90%

yield.[281,282] Michael addition also occurs between chalcone and Meldrum's acid on alumina, but only in poor yield, which is surprising given the high acidity of the carbon acid.[268]

In a broader sense, the amine additions to conjugated systems, catalyzed by solid bases, may also be considered to be Michael-type reactions.[9,283]

5.9 THE NEF REACTION

Illustrative of the wide frontiers of carbanion chemistry that are opened through the use of solid bases is the case of the Nef reaction. Hawthorne[284] published an interesting example of the base-catalyzed Nef reaction. In solution, the Nef reaction involves base-catalyzed formation of the nitronate salt of a primary or secondary nitroalkane, followed by hydrolysis of the salt with sulfuric acid to give aldehydes or ketones as the final products. The acidic decomposition involves successive protonation of the nitro group, attack of water on the imino C=N bond, followed by elimination of HNO. The slow step in the mechanism appears to be the attack of water on the aci form of the protonated nitro-substituted carbanion.

On silica gel, the Nef reaction takes place cleanly.[285] It has been proposed that formation of the nitro-substituted anion through reaction with basic sites on the silica is followed by reaction of the nitro group oxygen(s) with a surface silicon. Further reaction involves either adsorbed hydroxide ion on the surface or attack by a siloxide ($Si–O^-$) of the surface to yield the carbonyl product. Yields of 60–90% were reported.[284]

Clearly, further work remains to be done in order to completely elucidate the mechanism on silica.

5.10 PROSPECTS FOR DEVELOPMENT OF HETEROGENEOUS BASE-CATALYZED REACTIONS

The prospects appear limitless. In common with Laszlo, who has encouraged the development of solid catalysts for use in the synthetic laboratory,[10,11] we see the potential in the expansion of the types of solid basic catalysts, both from the point of view of tuning the basicity of the surface to the needs of a given reaction and from the point of view of achieving greater reaction selectivity (e.g., side-chain versus ring alkylation of arenes).

At the same time, however, it is apparent that many of the solid base systems that are synthetically useful have not been thoroughly examined mechanistically. We look forward to the application of kinetic isotope effects, substituent effects, and computational methods to the challenging problem of scrutinizing the reaction mechanisms on this important class of solid catalysts.

References

1. Corma, A. *Chem. Rev.* **1995**, *95*, 559.
2. Tanabe, K. *Solid Acids and Bases: Their Catalytic Properties*; Academic Press: New York, 1970.
3. Pines, H.; Stalick, W.M. *Base-Catalyzed Reactions of Hydrocarbons and Related Compounds*; Academic Press: New York, 1977.
4. Tanabe, K.; Misono, M.; Ono, Y.; Hattori, H. *New Solid Acids and Bases: Their Catalytic Properties*; Elsevier: Amsterdam, 1989.
5. Catalytica Associates. *Novel Solid Acid and Base Catalysts*; Multiclient Study 4182 AB, Catalytica Studies Division: Mountain View, CA., 1983.
6. Martens, J.A.; Jacobs, P.A. In *Theoretical Aspects of Heterogeneous Catalysis*; Moffatt, J.B., Ed.; Van Nostrand-Reinhold: New York, 1990; Chapter 2.
7. Rigby, A.M.; Kramer, G.J.; van Santen, R.A. *J. Catal.* **1997**, *170*, 1.
8. Pines, H.; Veseley, J.A.; Ipatieff, V.N. *J. Am. Chem. Soc.* **1955**, *77*, 6314.
9. Hattori, H. *Chem. Rev.* **1995**, *95*, 537.
10. Laszlo, P. *Acc. Chem. Res.* **1986**, *19*, 121.
11. Laszlo, P. *J. Phys. Org. Chem.* **1998**, *11*, 356.
12. Lednor, P.W.; de Ruiter, R. *J. Chem. Soc., Chem. Commun.* **1991**, 1625.
13. Tanabe, K. *Solid Acids and Bases: Their Catalytic Properties*; Academic Press: New York, 1970; p 2.
14. Tanabe, K. In *Catalysis by Acids and Bases*; Imelik, B.; Naccache, C.; Coudurier, G.; Ben Taarit, Y.; Vedrine, J.C., Eds.; Elsevier: Amsterdam, 1985; pp 1–14.
15. Barthomeuf, D. *Catal. Rev.—Sci. Eng.* **1996**, *38*, 521.
16. Tsuchiya, S.; Takase, S.; Imamura, H. *Chem. Lett.* **1984**, 661.
17. Noumi, H.; Misumi, T.; Tsuchiya, S. *Chem. Lett.* **1978**, 429.
18. Tanabe, K.; Misono, M.; Ono, Y.; Hattori, H. *New Solid Acids and Bases: Their Catalytic Properties*; Elsevier: Amsterdam, 1989; p 29.
19. Tanabe, K.; Misono, M.; Ono, Y.; Hattori, H. *New Solid Acids and Bases: Their Catalytic Properties*; Elsevier: Amsterdam, 1989; p 39.
20. Morris, R.M.; Klabunde, K.J. *Inorg. Chem.* **1983**, *22*, 682.
21. De Vleesschauwer, W.F.N.M. In *Physical and Chemical Aspects of Adsorbents and Catalysts*; Linsen, B.G., Ed.; Academic Press: New York, 1970; p 265.
22. Kawakami, H.; Yoshida, S. *J. Chem. Soc., Faraday Trans. 2* **1984**, *80*, 921.
23. Coluccia, S.; Tench, A.J. In *Proceedings of the 7th International Congress on Catalysis*; Seiyama, T.; Tanabe, K., Eds.; Elsevier: Amsterdam, 1980; pp 1154–1171.
24. Coluccia, S.; Bozzuzzi, F.; Ghiotti, G.; Morterra, C. *J. Chem. Soc., Faraday Trans.* **1982**, *78*, 2111.
25. Driscoll, D.J.; Martir, W.; Wang, J.X.; Lunsford, J.H. *J. Am. Chem. Soc.* **1985**, *107*, 58.
26. Utiyama, M.; Hattori, H.; Tanabe, K. *J. Catal.* **1978**, *53*, 237.
27. Garnett, J.L.; Long, M.A. In *Isotopes in the Physical and Biomedical Sciences, Vol. 1, Part A;* Buncel E.; Jones, J.R., Eds.; Elsevier: Amsterdam, 1987; Chapter 4, pp 86–121.
28. Hattori, H. *Stud. Surf. Sci. Catal.* **1984**, *21*, 319.
29. Hoq, M.F.; Nieves, I.; Klabunde, K.J. *J. Catal.* **1990**, *123*, 349.
30. Hoq, M.F.; Klabunde, K.J. *J. Am. Chem. Soc.* **1986**, *108*, 2114.
31. Hoq, M.F.; Klabunde, K.J. In *Acid–Base Catalysis. Proceedings of the International Symposium on Acid–Base Catalysis, Sapporo, 1988*; Tanabe, K.; Hattori, H.; Yamaguchi, T.; Tanaka, T., Eds.; Kodansha: Tokyo and VCH: New York, 1989; Chapter 2.1, pp 105–120.

32. Hattori, H. *Mater. Chem. Phys.* **1988**, *18*, 533.
33. Sumiyoshi, T.; Tanabe, K.; Hattori, H. *Bull. Jap. Petrol. Inst.* **1975**, *17*, 65.
34. Yamagata, N.; Owada, Y.; Ozaki, S.; Tanabe, K. *J. Catal.* **1977**, *47*, 358.
35. Wang, G.; Hattori, H.; Tanabe, K. *Bull. Chem. Soc. Jpn.* **1983**, *56*, 2407.
36. Ueda, W.; Yokoyama, T.; Moro-Oka, Y.; Ikawa, T. *J. Chem. Soc., Chem. Commun.* **1984**, 39.
37. Rouxlet, P.G.; Semples, R. *J. Chem. Soc., Faraday Trans. 1* **1974**, *70*, 2021.
38. Vinek, H.; Noller, H.; Ebel, M.; Schwartz, K. *J. Chem. Soc., Faraday Trans. 1* **1977**, *73*, 1977.
39. Tanabe, K.; Misono, M.; Ono, Y.; Hattori, H. *New Solid Acids and Bases: Their Catalytic Properties*; Elsevier: Amsterdam, 1989; p 113.
40. Tanabe, K.; Shimazu, K.; Hattori, H.; Shimazu, K. *J. Catal.* **1979**, *57*, 35.
41. Tanabe, K.; Hattori, H.; Sumiyoshi, T.; Tamaru, K.; Kondo, T. *J. Catal.* **1978**, *53*, 1.
42. Lercher, J.A.; Colombier, C.; Vinek, H.; Noller, H. In *Catalysis by Acids and Bases*; Imelik, B.; Naccache, C.; Coudurier, G.; BenTaarit, Y.; Vedrine, J.C., Eds.; Elsevier: Amsterdam, 1985; pp 25–31.
43. Take, J.; Kikuchi, N.; Yoneda, Y. *J. Catal.* **1971**, *21*, 164.
44. Kijenski, J.; Marczewski, M.; Malinowski, S. *React. Kinet. Catal. Lett.* **1977**, *7*, 151.
45. Malinowski, S. In *Catalysis by Acids and Bases*; Imelik, B.; Naccache, C.; Coudurier, G.; BenTaarit, Y.; Vedrine, J.C., Eds.; Elsevier: Amsterdam, 1985; pp 57–65.
46. Olah, G.A.; Prakash, G.K.S.; Sommer, J. *Superacids*; Wiley: New York, 1985.
47. Prakash, G.K.S.; Olah, G.A. In *Acid–Base Catalysis. Proceedings of the International Symposium on Acid–Base Catalysis, Sapporo, 1988*; Tanabe, K.; Hattori, H.; Yamaguchi, T.; Tanaka, T., Eds.; Kodansha: Tokyo and VCH: New York, 1989; Chapter 1.6, pp 59–78.
48. Kijenski, J.; Malinowski, S. *Bull. Acad. Polon. Sci.* **1977**, *25*, 669.
49. Tanabe, K. *In Proceedings of the 9th International Congress on Catalysis;* Phillips, M.J.; Ternan, M., Eds.; Kodansha: Tokyo, 1989; p 85.
50. Okuhara, T.; Mishimura, T.; Misono, M. In *11th International Congress on Catalysis—40th Anniversary. Part A*; Hightower, J.W.; Delgass, W.N.; Iglesia, E.; Bell, A.T., Eds.; Elsevier: Amsterdam, 1996; pp 581–590.
51. Kubelková, L.; Kotrla, J.; Florián, J.; Bolom, T.; Faissard, J.; Heeribout, L.; Doremieux-Morin, C. In *11th International Congress on Catalysis—40th Anniversary. Part B.;* Hightower, J.W.; Delgass, W.N.; Iglesia, E.; Bell, A.T., Eds.; Elsevier: Amsterdam, 1996; pp 761–770.
52. Caubère, P. *Chem. Rev.* **1993**, *92*, 2317.
53. Schlosser, M. *Mod. Synth. Methods* **1992**, *6*, 227.
54. Schlosser, M.; Maccaroni, P.; Marzi, E. *Tetrahedron* **1998**, *54*, 2763.
55. Lochmann, L.; Trekoval, J. *J. Organomet. Chem.* **1987**, *326*, 1.
56. Lochmann, L. *Eur. J. Inorg. Chem.* **2000**, 1115; and references therein.
57. Kowalik, J.; Tolbert, L.M. *J. Org. Chem.* **2001**, *66*, 3229.
58. Morton, A.A. *Ind. Eng. Chem.* **1950**, *42*, 1488.
59. Morton, A.A.; Claff, C.E., Jr.; Collins, F.W. *J. Org. Chem.* **1955**, *20*, 428.
60. Lochmann, L.; Pospisil, J.; Lim, D. *Tetrahedron Lett.* **1966**, 257.
61. Mordini, A. In *Advances in Carbanion Chemistry;* Sneickus, V., Ed.; JAI Press: Greenwich, CT, 1992; Vol. 1.
62. Bauer, W.; Lochmann, L. *J. Am. Chem. Soc.* **1992**, *114*, 7482.
63. Harder, S.; Streitwieser, A. *Angew. Chem., Int. Ed. Engl.* **1993**, *32*, 1066.
64. Harder, S.; Lutz, M.; Kremer, T. *Organometallics* **1995**, *14*, 2133.
65. Kijenski, J.; Malinowski, S. *Bull. Acad. Pol. Sci. Ser. Sci. Chim.* **1977**, *25*, 501.
66. Kijenski, J.; Malinowski, S. *J. Chem. Soc., Faraday Trans. 1* **1978**, *74*, 250.

67. Kijenski, J.; Brzozka, K.; Malinowski, S. *Bull. Acad. Pol. Sci. Ser. Sci. Chim.* **1978**, *26*, 271.
68. West, A.R. *Basic Solid State Chemistry*; Wiley: New York, 1988; Chapter 5, pp 206–255.
69. Klabunde, K.J.; Matsuhashi, H. *J. Am. Chem. Soc.* **1987**, *109*, 1111.
70. Goodman, D.W. *Chem. Rev.* **1995**, *95*, 523.
71. Breck, D.W. *Zeolite Molecular Sieves*; Wiley: New York, 1974.
72. Wilson, S.T.; Lok, B.M.; Messina, C.A.; Cannan, T.R.; Flanigen, E.M. *J. Am. Chem. Soc.* **1982**, *104*, 1146.
73. Lok, B.M.; Messina, C.A.; Patton, R.L.; Gajek, R.T.; Cannan, T.R.; Flanigen, E.M. *J. Am. Chem. Soc.* **1984**, *106*, 6092.
74. Tanabe, K.; Misono, M.; Ono, Y.; Hattori, H. *New Solid Acids and Bases: Their Catalytic Properties*; Elsevier: Amsterdam, 1989; p 142.
75. Barthomeuf, D. *J. Phys. Chem.* **1984**, *88*, 42.
76. Corma, A.; Martin-Aranda, R.M.; Sanchez, F. *J. Catal.* **1990**, *126*, 192.
77. Barr, T.L. *Zeolites* **1990**, *10*, 760.
78. Malinowski, S.; Szczepanska, S. *J. Catal.* **1963**, *2*, 310.
79. Khan, A.Z.; Ruckenstein, E. *J. Catal.* **1993**, *143*, 1.
80. Kasai, P.H.; Bishop, R.J., Jr. *J. Phys. Chem.* **1973**, *77*, 2308.
81. Trescos, E.; Rachdi, F.; de Menorval, L.C.; Fajula, F.; Nunes, T.; Feio, G. *J. Phys. Chem.* **1993**, *97*, 11855.
82. Smeulders, J.B.A.F.; Hefni, M.A.; Klaasen, A.A.K.; de Boer, E.; Westphal, U.; Geismar, G. *Zeolites* **1987**, *7*, 347.
83. Ushikabo, T.; Hattori, H.; Tanabe, K. *Chem. Lett.* **1984**, 649.
84. Martens, L.R.M.; Grobet, P.J.; Jacobs, P.A. *Nature* **1985**, *315*, 568.
85. Xu, B.; Kevan, L. *J. Chem. Soc., Faraday Trans.* **1992**, *88*, 1695.
86. Xu, B.; Kevan, L. *J. Phys. Chem.* **1992**, *96*, 2642.
87. Hathaway, P.E.; Davis, M.E. *J. Catal.* **1989**, *116*, 263.
88. Hathaway, P.E.; Davis, M.E. *J. Catal.* **1989**, *116*, 279.
89. Tsuji, H.; Yagi, F.; Hattori, H.; Kita, K. In *Proceedings of the 10th International Congress on Catalysis, Budapest*; Guczi, L.; Solymosi, F.; Tétenyi, P., Eds.; Akademiai Kiado: Budapest, 1993; p 1171.
90. Park, Y.S.; Lee, Y.S.; Yoon, K.B. *J. Am. Chem. Soc.* **1993**, *115*, 12220.
91. Bordiga, S.; Ferrero, A.; Giamello, E.; Spoto, G.; Zecchina, A. *Catal. Lett.* **1991**, *8*, 375.
92. Ozin, G.A. *Adv. Mater.* **1992**, *4*, 162.
93. Martens, L.R.M.; Grobet, P.J.; Vermeiren, W.J.M.; Jacobs, P.A. *Stud. Surf. Sci. Catal.* **1986**, *28*, 935.
94. Martens, L.R.M.; Vermeiren, W.J.M.; Grobet, P.J.; Jacobs, P.A. *Stud. Surf. Sci. Catal.* **1987**, *31*, 531.
95. Lorsbach, B.A.; Kurth, M.J. *Chem. Rev.* **1999**, *99*, 1549.
96. Tanabe, K.; Misono, M.; Ono, Y.; Hattori, H. *New Solid Acids and Bases: Their Catalytic Properties*; Elsevier: Amsterdam, 1989; p 173–175.
97. Astle, M.J.; Zaslowski, J.A. *Ind. Eng. Chem.* **1952**, *44*, 2867.
98. Ishikawa,Y; Kamio, S. *Tokyo Kogyo Shikensho Hokoku* **1963**, *58*, 40.
99. Bergmann, E.D.; Corett, R. *J. Org. Chem.* **1956**, *21*, 107.
100. Bergmann, E.D.; Corett, R. *J. Org. Chem.* **1958**, *23*, 1507.
101. Miller, J.M.; So, K.-H.; Clark, J.H. *J. Chem. Soc., Chem. Commun.* **1978**, 466.
102. Hodge, P.; Khoshdel, E.; Waterhouse, J. *J. Chem. Soc., Perkin Trans. 1* **1983**, 2205.
103. Komeili-Zadeh, H.; Dou, H.J.M.; Metzger, J. *J. Org. Chem.* **1978**, *43*, 156.
104. Tomoi, M.; Akada, Y.; Kakiuchi, M. *Makromol. Chem. Rapid Commun.* **1982**, *3*, 537.

105. Allmann, R.; Jepson, H.P. *Neues Jahrb. Mineral. Monatsch.* **1969**, 11, 544.
106. Allmann, R. *Acta Crystallogr. Sect. B.* **1968**, *24*, 972.
107. Reichle, W.T. *CHEMTECH* **1986**, *16*, 58.
108. Reichle, W.T. *J. Catal.* **1986**, *101*, 352.
109. Reichle, W.T. *J. Catal.* **1985**, *94*, 547.
110. Guida, A.; Lhouty, M.H.; Tichit, D.; Figueras, F.; Geneste, P. *Appl. Catal. A: General* **1997**, *164*, 251.
111. Kohjiya, S.; Sato, T.; Nakayama, T.; Yamashita, S. *Makromol. Chem. Rapid Commun.* **1981**, *2*, 231.
112. Miyata, S.; Kumura, T.; Hattori, H.; Tanabe, K. *Nippon Kagaku Kaishi* **1971**, *92*, 514; C.A. 75: 70781.
113. Bonneau, L.; Pezererat, H. *J. Chim. Phys. Phys.-Chim. Biol.* **1983**, *80*, 275.
114. Bonneau, L.; Suquet, H.; Mallard, C.; Pezererat, H. *Environ. Res.* **1986**, *41*, 251.
115. Chalais, S.; Laszlo, P.; Mathy, A. *Tetrahedron Lett.* **1985**, *26*, 4453.
116. Laszlo, P.; Pennetreau, P. *Tetrahedron Lett.* **1985**, *26*, 2645.
117. Csuk, R.; Glanzer, B.I.; Furstner, A. In *Advances in Organometallic Chemistry, Vol. 28*; Stone, F.G.A.; West, R., Eds.; Academic Press: San Diego, CA, 1996; p 85.
118. Lednor, P.W.; de Ruiter, R. *J. Chem. Soc., Chem. Commun.* **1989**, 320.
119. Massinon, A.; Guéguen, E.; Conanec, R.; Marchand, R.; Laurent, Y.; Grange, P. In *11th International Congress on Catalysis—40th Anniversary. Part B*; Hightower, J.W.; Delgass, W.N.; Iglesia, E.; Bell, A.T., Eds.; Elsevier: Amsterdam, 1996; pp 77–85.
120. Climent, M.J.; Corma, A.; Fornés, V.; Frau, A.; Guil-Lopez, R.; Iborra, S.; Primo, J. *J. Catal.* **1996**, *163*, 392.
121. Busca, G.; Lorenzelli, V.; Porcile, G.; Barton, M.I.; Quintard, P.; Marchand, R. *Mater. Chem. Phys.* **1986**, *14*, 123.
122. Bentley, T.W.; Jones, R.V.H.; Larder, A.H.; Lock, S.J. *J. Chem. Soc., Chem. Commun.* **1994**, 2309.
123. Fuentes, A.; Marinas, J.M.; Sinisterra, J.V. *Tetrahedron Lett.* **1987**, *28*, 4541.
124. Climent, M.J.; Corma, A.; Garcia, H.; Primo, J. *Appl. Catal. A.: General* **1989**, *51*, 113.
125. Varma, R.S.; Varma, M. *Tetrahedron Lett.* **1992**, *33*, 5937.
126. Walling, C. *J. Am. Chem. Soc.* **1950**, *72*, 1164.
127. Hammett, L.P.; Deyrup, A.J. *J. Am. Chem. Soc.* **1932**, *54*, 2721.
128. Paul, M.A.; Long, F.A. *Chem. Rev.* **1957**, *57*, 1.
129. Dolman, D.; Stewart, R. *Can. J. Chem.* **1967**, *45*, 911.
130. Cox, R.A.; Stewart, R. *J. Am. Chem. Soc.* **1976**, *98*, 488.
131. Buncel, E.; Wilson, H.A. *Adv. Phys. Org. Chem.* **1977**, *14*, 133.
132. Buncel, E., Symons, E.A.; Dolman, D.; Stewart, R. *Can. J. Chem.* **1970**, *48*, 3354.
133. Garcia, G.; Leal, J.M. *J. Phys. Org. Chem.* **1991**, *4*, 1991.
134. Cox, R.A. *Acc. Chem. Res.* **1987**, *20*, 27.
135. Bunnett, J.F.; Olsen, F.P. *Can. J. Chem.* **1966**, *44*, 1899.
136. Tanabe, K.; Misono, M.; Ono, Y.; Hattori, H. *New Solid Acids and Bases: Their Catalytic Properties*; Elsevier: Amsterdam, 1989; p 15.
137. Tanabe, K.; Misono, M.; Ono, Y.; Hattori, H. *New Solid Acids and Bases: Their Catalytic Properties*; Elsevier: Amsterdam, 1989; p 14.
138. Benesi, H.A. *J. Am. Chem. Soc.* **1956**, *78*, 5490.
139. Barbarin, V. Thesis, Paris, 1987.
140. Yashima, T.; Sato, K.; Hayasaki, H.T.; Hara, N. *J. Catal.* **1972**, *26*, 303.
141. Nagaraju, N.; Walvekar, S.P.; Nanje-Gowda, N.M. *Indian J. Technol.* **1990**, *28*, 59.
142. Przystajko,W.; Fiedorow, R.; Dalla-Lana, I.G. *Zeolites* **1987**, *7*, 477.
143. Yamanaka, T.; Tanabe, K. *J. Phys. Chem.* **1975**, *79*, 2409.

144. Yamanaka, T.; Tanabe, K. *J. Phys. Chem.* **1976**, *80*, 1723.
145. Farneth, W.E.; Gorte, R.J. *Chem. Rev.* **1995**, *95*, 615.
146. Tanabe, K. In *Catalysis: Science and Technology*; Anderson, J.R.; Boudart, M., Eds.; Springer Verlag: Berlin, 1986; Vol. 2, Chapter 5, p 231.
147. Auroux, A.; Gervasini, A. *J. Phys. Chem.* **1990**, *94*, 6371.
148. Fukuda, Y.; Tanabe, K. *Bull. Chem. Soc. Jpn.* **1972**, *46*, 1616.
149. Evans, J.V.; Whateley, T.L. *Trans. Faraday Soc.* **1967**, *63*, 2769.
150. Tanabe, K.; Misono, M.; Ono, Y.; Hattori, H. *New Solid Acids and Bases: Their Catalytic Properties*; Elsevier: Amsterdam, 1989; p 17.
151. Angell, C.L.; Howell, M.V. *Can. J. Chem.* **1969**, *47*, 3811.
152. Delaval, Y.; Cohen de Lara, E. *J. Chem. Soc., Faraday Trans. 1* **1981**, *77*, 869.
153. Mirodatos, C.; Abou-Kaïs, A.; Vedrine, J.C.; Pichat, P.; Barthomeuf, D. *J. Phys. Chem.* **1976**, *80*, 2366; and references therein.
154. Scokart, P.O.; Rouxhet, P.G. *J. Chem. Soc., Faraday Trans. 1* **1980**, *76*, 1476.
155. Scokart, P.O.; Rouxhet, P.G. *Bull. Soc. Chim. Belg.* **1981**, *90*, 983.
156. Barthomeuf, D. *J. Phys. Chem.* **1978**, *55*, 138.
157. Barthomeuf, D. *Stud. Surf. Sci. Catal.* **1991**, *65*, 157.
158. Barthomeuf, D.; de Mallmann, A. *Stud. Surf. Sci. Catal.* **1988**, *37*, 365.
159. Derouane, E.G.; Nagy, J.B. In *Catalytic Materials: Relationship between Structure and Reactivity*; Whyte, T.E., Jr.; Dalla Betta, R.A.; Baker, R.T.K., Eds.; ACS Symposium Series 248; American Chemical Society: Washington, DC, 1984; Chapter 7, pp 101–126.
160. Nagy, J.B.; Gigot, M.; Gourgue, A.; Derouane, E.G. *J. Mol. Catal.* **1977**, *2*, 265.
161. Nagy, J.B.; Abou-Kaïs, A.; Guelton, M.; Harmel, J.; Derouane, E.G. *J. Catal.* **1982**, *72*, 1.
162. Nelson, R.L.; Hale, J.W. *Discuss. Faraday Soc.* **1958**, *52*, 77.
163. Coluccia, S.; Tench, A.J.; Segall, R.L. *J. Chem. Soc., Faraday Trans. 1* **1978**, *75*, 1769.
164. Tench, A.J.; Pott, G.T. *Chem. Phys. Lett.* **1974**, *26*, 590.
165. Zecchina, A.; Lofthouse, M.G.; Stone, F.S. *J. Chem. Soc., Faraday Trans. 1* **1975**, *71*, 1476.
166. Timken, H.K.C.; Turner, G.L.; Gilson, J.P.; Welsch, L.B.; Oldfield, E. *J. Am. Chem. Soc.* **1986**, *108*, 7231.
167. Timken, H.K.C.; Janes, N.; Turner, G.L.; Lambert, S.L.; Welsch, L.B.; Oldfield, E. *J. Am. Chem. Soc.* **1986**, *108*, 7236.
168. Engelhardt, G.; Michel, D. *High Resolution Solid-State NMR of Silicates and Zeolites*; Wiley: Chichester, U.K., 1987.
169. Engelhardt, G. *Stud. Surf. Sci. Catal.* **1991**, *52*, 151.
170. Radeglia, R.; Engelhardt, G. *Chem. Phys. Lett.* **1985**, *114*, 28.
171. Noller, H.; Lercher, J.A.; Vinek, H. *Mater. Chem. Phys.* **1988**, *18*, 577.
172. Gasteiger, J.; Marsili, M. *Tetrahedron Lett.* **1978**, *34*, 3181.
173. Okamoto, Y.; Ogawa, M.; Maezawa, A.; Imanaka, T. *J. Catal.* **1988**, *112*, 427.
174. Barr, T.L.; Lishka, M.A. *J. Am. Chem. Soc.* **1986**, *108*, 3178.
175. Stoch, J.; Lercher, J.; Ceckiewicz, S. *Zeolites* **1992**, *12*, 81.
176. Huang, M.; Adnot, A.; Kaliaguine, S. *J. Catal.* **1992**, *137*, 322.
177. Kulkarni, S.J. *Indian J. Chem. Sect. A* **1990**, *29A*, 1125.
178. Edgell, M.J.; Paynter, R.W.; Mugford, S.C.; Castle, J.E. *Zeolites* **1990**, *10*, 51.
179. Cventanovic, R.J.; Amenomiya, Y. *Adv. Catal.* **1967**, *17*, 103.
180. Knötzinger, H. *Adv. Catal.* **1976**, *25*, 184.
181. Zhang, G.; Hattori, H.; Tanabe, K. *Appl. Catal. A: General* **1988**, *36*, 189.
182. Ito, T.; Murakami, T.; Tokuda, T. *J. Chem. Soc., Faraday Trans. 1* **1983**, *79*, 913.
183. Joshi, P.N.; Shiralkar, V.P. *J. Phys. Chem.* **1993**, *97*, 619.

184. Amari, D.; Lopez-Cuesta, J.M.; Nguyen, N.P.; Jerrentrup, R.; Ginoux, J.L. *J. Therm. Anal.* **1992**, *38*, 1005.
185. Bunnett, J.F. *Angew. Chem., Int. Ed. Engl.* **1962**, *1*, 971.
186. McLennan, D.J. *Q. Rev.* **1967**, *21*, 490.
187. Saunders, W.H. Jr.; Cockerill, A.F. *Mechanisms of Elimination Reactions*; Wiley: New York, 1973.
188. Saunders; W.H., Jr. *Acc. Chem. Res.* **1976**, *9*, 19.
189. Bordwell, F.G. *Acc. Chem. Res.* **1972**, *5*, 374.
190. Fry, A. *Chem. Soc. Rev.* **1972**, *1*, 163.
191. More O'Ferrall, R.A. In *The Chemistry of the Carbon–Halogen Bond, Part 2*; Patai, S.; Ed.; Wiley: New York, 1973; pp 630–640.
192. Sicher, J. *Angew. Chem., Int. Ed. Engl.* **1972**, *11*, 200.
193. Knötzinger, H. In *The Chemistry of the Hydroxyl Group*; Patai, S. Ed.; Wiley: New York,1971; Chapter 12.
194. Keeffe, J.R.; Jencks, W.P. *J. Am. Chem. Soc.* **1983**, *105*, 265.
195. Lundeen, A.J.; van Hoozen, J. *J. Org. Chem.* **1967**, *32*, 3386.
196. Yamaguchi, T.; Sasaki, H.; Tanabe, K. *Chem. Lett.* **1973**, 1017.
197. Yamaguchi, T.; Sasaki, H.; Tanabe, K. *Chem. Lett.* **1976**, 677.
198. Takahashi, K.; Hibi, T.; Higashio, Y.; Araki, M. *Shokubai (Catalyst)* **1993**, *35*, 12.
199. Yamadaya, M.; Shimomura, K.; Konoshita, T.; Uchida, H. *Shokubai (Catalyst)* **1965**, *7*, 313.
200. Knötzinger, H. *Angew. Chem., Int. Ed. Engl.* **1968**, *7*, 791.
201. Misono, M.; Yoneda, Y. *Dai 19-menkai Koenyoko-shu* **1966**, *12A*, 123.
202. Pines, H.; Pillai, C.N. *J. Am. Chem. Soc.* **1961**, *83*, 3270.
203. Pines, H.; Pillai, C.N. *J. Am. Chem. Soc.* **1960**, *82*, 2401.
204. Yashima, T.; Suzuki, H.; Hara, N. *J. Catal.* **1974**, *33*, 486.
205. Unland, M. L. *J. Phys. Chem.* **1978**, *82*, 580.
206. Yamawaki, J.; Kawate, T.; Ando, T.; Hanafusa, T. *Bull. Chem. Soc. Jpn.* **1983**, *56*, 1885.
207. Pines, H. *The Chemistry of Catalytic Hydrocarbon Conversions*; Academic Press: New York, 1981; Chaper 2.
208. Goldwasser, J.; Hall, W.K. *J. Catal.* **1981**, *71*, 53.
209. Hattori, H. *Stud. Surf. Sci. Catal.* **1993**, *78*, 35.
210. Bank, S.; Schriesheim, A.; Rowe, C.A., Jr. *J. Am. Chem. Soc.* **1965**, *87*, 3244.
211. Bank, S. *J. Am. Chem. Soc.* **1965**, *87*, 3245.
212. Tanabe, K.; Misono, M.; Ono, Y.; Hattori, H. *New Solid Acids and Bases: Their Catalytic Properties*; Elsevier: Amsterdam, 1989; pp 28–30.
213. Hoffmann, R.; Olofsen, R.A. *J. Am. Chem. Soc.* **1966**, *88*, 943.
214. Haag, W.O.; Pines, H. *J. Am. Chem. Soc.* **1960**, *82*, 387.
215. Tanabe, K.; Yoshii, N.; Hattori, H. *J. Chem. Soc., Chem. Commun.* **1971**, 464.
216. Hightower, J.W.; Hall, W.K. *J. Am. Chem. Soc.* **1967**, *89*, 778.
217. Satoh, A.; Hattori, H. *J. Catal.* **1976**, *45*, 36.
218. Lombardo, E.A.; Conner, W.C.; Madon, R.J.; Hall, W.K.; Karlamov, V.V.; Minachev, Zh.M. *J. Catal.* **1978**, *53*, 135.
219. Hino, M. *Hakodate Kogyo Koto Senmon Gakko Kiyo* **1983**, *17*, 95; C.A. 99: 138980.
220. Martens, L.R.M.; Grobet, P.J.; Jacobs, P.A. *Nature* **1985**, *315*, 568.
221. Reinacker, R.; Graefe, J. *Angew. Chem., Int. Ed. Engl.* **1985**, *24*, 320.
222. Hattori, A.; Hattori, H.; Tanabe, K. *J. Catal.* **1980**, *65*, 246.
223. Matsuhashi, H.; Hattori, H.; Tanabe, K. *Chem. Lett.* **1981**, 341.
224. Matsuhashi, H.; Hattori, H. *J. Catal.* **1984**, *85*, 457.
225. Lill-Nilsson, S.O.; Arvidsson, P.I.; Ahlberg, P. *Acta Chem. Scand.* **1998**, *52*, 280.

226. Hambling, J.K.; Alderson, G.W.; Yeo, A.A., U.K. Patent 958,161, 1964.
227. Meisinger, E.E.; Bloch, H.S., U.S. Patent 3,128,318, 1964.
228. Reichle, W.T., U.S. Patent 4,165,339, 1979.
229. Kricheldorf, H.R.; Berl, M.; Scharnagl, N. *Macromolecules* **1988**, *21*, 286.
230. Bush, L.W.; Breslow, D.S. *Macromolecules* **1968**, *1*, 189.
231. Malinowski, S.; Jedrzejewski, H.; Basinski, S.; Lipski, Z.; Moszczenska, J. *Roczniki Chem.* **1956**, *30*, 1129.
232. Malinowski, S.; Basinski, S.; Olsweska, M.; Zieleniewska, H. *Roczniki Chem.* **1957**, *31*, 123.
233. Malinowski, S.; Benbenek, S.; Pasynktewicz, I.; Wojciechowska, E. *Roczniki Chem.* **1958**, *32*, 1089.
234. Malinowski, S.; Kiewlicz, W.; Soltys, E. *Bull. Soc. Chim. Fr.* **1963**, 439.
235. Malinowski, S.; Basinski, S.; Szczepanska, S.; Kiewlicz, W. In *Proceedings of the 3rd International Congress on Catalysis, Amsterdam*; Sachtler, W.M.H.; Schmit, G.C.A.; Zeitering, P., Eds.; North-Holland: Amsterdam, 1964; pp 56–65.
236. Bram, G.; Galons, H.; Labidalle, S.; Loupy, A.; Miocque, M.; Petit, A.; Pigeon, P.; Sansoulet, J. *Bull. Soc. Chim. Fr.* **1989**, 247.
237. Sidorenko, Y.N.; Galich, P.N. *Neftkhimia* **1991**, *31*, 54.
238. Lacroix, C.; Deluzarche, A.; Keinnemann, A.; Boyer, A. *J. Chem. Phys.* **1984**, *81*, 473.
239. Lacroix, C.; Deluzarche, A.; Keinnemann, A.; Boyer, A. *J. Chem. Phys.* **1984**, *81*, 486.
240. Palomares, A.E.; Eder-Mirth, G.; Lercher, J.A. *J. Catal.* **1997**, *168*, 442.
241. Garces, J.M.; Vrieland, G.E.; Bates, S.I.; Scheidt, F.M. In *Catalysis by Acids and Bases*; Imelik, B.; Naccache, C.; Coudurier, G.; Ben Taarit, Y.; Vedrine, J.C., Eds.; Elsevier: Amsterdam, 1985; p 67.
242. Hathaway, P.E.; Davis, M.E. *J. Catal.* **1989**, *119*, 497.
243. Itoh, H.; Miyamoto, A.; Murakami, Y. *J. Catal.* **1980**, *64*, 284.
244. Corma, A.; Sastre, G.; Viruela, P. *Stud. Surf. Sci. Catal.* **1994**, *84C*, 2171.
245. Mielczarski, E.; Davis, M.E. *Ind. Eng. Chem. Res.* **1990**, *29*, 1579.
246. Sefcik, M.D. *J. Am. Chem. Soc.* **1979**, *101*, 2164.
247. Pines, H.; Shaap, L.A. *Adv. Catal.* **1960**, *12*, 120.
248. Suzukamo, G.; Fukao, M.; Hibi, T.; Chikaishi, K. In *Acid–Base Catalysis. Proceedings of the International Symposium on Acid–Base Catalysis, Sapporo, 1988*; Tanabe, K.; Hattori, H.; Yamaguchi, T.; Tanaka, T., Eds.; Kodansha: Tokyo and VCH: New York, 1989; p 405.
249. House, H.O. *Modern Synthetic Reactions*, 2nd ed.; Benjamin: Menlo Park, CA., 1972; Chapter 9.
250. Harris, J.M.; Wamser, C.C. *Fundamentals of Organic Reaction Mechanisms*; Wiley: New York, 1976; pp 213–236.
251. Fang, Y.; Zhang, S.; Xing, S.; Shi, H.; Zhao, W. *Gaodeng Xuexiao Huaxue Xuebao* **1988**, *9*, 601; C.A. 110: 153512.
252. Muzart, J. *Synth. Commun.* **1985**, 285.
253. Maple, S.R.; Allerhand, A. *J. Am. Chem. Soc.* **1987**, *109*, 6609.
254. Barot, B.C.; Sullins, D.W.; Eisenbraun, E.J. *Synth. Commun.* **1984**, *14*, 397.
255. Zhang, G.; Hattori, H.; Tanabe, K. *Appl. Catal.* **1988**, *36*, 189.
256. Scheidt, F.M. *J. Catal.* **1964**, *3*, 372.
257. Salvapati, G.S.; Ramanamurty, K.V.; Janardanarao, M. *J. Mol. Catal.* **1989**, *54*, 9.
258. Salvapati, G.S.; Ramanamurty, K.V.; Janardanarao, M.; Vaidyeswaran, R. *Appl. Catal.* **1989**, *48*, 223.
259. Fukui, K; Takei, M. *Bull. Inst. Chem. Res. Kyoto Univ.* **1951**, *26*, 85.

260. Campelo, J.M.; Garcia, A.; Luna, D.; Marinas, J.M. *Can. J. Chem.* **1983**, *62*, 638.
261. Tanabe, K.; Fukuda, Y. *React. Kinet. Catal. Lett.* **1974**, *1*, 21.
262. Kijenski, J.; Malinowski, S. *Bull. Acad. Polon. Sci. Ser. Chem.* **1978** *26*, 183.
263. Wierzchowski, P.T.; Zatorski, L.W. *Catal. Lett.* **1991**, *9*, 411.
264. Huang, N.; Kaliaguine, S. *Stud. Surf. Sci. Catal.* **1993**, *78*, 559.
265. Rode, E.J.; Gee, P.E.; Marquez, L.N.; Uemura, T.; Bazargani, M. *Catal. Lett.* **1991**, *9*, 103.
266. Corma, A.; Fornés, V.; Martin-Aranda, R.H.; Garcia, H.; Primo, J. *J. Appl. Catal.* **1990**, *59*, 537.
267. Texier-Boullet, F.; Foucaud, A. *Tetrahedron Lett.* **1982**, *23*, 4927.
268. Villemin, D.J. *J. Chem. Soc., Chem. Commun.* **1983**, 1092.
269. Cabello, J.A.; Campelo, J.M.; Garcia, A.; Luna, D.; Marinas, J.M. *J. Org. Chem.* **1984**, *49*, 5195.
270. Sinisterra, J.V.; Mouloungui, Z.; Marinas, M. *J. Colloid Interface Sci.* **1987**, *115*, 520.
271. Moison, H.; Texier-Boullet, F.; Foucaud, A. *Tetrahedron* **1987**, *43*, 537.
272. Siebenhaar, B.; Casagrande, B.; Studer, M.; Blaser, H.U. *Can. J. Chem.* **2001**, *79*, 566.
273. Rodriguez, I.; Cambon, H.; Brunel, D.; Lasperas, M.; Geneste, P. *Stud. Surf. Sci. Catal.* **1993**, *78*, 623.
274. Corma, A.; Fornés, V.; Martin-Aranda, R.M.; Garcia, H.; Primo, J. *Appl. Catal.* **1990**, *59*, 237.
275. Corma, A.; Martin-Aranda, R.M.; Sanchez, F. *Stud. Surf. Sci. Catal.* **1991**, *59*, 503.
276. Varma, R.S.; Kabalka, G.W.; Evans, L.T.; Pagni, R.M. *Synth. Commun.* **1985**, *15*, 279.
277. Foucaud, A. *Bull. Soc. Chim. Fr.* **1989**, 279.
278. Tanabe, K.; Saito, K. *J. Catal.* **1974**, *35*, 274.
279. Climent, M.J.; Corma, A.; Garcia, H.; Primo, J. *Appl. Catal.* **1989**, *51*, 113.
280. Morrison, R.T.; Boyd, R.N. *Organic Chemistry,* 5th ed.; Allyn & Bacon: Boston, MA, 1987; p 1087.
281. Inglesias, M.; Marinas, J.M.; Sinisterra, J.V. *Tetrahedron* **1987**, *43*, 2335.
282. Garcia-Raso, A.; Garcia-Raso, J.; Campaner, B.; Mestres, R.; Sinisterra, J.V. *Synthesis* **1982**, 1037.
283. Kakuno, Y.; Hattori, H.; Tanabe, K. *Chem. Lett.* **1982**, 2015.
284. Hawthorne, M. *J. Am. Chem. Soc.* **1957**, *79*, 2510.
285. Keinan, E.; Mazur, Y. *J. Am. Chem. Soc.* **1977**, *99*, 3861.

6

Enolates

6.1 INTRODUCTION

Enolates are central intermediates in a wide range of name reactions. These reactions pervade the introductory organic chemistry curriculum. Clearly, no study of the mechanistic and structural chemistry of carbanions would be complete without some discussion of enolates (see also the earlier monograph by Buncel).[1]

An enolate could be described as a carbanion that is stabilized by at least one adjacent carbonyl functionality, as shown in Scheme 6.1. Here, deprotonation of the α-carbon of the keto form of the carbonyl compound formally leads to formation of a carbanion, one of the canonical forms of the enolate anion. Conceptually, deprotonation of the enol form of the carbonyl compound results in formation of the oxyanion resonance structure of the enolate. The equilibrium between the keto and enol forms of the carbonyl compound is termed a tautomeric equilibrium and the two forms are more generally referred to as tautomers. (Tautomerism will be considered more generally in Chapter 7).

Scheme 6.1 emphasizes the ambident nature of the enolate anion. As we have seen in Chapter 1, alkylation, as but one example, can occur at either the carbon or oxygen centers, depending on the nature of the enolate, the alkylating agent, and, importantly, the nature of the reaction medium. It would appear, as well, that the ambident nature of enolate ions is tied to the keto–enol tautomerism, as Scheme 6.1 shows.

This is the starting point in our discussion of enolate anions: keto–enol tautomerism. However, in discussing this tautomerism we must echo the comment of

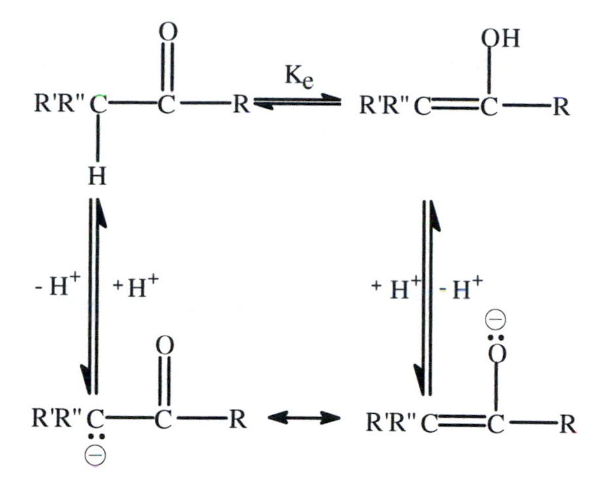

SCHEME 6.1

Toullec in his 1982 review that "several thousand papers have made enolization one of the best documented processes in organic chemistry".[2] Since then, several reviews[3-5] and an authoritative volume have appeared, devoted to enols and the process of enolization.[6] It would be presumptuous to expect a thorough accounting of the abundant activity in this area, let alone the broader field of enolate reactions,[7,8] including stereospecific enolate reactions.[9,10] Our aim, therefore, will be to consider a few key features of enolization, before proceeding to consideration of a model enolate reaction, the aldol condensation, and the nature of the transition state involved in this condensation, and, finally, examining briefly a few selected enolate rearrangements. (Note that rearrangements of other types of carbanions forms the topic of Chapter 7.)

6.2 KETO–ENOL TAUTOMERISM

6.2.1 Position of the Equilibrium

In principle, all compounds having an sp^3-hybridized carbon that is bonded both to a proton and to at least one adjacent carbonyl functionality, as in an alkanone, ester, alkanedione, and so forth, may undergo enolization. Generally, the keto–enol prototropic equilibrium lies well to the left (Scheme 6.1; K_e is small), that is, toward the keto form. Clearly, those cases in which the equilibrium favors the enol side are particularly instructive (see Table 6.1). The question then arises: How may the equilibrium be measured?

Classically, the two tautomers of ethyl acetoacetate (ethyl 3-oxobutanoate; equation 6.1) were isolated from the tautomeric mixture.[11] The keto form, with a melting point of −39 °C, was crystallized from solution at −78 °C, while the enol, which is liquid at −78 °C, was isolated after acidification of the enolate sodium salt with hydrogen chloride, also at −78 °C. At such low temperatures and in the absence of traces of acid or base, the two tautomers could be kept in pure form for extended periods. In examining equation 6.1, it is apparent that one way of describing the equilibrium constant for the tautomerization of a keto form to an enol form is in terms of "enol content", that is, the percentage of enol present in a given solution or in the neat alkanone/alkanal/ester (see Table 6.1). The greater the enol content, the higher the value of K_e.

$$CH_3C(=O)CH_2C(=O){-}OCH_2CH_3 \rightleftharpoons CH_3C(OH){=}CHC(=O){-}OCH_2CH_3 \qquad (6.1)$$

Since the enol contains an alkene functional group, the amount of enol tautomer of any alkanone or alkanal present, at room temperature, was initially believed to be accessible by titration with standard bromine,[12-16] but now such direct titrations are suspect, at least for simple enols. One factor that appears to be important is the perturbation of the keto–enol equilibrium by the formation of hydrogen bromide in the titration since HBr can act as a catalyst for keto–enol conversion. Spectroscopic methods, including UV[17-21] and IR spectrophotometry[22,23] and, particularly, ^{1}H NMR spectrometry,[24-27] but also ^{13}C NMR[28-30] and ^{17}O NMR,[31-33] often appear to be the methods of choice for direct determination of enol content (Table 6.1). Such methods usually require relatively high enol contents (> 1%), however, and for β-diketones and β-ketoesters, where the enol

Table 6.1. Equilibrium enol content in selected keto–enol tautomeric systems

Compound	Enol content (%)	Method of determination
Monoketones (solvent)		
Propanone (liquid)	1.5×10^{-4}	Bromine titration[a]
Propanone (water)	2.5×10^{-6}	Bromine titration[b]
Propanone (water)	9.0×10^{-7}	Bromine titration[c]
Butanone (liquid)	1.2×10^{-1}	Bromine titration[a]
Butanone (liquid)	$8 \pm 2 \times 10^{-1}$	Infrared spectroscopy[d]
Cyclopentanone (liquid)	8.8×10^{-2}	Bromine titration[a]
Cyclopentanone (water)	4.8×10^{-3}	Bromine titration[b]
Cyclopentanone (water)	1.3×10^{-3}	Bromine titration[c]
Cyclohexanone (liquid)	1.2	Bromine titration[a,e]
Cyclohexanone (liquid)	$1.5 \pm .2$	Infrared spectroscopy[d]
Cyclohexanone (water)	2×10^{-2}	Bromine titration[b]
Cyclohexanone (water)	4.1×10^{-4}	Bromine titration[c]
1,3-Diketones		
Ethyl-3-oxobutanoate (liquid)	7.5	Bromine titration[f]
2,4-Pentanedione (liquid)	78	Bromine titration[g]

SOURCES: [a]Data taken from reference 15; [b]Data taken from references 34–36; [c]Data taken from reference 16; [d]Data taken from reference 37; [e]Data taken from reference 38; [f]Data taken from references 12 and 13; [g]Data taken from reference 14.

content is generally high, the enol percentages determined by halogen titration and spectroscopic methods are often in reasonable agreement. The values determined by the titration method and spectroscopic methods are in worse agreement for the enols of simple alkanones and alkanals. As but one example, the enol content determined by halogen titration to be 1.2% for cyclohexanone (neat liquid),[38] has been criticized as being far too high on the basis of comparison with NMR data;[24] however, the value appears to be in good agreement with the enol content as determined by IR spectroscopy (Table 6.1). Note also the significant variation in enol contents for a given system, even where all values are determined by bromine titration, though to some degree this is a function of the improvement in the titration technique. These classical and spectroscopic methods of determining equilibrium constants (K_e, Scheme 6.1) for enolization have been critically reviewed by Toullec.[39]

Fast kinetic methods including temperature-jump[40] and flash photolysis[41-43] have also been applied to the problem of measuring the keto–enol equilibrium and, directly or indirectly, the relative enol content of various solutions.

An interesting example of the kinetic approach[44] is based on the flash-photolytic generation of the enol of ethanal (acetaldehyde), namely, vinyl alcohol.[42] The key to the approach is that the equilibrium constant (K_e, Scheme 6.1) can be determined from the rate constants for enolization [k(enol)] and for ketonization [k(keto)], where ketonization refers to formation of the keto form from the enol form], since $K_e = k(\mathrm{enol})/k(\mathrm{keto})$.

The rate constant for enolization may be obtained from kinetic measurements on enol trapping. The assumption here is that the rate of trapping of the enol, by addition of halogen, for example, is much greater than the rate of formation of the enol from the keto form. Therefore, the rate of enolization becomes rate limiting and the rate constant determined from trapping is actually the rate constant for enol formation. The further tacit assumption is that trapping occurs much faster than the reverse ketonization of the enol. Kresge has discussed[44] the validity of these assumptions under various conditions, as well as competing reactions (alternative sites of halogenation,[45] successive halogenation at the α-carbon under basic conditions, oxidation of aldehydes,[46] etc.). Regardless, if the assumptions hold then the rate constant for enolization, that is, k(enol), may be measured, as it was in the Kresge study.[42]

The other part of the kinetic method is the determination of the rate constant for ketonization, k(keto). While this may appear difficult, given our expectations concerning the stability of enols, particularly simple ones, many of these enols have in fact proven to be significantly persistent once generated independently rather than through tautomerization. In this regard, the simplest enol, vinyl alcohol, has been shown to have a half-life of 10 min (20 °C) in aqueous acetonitrile.[47] In the Kresge study, vinyl alcohol was formed photolytically in significantly greater concentration than would normally exist in tautomerization. One route explored was Norrish type II cleavage of 5-hydroxy-2-pentanone as shown in Scheme 6.2. Since both the enol of acetone and vinyl alcohol, the enol of ethanal, were formed in this photoelimination, the rate of ketonization of both could be monitored. (The ketonization of vinyl alcohol was found to proceed 2 orders of

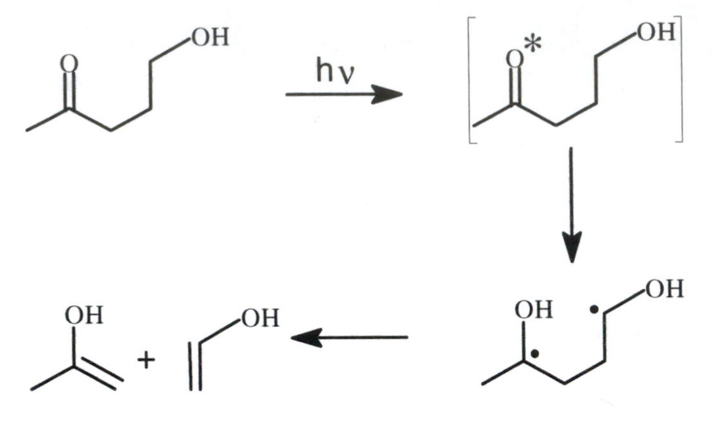

SCHEME 6.2

magnitude slower than that for the enol of acetone.) Nevertheless, in this way the rate constant, k(keto), could be determined for vinyl alcohol.

Combination of k(enol) and k(keto), determined under both acidic and basic conditions, yielded a value for K_e of 5.89 $(\pm 0.81) \times 10^{-7}$($pK_e = 6.23$) for the tautomeric equilibrium between ethanal and vinyl alcohol in water.[42] In the same study, the acid dissociation constant for vinyl alcohol was also determined: $K_a(\text{enol}) = 3.13 \times 10^{-11}$ [$pK_a(\text{enol}) = 10.50$]. From these values and the following thermodynamic cycle (Scheme 6.3), the acid dissociation constant for ethanal to give the corresponding ethanolate anion could be calculated: $K_a(\text{ethanal}) = 1.85 \times 10^{-17}$ or $pK_a(\text{ethanal}) = 16.73$, *in water*.[42] Thus, a study into keto–enol tautomerism leads to a value for the acid dissociation constant (K_a) that transforms an aldehyde into an enolate anion. Some enolization constant values (pK_e) for simple enols, as determined by this kinetic approach, are listed in Table 6.2.

Equilibrium constants for enolization may also be calculated from thermodynamic properties (ΔH and ΔS).[48,49] These may be accessible from group additivity schemes or from gas-phase measurements (which are then corrected for solvent effects).

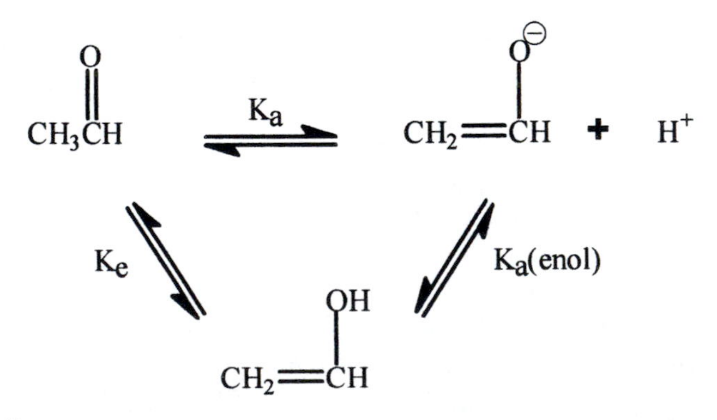

SCHEME 6.3

Table 6.2. Enolization constants (pK_e, aqueous, 25 °C) for selected carbonyl compounds determined by the kinetic method

Compound (in keto form)	pK_e
Acetaldehyde (ethanal)	6.23[a]
Cyclohexanone	6.39[b]
Mesityl acetophenone	6.92[c]
Acetophenone (ethanoylbenzene)	7.96[d]
Acetone (propanone)	8.33[e]

SOURCE: [a]Data taken from reference 42; [b]Data taken from reference 51; [c]Data taken from reference 52; [d]Data taken from reference 53; [e]Data taken from reference 54.

Calculational approaches to determining the stability of enol and keto pairs in the gas phase have been reviewed.[50] Methods of correlating gas-phase calculated enol acidities with solution acidities have been developed and thereby estimates of pK_a values may be obtained for other structurally related enols in solution.[55,56] This will likely continue to be an area of significant activity as the factors that stabilize (or destabilize) enols as compared with keto forms are probed.[57,58] In the following sections, we will briefly examine a few of these factors.

6.2.2 Factors That Influence Stability of Enols

It is beyond the scope of this brief review to consider each component that may influence the magnitude of the equilibrium constant for enolization; these factors that influence the position of equilibrium are a source of significant continuing research activity.[6] A few factors, however, are well established and can be enumerated here.

Enols may be stabilized by internal hydrogen-bonding. An example of this is provided by the enol generated from ethyl 3-oxobutanoate, **65** (where the internal hydrogen-bond is emphasized). Similar structures can be expected to stabilize the enols of other β-ketoesters and β-diketones. This hydrogen-bonding stabilization

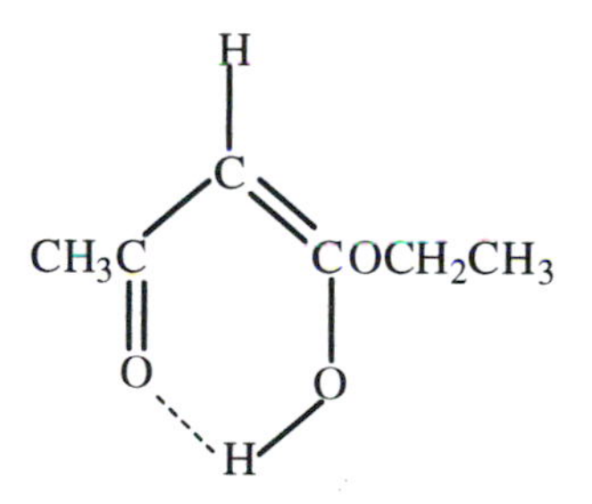

65

accounts for the relatively high values for enol content found for these enols. Thus, ethyl 3-oxobutanoate has been found to have an enol content of about 8% (Table 6.1, but a later reference[57] sets a value of 8.4% for the enol content). Similarly, the enol of 2,4-pentanedione, which would be stabilized by internal hydrogen-bonding, is found to be present in the tautomeric mixture to the level of about 80%.[14,57]

Another consequence of the internal hydrogen-bonding found in these systems is that the enol content is markedly dependent upon the solvent system. Solvents that may also hydrogen-bond with the enol and, so, partly disrupt the internal hydrogen-bonding or, alternatively, solvents that may preferentially stabilize the keto form, should reduce the enol content. This is reflected in the compilation of enol contents (Table 6.3), where, generally, the highest values for ethyl 3-oxobutanoate and 2,4-pentanedione are found in the gas phase and in low-polarity, non-hydrogen-bonding solvents such as tetrachloromethane, whereas the lowest values are found in water.

Enols formed by alkanones that have a high degree of α-fluorine substitution are apparently kinetically persistent.[63] These enols (e.g., equation 6.2) are not particularly stable, but the rate of ketonization found for such enols is slow as a result of the inductive electron-withdrawing effect of the fluorines (but see Section 2.3.3.2, concerning the effect of proximate electronegative substituents on carbanion stability).

$$F_2C{=}C(OH)CF_3 \xrightarrow{200^{o}C\text{ , }3h} HF_2CC({=}O)CF_3 \qquad (6.2)$$

Finally, the renewed interest in sterically hindered enols is noteworthy.[64,65] In these stable enols, the enolic double bond is substituted with at least two bulky and, typically, aryl groups. In honor of Fuson, who first isolated these stable enols (equation 6.3), these are often termed Fuson enols.[66]

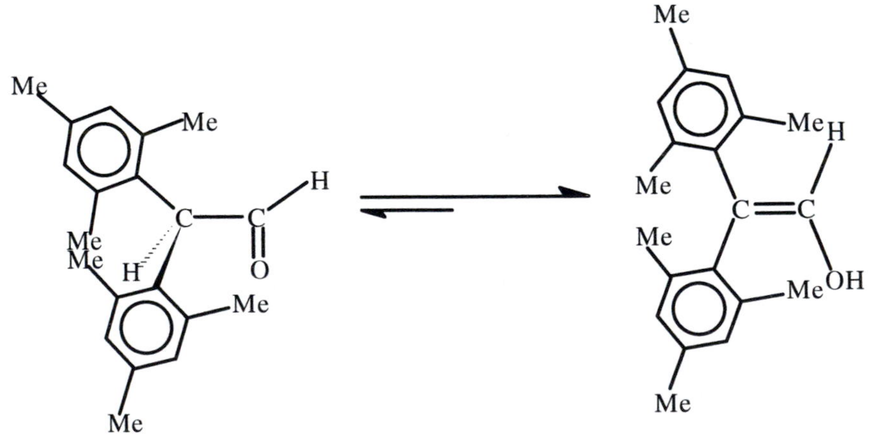

(6.3)

Table 6.3. Enol contents as a function of solvent for two β-dicarbonyl compounds

Compound	Solvent	Enol content (%)	Reference
Ethyl-3-oxobutanoate	Gas phase	53[a]	59
	CC_4	36.8[b]	60
	$(C_2H_5)_2O$	26[c]	61
	CH_3CN	4.9[c]	61
	Water	0.39[d]	62
2,4-Pentanedione	Gas phase	97.6[a]	59
	CCl_4	94.6[e]	23
	$(C_2H_5)_2O$	91.9[e]	23
	CH_3CN	52.9[e]	23
	Water	12.9[e]	23

SOURCES: [a]Data taken from reference 59; [b]Data taken from references 60; [c]Data taken from reference 61; [d]Data taken from reference 62; [e]Data taken from reference 23.

Essentially, in the keto form, the carbon bearing the bulky aryl groups is formally sp^3-hybridized and would be expected to have bond angles (Ar–C–Ar) of approximately 109.5°. By comparison, the enol, where the carbon bearing the aryl groups is sp^2-hybridized, contains less internal strain as a result of placing the bulky groups about 120° apart. In the case of the 2,2-bis(2′, 4′, 6′-trimethylphenyl)-ethenol (shown in equation 6.3), the enol content at equilibrium is 95%, reflecting the stability of the enol form.[5] The study of these sterically hindered enols is an active field and further factors affecting these stable, isolable enols have been detailed.[5,67]

Other factors, including increased delocalization in the enol, alkyl substitution, and other substituent effects on either keto or enol forms, also influence the magnitude of the enolization equilibrium constant. Analysis of these various ingredients is ongoing,[6,68,69] and we shall not consider these, in general terms, further, but will now turn to the mechanisms governing enolization.

6.2.3 Mechanisms of Enolization

Keefe and Kresge have reviewed the various mechanisms of enolization.[44] We will highlight the most common of these and the evidence in support of them.

6.2.3.1 Major Mechanisms

In principle, enolization requires deprotonation of the α-carbon and protonation of the oxygen center to give the enol. As might be expected, then, enolization is subject to general acid and general base catalysis,[70-73] that is, catalysis not only by hydronium ion and hydroxide ion, but also by other Brønsted acids and bases according to their acid or base strengths. In fact, enolization was one of the first reactions shown to be subject to general acid catalysis in aqueous solution.[74]

Later, the Stewart group determined and tabulated the catalytic coefficients for more than 100 general acids and bases in acetone enolization.[75-77]

Early kinetic isotope effect (KIE) studies also highlighted the importance of acid/base properties in the rate-determining transition state for enolization. In this regard, in 1937 Reitz found a primary deuterium isotope effect ($k_H/k_D = 8$) in the acid-catalyzed bromination of propanone as compared with perdeuteriopropanone.[78] Consistent with the observation of general acid and general base catalysis, Reitz and Kopp also reported a primary deuterium isotope effect in the acetate-ion-catalyzed bromination of propanone.[79] Since then, tritium KIEs have also been measured for such reactions, notably by the Jones group for enolization of substituted acetophenones.[80,81] These KIEs (k_H/k_T ca. 20) are again large and consistent with C–H bond breaking occurring in the rate-determining transition state in a system where internal return is not significant (see Section 2.1).

Taken together, the observation of general acid–general base catalysis and primary KIE, in reactions such as α-halogenation of a ketone, indicate that enolization is the usual rate-determining step in such reactions. However, a number of mechanisms can be suggested that fit these criteria. In essence, these mechanisms differ in the sequence of the two proton transfers. In the terminology of Bell,[82] we could term a mechanism *consecutive* if, in proceeding from keto form to enol form, either (1) protonation of the oxygen by acid, HA, and dissociation of the ion pair (i.e., oxonium ion and the conjugate base of the acid) precedes the abstraction of a proton from the α-carbon (equation 6.4a); or (2) deprotonation of the α-carbon center by general base, B:, and dissociation of the enolate ion–conjugate acid pair, occurs prior to protonation of the oxygen (equation 6.4b).

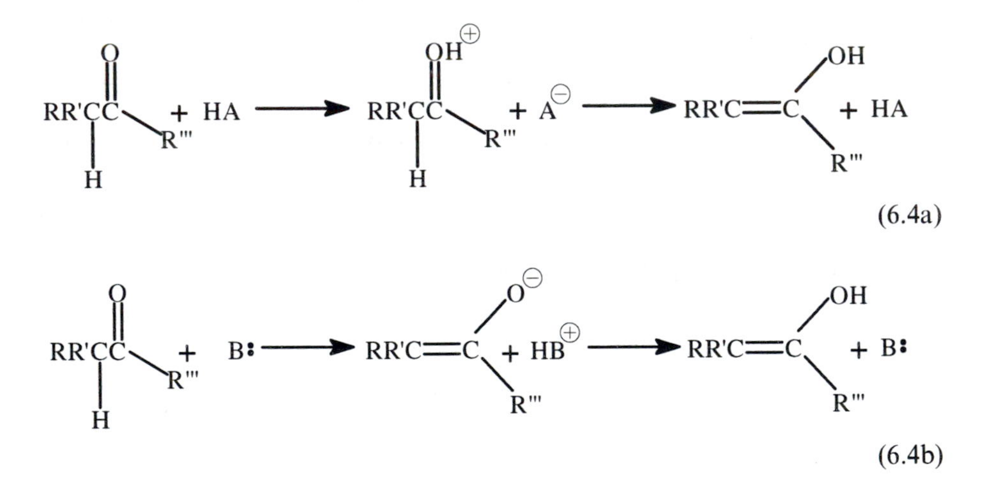

It follows that the transition states of these two consecutive processes contain only the substrate and a single acid or base catalyst molecule. In contrast, in what Bell refers to as a *concerted* process, the transition state will contain the substrate and two catalysts: the one protonating the carbonyl in equation 6.4a, as well as the one deprotonating the carbon acidic center in equation 6.4b (or vice versa).

Issues of which step is rate-determining (i.e., protonation of oxygen vs deprotonation of carbon) arise in consideration of the consecutive-type mechanism, whereas issues of synchronicity (degree of advancement of proton transfer to/from oxygen as compared with transfer to/from carbon) are inherent in consideration of the concerted-type mechanism. Our focus, at this point, will shift to base-catalyzed enolization, that is, the formation of the enolate anion.

(a) Base-Catalyzed Consecutive Mechanisms It is apparent (equation 6.4b) that in base-catalyzed consecutive enolization, deprotonation of the carbon site must occur prior to protonation of the oxygen center. However, the consecutive mechanism can be further subdivided on the basis of the nature of the rate-determining step. If the first step in equation 6.4b is rate-determining and presumably irreversible, then enolization will be governed by those factors that control formation of the enolate carbanion. In this case, protonation of the oxygen to give the enol is fast. Alternatively, deprotonation of the α-carbon could be rapid and reversible, in which case the reaction arrow in the first step in equation 6.4b should be replaced by an equilibrium sign. Here, proton transfer to the oxygen of the enolate to yield the enol becomes the rate-determining step.

It is important to emphasize that either of these alternatives would fit the criterion of general base catalysis, since in both mechanisms the transition state would incorporate a molecule of substrate and one of a general base catalyst. Determination of general base catalysis is, therefore, necessary, but does not distinguish between these two possibilities. At first glance, the observation of a 1° KIE would appear to favor the first mechanistic possibility, where carbon deprotonation is rate limiting. However, it must be pointed out that if the method of obtaining kinetic data in an enolization consists of trapping the enolate ion, perhaps by reaction with an alkylating agent, before the enolate ion either could be protonated (in the slow protonation suggested above) or could reketonize, then the kinetic method would be measuring the forward rate for deprotonation. In this case, a 1° KIE could still be observed, even though the rate-determining step of the mechanism does not involve C–H bond scission. The corollary of this argument is that enolization equilibrium constants (K_e) determined by the kinetic method under acid-catalyzed and base-catalyzed conditions will not be equivalent for any system that follows this second mechanism under alkaline conditions. It is important to note, then, that the systems that have been examined so far by the kinetic approach (*vide supra*) have yielded comparable K_e values determined under either acidic or basic catalysis.[43]

Other evidence, including interpretation of results from Brønsted plots (see Section 2.2) and arguments based on solvent isotope effects,[43] combines with the foregoing to suggest that the base-catalyzed enolization mechanism constituted by rapid and reversible proton transfer involving the carbon center with rate-controlling protonation of the oxygen is rare, at best. Consequently, for a consecutive process as defined by Bell, the most common mechanism, consistent with all facts, requires rate-limiting deprotonation to give the enolate anion, which is then rapidly protonated by the medium to give the enol.

(b) Concerted Mechanisms of Enolization A concerted mechanism for base-catalyzed enolization is something of a misnomer. For an enolization mechanism to be concerted, the rate-determining transition state should contain both a proton

acceptor and a proton donor or, in other words, both a base and an acid. In fact, there is evidence for such a concerted mechanism, as a minor contributor to the overall kinetics of enolization of propanone (acetone) in aqueous ethanoic (acetic) acid–ethanoate (acetate) ion buffer solutions.[83-85] Here, the observation of a third-order kinetic term (i.e., first-order dependence on the concentrations of keto form, acid form of the buffer, and base form of the buffer) implies the presence of all three species in the rate-limiting transition state and, therefore, evidence for the concerted mechanism. A third-order kinetic term has also been reported for the enolization of cyclohexanone in ethanoic acid–ethanoate buffers.[86]

While a role for a concerted component in enolization in these systems seems to be accepted, the nature of the transition state is still open to interpretation.[83,85,87,88] However, it should be emphasized that the concerted mechanism only appears to count to a minor degree in the enolization process. The major pathway, certainly in alkaline systems, is the consecutive mechanism with proton abstraction as the slow, rate-limiting step.

In water as solvent, it may be argued that water could play a role in enolization (as acid or base or both) and that this role could be obscured by the high concentration of water present.[44,89,90] Thus, a concerted mechanism may not be discernible from the kinetic form of the overall rate law for enolization. Capon's group, largely on the basis of comparison of rates of vinyl ether hydrolysis with those for ketonization,[91-93] have revived the concept of a concerted mechanism as a major route in enolization, even in systems where a third-order kinetic term is not present. This proposal is still controversial[44] and we look forward to future developments, especially given the obvious importance of the role of solvent in such concerted mechanisms.

In the preceding sections, we have considered the mechanisms for enolization, particularly those involving formation of the enolate anion under basic conditions. We have seen that the kinetic determination of enolization equilibrium constants can be combined with measured (or estimated) acid dissociation constants (pK_a values) for the enols to obtain acidities of carbon acids. In the next section, we will consider transition-state structures in some enolate reactions.

6.3 ENOLATE REACTIONS

Claisen, Claisen–Schmidt, Darzens, Dieckmann, Perkin, and Knoevenagel condensations: the number of "name" reactions in organic chemistry that are based upon enolates, as the central reactive intermediates, is large and growing. Many of these have been found useful, particularly in some stereoselective syntheses.

6.3.1 Enolate Alkylation

Alkylation of an enolate forms an important step in a number of well-known C–C bond-forming syntheses, including such classics as the malonic ester synthesis and the acetoacetate synthesis.[94] In these particular syntheses, the enolate may even be readily formed by such weak bases as potassium carbonate in aprotic solvents such as DMSO as a result of the relatively low pK_a of the carbon acid (pK_a ca.

9–11).[95] Reaction of the enolate as a nucleophile with an electrophilic site, typically of a *primary* haloalkane such as iodomethane or an alkyl arenesulfonate such as methyl tosylate, leads to alkylation; the alkylation steps are emphasized in the generalized Scheme 6.4 for ethyl acetoacetate (ethyl 3-oxobutanoate). In Scheme 6.4, the resonance-delocalized enolate ion, **66**, is formed by abstraction of the acidic proton (highlighted in the scheme) using ethoxide as base. (The counterion,

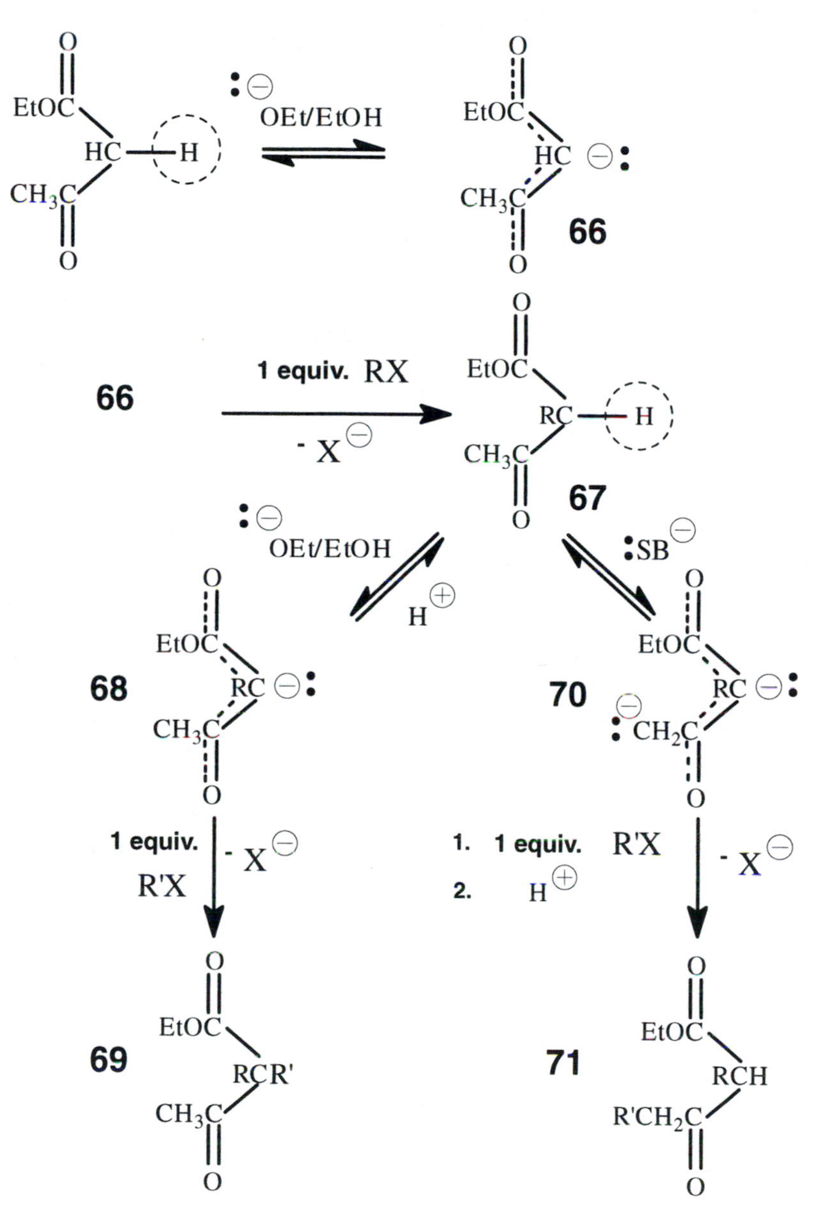

SCHEME 6.4

while important, has been omitted for the sake of simplicity.) Reaction with one equivalent of a general alkylating agent, R–X, leads to the initial product, **67**. For steric reasons, in practice, only alkylation with primary alkylating agents such as iodoethane, allyl chloride, or benzyl bromide is feasible.

However, Scheme 6.4 also illustrates some of the difficulties (or *opportunities*) inherent in enolate alkylation. First, the enolate **66**, of course, is ambident. In principle, alkylation may occur either at the former carbonyl oxygens or at the carbanion carbon. The factors that affect the site of alkylation have been discussed previously (Section 1.2.4) and will be discussed further below with respect to the *relative reactivity* of various aggregated lithium carbanides. Nonetheless, it should be emphasized that the nature of the counterion and solvent, and, hence, the degree and type of ion-pairing[96] or aggregation of the carbanion, are particularly important.[97,98] The second point is illustrated in Scheme 6.4 by the equilibrium between the first alkylation product **67** and the delocalized enolate ion **68** that arises from deprotonation of **67**. The pK_a value of the acidic proton of **67** highlighted in Scheme 6.4 is not very different from that of the starting material, ethyl 3-oxobutanoate, and, therefore, deprotonation of **67** by ethoxide base may occur readily under the reaction conditions. Addition of a new alkylating agent, R′X in Scheme 6.4, leads to product **69**. On the other hand, if the reaction is undertaken with two equivalents of the initial alkylating agent, then polyalkylation occurs. Finally, treatment of **67** with a stronger base, $^{-}$:SB, (or treatment of the initial ethyl 3-oxobutanoate with two equivalents of a strong base, such as LDA or *n*-butyllithium) results in formation of the dianion **70**. Alkylation of this dianion with R′X followed by acidic workup leads to **71**, where alkylation has occurred preferentially at the more basic carbanionic center. The selectivity illustrated for the conversion of **67** to **71** in Scheme 6.4 has been exploited successfully in a large number of cases.[99,100]

The acetoacetate synthesis proper continues from either alkylation products **67** or **69** with saponification of the ester moiety, followed by decarboxylation under acidic conditions to give alkanones as the final products.[94] The same decarboxylation step is also utilized in the malonic ester synthesis.[94] The mechanism of this decarboxylation has been reviewed and the interested reader is directed to the review for details.[101]

Alkylation at the carbon site is strongly influenced by the nature and degree of ion pairing/aggregation, which, in turn, is dependent upon the solvent and the counterion to the carbanion.[102-104] Again, using the example of alkylation of the sodium carbanide derived from ethyl 3-oxobutanoate, it has been found that the rate of alkylation is enhanced 15-fold when crown ether is added to the system; a 235-fold rate enhancement is noted when cryptand is introduced into the system. However, the amount of O-alkylation also increases significantly upon addition of cation-sequestering agents.[105] These results suggest that, at least in this system, the free (or approximately free) carbanion is the most reactive intermediate in this alkylation.[102,105] (Consideration of the most reactive species in alkylation will be made further below.)

Alkylation of carbanions derived from simple saturated ketones leads to a new set of challenges for the synthetic organic chemist. These have been discussed in detail in the volume edited by Augustine[106,107] and will only be briefly outlined here. Formation of such carbanions in low equilibrium concentration—for example, by deprotonation of the alkanone by metal alkoxides in the parent alcoholic solvent—results in condensation effectively competing with alkylation. Use of a strong base, such as potassium *t*-butoxide or, even more effectively, lithium diisopropylamide (LDA), can give quantitative formation of the enolate and, therefore, avoids competitive condensation reactions. In some cases, the reactive enolate may be isolated as a salt;[108] these enolates are referred to as "preformed".[109-111]

However, if the saturated alkanone is unsymmetrical, issues of regioselectivity (beyond O- vs C-alkylation) arise and mixtures of isomeric mono- and polyalkylation products typically result.[112] An oft-cited example is that provided by 2-methylcyclopentanone. In the study by House and Trost,[113] formation of the carbanion, **73**, from deprotonation of the least-substituted α-carbon is favored (72%) when 2-methylcyclopentanone, **72**, is added slowly to an excess of triphenylmethyllithium in DME; **73** is the product of kinetic control and alkylation of this ion leads to the kinetic alkylation product, **75** (Scheme 6.5, using iodoethane as the hypothetical alkylating agent). On the other hand, when excess 2-methylcyclopentanone is present in the system, equilibration occurs and the thermodynamically more stable enolate, **74**, predominates (94%);[113] alkylation of **74**, with iodoethane in this example, would give rise to the product of thermodynamic control, **76** (Scheme 6.5). Generally, in systems similar to that shown in Scheme 6.5, when the counterion for the enolate is potassium or sodium, equilibration of the enolates occurs more readily and monoalkylation would preferentially yield the thermodynamic product. Equilibration is also favored in protic solvents, notably in alkanols.[107] Where an equilibrium mixture is established, the more stable enolate may usually be predicted on the basis of known alkene stabilities; for example, enolates arising from deprotonation of the most highly substituted α-C are the most stable and *Z*-stereoisomeric enolates are more stable than their *E*-counterparts.[107] These conclusions are supported by the extensive studies of House and co-workers,[113-115] among others,[116,117] into the mixtures of enolates present in such systems under both kinetic and equilibrium conditions.

While the foregoing considerations give the synthetic organic chemist some control over the regioselectivity in the course of enolate C-alkylation reactions, these considerations are largely based on the view that the enolates involved are discrete species. However, in nonpolar (often ethereal) solvents, enolates are well known to be aggregated.[97,118,119] These pioneering studies of Zook and co-workers have been followed by the NMR studies of the Jackman group[104,120,121] among others.[122] The question currently being considered is: What is the role of these species—from monomer through tetrameric structure,[123,124] to aggregates, and including mixed aggregates—in the reactivity, as well as regio- and stereoselectivity, found in the C–C bond-forming reactions of enolates?[125-128]

In tetrahydrofuran solution, the lithium carbanide of 1-phenyl-2-methylpropanone (isobutyrophenone), **77**, appears to exist predominantly as tetramers, on the basis of NMR studies,[129,130] and the ratio of O- versus C-alkylation products was attributed to the intervention primarily of these tetramers[104] though some contribution from dimers was suggested subsequently.[121] Note that the structure of

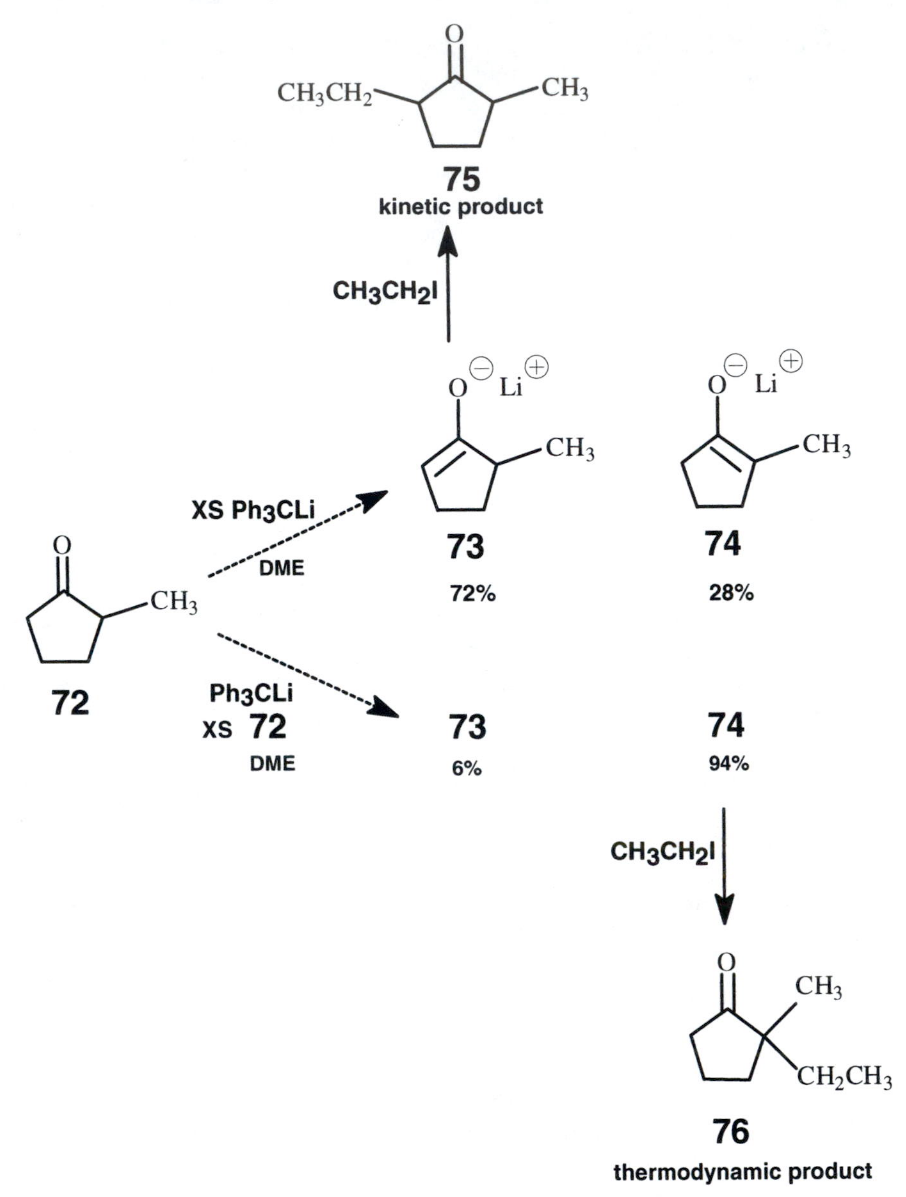

SCHEME 6.5

isobutyrophenone precludes formation of regioisomeric enolates. The Streitwieser group examined the concentration-dependent behavior of the UV spectra of enolates in THF and extracted both equilibrium constants for aggregation[131-133] and the kinetics for O- and C-alkylation by each species (e.g., monomer, dimer, etc.).[134-136] In the case of the cesium enolate of 1-biphenylyl-2-methylpropane (*p*-phenylisobutyrophenone) in THF, Streitwieser and co-workers[135] found that the enolate exists as a mixture of monomer, dimer, and tetramer; the equilibrium

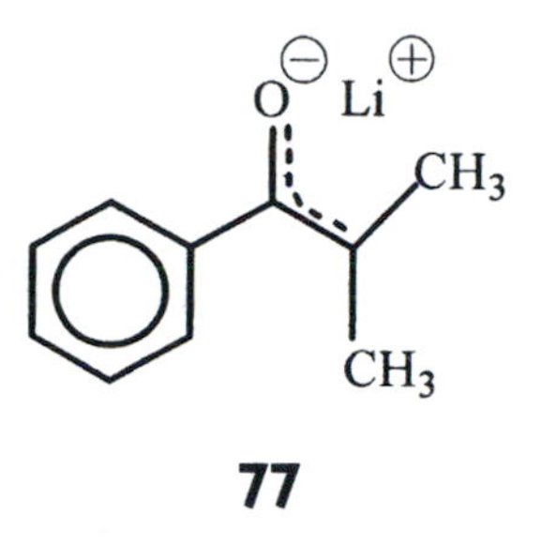

77

between monomer and dimer has an equilibrium constant of $2.89 \times 10^4\ M^{-1}$, whereas the equilibrium between monomer and tetramer has a significantly larger equilibrium constant of $7.78 \times 10^{12}\ M^{-3}$. However, it is the monomeric species that is the most reactive toward alkylation with methyl tosylate or *p-tert*-butylbenzyl chloride (25 °C, THF).[135] Interestingly, the cesium carbanide system displays a higher degree of aggregation than the corresponding lithium enolate system,[136] but, again, the most reactive species was shown to be the monomeric lithium carbanide ion pair. Similar results have also been reported by the Streitweiser group for alkylation of the lithium enolate of *p*-phenylsulfonylisobutyrophenone even where the aggregate was of the mixed kind (i.e., LiBr was also present in the THF solution).[132,133]

Returning to the cesium 1-biphenylyl-2-methylpropanide/THF system, only 3% of the alkylation products with *p-tert*-butylbenzyl chloride arose from O-attack, whereas yields of O- and C-alkylation products were almost equal (C:O ratio = 1.2) when methyl tosylate was used as the alkylating agent.[135] These results are consistent with the earlier work of Zook and co-workers[97] and can be explained by considering six-member transition states (TS), as suggested previously by Brändstrom.[137] As shown in Scheme 6.6, the only six-membered TS, **78**, that is feasible with the ion-paired enolate (shown as a simplified and generalized structure) and a haloalkane leads to exclusive C-alkylation, whereas two different TSs, **79** and **80**, may be proposed for reaction of the same ion-paired enolate with an alkyl sulfonate. Note that in both **79** and **80**, it is possible for the counterion of the enolate to be coordinated simultaneously to two oxygen centers. Since some O-alkylation does occur, clearly the TS for a haloalkane that corresponds to **80** is significantly higher in energy than the TS **78** to **80**.

Alkylation of enolates is a broad topic and we have not even touched upon the regioselectivity inherent in alkylation of unsaturated enolates (i.e., 1,2 vs 1,4 or Michael addition). Suffice it to say that in these systems, as well, the nature and degree of ion pairing and aggregation are also important, and further understanding of the regioselectivity involved will arise from consideration of the transition state structures[138-140] as outlined above for saturated enolates. Alkylation of allenyl enolates has also been reviewed elsewhere[141] and will not be discussed further here.

In the next section, we will consider the aldol condensation, which we have seen can compete effectively with alkylation under simple reaction conditions.

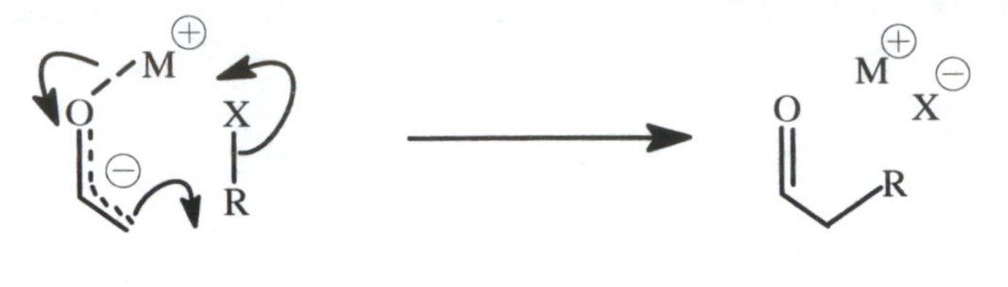

78

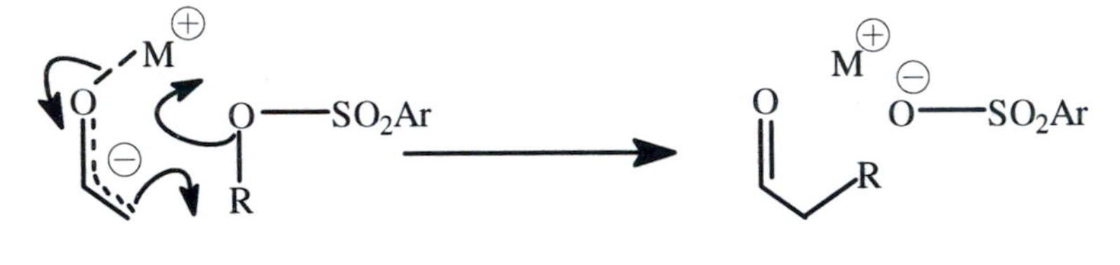

79

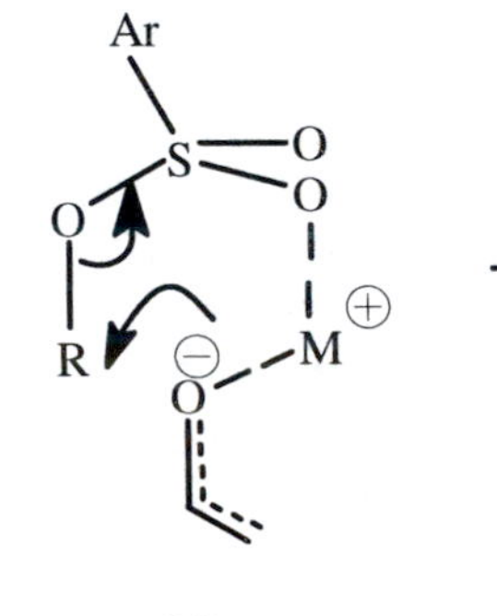

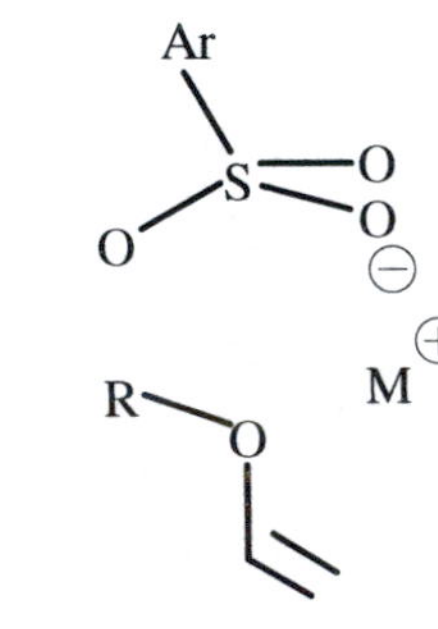

80

SCHEME 6.6

6.3.2 The Aldol Condensation

The aldol condensation, particularly the stereoselective aldol, has been shown to be an important reaction in the synthesis of carbon–carbon bonds. [8–10,142,143] As in other portions of this work, our information comes from using the tools of the physical organic chemist: product analysis, stereochemistry, kinetic isotope effects, substituent effects, and the insights gleaned from calculational studies.

In the aldol reaction[144,145] (and its related reactions) a C–C bond is formed between the carbon α to a carbonyl and the carbonyl carbon of another molecule of the same[146] or another compound. As an enolate reaction, the initial step is equilibrium formation of a low concentration of the enolate by deprotonation of the α-carbon, usually by hydroxide ion (or, in the case of ketones, by a stronger base, such as aluminum *tert*-butoxide). The reaction takes its name from the product. If, in equation 6.5, R‴ is H, then the product contains both an *ald*ehyde function and an alkan*ol* function and, hence, the name *aldol*.

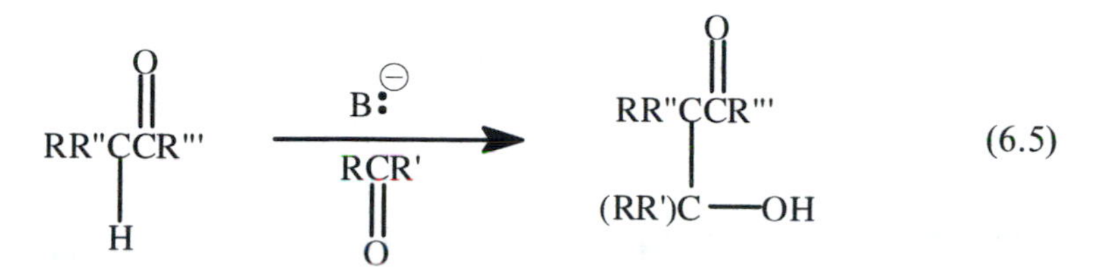

When R″ in the product is H, elimination may occur concurrently with alkylation and the product isolated is the β-ketoalkene. Each step in the reaction is reversible and, therefore, a retro-aldol reaction is also known, in which α, β-unsaturated ketones and aldehydes, as well as aldols, cleave under the action of base.

Choice of reaction conditions in clearly important. Significant effort has been expended to improve our understanding of the requirements necessary for stereoselectivity in the aldol condensation. It is not our intention to review in detail the many fine studies of stereoselectivity in these reactions. In fact, these have already been critically reviewed elsewhere.[144,147,148] Generally, modern aldol syntheses use preformed metal enolates in aprotic solvents and the products are formed under conditions of kinetic control, that is, short reaction times at low temperature. Our focus, then, will be on the transition states that have been advanced to explain the stereochemical results found in these reactions.

Studies in the laboratories of Dubois,[149-152] Heathcock,[147] Evans,[153] and House[154] have shown that under conditions of kinetic control, *E* enolates primarily yield aldols that have an *anti* conformation and, conversely, *Z* enolates react to give *syn* β-hydroxycarbonyl compounds, as shown in equations 6.6a and 6.6b.

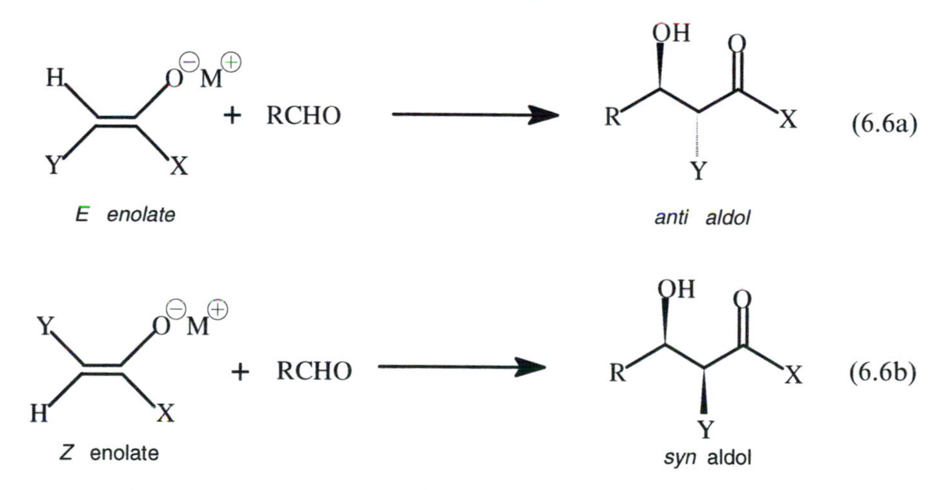

A few corroborating results for addition of *E*- and *Z*-lithium enolates to benzaldehyde are listed in Table 6.4. (Y = CH_3, X given in table).

As can be seen from Table 6.4 (as well as other examples),[147,148,156] *syn* aldols tend to form preferentially, notably when there is a bulky α-group in the enolate.

Table 6.4. Ratio of diastereomeric aldol products formed by addition of lithium ketone enolates to benzaldehyde

X group	Enolate stereochemistry (*E*:*Z*)	Aldol stereochemistry (*anti*:*syn*)
Ethyl	34:66	23:77
1-Adamantyl	2:98	2:98
Phenyl	2:98	12:88
Mesityl	13:87	12:88

SOURCE: Data taken from reference 155

It has also been found that even higher diastereoselectivities can be achieved with boron or zirconium enolates, namely, enolates whose counterion is zirconium or a boron-centered group. (For a review of stereoselectivity in reactions of zinc and boron enolates, see reference 157.) However, as Braun has commented[158] in a review of stereoselectivity in the aldol condensation: "The *Z-syn, E-anti* correlation appears to be a rule with many exceptions."

Nonetheless, the results that support the *Z-syn* rule have been rationalized on the basis of six-membered transition states originally proposed by Zimmerman and Traxler for the Ivanov condensation of phenylethanoic acid with benzaldehyde[159] but adapted by Dubois.[149-152] It must be recognized at the outset that transition states which include only a single lithium counterion may be an oversimplification; the propensity for lithium enolates to form aggregates is well known (see Chapter 4). On the other hand, to leave the counterion out of the transition state would be even more unreasonable. By way of justification, we have seen above in our consideration of alkylation of enolates that in a number of systems where higher-order aggregates are favored (e.g., tetramers), the monomeric ion pairs are the more reactive species.

The Zimmerman–Dubois transition states (Figure 6.1) have six-membered pseudo-cyclohexane structures. By analogy to substituted cyclohexanes, then, the transition state that places the bulky substituent of the aldehyde, R, in an equatorial-type position should be energetically preferred. In fact, for the *Z*-enolate, this transition state leads to formation of the *syn* aldol. Since the transition state leading to the *syn* aldol is stabilized relative to the transition

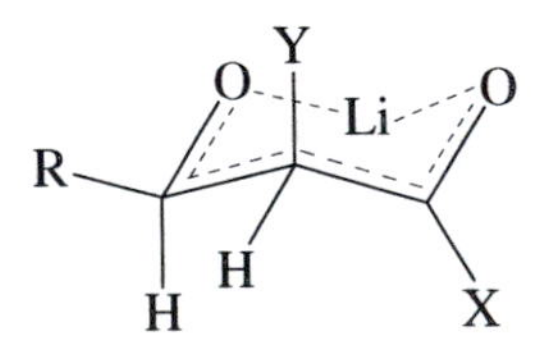

Figure 6.1. Zimmerman–Dubois transition state for preferred *syn* aldol formation.[147,148] In this rendering, the charges have been omitted for clarity; X and Y are general substituents.

state for *anti* aldol formation, the *syn* aldol will be formed more readily and will predominate under conditions of kinetic control. A similar argument may be made for the preferential reaction of the *E*-enolate to give the *anti* aldol. However, it has been pointed out by Heathcock that in the case of *E*-enolates any such chair-like transition state may entail unreasonably high steric congestion.[147]

It should be pointed out that, in general, calculational studies of the aldol condensation involving lithium enolates have predicted chair or half-chair transition-state structures in which the lithium is coordinated by two oxygens.[160,161]

An apparent weakness of the chairlike transition-state approach is that the model fails to explain the generally higher stereoselectivity exhibited by *Z*-enolates as compared with their *E*-stereoisomers. It has been argued, partly on these grounds, that boat- or skew-type transition states could be preferred in some systems.[162] Thus, a twist-boat transition state could be formed readily from the U-conformation of the enolate (equation 6.7), while a W-form would lead to the chair-type, transition state (equation 6.8).[163]

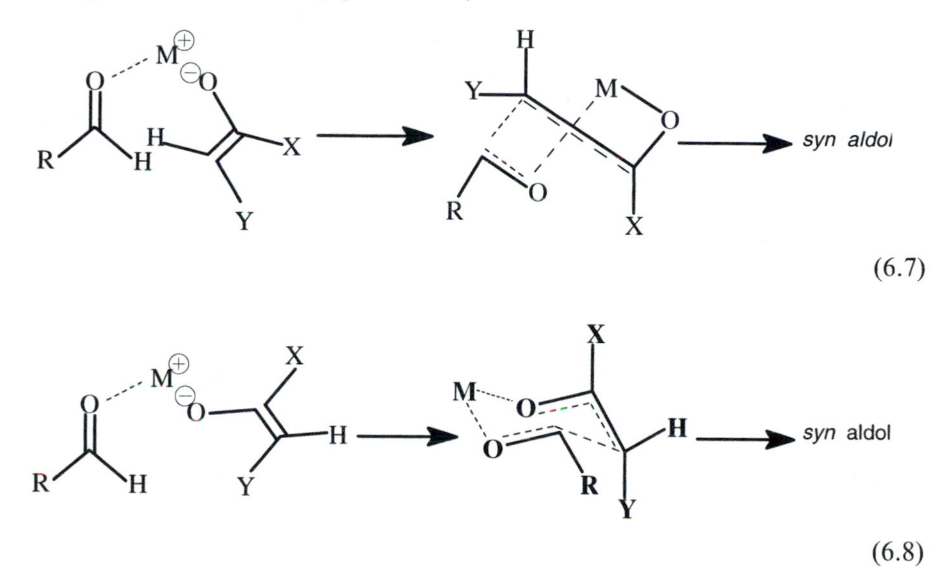

With *Z*-enolates, then, the chairlike transition state (equation 6.8) leads directly to the *syn* aldol. When bulky substituents are present on the metal, M (as is frequently the case with boron enolates), reaction of the *E*-enolate proceeds through the twist-boat transition state, shown in equation 6.8, leading again to the *syn* aldol. In the absence of bulky substituents on the chelating metal, competitive formation of the *anti* aldol occurs for the *E*-enolate, via a Zimmerman–Dubois chairlike transition state.

The Streitwieser group have examined the Aldol–Tishchenko reaction[164] of several lithium enolates with benzaldehyde.[165] In this modification of the aldol reaction, the initially formed aldol alkoxide reacts with another aldehyde (which generally lacks α-hydrogens) to give an ester, as shown in Scheme 6.7.

The final intramolecular hydride transfer step was determined to be rate limiting. The observed hydrogen–deuterium KIE_{ob} ($k_H/k_D = 2.1$) was found to be

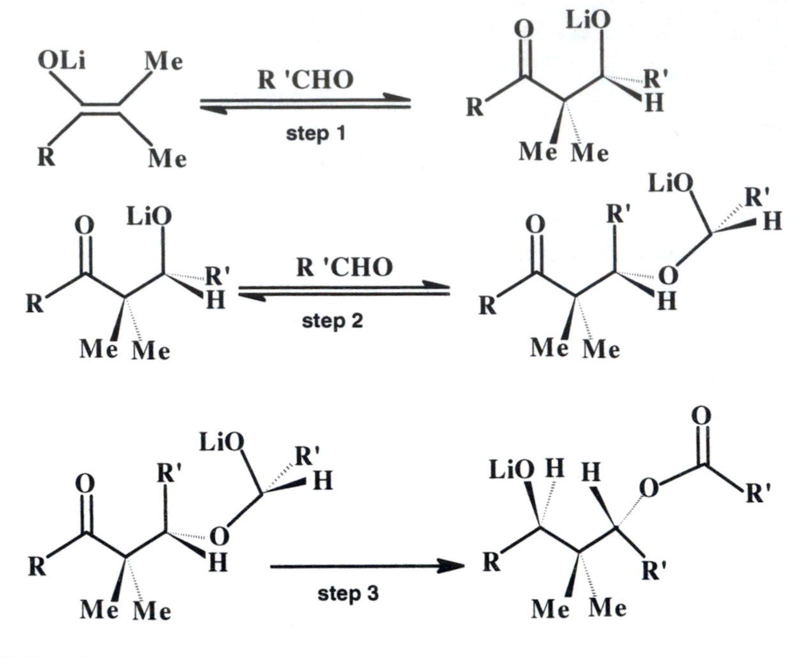

SCHEME 6.7

within the normal range for a primary effect, *where the equilibrium isotope effects (EIE) from the first two steps of the mechanism also contribute to the overall kinetic isotope effect*; that is, KIE_{ob} = KIE (step 3) × EIE (step 1) × EIE (step 2; Scheme 6.7). Interestingly, ab initio calculations nicely reproduced the observed kinetic isotope effect and predicted a six-membered chairlike transition state for the rate-determining hydride transfer step. Note that the stereochemistry of the Aldol–Tishchenko reaction is dictated by the irreversible hydride transfer step; here, the more stable transition state, which places all bulky groups equatorial, would lead to preferential formation of an *anti* product.

Clearly, the stereochemistry of the aldol reaction can be rationalized on the basis of six-membered transition states, and their stability can be analyzed by comparison with substituted cyclohexane conformers. Other possible transition states that explain specific stereochemical consequences in various enolate reactions have been reviewed by Heathcock.[147] Whether the current transition-state models will also accommodate more exotic enolates, notably those with counterions such as nickel and palladium,[166] is still open to question. However, the need for transition-state models that incorporate the counterion (or organometallic species) is apparent, given current work into the use of organolithium reagents and the interesting regio-, diastereo-, and enantioselectivity found in these systems.[167,168] There is also apparent need for transition-state models that can accommodate the effects of solvent on regioselectivity in carbanion reactions.

6.4 SELECTED ENOLATE REARRANGEMENTS

6.4.1 Coverage

In this section, we will concentrate our discussion on the Favorskii rearrangement and a few other enolate rearrangements. We recognize that enolates may undergo a range of rearrangements in structure, including homoenolate rearrangements, and these have been treated in some detail elsewhere.[169] Rearrangements of other types of carbanions are covered in Chapter 7. Therefore, in this chapter we will focus on the Favorskii rearrangement and a few other selected enolate rearrangements.

6.4.1.1 Favorskii Rearrangement

The reaction of bases with alkanones that are halogenated in the α-position results in the formation of carboxylic acids, esters, or amides (depending upon the base employed), and typically also includes structural reorganization of the carbon skeleton (equation 6.9). This is known as the Favorskii rearrangment.[170] With polyhalogenated ketones, the reaction yields unsaturated carboxylic acid derivatives.

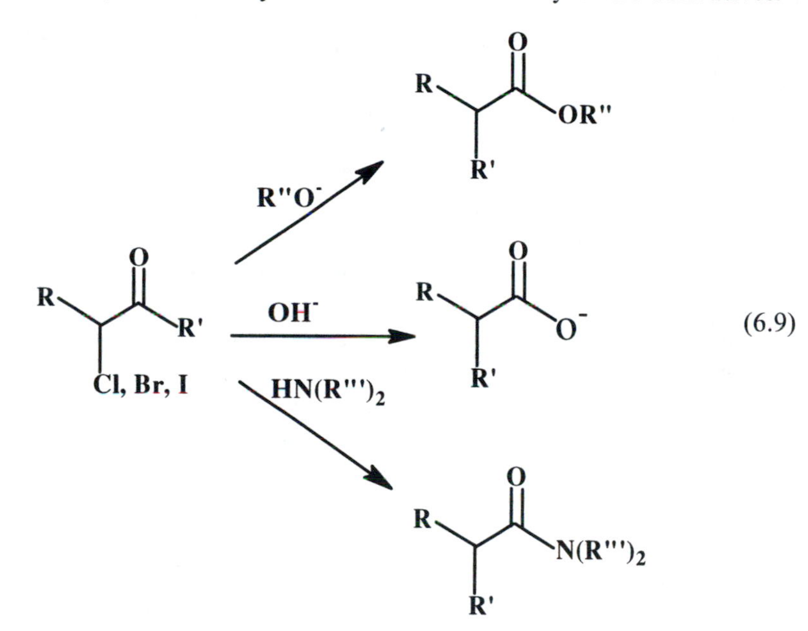

(6.9)

A review focuses on the synthetic utility of this rearrangement.[171] We shall highlight the mechanistic aspects of this rearrangement. In this regard, a number of possible mechanisms have been advanced for the Favorskii rearrangement.[172-174] Two main mechanistic pathways have emerged as the most likely. The first is termed the symmetrical mechanism or the cyclopropanone route, so named for the central cyclopropanone intermediate (Scheme 6.8, where chlorine is arbitarily shown as the halogen). The second is an unsymmetrical mechanism that may be termed the semibenzilic pathway.

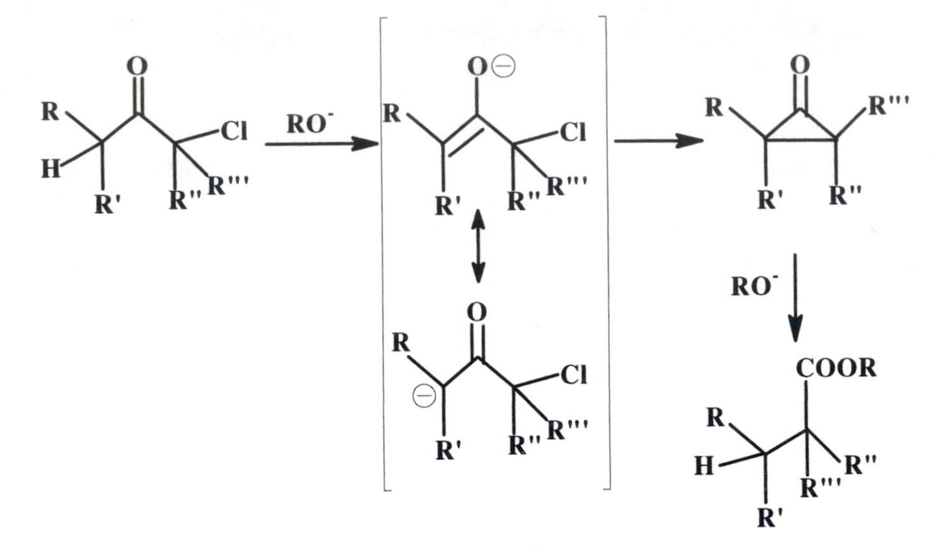

SCHEME 6.8

The first step in the cyclopropanone mechanism is formation of an enolate by base abstraction from the α-haloalkanone. An intramolecular S_N2 displacement results in formation of the cyclopropanone intermediate, which then ring opens to the rearrangement product. (Only one isomer is shown in Scheme 6.8.)

Evidence for the cyclopropanone mechanism comes from observation that both 1-bromo-3-methyl-2-butanone and 3-bromo-3-methyl-2-butanone react with sodium methoxide to give methyl 2,2-dimethylpropanoate, implying a common symmetrical intermediate: 2,2-dimethylcyclopropanone.[175] Significantly, 2,2-dimethylcyclopropanone was also prepared independently, and when treated with sodium methoxide it reacted to give the same methyl 2,2-dimethylpropanoate.[176]

Further support comes from the work of Loftfield.[177] Reaction of 2-chlorocyclohexanone, labeled with ^{14}C at the 1- and 2-positions, with sodium alkoxides yielded two ring-contracted products. As shown in equation 6.10, reaction of the labeled chlorocyclohexane (where the labeled carbons are denoted by asterisks) gave the cyclopentane ethyl ester product in which the carboxyl carbon contained 50% of the radioactive label and the *alpha* and *beta* carbons contained the remainder, equally divided between them as would be expected if a symmetrical intermediate were involved in the mechanism.

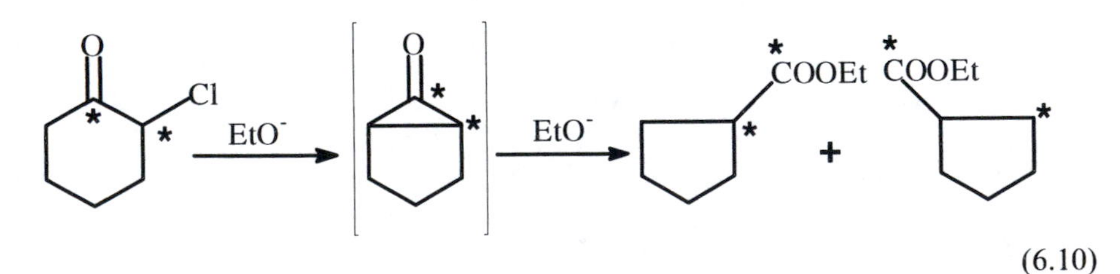

(6.10)

Thus, the location of the ^{14}C labels could be rationalized on the basis of a bicyclic intermediate containing the cyclopropanone moiety.[177] Partial reaction yielded starting material in which the label was undisturbed, indicating that no halogen migration had occurred prior to formation of the bicyclic intermediate.

It should also be noted that α, α′-dibromoketones undergo 1,3-elimination when treated with base. 1,3-Dibromo-1,3-diphenylpropanone (α, α′-dibromodibenzylketone) reacts with triethylamine to ultimately yield diphenylcyclopropenone as the stable product. Presumably, the 2-bromo-2,3-diphenylcyclopropanone, formed by initial 1,3-elimination, is an intermediate in the reaction sequence.[178]

Evidence has accumulated[179] to show that the cyclopropanone intermediate is likely in equilibrium with a zwitterionic intermediate, as shown in Scheme 6.9. In fact, in one system, products of both the cyclopropanone and the zwitterion were found, indicative of the competition between the two intermediates, as outlined in Scheme 6.9.[180] Product analysis in the reaction of sodium methoxide (in MeOH) with 1-chloro-1,1-diphenylpropanone and the isomeric 1-chloro-3,3-diphenylpropanone has also been taken to support the intermediacy of a dipolar intermediate in equilibrium with the cyclopropanone.[181] As may be expected, the formation of the zwitterionic intermediate is likely favored in hydroxylic and other ionizing solvents.

The alternative semibenzilic pathway[182] involves nucleophilic attack by the base, RO^-, on the carbonyl carbon of the haloalkanone, **81**; a tetrahedral alkoxide intermediate, **82**, results. An alkyl shift then occurs in tandem with loss of the halide, X, from **82** to yield the product, as shown in Scheme 6.10. The mechanism is termed *semibenzilic* because of its similarity to the benzilic acid

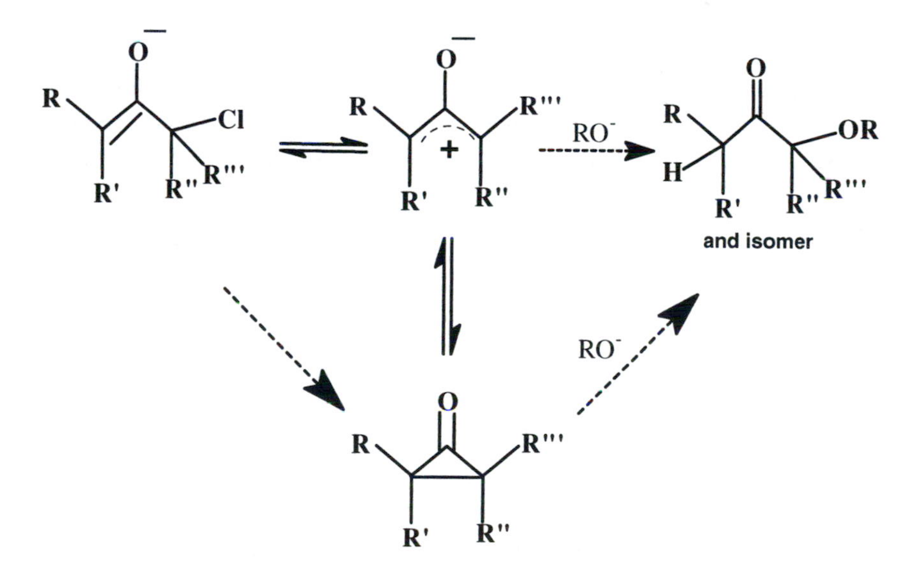

SCHEME 6.9

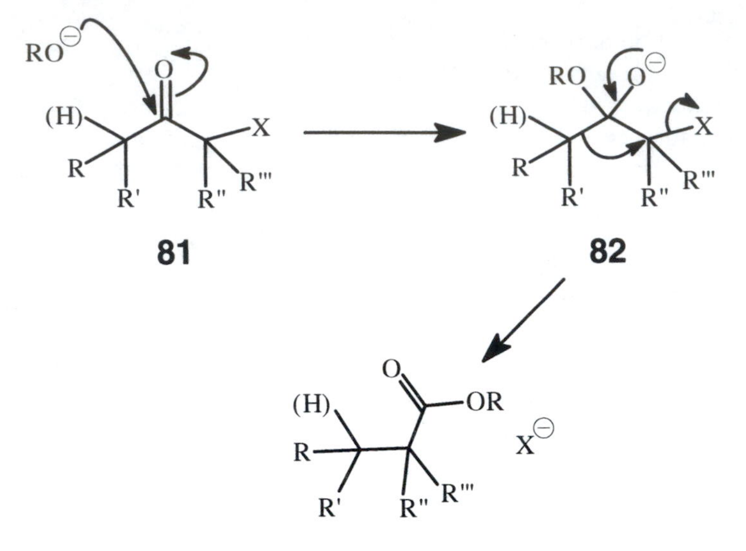

SCHEME 6.10

rearrangement (see Section 7.4.3). Such a mechanism requires inversion at the carbon center bearing the halogen, and in some systems this has, in fact, been reported.[183] It may be expected that in strained ring systems the cyclopropanone route, which would introduce further ring strain, may be disfavored and in this case the semibenzilic pathway may become dominant.[184] Finally, since the cyclopropanone mechanism requires abstraction of a proton from the α-position (i.e., H in parentheses in structure **81**, Scheme 6.10) to the activating carbonyl group, it might be expected that where these positions lack α-protons the semibenzilic mechanism would obtain. Alkanones of this type have been shown to undergo this variation on the Favorskii rearrangement.[185]

Other mechanisms[186,187] suggested for the Favorskii rearrangement have been covered in detail in the monograph by Buncel[188] and in the review by Hunter, Stothers, and Warnhoff.[189] In the next chapter we will examine carbanion rearrangements, other than enolate rearrangements, in some detail. Included in Chapter 7 is a discussion of the Ramberg–Bäcklund rearrangement of α-halogenosulfones. Some of the features of the mechanism of this rearrangement are quite similar to the cyclopropanone pathway found in certain of the Favorskii rearrangements.

6.4.1.2 Base-Promoted Rearrangements of α-Epoxy Ketones

Reaction of an α, β-unsaturated ketone with reagents such as peroxyalkanoic acids, including *m*-chloroperoxybenzoic acid (MCPBA), yields an α-epoxy ketone that can undergo rearrangement when treated with base (equation 6.11).

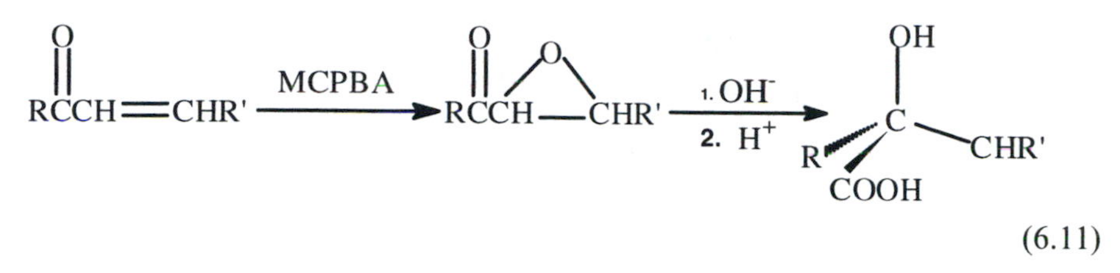

(6.11)

Clearly, equation 6.11 represents not only another example of a rearrangement that presumably involves an enolate ion, but it is also synthetically useful.[190]

The mechanistic pathway probably involves initial proton abstraction from the epoxy ketone to form the carbanion, followed by ring-opening to give the α-acyl enolate, as shown in Scheme 6.11. Reprotonation of the acyl enolate yields a benzil derivative. The remaining steps of the mechanism that convert the benzil to benzilic acid constitute the benzilic acid rearrangement.[191] These steps are discussed in Section 7.4.3.

6.4.1.3 α-Halo Amide Rrearrangements

By analogy with the Favorskii rearrangement, a number of a-halo amides react with base to give Favorskii-type rearrangment products.[192,193] The mechanism presumably proceeds via formation of the three-membered lactam or aziridone that is analogous to the cyclopropanone intermediate in the Favorskii rearrangement, as shown in Scheme 6.12 (R = *t*-butyl; R′OH = *t-butyl alcohol*).[192]

Evidence in support of the mechanism include isolation of the aziridone intermediate. Thus, reaction of *N*-*t*-butylphenylacetamide with *t*-butyl hypochlorite and potassium *t*-butoxide in toluene yielded the aziridone, when an excess of *t*-butoxide was maintained.[193] Other routes to α-lactams have since been reported.[194,195] The dichotomy of products that arise from ring-opening nucleophilic attack on the aziridone (as shown in Scheme 6.12) have been confirmed in

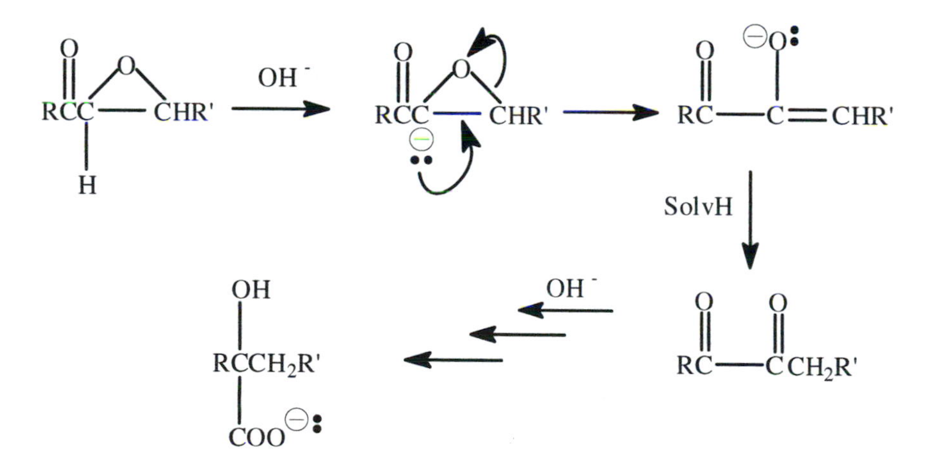

SCHEME 6.11

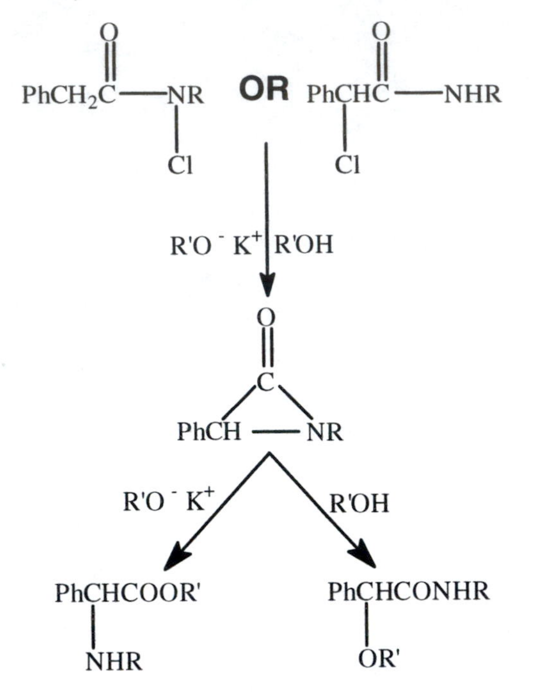

SCHEME 6.12

experiments on the isolated aziridones.[194-196] For example, 1-*t*-butyl-3,3-dimethylaziridone reacts with ionic nucleophiles solely by attack at the carbonyl center (acyl–nitrogen bond cleavage), while uncharged nucleophiles attack at the α-C (alkyl–nitrogen bond cleavage). However, the direction of the ring-opening reactions is sensitive to the structure of the substituents on the aziridone ring. Consequently, 1-*n*-propyl-3,3-dimethylaziridone reacts with *t*-butoxide to give products from both ring-opening pathways, while 1-*t*-butyl-3-methylaziridone reacts with the same nucleophile almost entirely by alkyl-nitrogen bond cleavage.[194-196]

In these rearrangement reactions, as in the Favorskii rearrangement, the key step is formation of the enolate anion. In the next chapter, our focus will be on rearrangements of carbanions more generally.

References

1. Buncel, E. *Carbanions: Mechanistic and Isotopic Aspects*; Elsevier: Amsterdam, 1975.
2. Toullec, J. *Adv. Phys. Org. Chem.* **1982**, *18*, 1.
3. Kresge, A.J. *Chemtech* **1986**, *16*, 250.
4. Capon, B.; Guo, B.X.; Kwok, F.C.; Siddhanta, A.K.; Zucco, C. *Acc. Chem. Res.* **1988**, *21*, 135.
5. Rappoport, Z.; Biali, S.E. *Acc. Chem. Res.* **1988**, *21*, 442.
6. Rappoport, Z., Ed.; *The Chemistry of Enols*; Wiley: Chichester, U.K., 1990.

7. Fraser, R.R. In *Comprehensive Carbanion Chemistry. Part B. Selectivity in Carbon–Carbon Bond Forming Reactions*; Buncel, E.; Durst, T., Eds.; Elsevier: Amsterdam, 1984; Chapter 2, pp 65–107.
8. Caine, D. In *Comprehensive Organic Synthesis. Selectivity, Strategy and Efficiency in Modern Organic Chemistry: Volume 3. Carbon–Carbon σ-Bond Formation*; Trost, B.M.; Fleming, I.; Pattenden, G.; Eds.; Pergamon Press: Oxford, U.K., 1991; Chapter 1.1, pp 1–65.
9. Heathcock, C.H. In *Asymmetric Synthesis, Volume 3*; Morrison, J.D., Ed.; Academic Press: New York, 1984.
10. Evans, D.A.; Nelson, J.V.; Taber, T.R. *Top. Stereochem.* **1982**, *13*, 1.
11. Knorr, L.; Rothe, O.; Averbeck, H. *Chem. Ber.* **1911**, *44*, 1138.
12. Meyer, K.H. *Justus Liebigs Ann. Chem.* **1911**, *380*, 212.
13. Meyer, K.H. *Chem. Ber.* **1914**, *47*, 826.
14. Schreck, R. *J. Am. Chem. Soc.* **1949**, *71*, 1881.
15. Gero, A. *J. Org. Chem.* **1954**, *19*, 1960.
16. Bell, R.P.; Smith, P.W. *J. Chem. Soc. B* **1966**, 241.
17. Bergon, M.; Calmon, J.P. *Bull. Soc. Chim. Fr.* **1972**, 1819.
18. Murthy, A.S.N.; Balasubramanian, A.; Rao, C.N.R.; Kasturi, T.R. *Can. J. Chem.* **1962**, *40*, 2267.
19. Mills, S.G.; Beak, P. *J. Org. Chem.* **1985**, *50*, 1216.
20. Rhoads, S.J.; Pryde, C. *J. Org. Chem.* **1965**, *30*, 3212.
21. Naoum, M.M.; Saad, G.R. *Indian. J. Chem.* **1987**, *26A*, 510.
22. Kulp, S.; Schmoyer, D.E.; Freeze, D.E.; Buzas, J. *J. Org. Chem.* **1975**, *40*, 453.
23. Emsley, J.; Freeman, N.J. *J. Mol. Struct.* **1987**, *161*, 193.
24. Allinger, N.L.; Chow, L.W.; Ford, R.A. *J. Org. Chem.* **1967**, *32*, 1994.
25. Reeves, L.W. *Can. J. Chem.* **1957**, *35*, 1351.
26. Ahlbrecht, A.; Funk, W.; Reiner, M.T. *Tetrahedron* **1976**, *32*, 479.
27. Bassetti, M.; Cerichelli, G.; Floris, B. *Tetrahedron* **1988**, *44*, 2997.
28. Olah, G.A.; Grant, J.L.; Westerman, P.E. *J. Org. Chem.* **1975**, *40*, 2102.
29. Lapachev, V.V.; Mainagashev, I.Ya.; Stekhova, S.A.; Fedotov, M.A.; Krivopalov, V.P.; Mamaev, V.P. *J. Chem. Soc., Chem. Commun.* **1985**, 494.
30. Kallury, K.R.; Krull, U.J.; Thompson, M. *J. Org. Chem.* **1988**, *53*, 1320.
31. Gorodetsky, M.; Luz, Z.; Mazur, Y. *J. Am. Chem. Soc.* **1967**, *89*, 1183.
32. Leffler, A.J.; Luz, Z.; Zimmerman, H. *J. Chem. Res., Synop.* **1991**, 262.
33. Geraldes, C.F.G.C.; Barros, M.T.; Maycock, C.D.; Silver, M.I. *J. Mol. Struct.* **1990**, *238*, 335.
34. Schwarzenbach, G.; Witwer, C. *Helv. Chim. Acta* **1947**, *30*, 659.
35. Schwarzenbach, G.; Witwer, C. *Helv. Chim. Acta* **1947**, *30*, 663.
36. Schwarzenbach, G.; Witwer, C. *Helv. Chim. Acta* **1947**, *30*, 669.
37. Domonkos, L.; Ratkovics, F. *Monatsh. Chem.* **1988**, *119*, 177.
38. Gero, A. *J. Org. Chem.* **1961**, *26*, 3156.
39. J. Toullec In *The Chemistry of Enols*; Rappoport, Z., Ed.; Wiley: Chichester, U.K., 1990; Chapter 6, pp 323–398.
40. Alcais, P.; Brouillard, R. *J. Chem. Soc., Perkin Trans. 2* **1972**, 1214.
41. Bakulev, V.A.; Chiang, Y.; Kresge, A.J.; Meng, Q.; Moreherin, Y.Y.; Popik, V.V. *J. Am. Chem. Soc.* **2001**, *123*, 2681.
42. Chiang, Y.; Hojatti, M.; Keeffe, J.R.; Kresge, A.J.; Schepp, N.P., Wirz, J. *J. Am. Chem. Soc.* **1987**, *109*, 4000.
43. Kresge, A.J. *Acc. Chem. Res.* **1990**, *23*, 43.
44. Keeffe, J.R.; Kresge, A.J. In *The Chemistry of Enols*; Rappoport, Z., Ed.; Wiley: Chichester, U.K., 1990; Chapter 7, pp 399–480.

45. Cox, R.A.; Smith, C.R.; Yates, K. *Can. J. Chem.* **1979**, *57*, 2952.
46. McTigue, P.T.; Sime, J.M. *Aust. J. Chem.* **1967**, *20*, 905.
47. Capon, B.; Rycroft, D.S.; Watson, T.W.; Zucco, C. *J. Am. Chem. Soc.* **1981**, *103*, 1761.
48. Guthrie, J.P. In *The Chemistry of Enols*; Rappoport, Z., Ed.; Wiley: Chichester, U.K., 1990; Chapter 2, pp 75–95.
49. Guthrie, J.P. *Can. J. Chem.* **1979**, *57*, 1177.
50. Apeloig, Y. In *The Chemistry of Enols*; Rappoport, Z., Ed.; Wiley: Chichester, U.K,. 1990; Chapter 1, pp 1–75.
51. Keeffe, J.R.; Kresge, A.J.; Schepp, N.P. *J. Am. Chem. Soc.* **1988**, *110*, 1993.
52. Kresge, A.J.; Schepp, N.P. *J. Chem. Soc. Chem. Commun.* **1989**, 1548.
53. Keeffe, J.R.; Kresge, A.J.; Toullec, J. *Can. J. Chem.* **1986**, *64*, 2470.
54. Chiang, Y.; Kresge, A.J.; Schepp, N.P. *J. Am. Chem. Soc.* **1989**, *111*, 3977.
55. Werstiuk, N.H. *Can. J. Chem.* **1988**, *66*, 2958.
56. Werstiuk, N.H.; Andrew, D. *Can. J. Chem.* **1990**, *68*, 1467,
57. Wong, W.W.; Wiberg, K.B.; Frisch, M.J. *J. Am. Chem. Soc.* **1992**, *114*, 1645.
58. Moriyasu, M.; Kato, A.; Hasimoto, Y. *J. Chem. Soc., Perkin Trans. 2* **1986**, 515.
59. Folkendt, M.M.; Weiss-Lopez, B.E.; Chauvel, J.P.; True, N.S. *J. Phys. Chem.* **1985**, *89*, 3347.
60. Bánkowska, Z.; Zadrożna, I. *Rocz. Chem.* **1968**, *42*, 1591.
61. Rogers, M.I.; Burdett, J.L. *Can. J. Chem.* **1965**, *43*, 1516.
62. Schwarzenbach, G.; Felder, E. *Helv. Chim. Acta* **1944**, *27*, 1701.
63. Bekker, R.A.; Knunyants, I.L. *Sov. Sci. Rev., Sect. B* **1984**, *5*, 145.
64. Nadler, E.B.; Rappoport, Z. *J. Am. Chem. Soc.* **1987**, *109*, 2112.
65. O'Neill, P.; Hegarty, A.F. *J. Chem. Soc. Chem. Commun.* **1987**, 744.
66. Fuson, R.C.; Southwick, P.L.; Rowland, B.I., *J. Am. Chem. Soc.* **1944**, *66*, 1109.
67. Hart, H.; Rappoport, Z.; Biali, S.E. In *The Chemistry of Enols*; Rappoport, Z., Ed.; Wiley: Chichester, U.K., 1990; Chapter 8, pp 481–591.
68. Pratt, D.V.; Hopkins, P.B. *J. Am. Chem. Soc.* **1987**, *109*, 5553.
69. Nadler, E.B.; Rappoport, Z.; Arad, D.; Apeloig, Y. *J. Am. Chem. Soc.* **1987**, *109*, 7873.
70. Bell, R.P. *Acid–Base Catalysis*; Oxford University Press: Oxford, U.K., 1941.
71. Jencks, W.P. *Catalysis in Chemistry and Enzymology*; McGraw-Hill: New York, 1969.
72. Jencks, W.P. *Chem. Rev.* **1972**, *72*, 705.
73. Gupta, K.S.; Gupta, Y.K. *J. Chem. Educ.* **1984**, *61*, 972.
74. Dawson, H.M.; Powis, F. *J. Chem. Soc.* **1913**, 2135.
75. Shelly, K.P.; Nagarajan, K.; Stewart, R. *Can. J. Chem.* **1987**, *65*, 1734.
76. Shelly, K.P.; Venimadhavan, S.; Nagarajan, K.; Stewart, R. *Can. J. Chem.* **1989**, *67*, 1274.
77. Venimadhavan, S.; Shelly, K.P.; Stewart, R. *J. Org. Chem.* **1989**, *54*, 2483.
78. Reitz, O. *Z. Phys. Chem A.* **1937**, *179*, 119.
79. Reitz, O.; Kopp, J. *Z. Phys. Chem. A* **1939**, *184*, 429.
80. Jones, J.R.; Marks, R.E.; Subba-Rao, S.C. *Trans. Faraday Soc.* **1967**, *63*, 111.
81. Jones, J.R.; Marks, R.E.; Subba-Rao, S.C. *Trans. Faraday Soc.* **1967**, *63*, 993.
82. Bell, R.P. *The Proton in Chemistry*, 2nd ed.; Cornell University Press: Ithaca, N.Y, 1973; pp 149–154.
83. Dawson, H.M.; Spivey, E. *J. Chem. Soc.* **1930**, 2180.
84. Bell, R.P.; Jones, P. *J. Chem. Soc. B* **1953**, 88.
85. Hegarty, A.F.; Jencks, W.P. *J. Am. Chem. Soc.* **1975**, *97*, 7188.
86. Hand, E.; Jencks, W.P. *J. Am. Chem. Soc.* **1975**, *97*, 6221.
87. Albery, W.J. *J. Chem. Soc., Faraday Trans. 1* **1982**, *78*, 1579.

88. Eliason, R.; Kreevoy, M.M. *J. Am. Chem. Soc.* **1978**, *100*, 7037.
89. Lowry, T.M. *J. Chem. Soc.* **1927**, 2254.
90. Swain, C.G.; Stivers, E.C.; Reuwer, J.F.; Schaad, L.J. *J. Am. Chem. Soc.* **1958**, *80*, 5885.
91. Capon, B.; Siddhanta, A.K. *J. Org. Chem.* **1984**, *49*, 255.
92. Capon, B.; Siddhanta, A.K.; Zucco, C. *J. Org. Chem.* **1985**, *50*, 3580.
93. Capon, B.; Zucco, C. *J. Am. Chem. Soc.* **1982**, *104*, 7564.
94. Fox, M.A.; Whitesell, J.K. *Organic Chemistry*, 2nd ed.; Jones & Bartlett: Sudbury, MA, 1997; pp 686–690.
95. Vaz de Araujo, A.C.; Vasconcelos de Almeida, F.; Bieber, L.W. *Quim. Nova* **1996**, *19*, 79; C.A. 124: 260343q.
96. Sun, X.; Collum, D.B. J. *Am. Chem. Soc.* **2000**, *122*, 2452; and references therein.
97. Zook, H.D.; Russo, T.J.; Ferrand, E.F.; Stotz, D.S. *J. Org. Chem.* **1968**, *33*, 2222.
98. LeNoble, W.J.; Morris, H.F. *J. Org. Chem.* **1969**, *34*, 1969.
99. Hauser, C.R.; Harris, T.M. *Org. React.* **1969**, *17*, 155; and references therein.
100. Kaiser, E.M.; Petty, J.D.; Knutson, P.L.A. *Synthesis* **1977**, 509.
101. Clark, L.W. In *The Chemistry of Carboxylic Acids and Esters*; Patai, S., Ed.; Wiley: New York, 1969, Chapter 12, p 589-620.
102. DePalma, V.M.; Arnett, E.M. *J. Am. Chem. Soc.* **1985**, *107*, 2091; and references therein.
103. Jackman, L.M.; Dunne, T.S. *J. Am. Chem. Soc.* **1985**, *107*, 2805.
104. Jackman, L.M.; Lange, B.C. *J. Am. Chem. Soc.* **1981**, *103*, 4494.
105. Cambillau, C.; Sarthou, P.; Bram, G. *Tetrahedron Lett.* **1976**, 281.
106. Augustine, R.L., Ed.; *Carbon–Carbon Bond Formation, Vol. 1*; Marcel Dekker: New York, 1979.
107. Caine, D. In *Carbon–Carbon Bond Formation, Vol. 1*; Augustine, R.L., Ed.; Marcel Dekker: New York, 1979; pp 86–326.
108. Kondo, Y.; Yano, K.; Urade, M.; Yamada, A.; Kouyama, R.; Takagi, T. *J. Chem. Soc., Perkin Trans 2* **1999**, 1181.
109. Stork, G.; D'Angelo, J. *J. Am. Chem. Soc.* **1974**, *96*, 7114.
110. Stork, G.; Kraus, G.A.; Garcia, G.A. *J. Org. Chem .* **1974**, *39*, 3460.
111. Auerbach, R.A.; Crumrine, D.S.; Ellison, D.L.; House, H.O. *Org. Synth.* **1975**, *54*, 49.
112. Conia, J.M. *Rec. Chem. Progr.* **1963**, *24*, 43.
113. House, H.O.; Trost, B.M. *J. Org. Chem.* **1965**, *30*, 1341 and 4395.
114. House, H.O.; Kramar, V. *J. Org. Chem.* **1963**, *28*, 3372.
115. House, H.O.; Czuba, L.J.; Gall, M.; Olmstead, H.D. *J. Org. Chem.* **1969**, *34*, 2324.
116. Stork, G.; Hudrlik, P.F. *J. Am. Chem. Soc.* **1968**, *90*, 4462.
117. Brown, C.A. *J. Org. Chem.* **1974**, *39*, 3913.
118. Zook, H.D.; Rellahan, W.L. *J. Am. Chem. Soc.* **1957**, *79*, 881.
119. Zook, H.D.; Russo, R.J. *J. Am. Chem. Soc.* **1960**, *82*, 1258.
120. Jackman, L.M.; Lange, B.C. *Tetrahedron* **1977**, *33*, 2737.
121. Jackman, L.M.; Dunne, T.S. *J. Am. Chem. Soc.* **1985**, *107*, 2805.
122. Sun, X.; Collum, D.B. *J. Am. Chem. Soc.* **2000**, *122*, 2459.
123. Amstutz, R.; Schweizer, W.B.; Seebach, D.; Dunitz, J.D. *Helv. Chim. Acta* **1981**, *64*, 2617.
124. Seebach, D.; Amstutz, R.; Dunitz, J.D. *Helv. Chim. Acta* **1981**, *64*, 2622.
125. Arnett, E.M.; Palmer, C.A. *J. Am. Chem. Soc.* **1990**, *112*, 7354.
126. Palmer, C.A.; Ogle, C.A.; Arnett, E.M. *J. Am. Chem. Soc.* **1992**, *114*, 5619.
127. Williard, P.G.; MacEwan, G.J. *J. Am. Chem. Soc.* **1989**, *111*, 7671.

128. Juaristi, E.; Beck, A.K.; Hansen, J.; Matt, T.; Mukhopadhyay, T.; Simson, M.; Seebach, D. *Synthesis* **1993**, 1271.
129. Jackman, L.M.; Haddon, R.C. *J. Am. Chem. Soc.* **1973**, *95*, 3687.
130. Jackman, L.M.; Szeverenyi, N. M. *J. Am. Chem. Soc.* **1977**, *99*, 4954.
131. Abbotto, A.; Streitwieser, A. *J. Am. Chem. Soc.* **1995**, *117*, 6358 and references therein.
132. Abu-Hasanayn, F.; Stratakis, M.; Streitwieser, A. *J. Org. Chem.* **1995**, *60*, 4688.
133. Abu-Hasanayn, F.; Streitwieser, A. *J. Am. Chem. Soc.* **1996**, *118*, 3186.
134. Krom, J.A.; Streitwieser, A. *J. Am. Chem. Soc.* **1992**, *114*, 8747.
135. Streitwieser, A.; Krom, J.A.; Kilway, K.V.; Abbotto, A. *J. Am. Chem. Soc.* **1998**, *120*, 10801.
136. Abbotto, A.; Leung, S.S.-W.; Streitwieser, A.; Kilway, K.V. *J. Am. Chem. Soc.* **1998**, *120*, 10807.
137. Brändstrom, A. *Ark. Kemi* **1953**, *6*, 155.
138. Reich, H.J.; Sikorski, W.H. *J. Org. Chem.* **1999**, *64*, 14.
139. Reich, H.J.; Sikorski, W.H.; Gudmundsson, B.O.; Dykstra, R.R. *J. Am. Chem. Soc.* **1998**, *120*, 4035.
140. Cohen, T.; Abraham, W.D.; Myers, M. *J. Am. Chem. Soc.* **1987**, *109*, 7923.
141. Fredrick, M.A.; Hulce, M. *Tetrahedron* **1997**, *53*, 10197.
142. Mateos, A.F.; Angel de la Fuente Blanco, J. *J. Org. Chem.* **1991**, *56*, 7084.
143. Masamune, S.; Choy, W.; Petersen, J.S.; Sita, R.L. *Angew. Chem., Int. Ed. Engl.* **1985**, *24*, 1.
144. Nielsen, A.T.; Houlihan, W.J. *Org. React.* **1968**, *16*, 1.
145. March, J. *Advanced Organic Chemistry*, 4th ed. Wiley: New York, 1992; pp 937–944.
146. Guthrie, J.P.; Wang, X.P. *Can. J. Chem.* **1991**, *69*, 339.
147. Heathcock, C.H. In *Comprehensive Carbanion Chemistry. Part B. Selectivity in Carbon–Carbon Bond Forming Reactions*; Elsevier: Amsterdam, 1984; pp 177–237.
148. Braun, M. In *Advances in Carbanion Chemistry. Vol. 3*; Snieckus, V., Ed.; JAI Press: Greenwich, CT, 1992; pp 177–247.
149. Dubois, J.E.; Dubois, M. *Tetrahedron Lett.* **1967**, 4215.
150. Dubois, J.E.; Dubois, M. *Bull. Soc. Chim. Fr.* **1969**, 3553.
151. Dubois, J.E.; Fellman, P. *C. R. Acad. Sci., Ser. C* **1972**, *274*, 1307.
152. Dubois, J.E.; Fellman, P. *Tetrahedron Lett.* **1976**, 1225.
153. Evans, D.A.; Takacs, J.M.; McGee, L.R.; Ennis, M.D.; Mathre, D.J.; Bartoli, J. *Pure Appl. Chem.* **1981**, *53*, 1109.
154. House, H.O.; Crumrine, D.S.; Teranishi, A.Y., Olmstead, H.D. *J. Am. Chem. Soc.* **1973**, *95*, 3310.
155. Heathcock, C.H.; Buse, C.T.; Kleschick, W.A.; Pirrung, M.C.; Sohn, J.E.; Lampe, J. *J. Org. Chem.* **1980**, *45*, 1066.
156. Majewsh, M.; Gleave, D.M. *Tetrahedron Lett.* **1989**, *42*, 5681.
157. Tagliavini, E.; Trombini, C.; Umani-Ronchi, A. In *Advances in Carbanion Chemistry. Vol. 2*; Snieckus, V., Ed.; JAI Press: Greenwich, CT, 1996; pp 111–146.
158. Braun, M. In *Advances in Carbanion Chemistry. Vol. 3.*; Snieckus, V., Ed.; JAI Press: Greenwich, CT, 1992; p 185.
159. Zimmerman, H.E.; Traxler, M. *J. Am. Chem. Soc.* **1957**, *79*, 1920.
160. Paddon-Row, M.N.; Rondan, N.G.; Houk, K.N. *J. Am. Chem. Soc.* **1982**, *104*, 7162.
161. Li, Y.; Paddon-Row, M.N.; Houk, K.N. *J. Org. Chem.* **1990**, *55*, 481.
162. Gennan, C.; Todeschini, R.; Beretta, M.G.; Favini, G.; Scholastico, C. *J. Org. Chem.* **1986**, *51*, 612.
163. Hoffmann, R.W.; Ditrich, K.; Froech, S.; Cremer, D. *Tetrahedron* **1985**, *41*, 5517.
164. Bodnar, P.M.; Shaw, J.T.; Woerpel, K.A. *J. Org. Chem.* **1997**, *62*, 5674.

165. Abu-Hasanayn, F.; Streitwieser, A. *J. Org. Chem.* **1998**, *63*, 2954.
166. Brukhardt, E.; Bergman, R.; Heathcock, C.H. *Organometallics* **1990**, *9*, 30.
167. Beak, P.; Basu, A.; Gallagher, D.J.; Park, Y.S.; Thayumanavan, S. *Acc. Chem. Res.* **1996**, *29*, 552.
168. Beak, P.; Meyers, A.I. *Acc. Chem. Res.* **1986**, *19*, 356.
169. Buncel, E. *Carbanions: Mechanistic and Isotopic Aspects*; Elsevier: Amsterdam, 1975; pp 116–138.
170. Favorskii, A. *J. Russ. Phys. Chem. Soc.* **1894**, *26*, 559.
171. Mann, J. In *Comprehensive Organic Synthesis: Selectivity, Strategy and Efficiency in Modern Organic Chemistry. Volume 3. Carbon–Carbon σ-Bond Formation*; Trost, B.M.; Fleming, I.; Pattenden, G.; Eds.; Pergamon Press: Oxford, U.K., 1991; Chapter 3.7, pp 839–859.
172. Kende, A.S. *Org. React.* **1960**, *11*, 261.
173. Turro, N.J. *Acc. Chem. Res.* **1969**, *2*, 25.
174. Bordwell, F.G. *Acc. Chem. Res.* **1970**, *3*, 281.
175. Turro, N.J.; Gagostian, R.B.; Rappe, C.; Knutsson, L. *J. Chem. Soc., Chem. Commun.* **1969**, 270.
176. Turro, N.J.; Hammond, W.B. *J. Am. Chem. Soc.* **1965**, *87*, 3528.
177. Loftfield, R.B. *J. Am. Chem. Soc.* **1951**, *73*, 4707.
178. Breslow, R.; Eicher, T.; Krebs, A.; Peterson, R.A.; Posner, J. *J. Am. Chem. Soc.* **1965**, *87*, 1320.
179. Baretta, A.; Waegell, B. In *Reactive Intermediates, Vol. 2*; Abramovitch, R.A., Ed.; Plenum Press: New York, 1982; p 527.
180. Finch, M.W.; Mann, J.; Wilde, P.D. *Tetrahedron Lett.* **1987**, *43*, 5431.
181. Bordwell, F.G.; Scamehorn, R.G. *J. Am. Chem. Soc.* **1971**, *93*, 3410.
182. Tchoubar, B.; Sackur, O. *C. R. Acad. Sci.* **1939**, *208*, 1020.
183. Baudry, D.; Begue, J.P.; Charpentier-Morize, M. *Bull. Soc. Chim. Fr.* **1971**, 1416.
184. Conia, J.M.; Salaum, J.R. *Bull. Soc. Chim. Fr.* **1961**, 1957.
185. Smissman, E.E.; Hite, G. *J. Am. Chem. Soc.* **1959**, *81*, 1201.
186. Warnhoff, E.W.; Wong, C.M.; Tai, W.T. *J. Am. Chem. Soc.* **1968**, *90*, 514.
187. Fong, W.C.; Thomas, R.; Scherer, K.V. Jr., *Tetrahedron Lett.* **1971**, 3789.
188. Buncel, E. *Carbanions: Mechanistic and Isotopic Aspects*; Elsevier: Amsterdam, 1975; pp 144–155.
189. Hunter, D.H.; Stothers, J.B.; Warnhoff, E.W. In *Rearrangements in Ground and Excited States*; de Mayo, P., Ed.; Academic Press: New York, 1980; pp 437–461.
190. Treves, G.R.; Stange, H.; Olofson, R.A. *J. Am. Chem. Soc.* **1967**, *89*, 6257.
191. Selman, S.; Eastham, J.F. *Q. Rev.* **1961**, *14*, 221.
192. Baumgarten, H.E.; Zey, R.L.; Krolls, U. *J. Am. Chem. Soc.* **1961**, *83*, 4469
193. Baumgarten, H.E. *J. Am. Chem. Soc.* **1962**, *84*, 4975.
194. Sheehan, J.C.; Beeson, J.H. *J. Am. Chem. Soc.* **1967**, *89*, 362.
195. Sheehan, J.C.; Beeson, J.H. *J. Am. Chem. Soc.* **1967**, *89*, 366.
196. Lengyel, I.; Sheehan, J.C. *Angew. Chem., Int. Ed. Engl.* **1967**, *7*, 25.

7

Carbanion Rearrangements

7.1 INTRODUCTION AND SCOPE

In a given reaction that proceeds by an established mechanism, the experienced chemist may reasonably expect to be able to predict the products. In fact, this predictive ability is a quality that we seek in our studies. However, in many systems, products may not accord with simple predictions. This is typically true when the product arises from the rearrangement of a reactive intermediate. On the other hand, observation of products that derive from rearrangement of an intermediate are, in themselves, diagnostic of the presence of these intermediates. The standard example in this case is the acid-catalyzed dehydration of a secondary alkanol where rearrangement products are often taken to indicate the intermediacy of carbocations.[1] As a corollary, the stereochemistry of rearrangement of double bonds on basic zeolites and other solid basic catalysts has been taken to imply the lack of intermediacy of carbocations and, instead, the intervention of carbanions. This begs the question: What is meant by rearrangement of carbanions?

One definition of carbanion rearrangement is a process that involves isomerization of the carbanion. The isomerization may be concerted, that is, bond formation and bond scission processes occur simultaneously (but not necessarily synchronously), or it may involve closely associated pairs of intermediates, for example, hydrogen-bonded carbanion ion pairs. The isomerization may result in conversion of functional groups, as in the Wittig rearrangement[2] where the carbanion of a benzylic ether undergoes a formal [1,2] shift of an alkyl, aryl, or allyl moiety and, so, transforms from a benzylic ether into an alkanol. Alternatively, the isomerization may be of the skeletal variety, as in a double-bond migration in an alkene or conjugated system.[3,4] Finally, there are related reactions that proceed via initial carbanionic rearrangement, but the final products arise from fragmentation of the rearranged intermediate, often with expulsion of a stable molecule. An example is offered by the Ramberg–Bäcklund rearrangement, where the carbanion of an α-halosulfone isomerizes, and yields a penultimate intermediate that expels sulfur dioxide to give the alkene product (but also see the related Favorskii rearrangement in Chapter 6).

As we have seen, in solution, carbanides exist in a diversity of environments from almost free carbanions to the more commonly found dimeric, tetrameric, and higher-order aggregates, depending on the nature of the medium, the counterion, temperature, and steric or other structural constraints at the anionic center.[5-7] In common with earlier reviews in aspects of this topic,[8-10] we will consider current understanding of rearrangements of these species and, where known, we will highlight the effect of carbanide structure in solution on the course of these isomerizations. However, we will take a more limited view of carbanionic rearrangements than that advocated by some authors.[10] Hence, we will not consider the conversion of a dimeric form of an alkyllithium into a tetramer as a rearrangement for purposes of discussion in this chapter, nor will we consider equilibration of *trans* alkenes and *cis* alkenes under basic conditions.

Reaction conditions can be the key determinant in the course of rearrangement. For example, when an alkanol α-substituted by a trialkyl- or triaryl-silyl group is treated with a *catalytic* amount of a strong base (RLi, Na–K, NaH, etc.), it undergoes a Brook rearrangement[11-13] to the isomeric alkoxysilane as shown in equation 7.1.

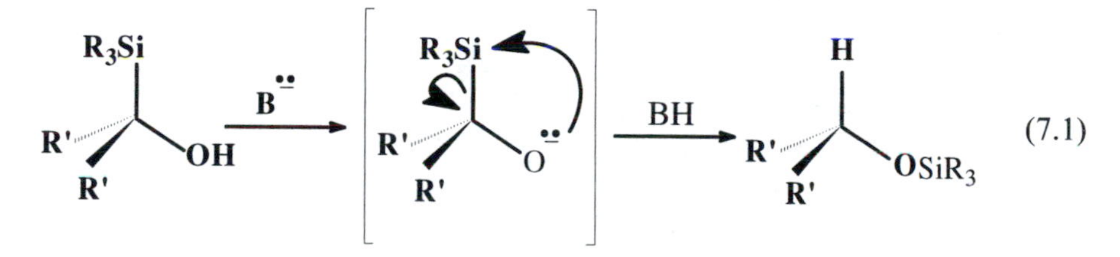

The Brook rearrangement, therefore, is an example of a [1,2] shift of an organosilyl group from carbon to an oxygen center. Conversely, a [1,2] migration of a trialkylsilyl group from oxygen to carbon occurs in the retro-Book (or silyl-Wittig or Wright–West) rearrangement[14-16] of a benzyloxysilane to a α-trialkylsilylalkanol, when the benzyloxysilane is treated with *excess* strong base (typically LDA or *t*-butyllithium):

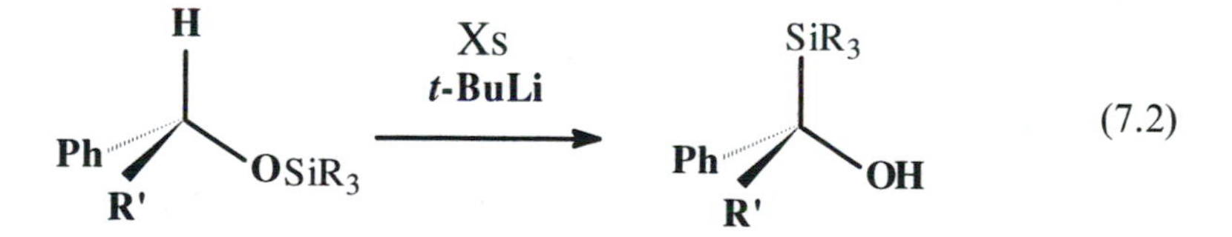

Thus, the direction of this 1,2-rearrangement (i.e., Brook vs retro-Brook) is dependent on the concentration ratio of strong base to substrate. Finally, it should be recognized that the Wittig and retro-Brook rearrangements are superficially similar in that both are members of the class of [1,2] isomerizations and in that the prototypical systems that undergo these rearrangements are isoelectronic.

For rearrangements proven (or, more often, presumed) to proceed in a concerted manner, conservation of orbital symmetry, first enunciated by Woodward and Hoffmann[17-19] and advanced in somewhat different form by Fukui,[20] Dewar,[21] and Zimmerman,[22,23] among others,[24,25] has provided a unifying principle for classification and description of such one-step isomerizations and their stereochemical outcomes. However, to date, there are no unequivocal examples of *concerted* [1,2] shifts of *alkyl fragments* (as opposed to phenyl) or [1,2] shifts of *hydrogen* along a saturated carbanionic skeleton. Neither concerted intramolecular nor intermolecular [1, *j*] alkyl migrations have been demonstrated beyond doubt.

These findings contrast markedly with the situation found for carbocations. Here, *concerted* [1,2] alkyl and hydrogen shifts to the cationic center are common. The explanation is rooted in the requirements of orbital symmetry[17-25] and will be discussed in some detail later in Section 7.3. (Notwithstanding the foregoing, Wittig rearrangements that involve an apparent [1,2] shift do occur. However, the best current evidence favors stepwise mechanisms in such systems.[26,27]) For an overview of symmetry control in carbanion rearrangements, including the basics of the Woodward–Hoffmann[17-19] and Zimmerman[22,23] approaches, the reader is referred to a monograph.[28]

Earlier research into specific carbanionic rearrangements has been ably discussed by Cram, particularly in the case of tautomerization of allylic and related systems,[29] and by Grovenstein, in the case of [1,2] aryl group migrations.[30,31] While we shall consider these systems, among others, in this chapter, the reader is also referred to these sources.

The content of this chapter falls into three broad categories: (1) tautomerization, which includes double-bond migration, but excludes keto–enol which was covered in the previous chapter; (2) orbital control in group migrations; and (3) *name* rearrangements, usually involving migration of heteroatom-centered groups, for example, Wittig, Brook, Stevens, and Grovenstein–Zimmerman rearrangements. Some enolate rearrangements, notably the Favorskii rearrangement, were dealt with in Chapter 6.

7.2 TAUTOMERISM AS A CARBANION REARRANGEMENT INVOLVING FORMAL PROTON MIGRATION

In a general sense, a tautomeric equilibrium implies a migration of a proton as shown in equation 7.3.

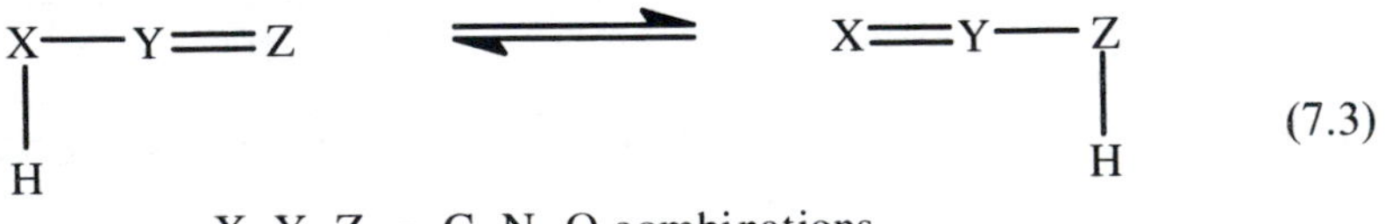

(7.3)

The two species shown in equation 7.3 are tautomers and the equilibrium is usually termed a tautomeric equilibrium, although emphasis may be placed on the fact that the process involves a proton shift by referring to the equilibrium as a prototropic one. Equation 7.3 clearly applies to the keto–enol tautomerism that is described in more detail in Chapter 6 (see also Scheme 6.1).

Equation 7.3 is meant to be general and, hence, while a formal [1,3] proton shift is depicted, the system, in principle, could be polyenic and the proton shift would be [1, *j*] in such a case. In a similar sense, X, Y, and Z may be any collection of atoms, although some combination of C, N, and O are the most common. In contrast to many of the other systems that we shall examine in this chapter, tautomeric species interconvert readily and, consequently, both tautomeric forms are usually present simultaneously in equilibrium.

Pertinent to our discussion is the role of carbanionic intermediates in tautomerization. As might be expected, a Brønsted base usually acts to generate an anionic intermediate as seen in equation 7.4a.

$$\text{B:}\ \ \underset{\displaystyle \text{H}}{\underset{|}{\text{X}}}\text{—Y=Z} \rightleftharpoons {}^{\oplus}\text{BH}\left[\text{X}\cdots\text{Y}\cdots\text{Z}\right]^{\ominus} \rightleftharpoons \text{X=Y—}\underset{\displaystyle \text{H}}{\underset{|}{\text{Z}}} \qquad (7.4a)$$

This foregoing equation may appear to imply that the Brønsted base, B:, acts to abstract a proton and yield a discrete anionic intermediate. Of course, proton transfer to B: could also occur at the same time as proton transfer from solvent or any general acid, HA, as shown in equation 7.4b.

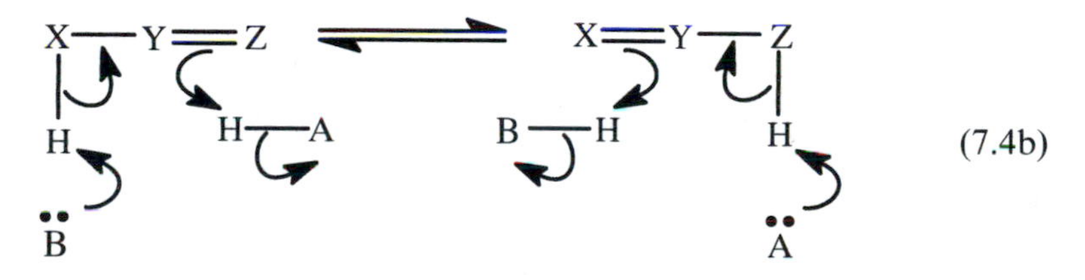

(7.4b)

Differentiating between the two mechanisms represented by equations 7.4a and 7.4b is not a trivial procedure; the process represented by Equation 7.4a becomes kinetically indistinguishable from the 7.4b process, where HA in equation 7.4b is also the solvent. In principle, two other possibilities are specifically tied to the structural features of either the catalytic species or the isomerization substrate (XYZ) itself. The first of these two possibilities corresponds to bifunctional catalysis if the catalytic species contains both acidic and basic centers (presumably with a suitable geometry for proton transfer) as illustrated by equation 7.4c.

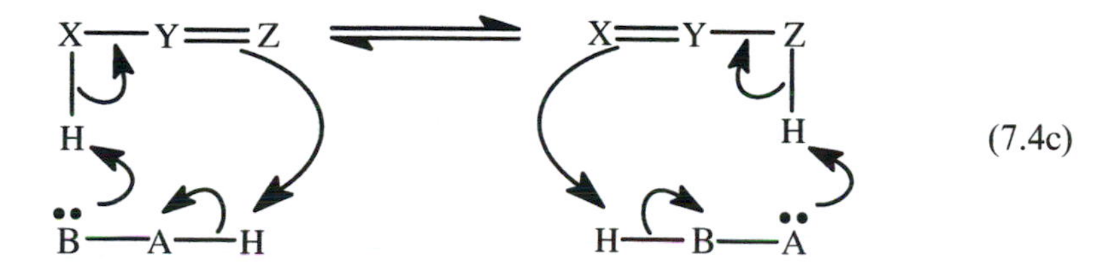

(7.4c)

Finally, the molecule may contain basic/acidic sites that would permit intramolecular proton transfer. Here, the molecule itself acts as the base catalyst.

7.2.1 Intramolecularity in Allylic 1,3 (and 1,5) Prototropic Rearrangements

We will now consider allylic rearrangement in light of the foregoing. The classic study of Cram is illustrative of prototropic rearrangement under H–D isotopic exchange conditions. In this regard, 3-phenyl-1-butene was found to rearrange to its conjugated isomer, 2-phenyl-2-butene (primarily as the *E*-stereoisomer), in O-deuterated *tert*-butyl alcohol with potassium *t*-butoxide as base (equation 7.5).[32,33]

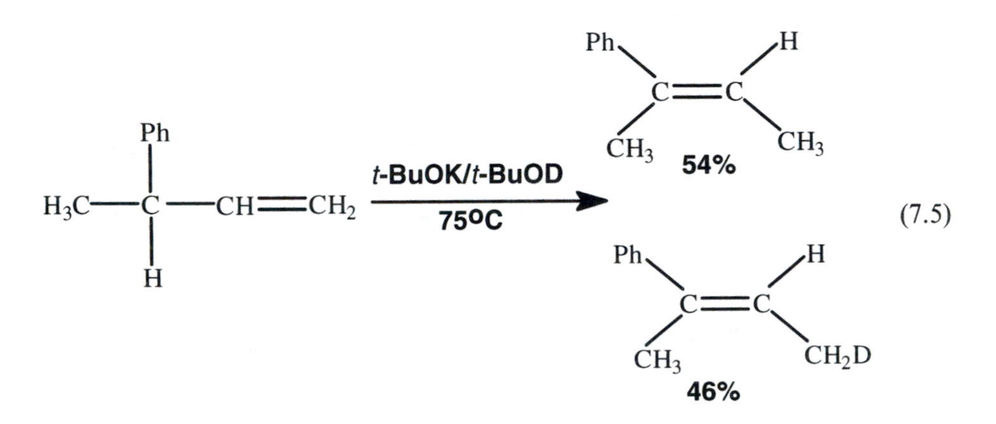

(7.5)

Even though the exchangeable deuterium pool was large (98% deuterated solvent), 54% of the starting material rearranged without incorporation of deuterium, that is, without proton loss to solvent. This result can be interpreted to mean that the rearrangement occurs with a degree of intramolecularity of 54%.

As proposed by Cram, the intramolecular mechanism would involve initial proton abstraction from the benzylic site to yield an allylic carbanion that can hydrogen-bond to the newly generated *tert*-butyl alcohol via two sites, shown in structure **83**. More generally, the *t*-butyl alcohol moiety in structure **83** may be replaced by the conjugate acid of the base used in the proton abstraction. It could

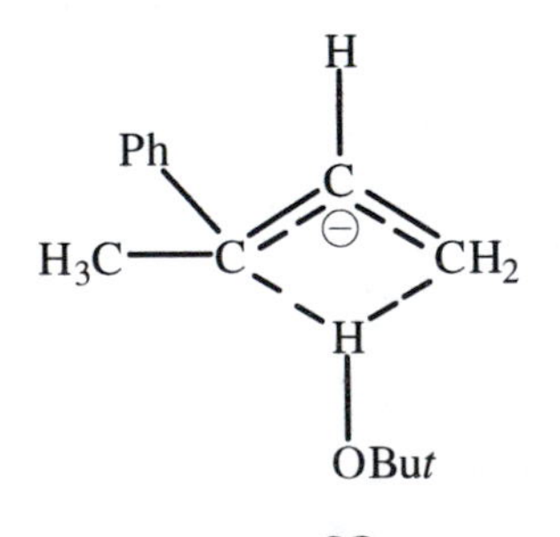

83

also be argued that the distances between the hydrogen of the alcohol and the anionic centers are too long for this to be a viable intermediate. In this case, **83** can be viewed as the average of a pair of rapidly equilibrating hydrogen-bonded structures. Regardless, collapse of this hydrogen-bonded carbanion (or carbanions) yields the rearranged product without isotopic exchange.

In the alternative intermolecular mechanism, the hydrogen-bonded allylic carbanion dissociates, at least to the extent that deuterated solvent replaces the *t*-butyl alcohol in the intermediate. Now, collapse of the anionic intermediate that includes the deuterated solvent in the delocalized structure leads to isotopic exchange along with rearrangement.

Since these pioneering studies, a full range of intramolecularities have been reported from a maximum of 100% found for the [1,3] proton shift in 3-deuterioindene (triethylamine: pyridine: 9M DO^- solvent system)[34] to a minimum of 0.3% in the 3-deuterio-1-(1,1-dimethylethyl)-1-methylindene:$KOCH_3$:CH_3OH reaction system.[35] Generally, some degree of intramolecularity is found in all alkene systems studied thus far, and Table 7.1 contains a selection of these prototropic rearrangements. Examination of the table, and related data for alkyne–allene[36,37] and imine rearrangements[38,39] compiled in the review by Hunter,[9] lead to a number of generalizations. First, intramolecularity in [1,3] and [1,5] proton shifts is enhanced by the use of tertiary amines as base catalysts as compared with alkoxides in alkanolic solvents. On the other hand, the identity of the alkoxide–alkanol system does not significantly influence the degree of intramolecularity.[9, 33] The higher degree of intramolecularity found with tertiary amine bases is consistent with strong ion pairing in the allylic carbanion-trialkylammonium ion-paired intermediate. (The secondary $^{11}C/^{14}C$ kinetic isotope effect in the 1,3-prototropic rearrangement of 1-methylindene to 3-methylindene catalyzed by DABCO,[43] as well as the Brønsted coefficient, 0.79, found for the same rearrangement with a series of structurally rigid tertiary amines[44] also supports significant ion pairing in these systems.) Secondly, the location of the isotopic label has a notable effect on the degree of intramolecularity. When the 3-deuterio-3-phenyl-1-butene systems are compared with the analogous undeuterated 3-phenyl-1-butene reactions,[32,33] a lower degree of intramolecularity is found for intramolecular [1,3] deuteron transfer than for the corresponding [1,3] proton transfer.

More recently, there has been interest in extended conjugated systems such as the α,ω-polyenes.[45,46] We look forward to a revival in interest in prototropic rearrangements involving such geometrically interesting systems. Will intramolecularity also obtain in these extended conjugated systems?

An important conclusion that may be drawn from these studies of allylic rearrangements is that concerted [1,3] (and [1,5]) proton shifts, which would show no intermolecularity, are rare, at the very least.

7.3 ORBITAL SYMMETRY CONSIDERATIONS IN CONCERTED CARBANION REARRANGEMENTS

What are the requirements for observation of *concerted* [1, *j*] rearrangements of carbanions? The concept of orbital symmetry control in concerted reactions invol-

Table 7.1. Intramolecularity in base-catalyzed rearrangement of selected alkenes systems

Reaction[a]	Conditions	% Intermolecularity
$H_3CC(Ph)(H)CH{=}CH_2 \rightarrow Ph(H_3C)C{=}C(H)CH_2{-}H$	*t*-BuOK-*t*-Bu**OD**, 75 °C	51[b]
	Et_3COK-Et_3C**OD**, 50 °C	56[b]
	Me_4N**OD**-*t*-Bu**OD**, 50 °C	45[b]
$H_3CC(Ph)(D)CH{=}CH_2 \rightarrow Ph(H_3C)C{=}C(H)CH_2{-}D$	*t*-BuOK-*t*-Bu**OH**, 75 °C	23[c]
	Et_3COK-Et_3C**OH**, 50 °C	17[c]
	Me_4N**OH**-*t*-Bu**OH**, 50 °C	6[c]
$CD_3(CD_2)_2CD{=}CD_2 \rightarrow CD_3CD_2CD{=}CDCD_3$	*t*-BuOK-*t*-Bu**OH**-CH_3SOCH_3, 55 °C	> 94[d]
7,7-dideuteriocycloheptatriene (D, D) $\rightarrow$ 7-D,7-H-cycloheptatriene with D moved onto the ring	Et_3COK-Et_3C**OD**, 50 °C	ca. 90[e]
	Et_3COK-Et_3C**OD**-CH_3SOCH_3, 50 °C	< 7[c]
$(CH_3)_2C(CO_2CH_3)$-cyclohexadiene(H)$=CPh_2 \rightarrow (CH_3)_2CCO_2CH_3$-$C_6H_4$-$CPh_2$-H	NaOMe-Me**OD**, 25 °C	47[f]
	t-BuOK-*t*-Bu**OD**, 25 °C	50[f]
	DABCO[g]-Et_3C**OD**, 75 °C	98[f]

[a]Intramolecularity is for the isomeric product shown.

Sources: [b]Data taken from reference 32; [c]Data taken from reference 33; [d]Data taken from reference 40; [e]Data taken from reference 41; [f]Data taken from references 37 and 42.

[g]DABCO = 1,4-diazabicyclo[2.2.2]octane.

ving π-systems has proven to be a powerful organizing principle.[17-25] Stated briefly, the principle of orbital symmetry holds that a concerted reaction will be favorable only if the participating reactant orbitals transform into the product orbitals, via a transition state, such that orbital symmetry is preserved throughout the process.[47] The key is that suitable orbital symmetry requires the maximum degree of bonding in proceeding from reactants to products.

While it is beyond the scope of this book to detail the background and general utility of this approach, we will briefly explore orbital symmetry as it applies to concerted rearrangements of carbanions (and carbanion-like species). For a fuller explanation of orbital symmetry, the reader is directed to the original articles[17-25] and books,[47,48] as well as the pertinent sections of a monograph.[28]

Consequently, in this section we will consider orbital symmetry only as it affects electrocyclic carbanion rearrangements, as typified by the cyclopropyl anion–propenyl anion rearrangement,[49,50] and in sigmatropic rearrangements in carbanions.

7.3.1.1 Carbanionic Electrocyclic Rearrangements

The process of ring closure of an unsaturated open-chain carbanion via formation of a single bond, or its reverse (ring opening of an unsaturated cyclic anion where a single bond in broken), may be termed a carbanionic electrocyclic rearrangement. This is illustrated generally by the equilibrium shown between the ring (**84**) and the open-chain structure (**85**).

Such rearrangements may be considered a subset of pericyclic reactions and would be expected to be governed by the rules of orbital symmetry. A priori, such processes may be *conrotatory or disrotatory*, as shown in equations 7.6a and 7.6b for the electrocyclization of 1,3-butadiene to cyclobutene, where the terminal hydrogens have been subscripted for clarity.

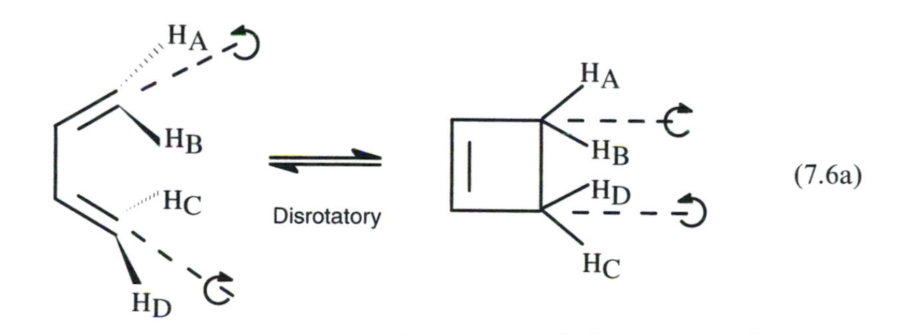

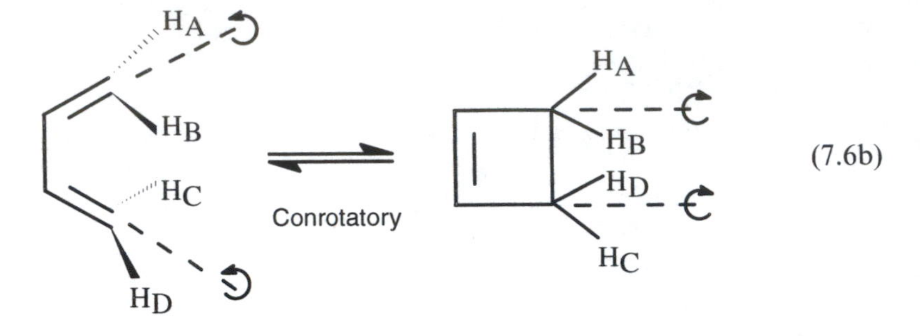

In their monograph, Woodward and Hoffmann elaborated on the process of electrocyclization for the 1,3-butadiene–cyclobutene system and showed the method for constructing orbital correlation diagrams for such systems, and thereby demonstrated that the ground-state path that would be preferred in this case is conrotatory.[47]

More generally, for a polyenic system with m π-electrons, the electrocyclic process that will occur most readily (i.e., with the lowest energy transition state) under thermal conditions will be disrotatory when m has a value equal to $4n+2$ and will be conrotatory when m has a value given by $4n$, where n is zero or an integer, (i.e. n = 0, 1, 2, etc.). For the electrocyclization of 1,3-butadiene to cyclobutene, $m = 4$ (for the four π-electrons of the double bonds of cyclobutadiene), as expected when $n = 1$ in the $4n$ formula, and so the conrotatory process is the favored in the ground-state concerted process. Note that an electrocyclization that is predicted to proceed in a conrotatory fashion under thermal conditions is expected to proceed via the disrotatory route under photochemical conditions. The processes predicted by orbital symmetry for some carbanionic systems are given in Table 7.2. It should be noted that processes that are considered favorable according to orbital symmetry are usually described as "allowed", while those paths that are high in energy and, therefore, unfavorable are termed "forbidden". Halevi, among others, has commented on this usage and suggested that more appropriately these terms be replaced with "facile" and "difficult".[51] Processes that are "forbidden" may, in fact, be observed, particularly where steric or other structural features raise the barrier to the thermally allowed pathway.

A significant corollary of orbital symmetry conservation is that the various modes of rearrangement—that is conrotatory versus disrotatory—can be detected experimentally from the stereochemistry found in reactions of *suitably substituted reactants*. On the contrary, bulky substituents may not only act as markers for the process of electrocyclization, but may also change the energetics involved.

In the following subsections, we will consider some carbanionic systems that can be described according to the principles outlined.

7.3.1.2 The Cyclopropyl-Propenyl Anion Rearrangement

The carbanion electrocyclic rearrangement involving the structurally simplest carbanion, namely, the ring opening of the cyclopropyl anion (and analogues) to give the corresponding propenyl (allyl) anion, was also the last to be confirmed

Table 7.2. Electrocyclization in some carbanionic systems

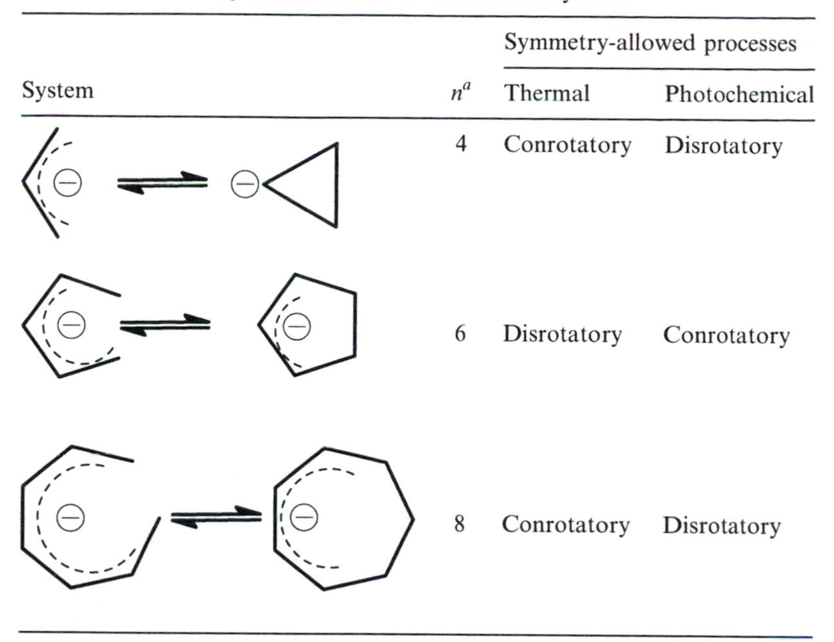

System	n^a	Symmetry-allowed processes	
		Thermal	Photochemical
	4	Conrotatory	Disrotatory
	6	Disrotatory	Conrotatory
	8	Conrotatory	Disrotatory

SOURCE: Data taken from reference 47.
[a]The number of participating electrons is given as *n*.

as an orbital-symmetry-controlled process. As a 4π-electron system, the propenyl carbanion was predicted to undergo a conrotatory electrocyclization to the cyclopropyl anion (and vice versa).[52] However, it has proven difficult to obtain unequivocal evidence in support of the prediction based on conservation of orbital symmetry.[53]

There is no doubt that cyclopropyl anions do undergo ring-opening rearrangement. For example, 1,2,3-triphenylcyclopropane under basic conditions (*n*-butyllithium and tetramethylethylene diamine), followed by D_2O quenching, leads to the *E*- and *Z*- stereoisomers of 3-deuterio-1,2,3-triphenylpropene as products (equation 7.7)[55] (see also reference 54).

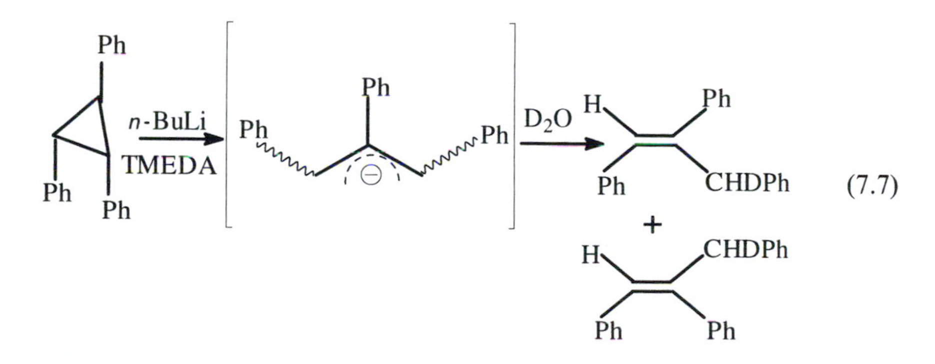

(7.7)

In the following system (equation 7.8), Londrigan and Mulvaney[56] found the rearrangement to proceed in an overall disrotatory fashion; steric factors may have intervened to raise the barrier to conrotatory ring opening.

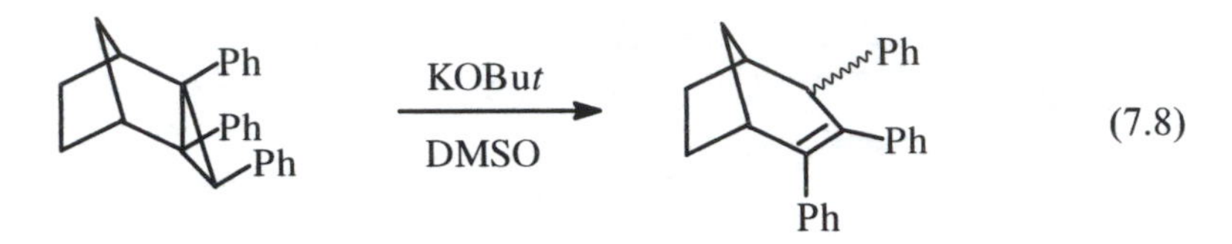

Treatment of 1-methyl-3,3-diphenylcyclopropane-1,2-*trans*-dimethyl dicarboxylate with sodium methoxide in methanol (60 °C) leads to equilibration of the isomeric cyclopropanes, presumably via the cyclopropyl carbanion as shown in equation 7.9.

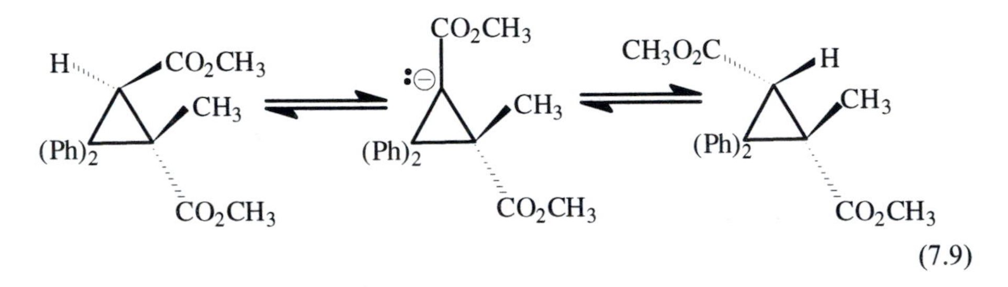

However, reaction of the same cyclopropane derivative with sodium hydride in dimethylformamide led to ring-opening isomerization to the propenyl anion that could be trapped by alkylation with bromomethane (or quenched with methanolic hydrogen chloride) as shown in Scheme 7.1. The allylic carbanion could also be generated independently via deprotonation of the dimethyl 1,1-diphenyl-1-but-

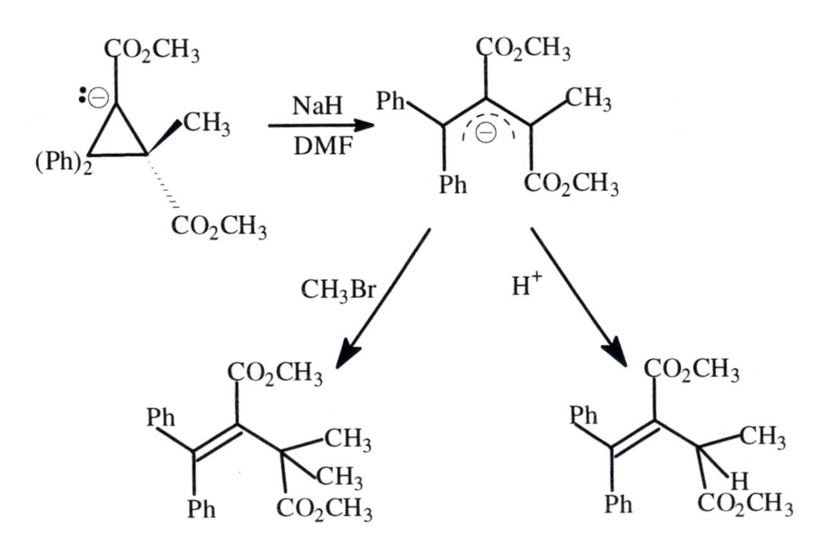

SCHEME 7.1

Table 7.3. Experimental gas-phase activation barriers for the cyclopropyl anion–allylic anion rearrangement

X	Y	Activation energy (kcal/mol)[a]
CO_2Me	CO_2Me	19
H	Ph	26
H	CHO	29
H	CN	36

SOURCE: Data taken from reference 50.
[a]Estimated error is ± 4 kcal/mol.

enedioate with NaH in DMF. In this case, the ring-opening isomerization was found to be irreversible.[57]

The foregoing examples are mechanistically ambiguous, at best, concerning the conrotatory nature of the rearrangement. In this regard, two main problems have been identified.[10] First, the conrotatory ring opening rearrangement appears to be slower than the isomerization of the resultant allylic anion to its most stable isomeric form. Consequently, stereochemical information that would confirm the conrotatory route is lost. Secondly, the identity of reacting species is not always clear-cut. Although free carbanions have been presumed in the examples shown, strongly ion-paired species may actually be involved, depending on the reaction system. Thus, the parent cyclopropyllithium,[58] and the 1-cyano-2,2-diphenylcyclopropyllithium[59] do not undergo the ring-opening rearrangement, nor does 1-phenylcyclopropylpotassium (in refluxing hexane),[60] whereas 1,2,3-triphenylcyclopropane *treated with n-butyllithium and TMEDA* does.[55] On the other hand, in the gas phase, a series of substituted cyclopropanes undergo deprotonation and ring-opening isomerization to the respective allyl anions (Table 7.3),[50] whereas the bicyclobutyl anion is stable in the gas phase to 300 °C.[61] The relative ease of the cyclopropyl anion–allylic anion rearrangement as compared with the stability of bicyclobutyl anion was considered consistent with the conrotatory nature of the cyclopropyl anion ring openings and with the effect of resonance-delocalizing substituents on the degree of pyramidalization of the cyclopropyl anions.[50] The argument here is that the conrotatory electrocyclization would be assisted by a planar anion in which overlap between the anionic orbital and the breaking carbon–carbon σ-bond would be maximized (Figure 7.1). Resonance-delocalizing groups at the anionic center (Y) are known to decrease the barrier to inversion in substituted cyclopropyl carbanions.[50,62]

Significant support for the concerted conrotatory process in the cyclopropyl anion ring opening is found in various calculational studies.[63-67] However, the solution-phase experimental evidence consists primarily of comparative kinetic

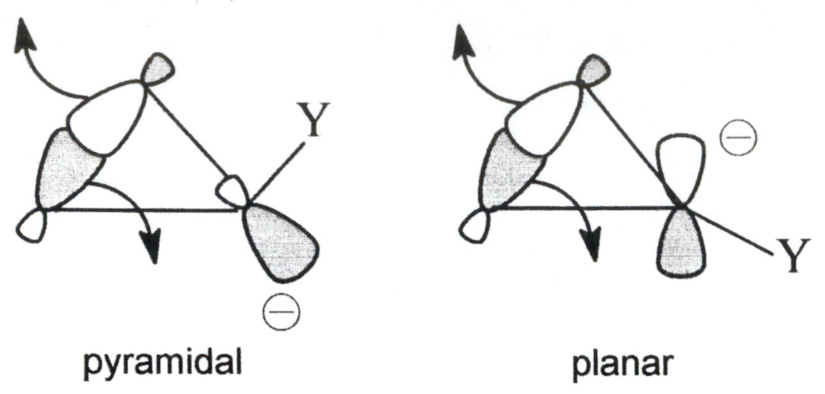

Figure 7.1. Conrotatory ring opening of a substituted cyclopropyl anion is favored by substituents, Y, that impose planarity on the carbanionic carbon. In the planar anion, the anionic orbital (shown as an approximately pure p-orbital) can overlap with an orbital of the breaking C–C σ-bond, as shown. Such overlap is not possible for the pyramidal anion.

studies and examination of an analogous *N*-lithioaziridine system. On the other hand, Boche and co-workers[68,69] examined the ring-opening rearrangement of *trans* 1-cyano-2,3-diphenyllithiums, where conrotatory ring opening of the *trans* lithium carbanide can give two isomeric allylic anions. (The lithium counterion has been omitted from Scheme 7.2.)

The situation is complicated by the rapid isomerization of the initially formed propenyl anions to the more stable isomeric form that would arise directly from the cyclopropanide anion only via the unfavorable disrotatory ring opening. In

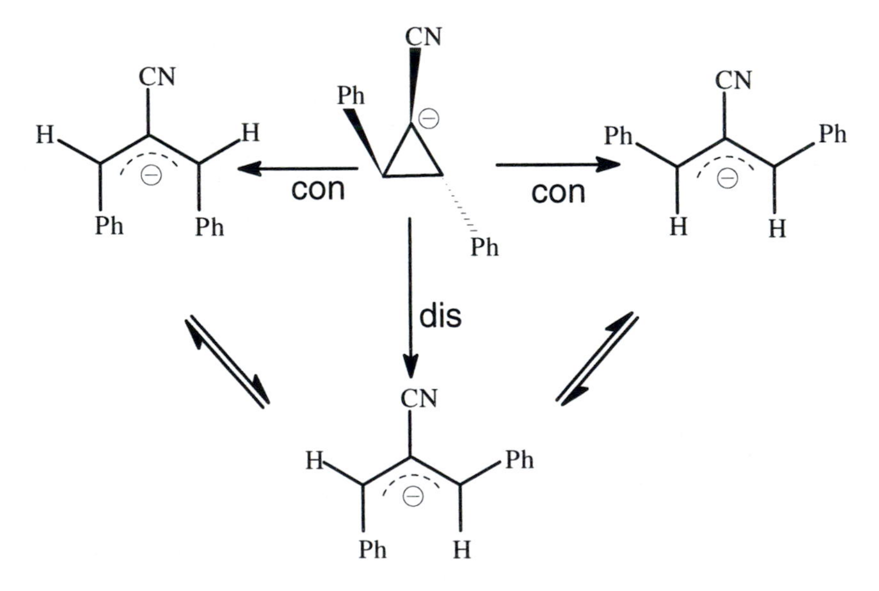

SCHEME 7.2

fact, the ring-opening process was found to be approximately 1500-fold slower than the allylic isomerization. (A similar result was obtained from the *cis*-1-cyano-2,3-diphenylcyclopropanide.[69]) Comparison can be made, however, to the "thermally forbidden" disrotatory ring opening found by Wittig and co-workers,[70] as shown in equation 7.10.

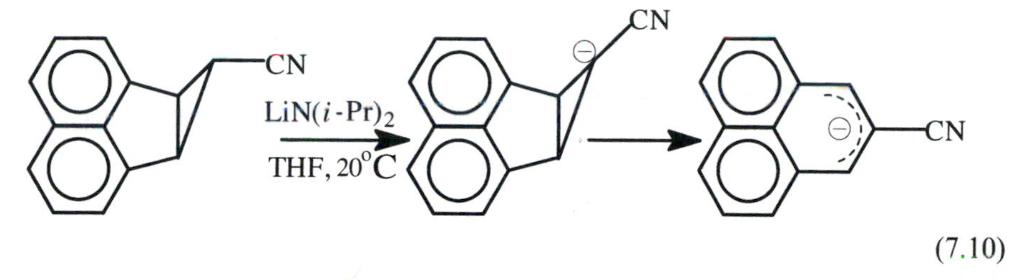

(7.10)

The lithium *trans*-1-cyano-2,3-diphenylcyclopropanide was shown to isomerize at a rate about 11,000 times faster than the disrotatory ring opening found by Wittig.[66,71] Therefore, the much more favorable ring-opening rearrangement in the 1-cyano-2,3-diphenylcyclopropyl anion system would be indicative of the preferred conrotatory process.

Definitive experimental evidence for the stereochemical course of the ring-opening isomerization has been provided by *N*-lithio-2,3-*cis*-diphenylaziridine, which is isoelectronic with the corresponding cyclopropyl anion. When an initially colorless THF solution of the lithium salt, under nitrogen, is warmed (40–60 °C), it turns red, indicative of the presence of the 1,3-*trans*-diphenyl-2-azapropenyl anion arising from conrotatory ring opening. Trapping of this *trans* azaallyl

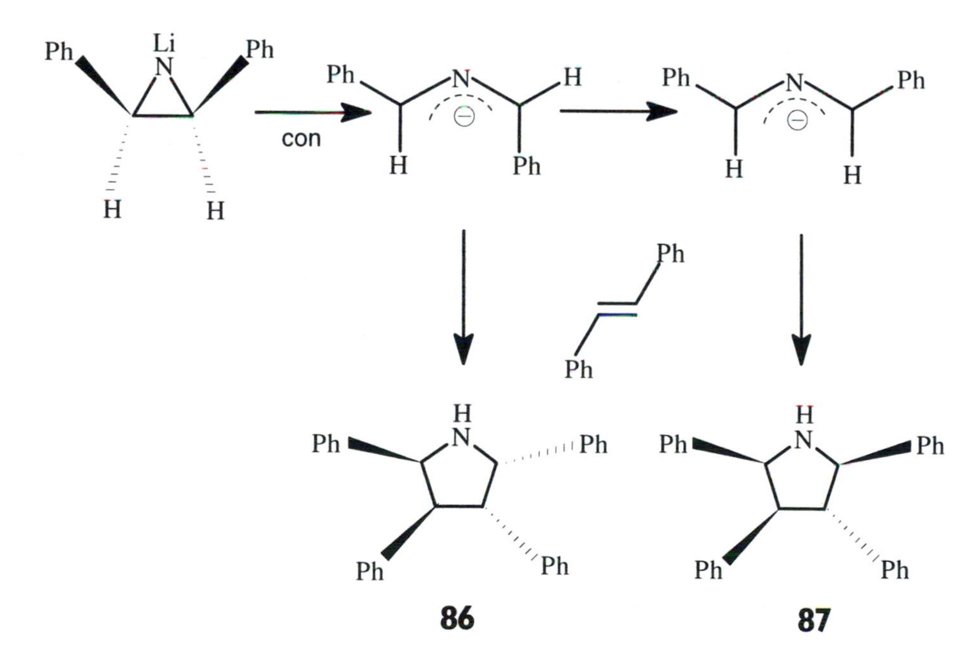

SCHEME 7.3

anion with *trans*-stilbene (*E*-1,2-diphenylethene) in an allowed 4 + 2 cycloaddition yielded as the major product (73%, **86**) the tetraphenylpyrrolidine in which the azaallyl moiety was shown by NMR to retain its *trans* orientation. The minor tetraphenylpyrrolidine (11%, **87**) results from trapping of the isomerized *cis* azaallyl anion (Scheme 7.3).[72]

The conclusion is that conrotatory ring opening is the preferred pathway in cyclopropyl anion–allylic anion rearrangements.

7.3.1.3 Electrocyclization of Dienyl Anions

Conjugated dienyl anions are 6π-electron systems and would be expected to undergo electrocyclization via the disrotatory mode (Table 7.2). In fact, the closure of the cyclooctadienyl anion to the bicyclo[3.3.0]octenyl anion may be achieved by treating either 1,3-cyclooctadiene or 1,4-cyclooctadiene with a variety of strong bases (phenylpotassium,[73] or sodium hydride[74] in heptane, cumyl potassium[75] or *n*-butyllithium[76] in THF, or potassium amide in liquid ammonia[77]).

The study by Bates and McCombs[76] is particularly pertinent to our discussion. Here, reaction of the cycooctadiene with *n*-butyllithium gave rise to the delocalized cyclooctadienyl anion, identified by its NMR spectrum (at −78 °C). As the temperature was raised (to −20 °C), reversible changes in the NMR spectrum were noted that may be ascribed to conformational changes in this nonplanar anion. At room temperature, the conversion of the cyclooctadienyl carbanion into the bicyclooctenyl anion can be monitored; a first order rate constant of $1.5 \times 10^{-4}\ s^{-1}$ ($t_{1/2} = 1.3$ h) was determined (at 35 °C).The process is readily described as a disrotatory electrocyclization (equation 7.11).

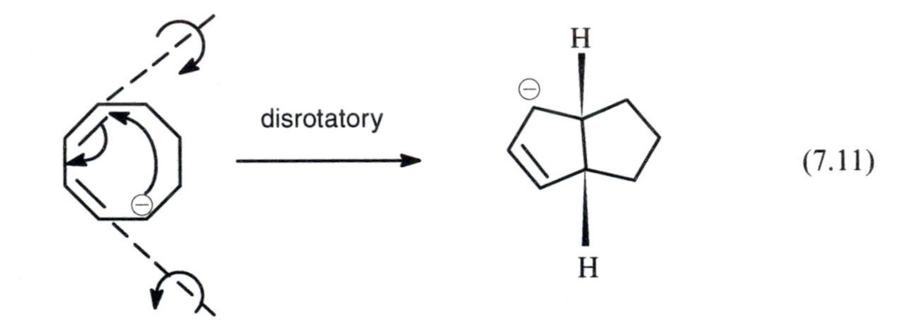

(7.11)

In the Woodward–Hoffmann correlation diagram for a generalized pentadienylic–pentenylic system (Figure 7.2),[28] the requirement that a plane of symmetry be maintained throughout a disrotatory ring closure leads to correlation of the bonding orbitals of the pentadienyl anion with the bonding molecular orbitals of the product cyclopentenyl anion, that is equivalent to the superposition of a σ-bond and an allylic π-system. No such correlation of bonding molecular orbitals is possible for the conrotatory process and, consequently, the disrotatory route is energetically facile.

A number of other pentadienyl anion systems have been examined whose rearrangements may be interpreted by electrocyclic mechanisms. For example, 6,6-diphenylbicyclo[3.1.0]hex-2-ene also isomerizes when heated with potassium

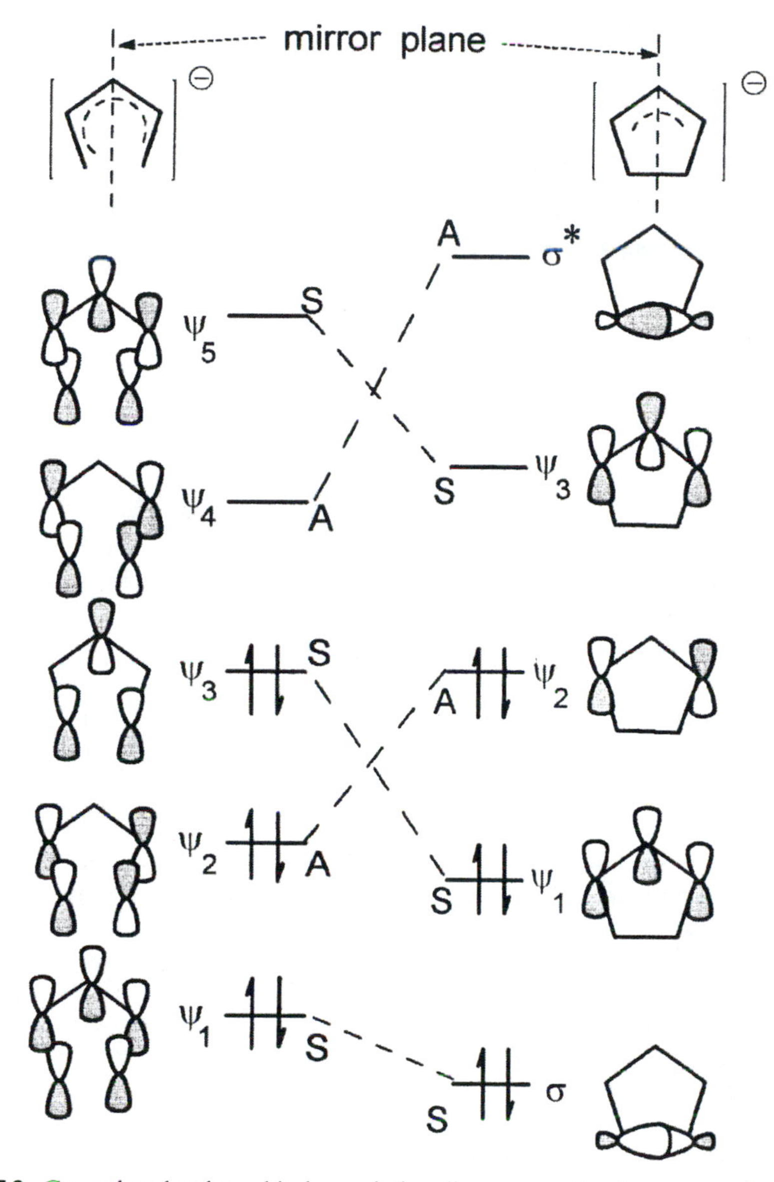

Figure 7.2. General molecular orbital correlation diagram for disrotatory ring closure of a pentadienyl carbanion to a cyclopentyl anion (and the reverse process). Letters S and A refer to the symmetry (i.e., "symmetric" or "antisymmetric" respectively) with respect to the mirror plane.

t-pentoxide in *t*-pentyl alcohol to give isomeric diphenylcyclohexadienes (Scheme 7.4).[79] Here, the initially formed bicyclohexadienyl anion opens in a disrotatory fashion to give the diphenylhexadienyl carbanion that is protonated by the alkanolic solvent under the reaction conditions. Interestingly, the structurally related

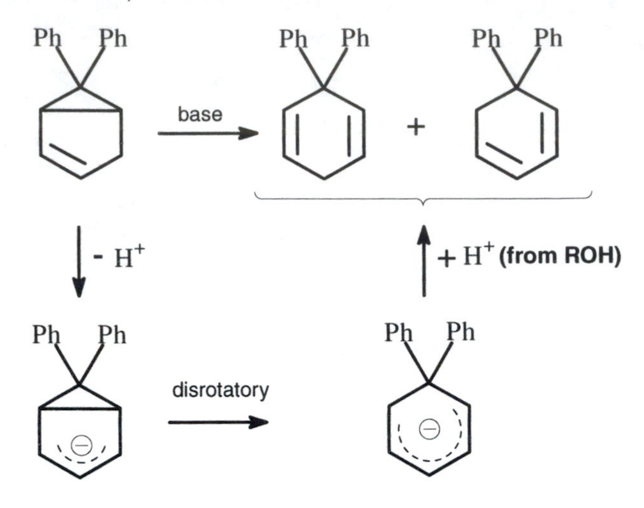

SCHEME 7.4

6-methyl-6-phenylcyclohexadienyl anion yields products of ring contraction in liquid ammonia (equation 7.12).[80]

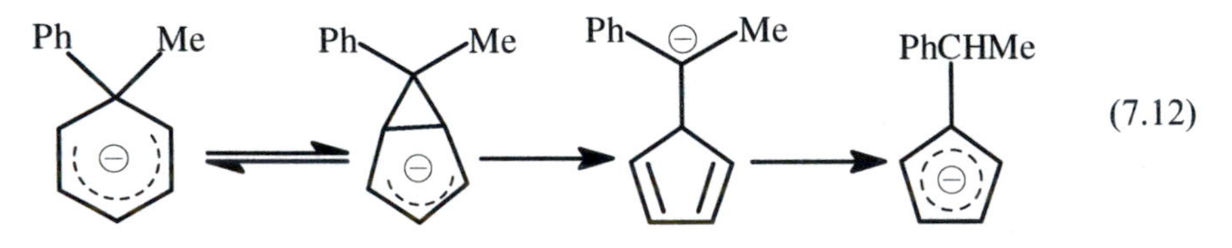

(7.12)

By analogy to the anion of 6,6-diphenylbicyclo[3.1.0]hex-2-ene, that ring opens to the corresponding cyclohexadienyl anion according to the disrotatory route, it is reasonable to expect the cyclohexadienyl anion to ring close by the disrotatory mode. However, in both systems the bicyclic anion is relatively unstable as a result of ring strain and, in this second example, the three-membered ring reopens, leading to overall ring contraction.[79] Quenching the reaction mixture with water regenerates the starting 6-methyl-6-phenylcyclohexadiene and gives 1-(1-phenylethyl)-1,4- and 1,3-cyclopentadienes via the cyclopentadienyl anion.

Although electrocyclization of the unsubstituted pentadienyl anion has not been reported,[81] Hunter and Sim[82,83] have described the formation and ring closure of the diazaanion formed as shown below. Deprotonation of the hydrobenzamide, **88**, with phenyllithium in THF yields a blue solution consistent with formation of the delocalized diazaanion, **89**, or a conformer. When this solution is warmed to room temperature, the solution changes color; it becomes a pale yellow. Following the left-hand side of Scheme 7.5, the yellow solution contains the cyclodiazapentyl anion, **90**, formed by the expected disrotatory ring-closing path. This is confirmed by quenching with ethanoic acid; only the *cis* stereoisomer (*cis*-amarine, **91**, shown in Scheme 7.5) is formed. Significantly, equilibration of **91** with *t*-butoxide in *t*-butyl alcohol yields a mixture of *cis:trans* product in a ratio of 4:96. Thus, formation of the *less stable isomer* under the reaction conditions

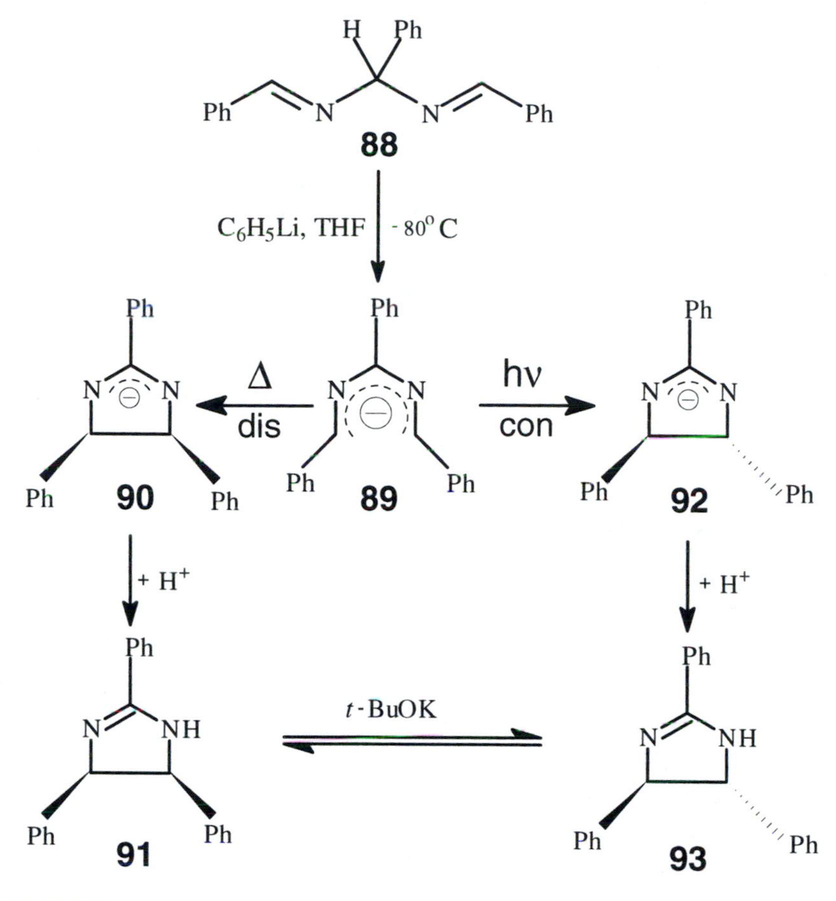

SCHEME 7.5

emphasizes the controlling influence of the thermally allowed disrotatory route on the course of the reaction. The right-hand side of Scheme 7.5 (**89** →**92** → **93**) is equally important in establishing orbital symmetry control in this (and related) system(s). Irradiation of the initial alkaline solution of **89** (577–579 nm) followed by acetic acid quenching produced the *trans*-isomarine, **93**, in increasing proportions as the temperature of reaction was lowered. Again, the stereochemistry is consistent with orbital symmetry arguments; upon irradiation, the diazaanion undergoes conrotatory electrocyclization to **92** and this stereochemistry is preserved in the *trans* isomarine product, **93**.[82,83]

7.3.1.4 Cycloheptatrienyl Carbanion Electrocyclization

As an 8π-electron system, the heptatrienyl carbanion–cycloheptatrienyl anion electrocyclization would be expected to proceed via the thermally allowed conrotatory route,[47] although experimental confirmation is limited.[48] As but one example[84,85] of many,[86-89] low-temperature reaction of either 1,3,5-heptatriene or 1,3,6-heptatriene with strong bases such as KNH_2 in NH_3 leads initially to formation of the delocalized carbanion (equation 7.13).

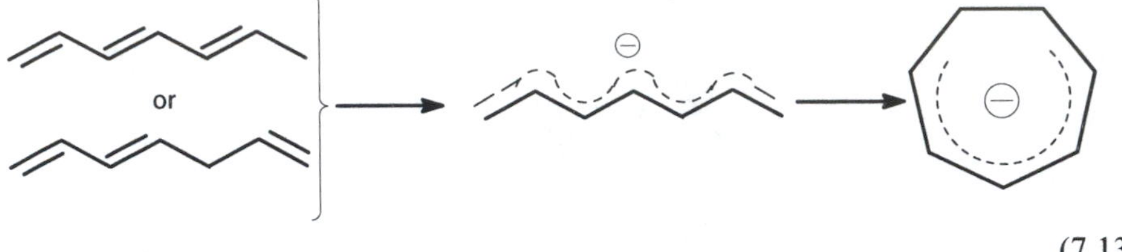

(7.13)

With increase in temperature, the formation of the cycloheptadienyl carbanion was observed by NMR and a half-life in liquid ammonia of 1 h assigned (0 °C). It is interesting to note that for electrocyclization to occur in the concerted manner, the extended-W (*all-trans*) form of the heptatrienyl anion, which is the more stable form for this anion, must first convert to a helical U-shaped (all-*cis*) conformation.[90] However, firm stereochemical evidence for the preferred conrotatory mode of electrocyclization in these systems remains elusive.[91]

We now turn our attention to the consideration of orbital symmetry predictions in regard to sigmatropic rearrangements of carbanionic systems.

7.3.2 [1, *j*] Sigmatropic Rearrangements of Carbanions

A sigmatropic rearrangement is one in which a σ-bond migrates to a new position within a molecule. Such migrations, if concerted, require one or more flanking π-systems. In common with the electrocyclizations discussed earlier, selection rules have been developed for thermally and photochemically allowed sigmatropic rearrangements.[47,92] Our discussion will be limited largely to the migration of hydrogen, alkyl groups, and unsaturated functions in unsaturated carbanionic systems. Rearrangements of anionic systems that contain heteroatoms, and where the migration generally involves the heteroatom(s), will be discussed in the following sections that concern "name rearrangements".

7.3.2.1 [1, *j*] Hydrogen Migrations

The selection rules for thermally allowed migrations of hydrogen from position 1 to position *j* (i.e., [1, *j*] migrations) in a carbanionic polyene are listed in Table 7.4. (In the designation of the migration, [*i*, *j*], the first number refers to the site at which the migrating atom or group is bonded; by convention, *i* is always numerically less than *j*). As can be seen (Table 7.4), for the simplest migration, a [1,2] proton shift in a four-electron system, the preferred process is antarafacial; that is, for the migration to occur in a concerted fashion, the hydrogen atom would pass from the top face of one carbon terminus to the bottom face of the other as shown in structure **94**. On the other hand, in the thermally disfavored [1,2] suprafacial process, the hydrogen migration from position 1 to carbon 2 would occur with the hydrogen always associated with the same face of the π-electron system.

Table 7.4. Selection rules for thermally allowed [1, *j*] sigmatropic hydrogen migrations

Number of participating electrons	Anionic polyene [1, *j*] migration	Stereochemical course
4	[1,2]	Antarafacial
6	[1,4]	Suprafacial
8	[1,6]	Antarafacial
$4n + 2$	[1,4*n*]	Suprafacial
$4n$	[1,4*n* − 2)]	Antarafacial

Source: Data taken from reference 47 and 92.

In comparison, a [1,2] shift is predicted to occur according to the suprafacial mode for a carbocationic polyenic system (with two electrons participating in the migration) as illustrated by structure **95**. Since this suprafacial mode of migration entails less steric hindrance than the antarafacial path, it is not surprising that concerted [1,2] hydrogen shifts are virtually a hallmark of carbocation chemistry, but unknown in the chemistry of carbanions.[8] In a standard curvy arrow presentation, the [1,2] shift of a generalized carbanion and carbocation can be illustrated thus in structures **94** and **95**.

Calculational studies on the ethyl anion support the view that the concerted [1,2] migration of hydrogen in a carbanion is an unlikely process.[93] A high-level calculational study of [1,2] migration in substituted ethynide (acetylide) anions suggests that activation barriers for rearrangement in these species are low when the migrating group is SiH_3 or AlH_2 (ca. 14 and 0.5 kcal/mol, respectively) and high when the migrating group is a second period element (or the center of initial attachment is a second period element), such as F or groups such as CH_3 and OH (barrier of ca. 40 kcal/mol). Hydrogen was found to migrate with a barrier of between 20 and 25 kcal/mol, depending on the level of sophistication of the calculation.[94]

The thermally allowed suprafacial [1,4] shift of hydrogen corresponds to prototropic rearrangement of an allylic-type ion and the favorable antarafacial [1,6] shift of hydrogen would be expected in pentadienyl derivatives, as illustrated in structures **96** and **97**.

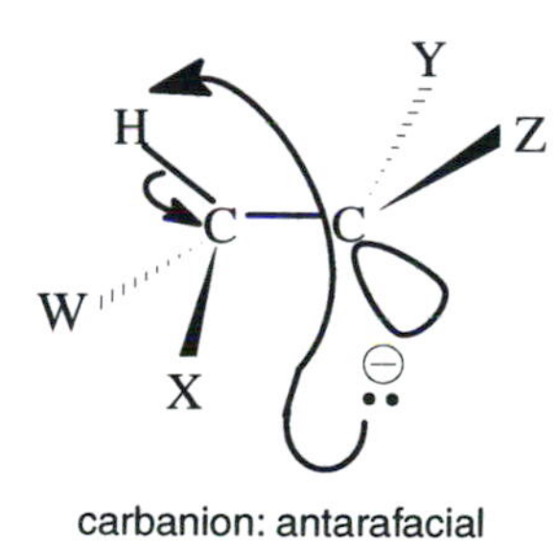

94

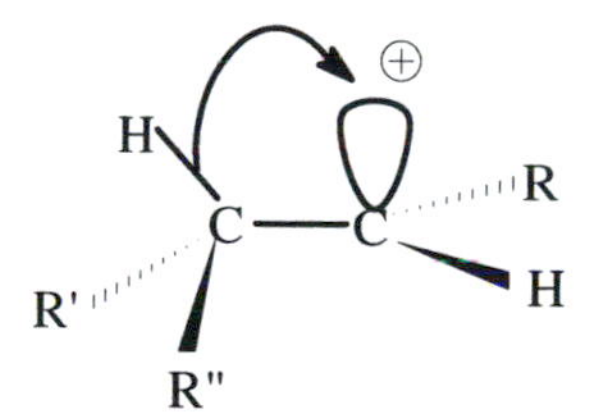

95

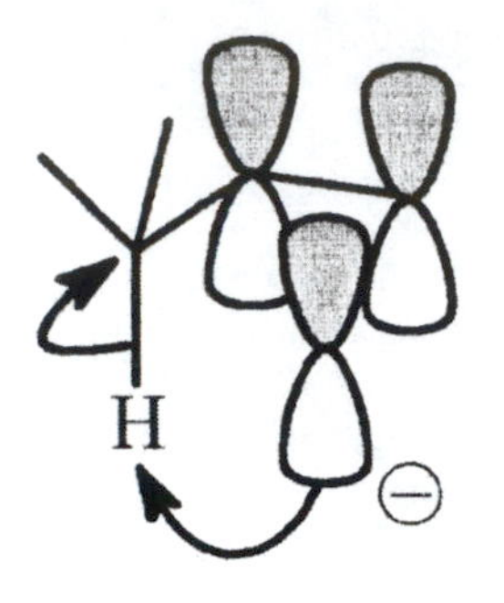

allylic: suprafacial

96

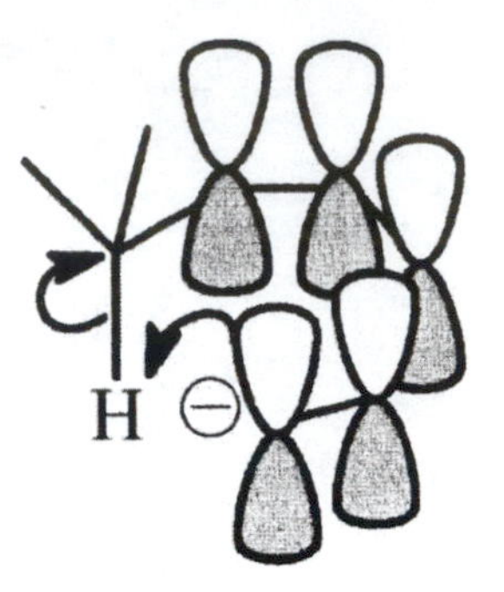

pentadienyl: antarafacial

97

Even though steric hindrance is alleviated in the suprafacial transfer of hydrogen in allylic carbanions, this [1,4] shift has not been observed, although examples have been sought. In this regard, deprotonation of 3,4-diphenyl-1-butene with strong base (*n*-BuLi in THF–hexane or hexane–TMEDA) at 0 °C yielded a red solution consistent with the presence of the corresponding allylic anion.[95] When the reaction was quenched, the product mixture consisted of 17% starting 3,4-diphenyl-1-butene with 78% 1,2-diphenyl-(2Z)-butene, but no 1,2-diphenyl-(1Z)-butene that would have arisen from the rearranged allylic carbanion. (Note that 5% of the products were not characterized.)[95] Significantly, the preferred structure of the initially formed allylic carbanion would provide a suitable geometry for the suprafacial [1,4]-H migration (Scheme 7.6). One possibility that could account for the apparent lack of the [1,4] shift here is that the rearranged allylic anion might form, but in equilibrium with the normal allylic carbanion, which

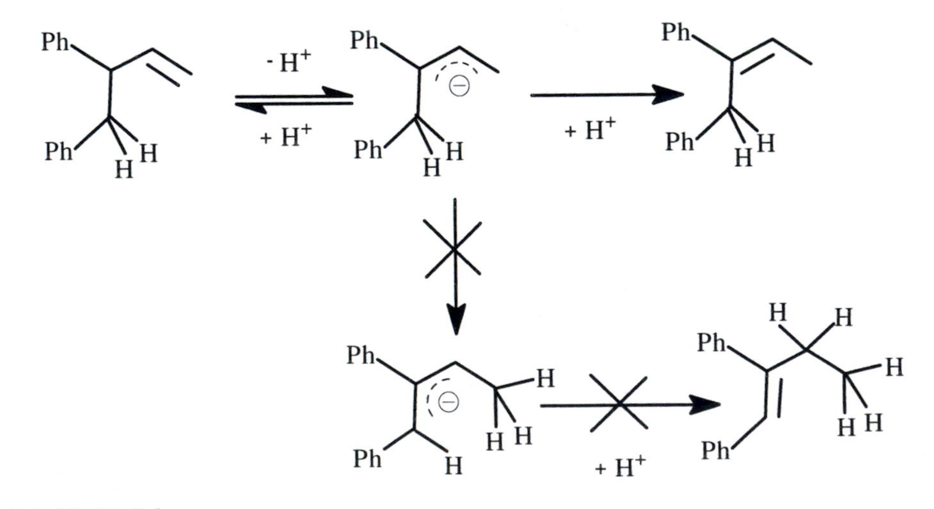

SCHEME 7.6

could be favored. This possibility was tested using the 1,1-deuterium-labeled 3,4-diphenyl-1-butene; treatment with strong base would be expected to lead to deuterium scrambling in the final product, but no such scrambling was found under a variety of reaction conditions.[95]

The products of carbonation with carbon dioxide or methylation of the carbanion formed from the methyl ether of linoleyl alcohol upon treatment with *n*-butyllithium (in Et_2O at room temperature) have been interpreted as resulting from a series of intramolecular [1,6]-H migrations.[96,97] Deprotonation of 5-methyl-1,4-hexadiene (with *n*-BuLi in THF–hexane) was shown by NMR to lead to 52% rearranged dienyl anion[98] (shown in equation 7.14 as the charge-localized species for simplicity, with the migrating H highlighted).

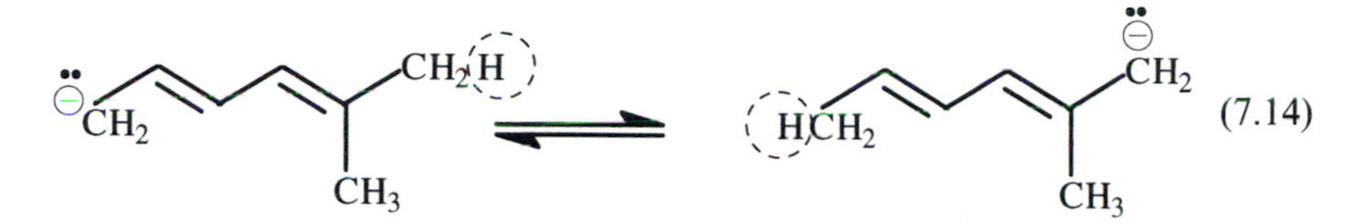

(7.14)

Deuterium labeling experiments demonstrate the intramolecularity of this [1,6]-H migration. The same study demonstrated [1,6]-H shifts in related heptadienyl carbanions.

Finally, while [1,6]-hydrogen shifts are predicted to proceed according to the antarafacial mode under thermal conditions (Table 7.4), the opposite is the case under photochemical conditions. Here, the migration should follow the suprafacial route. True to prediction, irradiation of a solution containing the cycloheptadienyl anion (THF–hexane at 0 °C) results in quantitative conversion to the rearranged anion (equation 7.15).

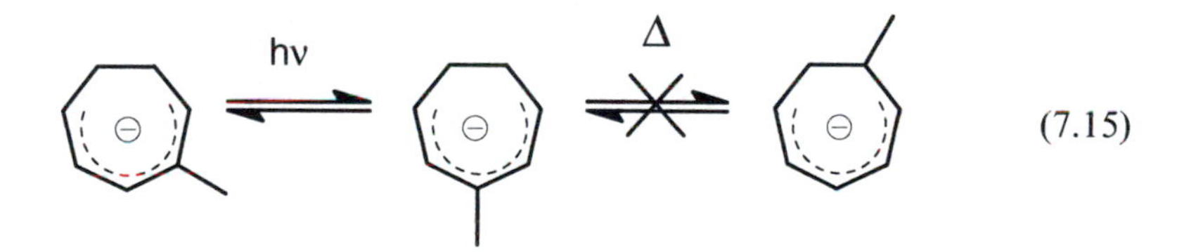

(7.15)

Conversely, rearrangement does not occur upon heating the solution (150 °C). Clearly, the ring system prevents operation of the thermal antarafacial mode but permits the photochemical suprafacial sigmatropic rearrangement.[98]

7.3.2.2 [i, j] Alkyl, Alkenyl, and Aryl Migrations

Sigmatropic rearrangement of carbon-centered groups falls into two categories: migration of saturated carbon groups and migration of unsaturated moieties. Rearrangement of alkyl groups may be analyzed in the same manner as hydrogen shifts, with one modification: for a migrating hydrogen, the bond between the carbon terminus and the 1s orbital is broken and then reformed. In short, the orbital involved in the sigmatropic shift of a hydrogen is the spherically symmetrical 1s orbital, whereas an alkyl carbon fragment would plausibly utilize a p-orbital (or hybrid orbital) with its corresponding nodal property. This can be

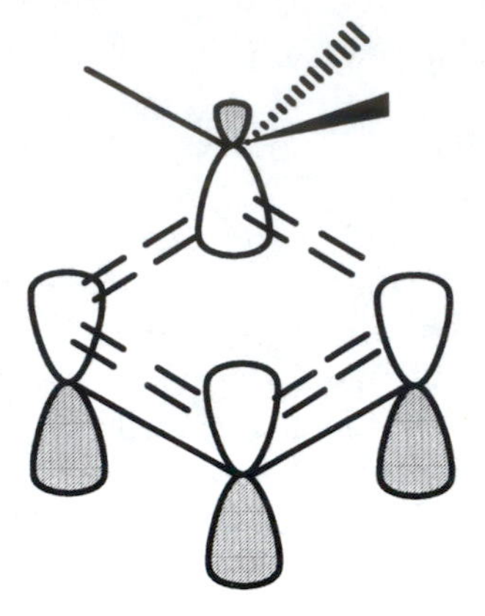

Figure 7.3. Hückel-type transition state.

illustrated for [1,3]-alkyl migration in a propenyl system (i.e., R–C–C = C → C = C–C–R). In Zimmerman's terminology,[22,23,92] under thermal conditions, the suprafacial [1,3] shift would involve a Hückel-type transition state (Figure 7.3); this transition state would contain only $4n$ electrons and, so, would be disallowed. Note that such a suprafacial [1,3] alkyl shift would occur with retention of absolute stereochemistry at the migrating carbon center. On the other hand, a suprafacial [1,3] shift is possible in a 4π-electron transition state involving one node, namely, a Möbius-type transition state,[92] as shown in Figure 7.4 using a p-orbital for the alkyl group. A consequence of this type of alkyl suprafacial shift is that inversion of configuration occurs at the carbon center of the migrating carbon fragment.

Let us now consider the [1,2] shift of an alkyl group. In the case of a carbocation, the suprafacial concerted transfer with retention of configuration at the migrating center follows from a Hückel transition state. In this 2π-electron (i.e., $4n + 2$ electron) system, the suprafacial [1,2] shift can be seen to be favorable. For

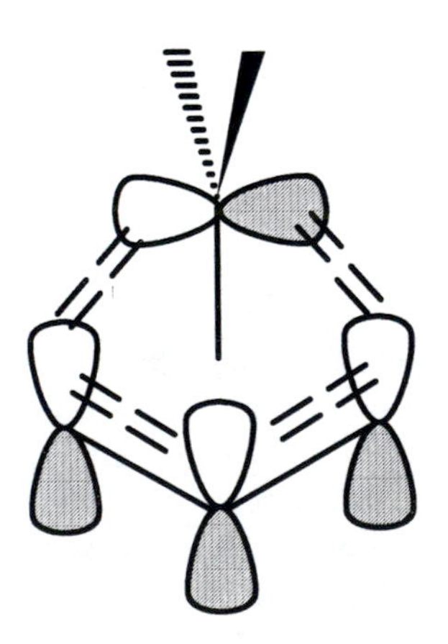

Figure 7.4. Möbius-type transition state.

a carbanion, a 4π-electron (i.e., 4*n* electron) system, the suprafacial [1,2]-shift with retention would be disallowed. While the suprafacial [1,2] shift of an alkyl group with inversion would be permissible on orbital symmetry grounds, the process entails such geometrical constraints as to be energetically difficult. (In contrast, Berson has demonstrated that structurally related transition states are accessible for neutral species.)[99] The antarafacial [1,2] shift of an alkyl group with retention would also generate a thermally allowed Möbius-type transition state, but, again, steric considerations suggest that such a concerted process would be energetically prohibitive, as was previously concluded by Zimmerman and Zweig.[100]

The consequence that follows from the limited number of orbital-symmetry-allowed thermal pathways and the sterically challenging nature of those paths that are favorable according to orbital symmetry considerations is as follows. No unambiguous examples of *concerted* [1,2]- or [1,3]-alkyl group migration are known.[9,94,101] In fact, no clear examples of concerted [1,4] or [1,6] shifts of saturated carbon moieties have been reported, thus far.

It may well be argued that if relatively few thermally allowed pathways are available and, further, if those that are available are precluded on steric grounds, then any simple [1, *j*]-alkyl migration that is observed may proceed via a lower-energy, but nonconcerted route. In this regard, a Grovenstein–Zimmerman rearrangement (Section 7.4.4, *vide infra*) in the 2,2,3-triphenylpropyl system that involves the formal [1,2] shift of a benzyl group appears to follow an elimination–addition (stepwise) mechanism, as shown in Scheme 7.7.[100,102-105] Importantly, the putative intermediate in this system (1,1-diphenylethene) was diverted by trapping with a range of reagents including benzyllithium isotopically labeled at the α-position with ^{14}C.[101,102] (Other examples of the Grovenstein–Zimmerman rearrangement, as well as an alternative mechanism, will be considered in Section 7.4.4.).

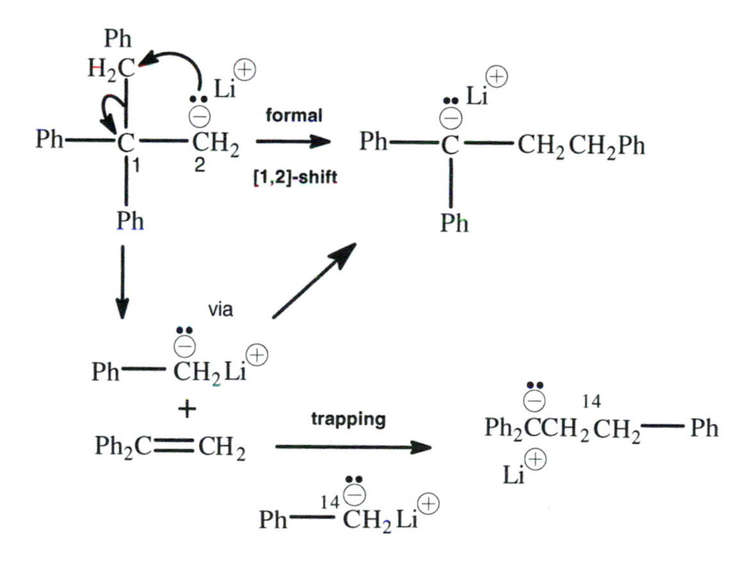

SCHEME 7.7

A [1,2]-sigmatropic shift of an *unsaturated* carbon fragment along a C–C chain is constrained to proceed with retention and, therefore, according to the antarafacial mode. The steric costs are such that this concerted process has not been observed.

A formal [1,2] shift of an allyl group (note numbering: 1′, 2′ in Scheme 7.8) has been reported;[106] the product may arise by either an elimination–addition mechanism or by a sigmatropic rearrangement of the order [2,3]. A similar competition between [2,3]-sigmatropic rearrangement and the stepwise elimination mechanism has been reported;[107] it involves a fluorenyl methide derivative as shown in Scheme 7.9.

As we have seen, there are relatively few examples of concerted sigmatropic rearrangements involving simple carbanions.[95-108] However, a significant number of name rearrangements, which generally involve migrations of heteroatom-centered moieties, may proceed, at least partly, by concerted sigmatropic routes. Where appropriate, therefore, orbital symmetry considerations will be discussed in the following sections.

7.4 NAME REARRANGEMENTS INVOLVING CARBANION INTERMEDIATES

A number of different rearrangements have been investigated to such a depth or have shown such synthetic utility that they have come to be recognized by name or, rather, by the name of the initial discoverer(s) or investigator(s) of these rearrangements. In the following sections, we will examine some of these anionic name rearrangements.

In discussing these name rearrangements, our focus will be on the mechanistic aspects. Nonetheless, it is important to recognize that many of the rearrangements have been found to be of significant synthetic utility.[109] As but one example,[110] a review shows that [2,3]-Wittig rearrangements[111-114] have taken their place in the arsenal of synthetic organic chemistry.

7.4.1 Wittig, Brook, and Retro-Brook Rearrangements

Rearrangement of ethers in the presence of alkyllithiums to give lithium alkoxides is termed the [1,2]-Wittig rearrangement.[2,115,116] Generally, the initially formed lithium carbanide is stabilized by an α-aryl group. The rearrangement is, therefore, a formal [1,2] shift of a group (originally R = Me or CH_2Ph)[115] from an oxygen center to the carbanionic center (equation 7.16).

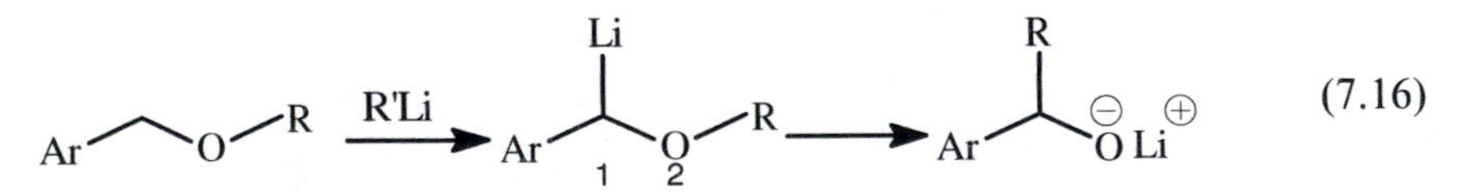

(7.16)

In equation 7.16, the migrating R group may be alkyl, aryl, or alkenyl; migratory tendency follows the order allylic, benzylic > ethyl > methyl > phenyl.[2,117] The most common mechanism for the rearrangement when R is alkyl or aryl involves

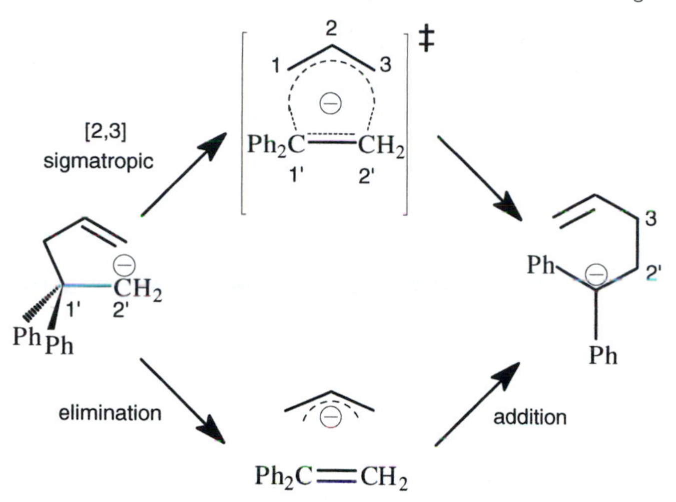

SCHEME 7.8

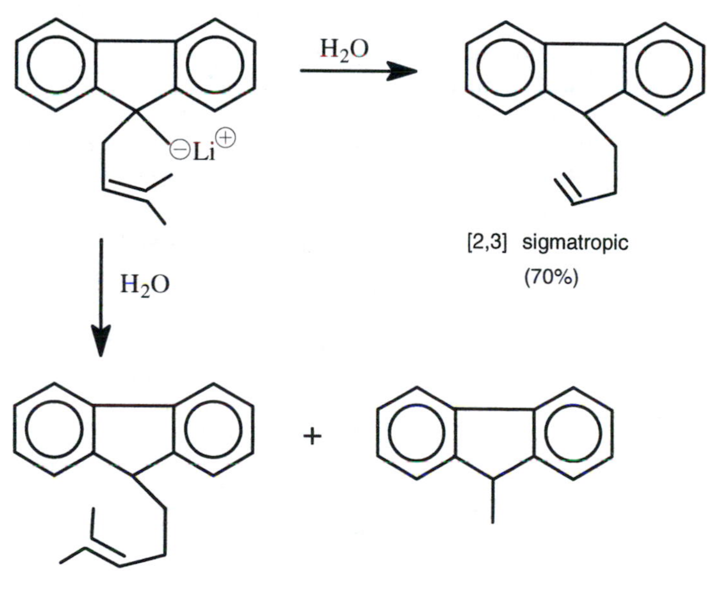

SCHEME 7.9

initial homolysis of the carbanide to yield a radical–radical anion pair that then recombines, presumably in a solvent cage, to yield the rearranged alkoxide product,[118,119] as shown in equation 7.17.

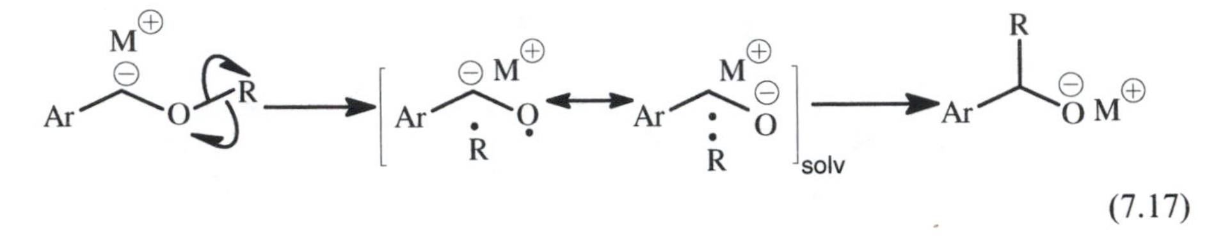

(7.17)

Note that the radical anion of the radical–radical anion pair in equation 7.17 is of the well-known ketyl type. Further, the migratory aptitudes cited are more consistent with the known order of radical stability than that of carbanion stability, which therefore supports the view that radical species are involved.[26] Cross-over products have been obtained in some cases, indicative of a stepwise mechanism rather than a concerted process.[120] Significantly, similar products result when ketyl radical anions and radicals derived independently from other precursors are allowed to react together.[121] Inasmuch as this radical anion–radical pair mechanism requires coupling of the radical species in a solvent cage, the nature of the counterion (M^+ above) and the solvent system are clearly important in determining the yield and purity of the products.[122,123]

Ab initio calculations on the prototypical system for a [1,2]-Wittig migration of a methyl group (i.e., $^-CH_2OCH_3$) have supported the view that a concerted [1,2]-sigmatropic shift is not energetically feasible. However, this initial study suggested a *heterolytic* rather than homolytic dissociation of the carbanion as the preferred route in the gas phase,[124] in agreement with experimental gas-phase studies of deprotonated benzyl and allyl ethers.[125] (Interestingly, rearrangement of the $^-CH_2OSiH_3$ anion was predicted to follow a process involving a pentacoordinated silicon in a cyclic intermediate.[126]) These gas-phase calculations ("complete active space SCF" method using 6-31G(d) basis sets)[127,128] have been refined to include both the lithium counterion and some solvent molecules and, so, they more closely approximate the situation in solution.[128] The conclusions of this study are: (1) in the gas phase, the free carbanion, $^-CH_2OCH_3$, most likely dissociates in a heterolytic manner to give formaldehyde and the methide anion, though a homolytic dissociation could occur in solution, depending upon the tightness of a presumed solvent cage; (2) lithium ion in the $Li^+\,^-CH_2OCH_3$ system associates with the oxygen of—and stabilizes—the incipient ketyl radical anion and thereby encourages the homolytic pathway; and (3) with lithium cation and three dimethyl ether molecules to complete coordination to the lithium, the preferred pathway is still homolytic.[126] These calculations lead us to conclude that methyl (and by extension other alkyl groups) in the [1,2]-Wittig rearrangement migrates via the homolytic elimination–addition mechanism. The significance of solvent and of the role of the counterion in these systems, however, should not be minimized.

A particularly interesting example of the role of the counterion in the [1,2]-Wittig rearrangement is provided by the following enantiomeric ethers (equations 7.18a and 7.18b):[129]

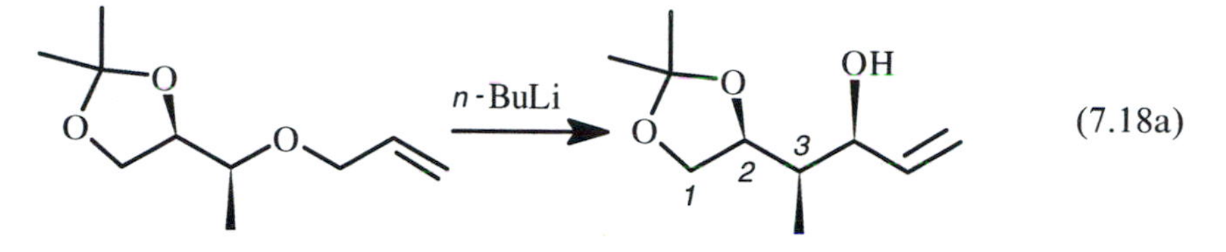

In the [1,2] rearrangement shown in equation 7.18a, 94% retention was reported at the migrating center (C-3 in equation 7.18a). Similarly, rearrangement of the stereoisomeric allyl ether occurred with 93% retention at the migrating center:

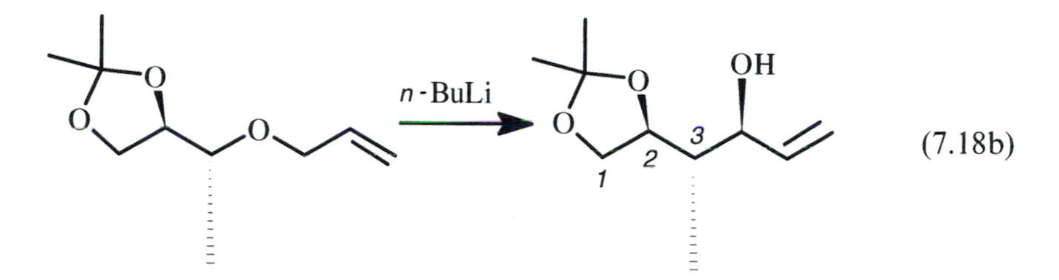

The retention of configuration at the migrating center, rather than inversion, is consistent with the nonconcerted mechanism, in that inversion would be expected in the thermally allowed process. However, the high degree of retention (rather than racemization) was rationalized on the basis that the lithium counterion could coordinate to both the carbanionic center and an oxygen of the acetal to preserve structure during the homolytic dissociation and radical recombination (Scheme 7.10).[129]

Unlike the [1,2]-Wittig alkyl and aryl rearrangements that generally proceed by a nonconcerted route,[130,131] allylic migrations can show a range of concerted and nonconcerted behavior. An excellent example is provided by the carbanion formed from the bis-γ,γ-(dimethyl)allyl ether[122,123] shown in equation 7.19.

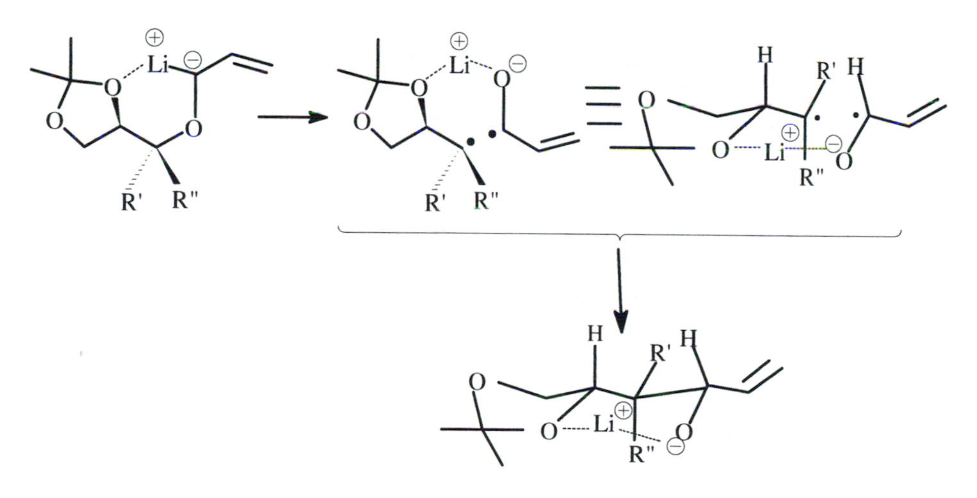

SCHEME 7.10

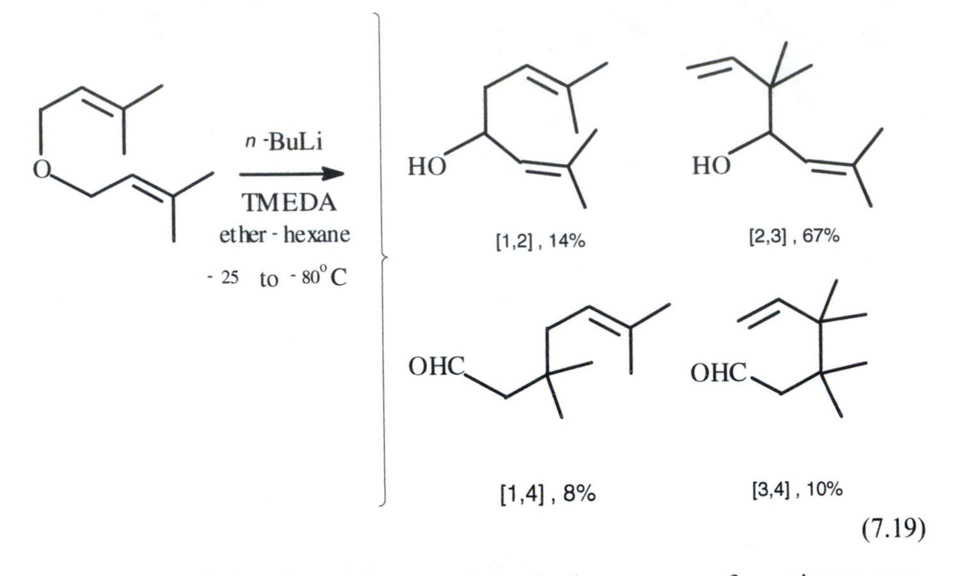

(7.19)

Here, treatment of the ether with strong base, in the presence of a cation sequestering agent, yields products of the stepwise pathways for both the [1,2] and [3,4] migrations. These stepwise rearrangement products are formed in relatively low yield, along with the products of the concerted thermally allowed [2,3]- and [1,4]-sigmatropic Wittig rearrangements. The aldehydes (in the [1,4] and [3,4] rearrangements) form via tautomerization of the initial enolate anions. In the related allyl cyclohexenyl system, the orbital-symmetry-required retention of sterochemistry has been demonstrated for the [1,4]-sigmatropic rearrangement (as shown in equation 7.20).

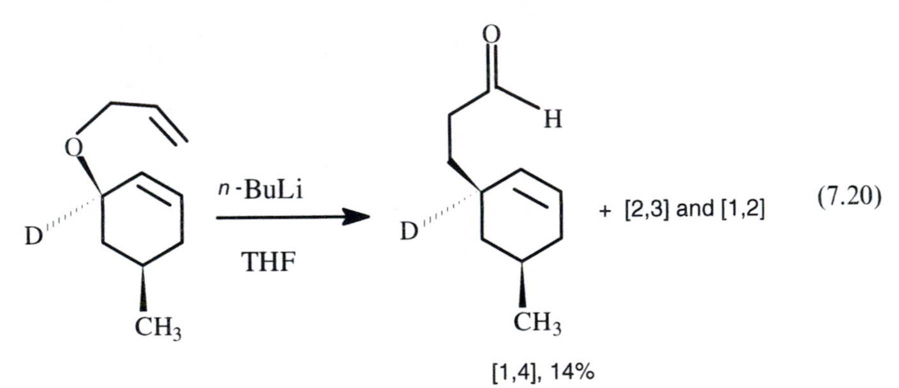

(7.20)

In general, the [2,3]-Wittig rearrangement of allyl ether derivatives occurs under milder conditions than the [1,2] migration. (Note, however, that the search continues for even milder conditions to induce this rearrangement.[132]) Higher yields may also be realized. As a result, the synthetic utility of this rearrangement has been recognized.[133] The thorough review by Marshall[110] details many interesting examples of syntheses that rely on the rearrangement, including its use in the total synthesis of the sesquiterpene antibiotic, (−)-punctatin A.[134] Particular attention has also been drawn to the possibility of linking the [2,3]-Wittig rearrangement to other sigmatropic rearrangements, occuring either in tandem or sequentially. The application of the [2,3]-Wittig rearrangement along with [3,3]-sigmatropic rearrangements, such as the [3,3]-oxy-Cope rearrangement,[111-113] have also been covered in the review by Marshall.[110]

The concerted [2,3]-Wittig rearrangement has been suggested to occur through a five-membered transition state with an envelope conformation.[135] However, a somewhat different transition-state structure has been advanced on the basis of different ab initio calculations,[136] and a later study suggested a dependence of transition-state structure on the stabilization afforded to the carbanion by substituents.[137] It would appear premature, therefore, to describe the transition state for this rearrangement in any further detail.

Alkyl allyl sulfides, as analogues of the ethers discussed above, also undergo the [2,3]-Wittig rearrangement in the presence of strong bases, although rates of reaction tend to be slower for the sulfides.[138] Carbanions substituted in the α-position by various amino groups also undergo the [2,3]-Wittig rearrangement, now termed the [2,3]-*aza*-Wittig rearrangement.[139-142] (The rearrangement of nitrogen ylides or α-ammonium carbanions, the Stevens rearrangement, will be discussed in the next section). Further, the Wittig rearrangement has also been demonstrated with dialkyl selenides.[143]

Organosiloxanes also undergo the related retro-Brook rearrangement (and its reverse, the Brook rearrangement). We now turn our attention to the behavior of these organosiloxanes under basic conditions.

Like the related Wittig rearrangements, the Brook and retro-Brook rearrangements have demonstrated synthetic utility.[11-15,144,145] In the Brook rearrangement, a silyl alkanol rearranges to an alkoxysilane under the influence either of small amounts of strong bases (e.g., NaH, $NaNH_3$, KH, and RLi) or even of milder bases (pyridine and Et_3N).[11] As shown in equation 7.21, the process is again formally a [1,2] shift; in this case, a silicon-centered group rearranges.

$$R_3Si\overset{1}{—}CR'_2(\overset{2}{OH}) \xrightarrow[\text{catalytic}]{B:} R_3Si—O—CHR'_2 \qquad (7.21)$$

As demonstrated by Brook and co-workers, a chiral silicon group rearranges with retention of configuration.[12] Hammett substituent constant studies show that significant negative charge develops at the benzylic carbon in the rate-determining transition state and that the transition state is stabilized by electron-withdrawing substitutents.[146] The proposed mechanism (Scheme 7.11) involves initial equilibration of the catalytic amount of strong base with the trialkylsilyl alkanol to give the alkoxide. The alkoxy anion attacks at the silicon center possibly giving rise to a pentacoordinate silicon intermediate[126,147] (or a transition state with pentacoordinate silicon character) with concomitant cleavage of the Si–C bond. In the final step, the much more basic carbanion re-equilibrates to yield the isolated alkoxysilane. The driving force for the Brook rearrangement appears to be the formation of the strong Si–O bond (ca. 120–130 kcal/mol) as compared with the relatively weak Si–C bond (ca. 75–85 kcal/mol) that was broken.

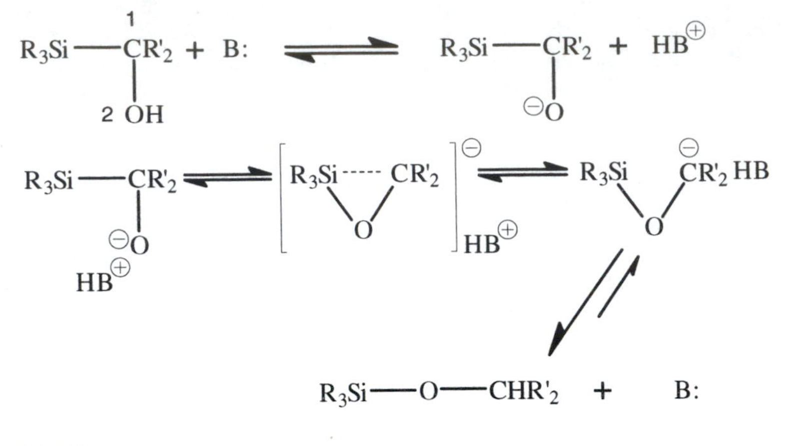

SCHEME 7.11

The retro-Brook rearrangement, that is, [1,2] shift of the silicon group from oxygen to carbon, is strictly comparable to the Wittig rearrangement and is often termed a silyl-Wittig rearrangement. (Note that a distinction should be made between the silyl-Wittig rearrangement and the [2,3]-*sila*-Wittig rearrangement, the latter of which involves rearrangement of an (allyloxy)silyllithium, a silyl anion, to a lithium allylsiloxide).[148] In the retro-Brook rearrangment (equation 7.22), an alkoxysilane, typically a benzyloxytrialkylsilane, is treated with equimolar or greater amounts of strong base (e.g., RLi with TMEDA),[14,15] and a benzylic carbanion is generated, followed by [1,2] migration of the trialkylsilyl group from the oxygen to the carbon center.[149]

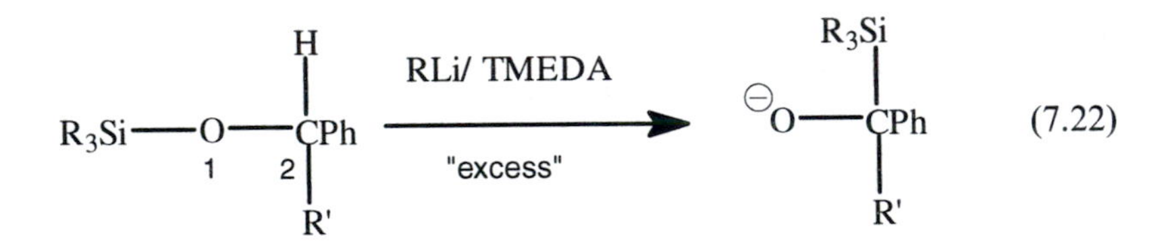

Protonation in workup would yield the silyl alkanol, the product of the retro-Brook, the starting material for the Brook rearrangement. The preference of the negative charge to reside on the more electronegative oxygen rather than on the carbon favors this rearrangement in the more highly basic media employed.

Clearly, reaction conditions are important in these rearrangements. Not only will an excess of strong base favor the retro-Brook rearrangement over the Brook rearrangement, but also, depending on the nature of the base, competitive formation of other products may occur. In this regard, Brook and Fieldhouse[150] reported that in the reaction of benzoyltriphenylsilane with the Wittig reagent, triphenyl phosphonium ethanide, the Brook rearrangement was followed by elimination to give the enol silyl ether, as shown in equation 7.23.

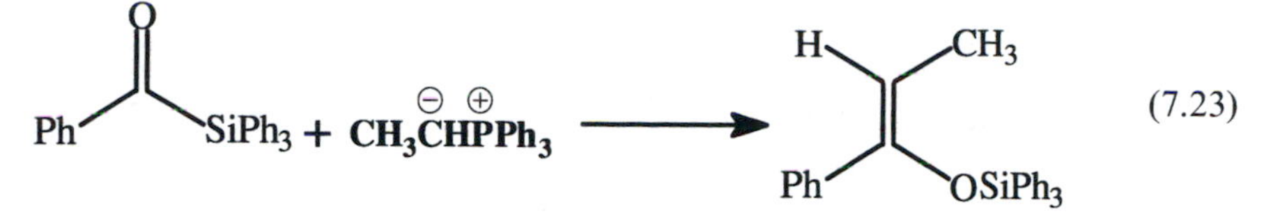

(7.23)

Whereas in mechanistic studies, such side reactions may "muddy the waters", it should be recognized that the linkage of several reactions, such as the Brook rearrangement–elimination cited here, may be exploited synthetically.[151,152]

We now turn our attention to the Stevens rearrangement and related reactions.

7.4.2 Stevens and Related Rearrangements

The Stevens[153] and related rearrangements, such as the Sommelet–Hauser[154] and Meisenheimer rearrangements,[155] involve ylides or comparable zwitterionic species. As we have seen (Chapter 3), ylides may be treated as carbanions that are strongly stabilized by an adjacent positively charged and, in the case of groups that contain a period three or greater central atom, highly polarizable substituent. In the Stevens rearrangement (equation 7.24), then, allyl, benzyl, propargyl, and phenacyl groups undergo a [1,2] (or [2,3] in the case of allyl moieties) shift from an ammonium ion center to a carbanion center in the ammonium (and later extended to the sulfonium)[156] ylide:

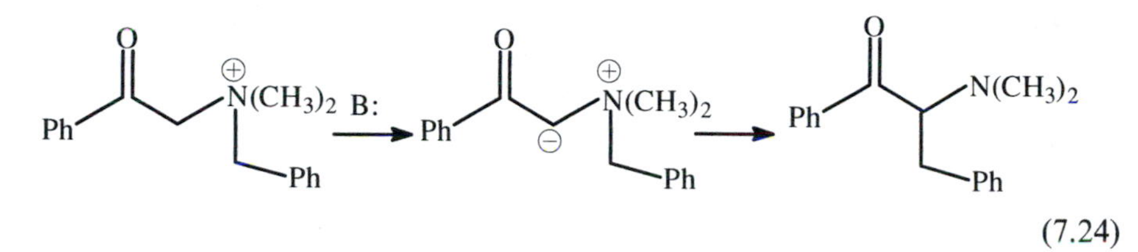

(7.24)

Mechanistically, substituent effect studies (4-X groups on the phenyl ring of the migrating benzyl group) suggest the intermediacy of the ylide.[157] In fact, in appropriate cases the ylide can be isolated.[158] Further, it was shown by use of isotopic labeling (^{14}C starting material) and by cross-over experiments that the rearrangement is intramolecular.[159] That this [1,2]-alkyl shift from the ammonium nitrogen to the carbanionic center occurs with retention at the carbon center was shown by Kenyon and co-workers[160] and, since a suprafacial [1,2]-sigmatropic shift with retention is symmetry forbidden and an antarafacial migration is energetically prohibitive on steric grounds, a concerted intramolecular rearrangement may be ruled out. (Equally clearly, a concerted sigmatropic shift of the [2,3] kind can operate in those systems where an allylic fragment migrates, as found early for the Wittig rearrangement.)

At the time of writing, the accepted mechanism relies on the concept of a solvent cage within which the short-lived intermediates may recombine without inversion. This mechanism (Scheme 7.12)[161] is reminiscent of that generally accepted for the [1,2]-Wittig rearrangement,[117,118] that is, homolysis of the ylide followed by recombination in the solvent cage. Evidence for this radical ion–radical pair mechanism comes from observation of the CIDNP (chemically

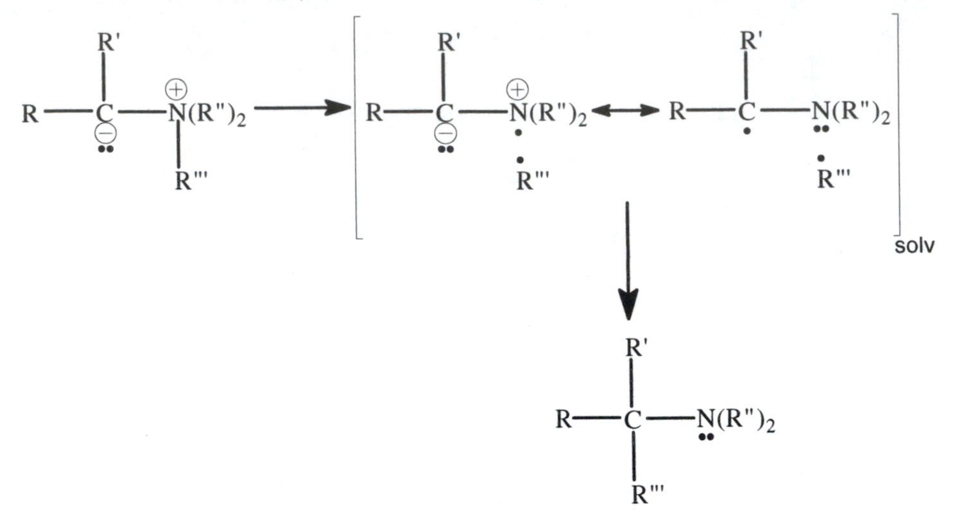

SCHEME 7.12

induced dynamic nuclear polarization) phenomenon,[162] which generally supports the existence of free radicals in a mechanism.[163,164] Further evidence comes from identification of small amounts of radical coupling product (R'''-R''') in some reactions; such products could arise from radicals that had "escaped the solvent cage".[165] Finally, a similar mechanism[166] has also been suggested for the related Meisenheimer rearrangement,[167] whereby some tertiary amine oxides undergo an apparent [1,2] shift of an allylic or benzylic group (i.e., $R' =$ allylic or benzylic) from the positively charged nitrogen to the anionic oxygen center to yield an alkoxyamine, as shown in equation 7.25.

$$(R,R'')\underset{1}{\overset{\oplus}{N}}(R')-\underset{2}{\overset{\ominus}{O}}\xrightarrow{\Delta}(R,R'')N-OR' \qquad (7.25)$$

The synthetic utility of the Stevens rearrangement is somewhat hampered by the tendency of the starting ammonium salts (and, to a lesser extent, sulfonium salts) to undergo Hofmann elimination rather than rearrangement upon treatment with base. Moreover, in formation of the ammonium or sulfonium salt by alkylation of a tertiary amine or a dialkyl sulfide with a haloalkane, for example, the halide anion may, in turn, act as a nucleophile to dealkylate the ammonium or sulfonium salt. Finally, in the case of a benzyl trialkylammonium ylide equilibration may occur to produce an alternative ylide that then undergoes the Sommelet–Hauser rearrangement (as shown in Scheme 7.13).[168-170] Note that the final step in the above sequence is a 1,3-prototropic rearrangement with concomitant rearomatization. The requirement for this step in Sommelet–Hauser rearrangement means that it competes with the Stevens rearrangement; this may be suppressed

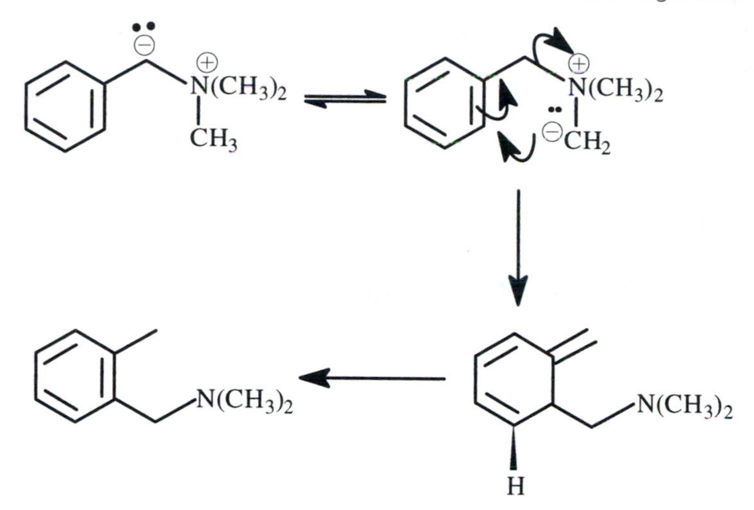

SCHEME 7.13

under aprotic reaction conditions and, conversely, encouraged under protic alkaline conditions.[171,172]

Furthermore, the use of alkylating agents whose counterions are non-nucleophilic,[173] or more novel methods of preparing the precursor salts to the ylides,[174–177] or direct formation of the ylide by addition of a carbene or benzyne to a sulfide or amine,[178,179] have increased the utility of the Stevens rearrangement in synthesis by suppressing competitive dealkylation and Hofmann elimination. In particular, the [2,3]-Stevens rearrangement of allylic nitrogen- and sulfur-centered ylides has proven to be synthetically useful. The early successful use of a [2,3]-Stevens rearrangement of a sulfonium ylide in the synthesis of squalene[180] has been joined by numerous examples that have been compiled in review by Markó.[181] We can only look forward to the increasing popularity of these rearrangements (and those outlined in the previous sections) in organic synthesis, particularly where stereochemical control is important.

7.4.3 Benzil–Benzilic Acid Rearrangement

As early as 1838, von Liebig[182] described the reaction of the α-diketone, benzil, with hydroxide, to give the salt of the α-hydroxy acid, namely benzilic acid, as shown in equation 7.26.

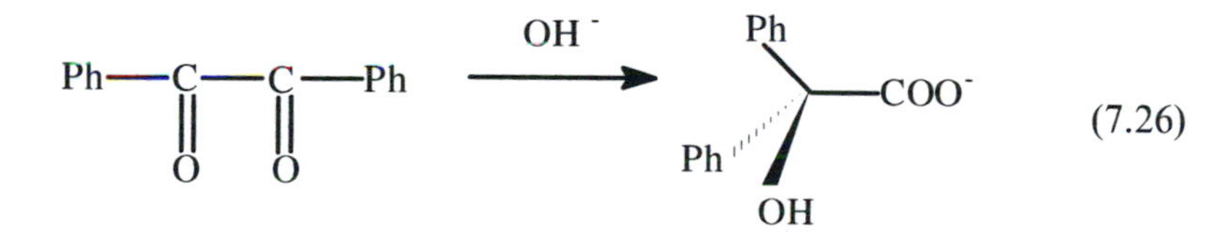

(7.26)

Benzil or 1,2-diphenyldione is the parent compound for the benzil–benzilic acid rearrangement, but this [1,2] shift may involve aryl groups and, more generally, alkyl,[183] acyl, and nitrogen-centered moieties.[184] (Also see Section 6.4.1.2 for the related base-promoted rearrangement of α-epoxy ketones, as well as the semibenzilic mechanism found in some examples of the Favorskii enolate rearrangement.)

The Ingold mechanism[185] involves initial and reversible nucleophilic attack of hydroxide (or alkoxide) on a carbonyl group, followed by slow [1,2] shift of the aryl (or other group, given here as R′) to the adjacent carbonyl, with loss of C = O π-bonding, and then followed by equilibrium protonation, as shown in Scheme 7.14.

Evidence in favor of the mechanism comes from a number of studies. First, when the diketone is an unsymmetrical diaryl one, attack of hydroxide occurs preferentially at the more electrophilic carbonyl, namely the one substituted by a phenyl ring bearing more of the more electron-withdrawing substituents.[186] Barring steric difficulties, then, it is also the more electron-deficient group that undergoes the [1,2] migration in the rate-determining irreversible step. Isotopic labeling studies have been used to determine migratory tendencies for groups (R′); generally, preferential migration occurs, in order, for the following pairs: H > Ph, $CONH_2$ > Ph, Ph > Me, CO_2R > COO^- and cyclopropyl.[187,188]

The benzil–benzilic acid rearrangement also occurs in aprotic solvents with non-oxygen-containing bases such as sodium amide, followed by a mild aqueous workup.[189] Under these conditions, it might be expected that the product would not be benzilic acid, but benzilamide. The conditions are clearly too mild for any benzilamide, once formed, to hydrolyze. One possibility[186] is that one molecule acts as the oxygen source for another molecule of the diketone, as shown in Scheme 7.15. The penultimate product could then be described as a hemiaminal and, consequently, would be susceptible to mild hydrolysis.

While the Ingold mechanism is widely accepted, there is also evidence for free-radical routes, perhaps involving single electron transfer (SET).[190,191]

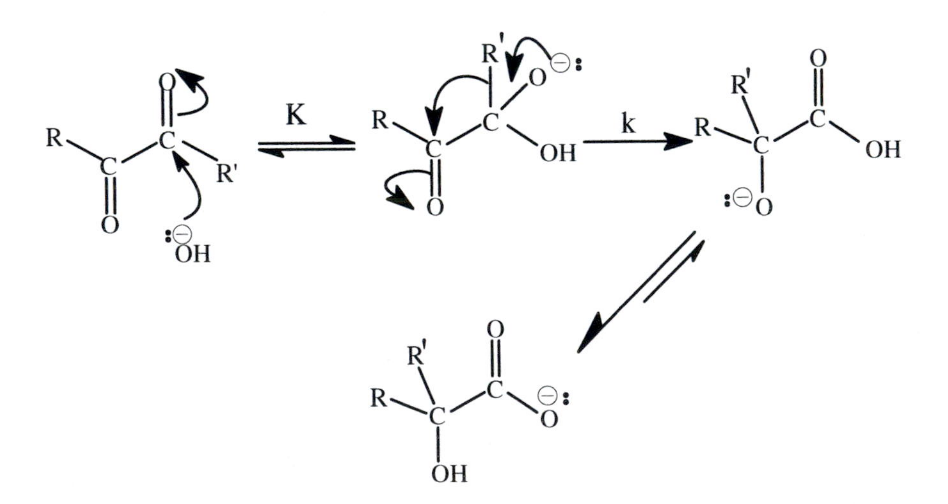

SCHEME 7.14

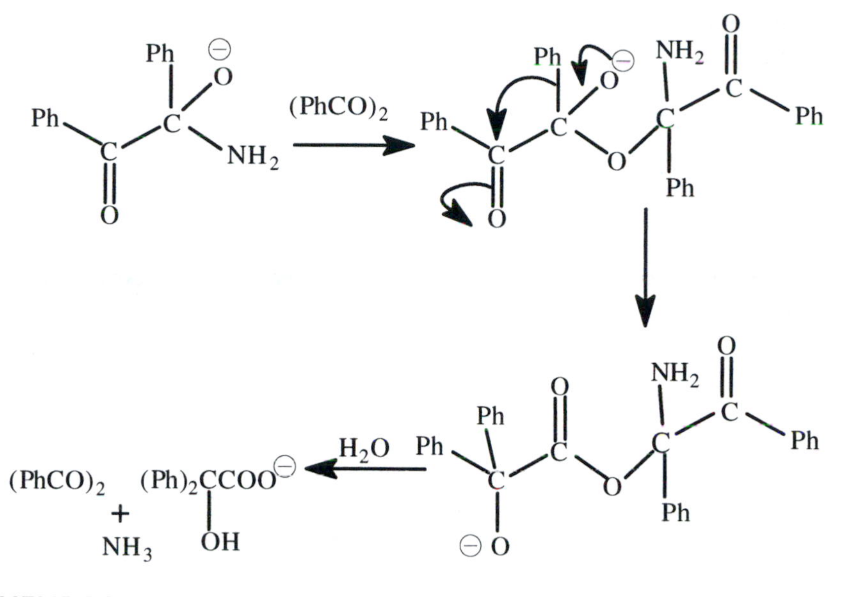

SCHEME 7.15

Regardless of mechanism, however, it is clear that the benzil–benzilic acid (and related benzil–benzilic ester) rearrangements do not directly involve carbanions (except as illustrated in Section 6.4.1.2), although they occur under basic conditions and carbanions, such as acetylide,[192] have been shown to initiate them. On the other hand, the rearrangement does represent an apparent [1,2] shift of aryl groups to a neutral (or, at least, uncharged, albeit electrophilic, carbon), and, as such, it is a useful introduction to the Grovenstein–Zimmerman rearrangement that we will discuss in the next section. The review by Gill provides further information on the benzil–benzilic acid rearrangement, including examples of the application of this rearrangement to synthesis.[193]

7.4.4 Grovenstein–Zimmerman Rearrangement

The Grovenstein–Zimmerman rearrangement[100-106] comprises the [1,2] migration of an aryl group in a carbanion (or metal carbanide). The first, and most studied, example of the rearrangement was that found by Grovenstein,[194] and independently and almost simultaneously by Zimmerman and Smentowski:[195] 2-chloro-1,1,1-triphenylethane treated with sodium metal in ethereal solvents yields, after protonation in workup, 1,1,2-triphenylethane. Significantly, reactions with the sodium carbanide in the presence of a small quantity of *tert*-pentyl alcohol generated only 1,1,1-triphenylethane, the product of protonation of the initially formed carbanion without rearrangement. This last observation is supported by the study of Dixon and Streitwieser into the kinetic acidity of 1,1,1-triphenylethane; 1,1,1-triphenylethane undergoes exchange with cesium cyclohexylamide much faster than rearrangement.[196]

In a general form, the Grovenstein–Zimmerman rearrangement may be described by equation 7.27.

$$\underset{1}{\mathrm{RR'C}}(\mathrm{Ar})\text{—}\underset{2}{\mathrm{CH_2R''}} \xrightarrow[\text{2. } \mathrm{H^+}]{\text{1. R'''Li}} \mathrm{RR'CH}\text{—}\mathrm{CHR''}(\mathrm{Ar}) \qquad (7.27)$$

Ease of rearrangement is clearly dependent on the cation associated with the carbanide and the solvent. Thus, the lithium 1,1-dideuterio-2-(4′-phenylphenyl)-ethanide fails to rearrange (0 °C, Et_2O), whereas 50% rearrangement is found with the potassium carbanide (THF).[197] Similarly, 2,2,2-triphenylethyllithium is stable in THF at low temperature, but rearranges at 0 °C to the 1,1,2-triphenylethyllithium, and rearrangement occurs immediately at 50 °C with 2,2,2-triphenylethylpotassium in dimethoxyethane.[198] Further, addition of potassium *tert*-butoxide to 2,2,2-triphenylethyllithium at -75°C (THF) induces complete rearrangement within 10 minutes; presumably the reactive species is the potassium carbanide.[199]

Although a concerted mechanism may be ruled out on a combination of orbital symmetry and steric grounds (Section 7.3.2.2), two possible step-wise mechanisms have been advanced for the Grovenstein–Zimmerman rearrangement, and it is possible that either can operate, depending on the system under consideration. As we have seen above (Section 7.3.2.2), in the 1,2-benzyl migration in the 2,2,3-triphenylpropyl carbanion system, the intermediate 1,1-diphenylethene that would arise from initial elimination (in an elimination–addition process) could be trapped with [α-^{14}C]benzyllithium.[100-102] On the other hand, such trapping failed in the 2,2,2-triphenylethyllithium system.[200] A reconciliation of these results would require a much faster collapse of the caged-ion–alkene pair in the latter case than in the former.

However, the lithium carbanide of the following fluorenyl derivative did not rearrange[201] with the expected ring expansion as shown in equation 7.28.

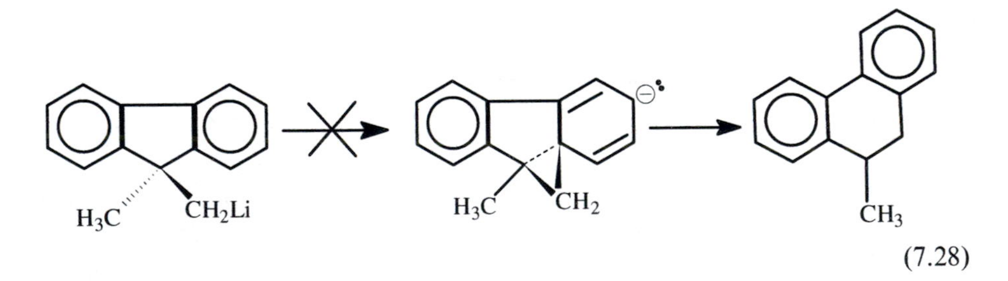

(7.28)

[With the related (9-*tert*-butyl-9-fluorenyl)methyllithium, Grovenstein and co-workers have reported loss of the *tert*-butyl group via an internal elimination mechanism.[202]] Since an elimination–addition mechanism should provide no barrier for the proposed rearrangement, the implication is that the rearrangement proceeds via an alternative mechanism. This alternative mechanism could involve the bridged species, which in this system would, in fact, be a strained spiro intermediate. The strain involved in formation of the intermediate could account for the failure of the rearrangement.

Migratory tendencies also support the proposal that the mechanism involves a bridged intermediate (or transition state modeled on this bridged intermediate).[103] For example, the tendency of a (4′-phenyl)phenyl group to migrate preferentially over a (3′-phenyl)phenyl group can be attributed to the ability of the former substituent to effectively delocalize negative charge in the spiro intermediate.

The spiro anion formed by dechlorination of 1-chloro-4-(4′-phenylphenyl)butane with cesium or potassium in THF (equation 7.29) was observed by NMR.[203]

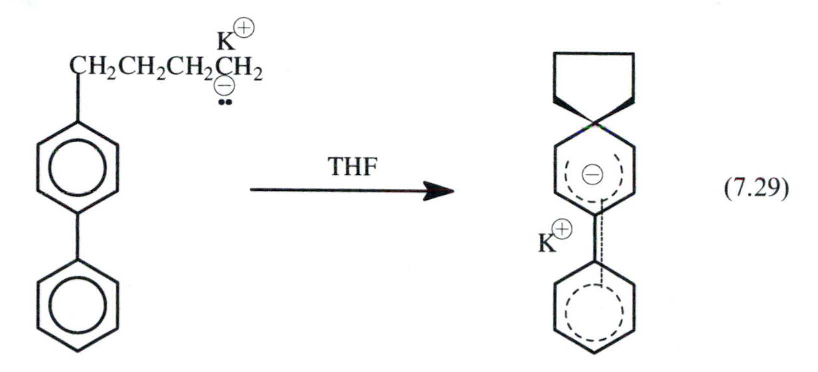

This evidence in favor of a spiro anion intermediate in the Grovenstein–Zimmerman rearrangement was further strengthened by the finding that reaction of 2-(4′-phenylphenyl)-3-chloro-2,3-dimethylbutane with Cs–K–Na alloy (THF, −75 °C) yielded, upon carbonation, a product likely formed from the spiro carbanion as well as the carboxylic acid derived from an open-chain carbanion (Scheme 7.16).[204,205] On these grounds, it appears that the most general mechanism for the Grovenstein–Zimmerman rearrangement involves a three-membered spiro carbanion (or a transition state similar to it).[203-206]

It should be pointed out, however, that in some systems highly reactive bases such as cesium–potassium–sodium alloy lead to formation of dianions and, hence, side reactions.[207,208] Nonetheless, the Grovenstein–Zimmerman rearrangement provides an alternative and likely nonconcerted route to [1,2] migration of aryl and vinyl groups in carbanions.

7.4.5 An Anionic Equivalent of the Fries Rearrangement: The Snieckus Rearrangement

Lithiation of an aromatic ring *ortho* to a directing group (methoxyl, carbamoyl, carboxamido, etc.) has proven itself to be a very useful synthetic methodology.[209-211] *Ortho* lithiation has also been combined with the carbanionic [1,2]-Wittig rearrangement to give an efficient synthesis of 2-aryl-2-hydroxypropanamides, analogues of anti-inflammatory drugs such as ibuprofen.[212] Stereochemical and regiochemical control in many of these reactions has been usefully rationalized by involving complex induced proximity effects (CIPE).[213]

In 1983, both Sneickus[214] and Chenard[215] reported a new carbanionic rearrangement involving initial *ortho* lithiation. The original paper by Sibi and

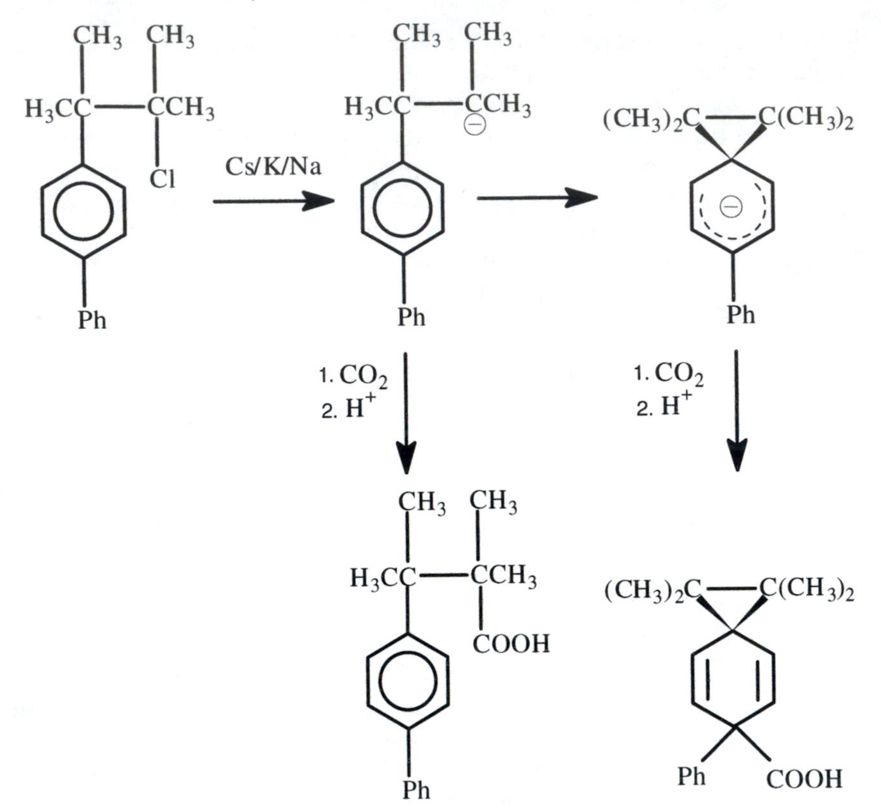

SCHEME 7.16

Sneickus[214] described the [1,3] shift of a carbamoyl group from an oxygen to the carbanide center (equation 7.30) as the anionic equivalent of a Fries rearrangement.[214]

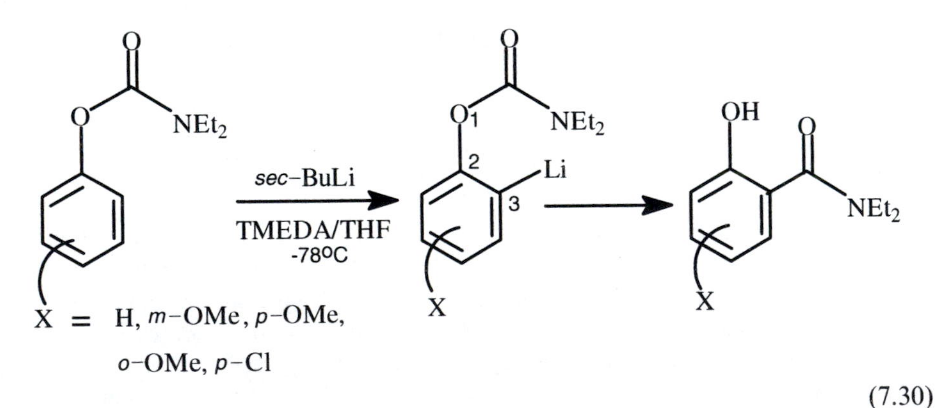

(7.30)

The Chenard paper[215] described a similar [1,3] rearrangement, but here the starting materials were 2-(arylthio)-4,4-dimethyloxazolines and, hence, the rearrangement involved a migration of oxazoline group from sulfur to the ring carbon (equation 7.31).

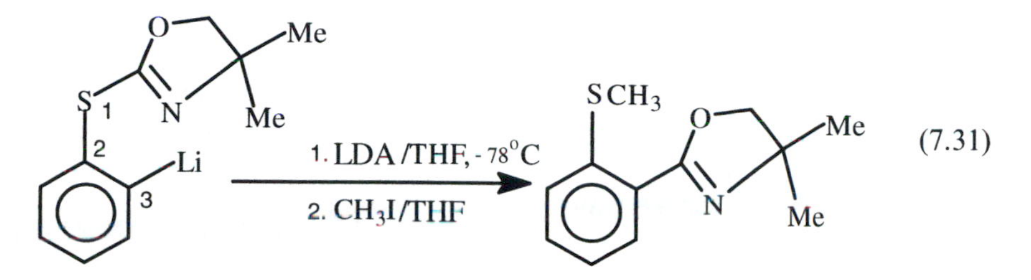

In this case, the rearrangement appears to be a carbanionic analogue of the Smiles rearrangement.[216] Interestingly, ring lithiation and rearrangement did not occur in this case with *n*-butyllithium (with or without TMEDA) or with phenyllithium; lithium diisopropylamide (LDA) was required to effect the rearrangement. Moreover, the rearrangement under these vigorous conditions also produced a number of side products, depending on the nature of the other substituents in the aryl ring.

Since these initial reports, a number of other investigations have demonstrated that the original anionic rearrangement has some generality,[217] and in a "remote" version it has clear synthetic utility as well.[218]

The mechanism of this rearrangement, which we will term the Snieckus rearrangement, is, as yet, unknown. While Chenard suggested a ring-closure/ring-opening process, based on the susceptibility of the oxazoline ring in his system to nucleophilic attack, it appears unlikely that the remote version of the rearrangement reported by the Snieckus group[218,219] proceeds via an intramolecular process. It is also possible that different mechanisms will apply in the proximate and remote versions of the Snieckus rearrangement.

At the time of writing, it appears prophetic that Snieckus has described this rearrangement as an "anionic-equivalent of the Fries" rearrangement, considering that the mechanism of the standard (cationic, Lewis acid-catalyzed) Fries rearrangement is still not fully understood.[220] Nonetheless, efforts of the Snieckus group, among others, have shown a parallelism between aromatic cationic chemistry (e.g., Friedel–Crafts acylation) and aromatic carbanionic chemistry,[211,221] and new rearrangements based on this connection may be expected to be found in the future.

7.4.6 Ramberg–Bäcklund Rearrangement

Unlike the Snieckus rearrangement, the mechanism of the Ramberg–Bäcklund rearrangement appears to be well understood. In fact, a review by Clough notes: "Very little additional work concerning the mechanism of the reaction has been published since 1975."[222]

In the Ramberg–Bäcklund rearrangement, an α-halosulfone reacts under alkaline conditions to lose sulfur dioxide and yield an alkene, often with the less stable *Z*-stereoisomer predominating (equation 7.32).[223, 224]

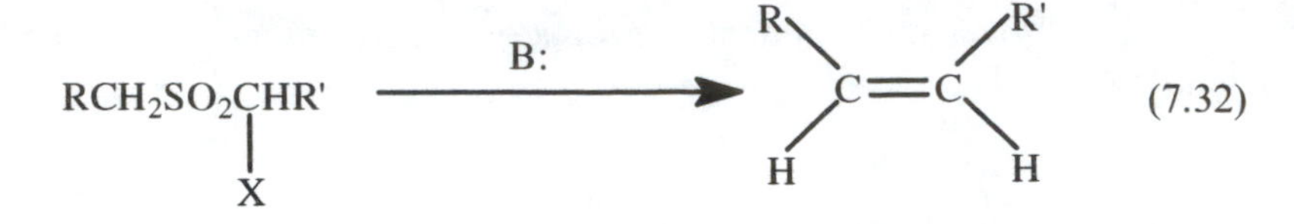

(7.32)

Halogen reactivity follows a standard order for nucleophilic displacements, namely I > Br > > Cl, but other leaving groups, such as sulfonates, and sulfinates can also be used.[225,226]

In the accepted mechanism,[222,224] the starting α-halosulfone is reversibly deprotonated; the reversibility was demonstrated by carrying out the reaction for a short period of time in methoxide/methanol-O-*d*, where deuterated starting material could be recovered. Deprotonation likely occurs preferentially from conformations that lead to stereoelectronic ($n \rightarrow \sigma^*$) stabilization of the resultant carbanion, that is from conformations that place the proton in between the two oxygens of the sulfone (cf. Sections 3.3.2 and 3.3.3). In principle, equilibration may take place between α and α′ carbanions (formed via deprotonation at each of the sites α to the sulfone), but only the α-carbanion depicted is aligned so as to permit the necessary internal displacement of the halide and ring closure to the thiirrane 1,1-dioxide (episulfone). (Choice of carbanion stabilizing substituents at a given position α to the sulfone would also provide some selectivity in the deprotonation step.) Subsequent loss of sulfur dioxide from the episulfone yields the alkene(s), as shown in Scheme 7.17.

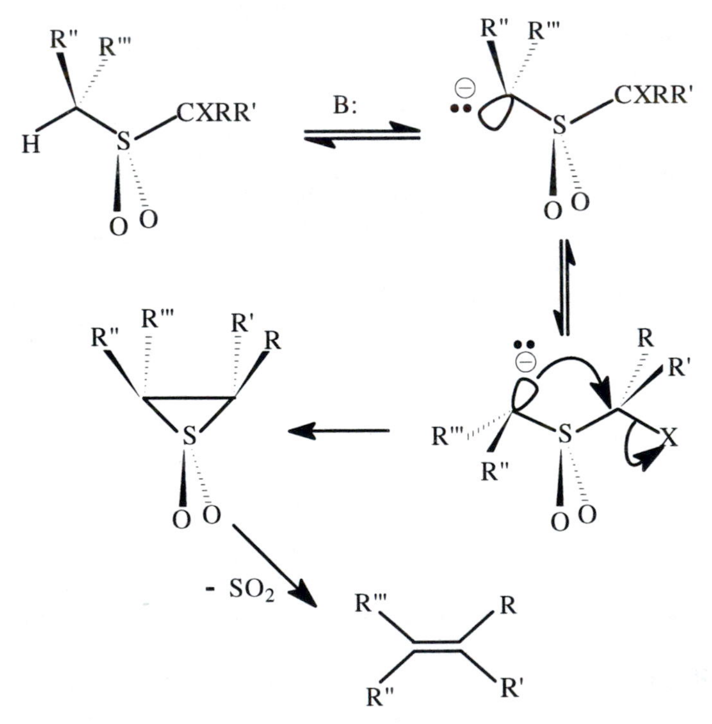

SCHEME 7.17

The mechanism accounts for the inversion of configuration found at both carbon centers;[227,228] inversion at the carbon bearing the leaving group as a result of backside attack by the carbanion leading to ring closure. The inversion at the carbanionic center is not so clear-cut. Based on calculations by Wolfe,[229] the typical (but unstabilized) carbanion adjacent to a sulfone should be intrinsically planar and, therefore, a process of inversion of the carbanion, as depicted in Scheme 7.17, may not be required. Regardless, inversion at the carbanionic center is also imposed by the requirements of the ring-closing nucleophilic attack.

The final step of the mechanism is the loss of sulfur dioxide and formation of the alkene. Episulfones have been prepared independently and shown to extrude sulfur dioxide, either when heated or when treated with base to form alkenes, but at a rate faster than that found in the Ramberg–Bäcklund rearrangement,[230,231] indicating that loss of sulfur dioxide is not rate determining in the rearrangement. Subsequently, an episulfone intermediate was isolated from the reaction.[232]

Concerted loss of sulfur dioxide under thermal conditions to give an alkene constitutes one of the chelotropic reactions governed by orbital symmetry.[233] The symmetry-allowed pathway for chelotropic extrusion of sulfur dioxide, in this case, is antarafacial, an unlikely proposition for a three-membered ring that is constrained to be planar. Consequently, the loss of sulfur dioxide in this rearrangement reasonably occurs by a stepwise mechanism. However, this leaves the problem of the observed stereoselectivity in the reaction, that is, that generally the less stable *Z*-isomers of the alkenes are preferred. There appears to be no satisfactory answer to this problem at this time.

Synthetically, the Ramberg–Bäcklund rearrangement has proven to be quite useful in the preparation of various cycloalkenes,[234,235] including the strained fused cyclobutene shown[236] in equation 7.33.

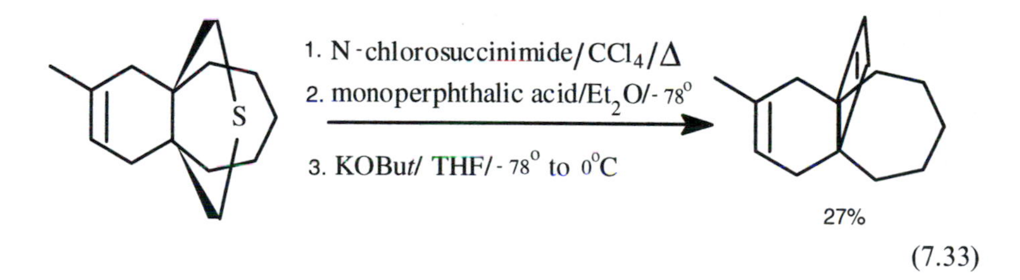

(7.33)

Another area of synthetic interest is the use of the Ramberg–Bäcklund rearrangement in the preparation of various cyclophanes and cyclophanedienes.[237,238] Although other synthetic methods can give better yields of these phanes,[239,240] the Ramberg–Bäcklund rearrangement can offer advantages in terms of control of regiochemistry in some cases. In this regard, the [3.3]paracyclophanediene, **98**, could be prepared in 2% yield from the precursor bis-sulfone, **99**, (equation 7.34), but only this single regioisomer was obtained.

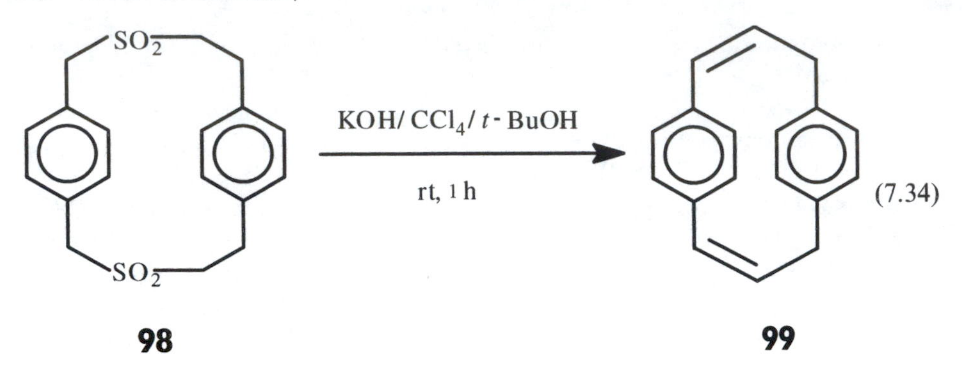

Other methods give higher yields, but also mixtures of regioisomeric cyclophanedienes.[241] Numerous other examples of the synthetic applications of this reaction are compiled in the review by Clough.[222]

Finally, we note the similarities between the Ramberg–Bäcklund reaction and the Favorskii rearrangement of enolates (Chapter 6). In both, a carbanion, stabilized by an adjacent group (carbonyl in the Favorskii case and sulfonyl in the Ramberg–Bäcklund reaction), attacks an sp^3-hybridized carbon in an intramolecular S_N2-type displacement to give a three-membered ring intermediate. In the Favorskii rearrangement, this intermediate does not undergo extrusion of a neutral molecule, nor is there evidence in the Ramberg–Bäcklund reaction for equilibration between the three-membered ring intermediate and an open dipolar intermediate, as is found in the major mechanism of the Favorskii rearrangement. Further, in the Ramberg–Bäcklund rearrangement, carbanion formation is rarely rate determining, unlike the situation in the Favorskii rearrangement. In fact, in the Ramberg–Bäcklund reaction, hydrogen–deuterium exchange has been found to occur prior to ring closure to the episulfone.[242] Thus, while there are superficial similarities between these two rearrangements, there are also significant differences that partly reflect the differences in stabilization of the initial carbanion afforded by a carbonyl group in the Favorskii case and by a sulfonyl group in the Ramberg–Bäcklund reaction.

7.5 CONCLUSIONS AND PROSPECTS

In this final chapter, we have attempted to provide an overview of rearrangements that involve carbanions. In so doing, we have attempted to cover the high points, while recognizing the impossibility of exhaustively covering this rapidly growing aspect of carbanion chemistry.

Rearrangements involving silicon centers that lead to elimination have been reported.[243,244] Dianionic rearrangements, such as those of alkoxyl benzyl dianions reported by Bates,[245] have not been considered in this review, nor has the interesting work of the Erickson group into ring expansion rearrangement with halomethylenecycloalkanes, under basic conditions, been examined here.[246,247] Rearrangements of homoenolate anions, which have been reviewed previously,[248,249] have also been omitted. The rearrangement of α-hydroxyphosphonates to phosphates may involve carbanions in some cases,[250,251] though not detectably in other related systems;[252] this rearrangement has not been considered here. Our desire to be comprehensive is, by necessity, tempered by the tremendous growth in this field of carbanion chemistry.[253,254]

One area particularly deserves mention, however briefly, and that is the area of carbanion-accelerated rearrangements. These are not, strictly speaking, rearrangements of carbanions. Rather, these are rearrangements whose rates are accelerated because a carbanionic center is built into the molecular structure. Since it is well known that π-electron-donating groups (Z) hasten the Claisen rearrangement (equation 7.35) of allyl vinyl ethers,[255,256] the Denmark group developed a carbanion-accelerated Claisen rearrangement, where the carbanion (or metalated carbon center) apparently acts as the electron donating group.[257,258]

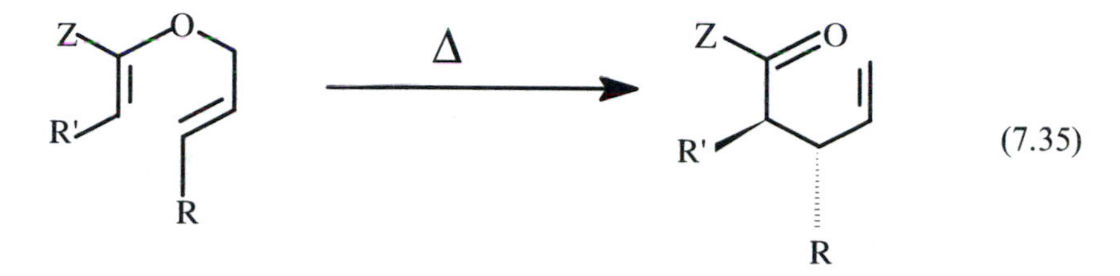

In this regard, the phenylsulfonylmethyl-substituted vinyl allyl ether (Scheme 7.18) does not undergo Claisen rearrangement at 50 °C (HMPT 3.5 h); that is, **100** does not rearrange directly to **103**. However, in the presence of KH, the phenylsulfonyl-substituted carbanion, **101**, is generated, which rearranges to the enolate, **102**, which, when protonated, gives the Claisen rearrangement product in 78% yield. Therefore, the rearrangement is achieved via the carbanionic pathway: **100** → **101** → **102** → **103** (Scheme 7.18).

Since these studies, a wide range of carbanion-accelerated rearrangements have been reported, including the carbanion-promoted hetero-Cope rearrangement,[259]

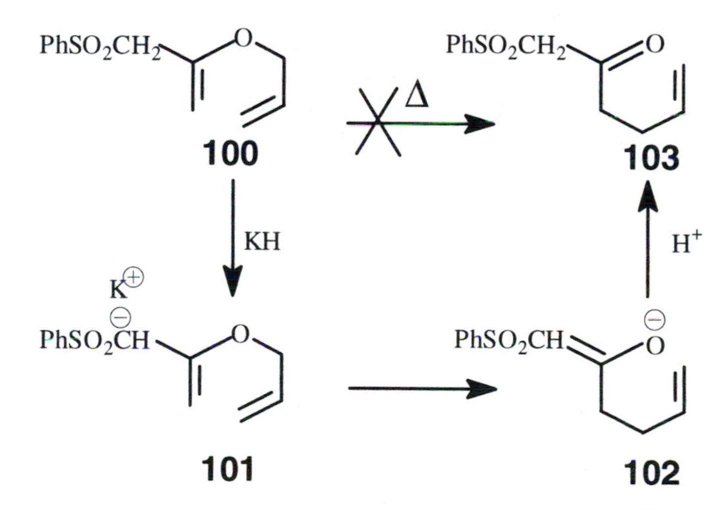

SCHEME 7.18

vinylcyclopropane–cyclopentene rearrangement,[260] as well as further examples of the Claisen rearrangement,[261] among others.[262-264]

There can be no doubt that the synthetic utility of these and the other rearrangements reviewed in this chapter will lead to further development of this aspect of carbanion chemistry. As in all of the studies we have considered in this work, our understanding of mechanism, and, where pertinent, different competing mechanisms has been highlighted. Insights that will be provided by examination of solvent effects, stereochemistry, and isotope effects, will "walk hand in hand" with our ability to exploit carbanion chemistry in organic synthesis.

References

1. Lowry, T.H.; Richardson, K.S. *Mechanism and Theory in Organic Chemistry,* 3rd ed.; Harper-Collins: New York, 1987; Chapter 5, pp 425–505.
2. Wittig, G. *Angew. Chem., Int. Ed. Engl.* **1954**, *66*, 10.
3. Pines, H.; Stalick, W.M. *Base-catalyzed Reactions of Hydrocarbons and Related Compounds*; Academic Press: New York, 1977.
4. Hattori, H. *Chem. Rev.* **1995**, *95*, 537.
5. Xu, F.; Redmer, R.A.; Tillyer, R.; Cummins, J.M.; Grabowski, E.J.J.; Reider, P.J.; Collum, D.B.; Huffman, J.C. *J. Am. Chem. Soc.* **2000**, *122*, 11212.
6. Remenar, J.F.; Lucht, B.L.; Kruglyak, D.; Romesberg, F.E.; Gilchrist, J.H.; Collum, D.B. *J. Org. Chem.* **1997**, *62*, 5748; and references therein.
7. Abbotto, A.; Streitwieser, A. *J. Am. Chem. Soc.* **1995**, *117*, 6358.
8. Hunter, D.H.; Stothers, J.B.; Warnhoff, E.W. In *Rearrangements in Ground and Excited States*; de Mayo, P., Ed.; Academic Press: New York, 1980; Vol. 1; Chapter 6, p 391.
9. Hunter, D.H. In *Isotopes in Organic Chemistry*; Buncel, E.; Lee, C.C., Eds.; Elsevier: Amsterdam, 1980; Vol. 1, p. 135.
10. Boche, G. In *Physical Organic Chemistry. Topics in Current Chemistry*; Springer-Verlag: Berlin, 1988; Vol. 146, pp 1–56.
11. Brook, A.G. *Acc. Chem. Res.* **1974**, *7*, 77.
12. Brook, A.G.; Warner, C.M.; Limburg, W.W. *Can. J. Chem.* **1967**, *45*, 1231.
13. Brook, A.G.; Bassindale, A.R. In *Rearrangements in Ground and Excited States*; de Mayo, P., Ed.; Academic Press: New York, 1980; Vol. 2, p 42.
14. West, R.; Lowe, R.; Stewart, H.F.; Wright, A. *J. Am. Chem. Soc.* **1971**, *93*, 282.
15. Wright, A.; West, R. *J. Am. Chem. Soc.* **1974**, *96*, 3227.
16. Lautens, M.; Delange, P.H.M.; Goh, J.G.; Zhang, C.H. *J. Am. Chem. Soc.* **1992**, *57*, 3270.
17. Woodward, R.B.; Hoffmann, R. *J. Am. Chem. Soc.* **1965**, *87*, 395.
18. Woodward, R.B.; Hoffmann, R. *J. Am. Chem. Soc.* **1965**, *87*, 4388.
19. Woodward, R.B.; Hoffmann, R. *Angew. Chem., Int. Ed. Engl.* **1969**, *8*, 781.
20. Fukui, K. *Acc. Chem. Res.* **1971**, *4*, 57.
21. Dewar, M.J.S. *Angew. Chem., Int. Ed. Engl.* **1971**, *10*, 761.
22. Zimmerman, H.E. *Acc. Chem. Res.* **1971**, *4*, 272.
23. Zimmerman, H.E. *Acc. Chem. Res.* **1972**, *5*, 393.
24. Salem, L. *J. Am. Chem. Soc.* **1968**, *90*, 553.
25. Epiotis, N.D. *J. Am. Chem. Soc.* **1973**, *95*, 1214.
26. Schäfer, H.J.; Schöllkopf, U.; Walter, D. *Tetrahedron Lett.* **1968**, 2809.

27. Marshall, J.A. In *Comprehensive Organic Synthesis: Selectivity, Strategy and Efficiency in Modern Organic Chemistry. Volume 3. Carbon–Carbon σ-Bond Formation*; Trost, B.M.; Fleming, I.; Pattenden, G.; Eds.; Pergamon Press: Oxford, U.K., 1991; Chapter 3.11, p 975.
28. Buncel, E. *Carbanions: Mechanistic and Isotopic Aspects*; Elsevier: Amsterdam, 1975. Chapter 6, pp 171–203.
29. Cram, D.J. *Fundamentals of Carbanion Chemistry*; Academic Press: New York, 1965; pp 176–182.
30. Grovenstein, E. *Angew. Chem., Int. Ed. Engl.* **1978**, *17*, 313.
31. Grovenstein, E.; Stevenson, R.W. *J. Am. Chem. Soc.* **1959**, *81*, 4850.
32. Cram, D.J.; Uyeda, R.T. *J. Am. Chem. Soc.* **1962**, *84*, 4358.
33. Cram, D.J.; Uyeda, R.T. *J. Am. Chem. Soc.* **1964**, *86*, 5466.
34. Bergson, G. *Acta Chem. Scand.* **1963**, *17*, 2691.
35. Almy, J.; Garwood, D.C.; Cram, D.J. *J. Am. Chem. Soc.* **1970**, *92*, 4321.
36. Klein, J.; Brenner, S. *J. Chem. Soc. Chem., Commun.* **1969**, 1020.
37. Cram, D.J.; Willey, F.; Fischer, H.P.; Relles, H.M.; Scott, D.A. *J. Am. Chem. Soc.* **1966**, *88*, 2759.
38. Guthrie, R.D.; Hedrick, J.L. *J. Am. Chem. Soc.* **1973**, *95*, 2971.
39. Hunter, D.H.; Kim, S.K. *Can. J. Chem.* **1972**, *50*, 678.
40. Bank, S.; Rowe, C.A.; Schriesheim, A. *J. Am. Chem. Soc.* **1963**, *85*, 2115.
41. Doering, W. von E.; Gaspar, P.P. *J. Am. Chem. Soc.* **1963**, *85*, 3043.
42. Cram, D.J.; Willey, F.; Fischer, H.P.; Scott, D.A. *J. am. Chem. Soc.* **1964**, *86*, 5370.
43. Axelsson, B.S.; Engdahl, K.A.; Långström, B.; Matsson, O. *J. Am. Chem. Soc.* **1990**, *112*, 6656.
44. Meurling, L. *Chem. Scr.* **1975**, *7*, 23.
45. Tolbert, L.M.; Ogle, M.E. *J. Am. Chem. Soc.* **1989**, *111*, 5958.
46. Tolbert, L.M.; Ogle, M.E. *J. Am. Chem. Soc.* **1990**, *112*, 9519.
47. Woodward, R.B.; Hoffmann, R. *The Conservation of Orbital Symmetry*; Verlag Chemie: Weinheim, Germany, 1970.
48. Marchand, A.P.; Lehr, R.E., Eds.; *Pericyclic Reactions*; Academic Press: New York, 1977; Vols. 1 and 2.
49. Patel, D.J.; Hamilton, C.L.; Roberts, J.D. *J. Am. Chem. Soc.* **1960**, *82*, 2646.
50. Chou, P.K.; Dahlke, G.D.; Kass, S.R. *J. Am. Chem. Soc.* **1993**, *115*, 315.
51. Halevi, E.A. *Orbital Symmetry and Reaction Mechanism. The OCAMS View*; Springer-Verlag: Berlin, 1992; p 131.
52. Woodward, R.B.; Hoffmann, R. *J. Am. Chem. Soc.* **1965**, *87*, 395.
53. Boche, G.; Walborsky, H.M. In *The Chemistry of the Cyclopropyl Group*; Rappoport, Z., Ed.; Wiley: New York, 1987; Chapter 2.
54. Hoffmann, R.W.; Koberstein, R. *J. Chem. Soc., Perkin Trans. 2* **2000**, 595.
55. Mulvaney, J.E.; Savage, D. *J. Org. Chem.* **1971**, *36*, 2592.
56. Londrigan, M.E.; Mulvaney, J.E. *J. Org. Chem.* **1972**, *37*, 2832.
57. Huisgen, R.; Eberhard, P. *J. Am. Chem. Soc.* **1972**, *94*, 1346.
58. Seyferth, D.; Cohen, H.M. *J. Organomet. Chem.* **1963**, *1*, 15.
59. Motes, J.M.; Walborsky, H.M. *J. Am. Chem. Soc.* **1970**, *92*, 3697.
60. Ogle, C.A.; Riley, P.A.; Dorchak, J.J.; Hubbard, J.L. *J. Org. Chem.* **1988**, *53*, 4409.
61. Chou, P.K.; Kass, S.R. *Org. Mass Spectrom.* **1991**, *26*, 1039.
62. Wiberg, K.B.; Castejon, H. *J. Org. Chem.* **1995**, *60*, 6327.
63. Froelicher, S.W.; Freiser, B.S.; Squires, R.R. *J. Am. Chem. Soc.* **1986**, *108*, 2853.
64. Hopkinson, A.C.; McKinney, M.A.; Lien, M.H. *J. Comput. Chem.* **1983**, *4*, 513.
65. Dewar, M.J.S.; Nelson, D.J. *J. Org. Chem.* **1982**, *47*, 2614.

66. Boche, G.; Buckl, K.B.; Martens, D.; Schneider, D.R.; Wagner, H.U. *Chem. Ber.* **1979**, *112*, 2961.
67. Clark, T.; Rohde, C.; Schleyer, P. von R. *Organometallics* **1983**, *2*, 1344.
68. Boche, G.; Martens, D. *Angew. Chem., Int. Ed. Engl.* **1972**, *11*, 724.
69. Boche, G.; Buckl, K.; Martens, D.; Schneider, D.R. *Liebigs Ann. Chem.* **1980**, *729*, 1135.
70. Wittig, G.; Rautenstrauch, V.; Wingler, F. *Tetrahedron* **1966**, (Suppl. 7) 189.
71. Newcomb, M.; Ford, W.T. *J. Am. Chem. Soc.* **1974**, *96*, 2968.
72. Kaufmann, T.; Habersaat, K.; Koppelmann, E. *Angew. Chem., Int. Ed. Engl.* **1972**, *11*, 291.
73. Stapp, P.R.; Kleinschmidt, R.F. *J. Org. Chem.* **1965**, *30*, 3006.
74. Slaugh, L.H. *J. Org. Chem.* **1967**, *32*, 108.
75. Tardi, M.; Vairon, J.P.; Sigwalt, P. *Bull. Soc. Chim. Fr. 1972*, 1791.
76. Bates, R.B.; McCombs, D.A. *Tetrahedron Lett.*, **1969**, 977.
77. Kloosterziel, H.; van Drunen, J.A.A. *Recl. Trav. Chim. Pays-Bas* **1970**, *89*, 368.
78. Buncel, E. *Carbanions: Mechanistic and Isotopic Aspects*; Elsevier: Amsterdam, 1975; Chapter 6, p 181.
79. Atkinson, D.J.; Perkins, M.J.; Ward, P. *J. Chem. Soc., Chem. Commun.* **1969**, 1390.
80. Staley, S.W.; Erdman, J.P. *J. Am. Chem. Soc.* **1970**, *92*, 3832.
81. Bates, R.B.; Gosselink, D.W.; Kaczynski, J.A. *Tetrahedron Lett.* **1967**, 205.
82. Hunter, D.H.; Sim, S.K. *Can. J. Chem.* **1972**, *50*, 669.
83. Hunter, D.H.; Sim, S.K. *Can. J. Chem.* **1972**, *50*, 678.
84. Kloosterziel, H.; van Drunen, J.A.A.; Galama, P. *J. Chem. Soc. Chem. Commun.* **1969**, 885.
85. Bates, R.B.; Kroposki, L.M.; Potter, D.E. *J. Org. Chem.* **1972**, *37*, 560.
86. Bates, R.B.; Deines, W.H.; McCombs, D.A.; Potter, D.E. *J. Am. Chem. Soc.* **1969**, *91*, 4608.
87. Barjot, J.; Bony, G.; Dauphin, G.; Duprat, P.; Kergomard, A.; Veschambre, H. *Bull. Soc. Chim. Fr.* **1973**, 3187.
88. David, L.; Kergomard, A. *Tetrahedron* **1971**, *27*, 653.
89. Dauphin, G. *Bull. Soc. Chim. Fr.* **1975**, 1208.
90. Hoffmann, R.; Olofson, R.A. *J. Am. Chem. Soc.* **1966**, *88*, 943.
91. Staley, S.W.; Heyn, A.S. *J. Am. Chem. Soc.* **1975**, *97*, 3852.
92. Zimmerman, H.E. In *Pericyclic Reactions*; Marchand, A.P.; Lehr, R.E., Eds.; Academic Press: New York, 1977; Vol. 1, pp 90–101.
93. Li, W.K.; Nobes, R.H.; Radom, L. Theochem **1987**, *149*, 67.
94. Borosky, G.L. *J. Org. Chem.* **1998**, *63*, 3337.
95. Magid, R.L.; Wilson, S.E. *Tetrahedron Lett.* **1971**, 19.
96. Klein, J.; Gilly, S.; Kost, D. *J. Org. Chem.* **1970**, *35*, 1281.
97. Klein, J.; Gilly, S. *Tetrahedron* **1971**, *27*, 3477.
98. Bates, R.B.; Brenner, S.; Deines, W.H.; McCombs, D.A.; Potter, D.E. *J. Am. Chem. Soc.* **1970**, *92*, 6345.
99. Berson, J.A. In *Rearrangements in Ground and Excited States*; de Mayo, P., Ed.; Academic Press: New York, 1980; Vol. 1, pp 368–390.
100. Zimmerman, H.E.; Zweig, A.Z. *J. Am. Chem. Soc.* **1961**, *83*, 1196.
101. Staley, S.W.; Cramer, G.M.; Kingsley, W.G. *J. Am. Chem. Soc.* **1973**, *95*, 5052.
102. Grovenstein, E.; Wentworth, G. *J. Am. Chem. Soc.* **1967**, *89*, 1852.
103. Grovenstein, E.; Wentworth, G. *J. Am. Chem. Soc.* **1967**, *89*, 2348.
104. Grovenstein, E. *Adv. Organomet. Chem.* **1977**, *16*, 167.
105. Grovenstein, E. *Angew. Chem., Int. Ed. Engl.* **1978**, *17*, 313.
106. Grovenstein, E.; Cottingham, A.B. *J. Am. Chem. Soc.* **1977**, *99*, 1881.

107. Baldwin, J.E.; Urban, F.J. *J. Chem. Soc., Chem. Commun.* **1970**, 165.
108. Scheffer, J.R.; Gayler, R.A.; Zakouras, T.; Dzakpasu, A.A. *J. Am. Chem. Soc.* **1977**, *99*, 7726.
109. Trost, B.M.; Fleming, I., Eds.; *Comprehensive Organic Synthesis. Vol. 3. Carbon–Carbon σ-Bond Formation*; Pergamon Press: Oxford, U.K., 1991; pp 821–1015
110. Marshall, J.A. In *Comprehensive Organic Synthesis. Vol. 3. Carbon–Carbon σ-Bond Formation*; Trost, B.M.; Fleming, I., Eds.; Pergamon Press: Oxford, U.K., 1991; Chapter 3.11, pp 975–1015.
111. Greeves, N.; Lee, W.M. *Tetrahedron Lett.* **1997**, *38*, 6445.
112. Greeves, N.; Lee, W.M. *Tetrahedron Lett.* **1997**, *38*, 6449.
113. Greeves, N.; Lee, W.M.; Barkley, J.V. *Tetrahedron Lett.* **1997**, *38*, 6453.
114. Enders, D.; Backhaus, D.; Runsink, J. *Tetrahedron* **1996**, *52*, 1503.
115. Wittig, G.; Löhmann, L. *Liebigs Ann. Chem.* **1942**, *550*, 260.
116. Tomooka, A.; Yamamoto, H.; Nakai, T. *Liebigs Ann. Chem.* **1997**, *815*, 1275.
117. Solov'yanov, A.A.; Ahmed, E.A.A.; Beletskaya, I.P.; Reutov, O.A. *J. Chem. Soc., Chem. Commun.* **1987**, *23*, 1232.
118. Schöllkopf, U. *Angew. Chem., Int. Ed. Engl.* **1970**, *9*, 763.
119. Nakai, T.; Mikami, K. *Chem. Rev.* **1986**, *86*, 885.
120. Lansbury, P.T.; Pattison, V.A. *J. Am. Chem. Soc.* **1962**, *84*, 4295.
121. Garst, J.F.; Smith C.D. *J. Am. Chem. Soc.* **1973**, *95*, 6870.
122. Crombie, L.; Darnborough, G.; Pattenden, G. *J. Chem. Soc., Chem. Commun.* **1976**, 684.
123. Rautenstrauch, V. *J. Chem. Soc., Chem. Commun.* **1970**, 4.
124. Antoniotti, P.; Tonachini, G. *J. Org. Chem.* **1993**, *58*, 3622.
125. Eichinger, P.C.H.; Bowie, J.H. *J. Chem. Soc., Perkin Trans. 2* **1988**, 497.
126. Antoniotti, P.; Canepa, C.; Tonachini, G. *J. Org. Chem.* **1994**, *59*, 3952.
127. Roos, B. In *Ab Initio Methods in Quantum Chemistry II*; Lawley, K.P., Ed.; Wiley: New York, 1987.
128. Antoniotti, P.; Tonachini, G. *J. Org. Chem.* **1994**, *59*, 9756; and references therein.
129. Schreiber, S.L.; Goulet, M.T. *Tetrahedron Lett.* **1987**, *28*, 1043.
130. Garst, J.F.; Smith, C.D. *J. Am. Chem. Soc.* **1976**, *98*, 1526.
131. Hebert, E.; Welvart, Z.; Ghelfenstein, M.; Szwarc, H. *Tetrahedron Lett.* **1983**, *23*, 1381.
132. Kunishima, M.; Hioki, K.; Kono, K.; Kato, A.; Tani, S. *J. Org. Chem.* **1997**, *62*, 7542.
133. Brckner, R. *Kontacke (Darmstadt)* **1991**, 3; C.A. 115: 278990x.
134. Paquette, L.; Sugimura, T. *J. Am. Chem. Soc.* **1986**, *108*, 3841.
135. Wu, Y.D.; Houk, K.N.; Marshall, J.A. *J. Org. Chem.* **1990**, *55*, 1421.
136. Takahashi, T.; Nemoto, H.; Kanda, Y.; Tsuji, J.; Furazasawa, Y.; Okajima, T.; Fujise, Y. *Tetrahedron* **1987**, *43*, 5499.
137. Mikami, K.; Tomoya, U.; Hirano, T.; Wu, Y.D.; Houk, K.N. *Tetrahedron* **1994**, *50*, 5917.
138. Snider, B.B.; Hrib, N.J.; Fuzesi, L. *J. Am. Chem. Soc.* **1976**, *98*, 7115.
139. Durst, T.; Elzen, R.V.D.; LeBelle, M. *J. Am. Chem. Soc.* **1972**, *94*, 9261.
140. Ahman, J.; Somfai, P. *Tetrahedron Lett.* **1995**, *36*, 303; and references therein.
141. Coldham, I.; Collis, A.J.; Mould, R.J.; Rathmell, R.E. *J. Chem. Soc., Perkin Trans. 1* **1995**, 2739.
142. Anderson, J.C.; Siddons, D.C.; Smith, S.C.; Swarbrick, M.E. *J. Org. Chem.* **1996**, *61*, 4820.
143. Duchenet, V.; Pelloux, N.; Vallée, Y. *Tetrahedron Lett.* **1994**, *35*, 2005.

144. Lindermann, R.J.; Ghannam, A. *J. Am. Chem. Soc.* **1990**, *112*, 2392; and references therein.
145. Koreeda, M.; Koo, S. *Tetrahedron Lett.* **1990**, *31*, 831.
146. Brook, A.G.; LeGrow, G.E.; MacRae, D.M. *Can. J. Chem.* **1967**, *45*, 239.
147. Biernbaum, M.S.; Mosher, H.S. *Tetrahedron Lett.* **1968**, 5289.
148. Kawachi, A.; Doi, N.; Tamao, K. *J. Am. Chem. Soc.* **1997**, *119*, 233.
149. Wright, A.; West, R. *J. Am. Chem. Soc.* **1974**, *96*, 3214.
150. Brook, A.G.; Fieldhouse, S.A. *J. Organomet. Chem.* **1967**, *10*, 235.
151. Reich, H.J.; Holtan, R.C.; Bolm, C. *J. Am. Chem. Soc.* **1990**, *112*, 5609.
152. Jankowski, P.; Marczak, S.; Masnyk, M.; Wicha, J. *J. Organomet. Chem.* **1991**, *403*, 49.
153. Stevens, T.S.; Creighton, E.M.; Gordon, A.B.; MacNicol, M. *J. Chem. Soc.* **1928**, 3193.
154. Pine, S.H. *J. Chem. Educ.* **1971**, *48*, 99.
155. Johnstone, R.A. *Mech. Mol. Migr.* **1969**, *2*, 249.
156. Thomson, T.; Stevens, T.S. *J. Chem. Soc.* **1932,** 69.
157. Dunn, J.L.; Stevens, T.S. *J. Chem. Soc.* **1932**, 1926.
158. Jemison, R.W.; Mageswaran, S.; Ollis, W.D.; Potter, S.E.; Pretty, A.J.; Sutherland, I.O.; Thebtaranonth, Y. *J. Chem. Soc., Chem. Commun.* **1970**, 1201.
159. Johnstone, R.A.; Stevens, T.S. *J. Chem. Soc.* **1955**, 4487.
160. Campbell, A.; Houston, A.H.J.; Kenyon, J. *J. Chem. Soc.* **1947**, 93.
161. Ollis, W.D.; Rey, M.; Sutherland, I.O. *J. Chem. Soc., Perkin Trans. 1* **1983**, 1009.
162. Lepley, A.R. *J. Am. Chem. Soc.* **1969**, *91*, 1237.
163. Lepley, A.R. In *Chemically Induced Magnetic Polarization*; Lepley, A.R.; Closs,G.L., Eds.; Wiley: New York, 1973; pp 323–384.
164. Barclay, L.R.C.; Dust, J.M. *Can. J. Chem.* **1982**, *60*, 607.
165. Hennion, G.F.; Shoemaker, M.J. *J. Am. Chem. Soc.* **1970**, *92*, 1769.
166. Lorand, J.P.; Grant, R.W.; Samuel, P.A.; O'Connell, E.; Zaro, J. *Tetrahedron Lett.* **1969**, 4087.
167. Khuthier, A.; Al-Mallah, K.Y.; Hanna, S.Y.; Abdulla, N.I. *J. Org. Chem.* **1987**, *52*, 1710.
168. Sommelet, M. *C. R. Hebd. Seances Acad. Sci.* **1937**, *205*, 56.
169. Kantor, S.W.; Hauser, C.R. *J. Am. Chem. Soc.* **1951**, *73*, 4122.
170. Puterbaugh, W.H.; Hauser, C.R. *J. Am. Chem. Soc.* **1964**, *86*, 1108.
171. Schöllkopf, U.; Ostermann, G.; Schossig, J. *Tetrahedron Lett.* **1969**, 2619.
172. Ratts, K.W.; Yao, A.N. *J. Org. Chem.* **1968**, *33*, 70.
173. Vedejs, E.; Engler, D.A.; Mullins, M.J. *J. Org. Chem.* **1977**, *42*, 3109.
174. Vedejs, E.; Martinez, G.R. *J. Am. Chem. Soc.* **1979**, *101*, 6452.
175. Vedejs, E.; West, F.G. *J. Org. Chem.* **1983**, *48*, 4773.
176. Sato, Y.; Sakakibara, H. *J. Organomet. Chem.* **1979**, *166*, 303.
177. Sato, Y.; Yagi, Y.; Koto, M. *J. Org. Chem.* **1980**, *45*, 613.
178. Ando, W. *Acc. Chem. Res.* **1977**, *10*, 179.
179. Doyle, M.P. *Acc. Chem. Res.* **1986**, *19*, 348.
180. Blackburn, G.M.; Ollis, W.D.; Smith, C.; Sutherland, I.O. *J. Chem. Soc., Chem. Commun.* **1969**, 99.
181. Markó, I.E. In *Comprehensive Organic Synthesis. Vol. 3. Carbon–Carbon σ-Bond Formation*; Trost, B.M.; Fleming, I., Eds.; Pergamon Press: Oxford, U.K.; 1991. Chapter 3.10, pp 913–974.
182. von Liebig, J. *Justus Liebigs Ann. Chem.* **1838**, *25*, 27.
183. Schaltegger, A.; Bigler, P. *Helv. Chim. Acta* **1986**, *69*, 1666.
184. Wöhler, F.; von Liebig, J. *Justus Liebigs Ann. Chem.* **1838**, *26*, 241.

185. Ingold, C.K. *Annu. Rep. Prog. Chem.* **1928**, *25*, 124.
186. Selman, S.; Eastham, J.F. *Q. Rev. Chem. Soc.* **1960**, *14*, 221.
187. Kwart, H.; Sarasohn, I.M. *J. Am. Chem. Soc.* **1961**, *83*, 909.
188. Dahn, H.; Dao, L.H.; Hunma, R. *Helv. Chim. Acta* **1982**, *65*, 2458.
189. Kasiwagi, I. *Bull. Chem. Soc. Jpn.* **1926**, *1*, 66.
190. Screttas, C.G.; Micha-Screttas, M.; Cazianis, C.T. *Tetrahedron Lett.* **1983**, *24*, 3287.
191. Thierrichter, B.; Junek, H. *Monatsch. Chem.* **1979**, *110*, 729.
192. Cymerman-Craig, J.; Moyle, M.; Rowe-Smiith, P.; Wailes, P.C. *Aust. J. Chem.* **1956**, *9*, 391.
193. Gill, B.G. In *Comprehensive Organic Synthesis. Vol. 3. Carbon–Carbon σ-Bond Formation*; Trost, B.M.; Fleming, I., Eds.; Pergamon Press: Oxford, U.K., 1991; Chapter 3.6, pp 821–838.
194. Grovenstein, E. *J. Am. Chem. Soc.* **1957**, *79*, 4985.
195. Zimmerman, H.E.; Smentowski, F.J. *J. Am. Chem. Soc.* **1957**, *79*, 5455.
196. Dixon, R.E.; Streitwieser, A. *J. Org. Chem.* **1992**, *57*, 6125.
197. Grovenstein, E.; Cheng, Y.M. *J. Am. Chem. Soc.* **1972**, *94*, 4971.
198. Grovenstein, E.; Williams, L.P. *J. Am. Chem. Soc.* **1961**, *83*, 412.
199. Grovenstein, E.; Williamson, R.E. *J. Am. Chem. Soc.* **1975**, *97*, 646.
200. Waack, R.; Doran, M.A. *J. Am. Chem. Soc.* **1969**, *91*, 2456.
201. Eisch, J. *Ind. Eng. Chem. Prod. Res. Dev.* **1975**, *14*, 11.
202. Grovenstein, E.; Singh, J.; Patil, B.B.; Van Derveer, D. *Tetrahedron* **1994**, *50*, 5971.
203. Grovenstein, E.; Akabori, S. *J. Am. Chem. Soc.* **1975**, *97*, 4620.
204. Bertrand, J.A.; Grovenstein, E.; Liu, P.C.; Van Derveer, D. *J. Am. Chem. Soc.* **1976**, *98*, 7835.
205. Grovenstein, E.; Lu, P.C. *J. Am. Chem. Soc.* **1982**, *104*, 6681.
206. Grovenstein, E.; Black, K.W.; Goel, S.C.; Hughes, R.L.; Northrop, J.H.; Streeter, D.I.; Van Derveer, D. *J. Org. Chem.* **1989**, *54*, 1671.
207. Grovenstein, E.; Bhatti, A.M.; Quest, D.E.; Sengupta, D.; Van Derveer, D. *J. Am. Chem. Soc.* **1983**, *103*, 6290.
208. Grovenstein, E.; Bhatti, A.M.; Plagge, F.A.; Heinrich, Y.M.; Longfield, T.H.; Singh, J.; Van Derveer, D. *Organometallics* **1990**, *9*, 2587.
209. Snieckus, V. *Heterocycles* **1980**, *14*, 1649.
210. Beak, P.; Snieckus, V. *Acc. Chem. Res.* **1982**, *15*, 306.
211. Snieckus, V. In *Chemical Synthesis. Gnosis to Prognosis*; Chatgilialoglu, C.; Snieckus, V., Eds.; Kluwer: Boston, M, 1994; pp 191–221.
212. Superchi, S.; Sotomayor, N.; Miao, G.; Joseph, B.; Campbell, M.G.; Snieckus, V. *Tetrahedron Lett.* **1996**, *37*, 6061.
213. Beak, P.; Meyers, A.I. *Acc. Chem. Res.* **1986**, *19*, 356.
214. Sibi, M.; Snieckus, V. *J. Org. Chem.* **1983**, *48*, 1937.
215. Chenard, B.L. *J. Org. Chem.* **1983**, *48*, 2610.
216. A. Streitwieser, Personal communication 1997.
217. Dankwardt, J.W. *J. Org. Chem.* **1998**, *63*, 3753.
218. Gray, M.; Chapell, B.J.; Taylor, N.J.; Snieckus, V. *Angew. Chem., Int. Ed. Engl.* **1996**, *35*, 1558.
219. Beaulieu, F.; Snieckus, V. *J. Org. Chem.* **1994**, *59*, 6508.
220. Superchi, S.; Sotomayor, N.; Miao, G.; Joseph, B.; Snieckus, V. *Tetrahedron Lett.* **1996**, *37*, 6057.
221. March, J. *Advanced Organic Chemistry*, 4th ed.; Wiley: New York, 1992; pp 555–556.
222. Clough, J.M. In *Comprehensive Organic Synthesis. Vol. 3. Carbon–Carbon σ-Bond Formation*; Trost, B.M.; Fleming, I., Eds.; Pergamon Press: Oxford, U.K., 1991; Chapter 3.8, pp 861–886.

223. Ramberg, L.; Bäcklund, B. *Ark. Kemi. Mineral. Geol.* **1940**, *13A*, 1.
224. Bordwell, F.G. *Acc. Chem. Res.* **1970**, *3*, 281.
225. Meyers, C.Y.; Hua, D.H.; Peacock, N.J. *J. Org. Chem.* **1980**, *45*, 1719.
226. Hendrickson, J.B.; Boudreaux, G.J.; Palumbo, P.S. *J. Am. Chem. Soc.* **1986**, *108*, 2358.
227. Bordwell, F.G.; Jarvis, B.B.; Corfield, P.W.R. *J. Am. Chem. Soc.* **1968**, *90*, 5298.
228. Bordwell, F.G.; Doomes, E.; Corfield, P.W.R. *J. Am. Chem. Soc.*. **1970**, *92*, 2581.
229. Wolfe, S. In *Organic Sulfur Chemistry. Theoretical and Experimental Advances*; Bernardi, F.; Czismadia, I.G.; Mangini, A., Eds.; Elsevier: Amsterdam, 1985; pp 133–190.
230. Bordwell, F.G.; Williams, J.M. *J. Am. Chem. Soc.* **1968**, *90*, 435.
231. Bordwell, F.G.; Williams, J.M.; Hoyt, E.B.; Jarvis, B.B. *J. Am. Chem. Soc.* **1968**, *90*, 429.
232. Sutherland, A.G.; Taylor, R.J.K. *Tetrahedron Lett.* **1989**, *30*, 3267.
233. Woodward, R.B.; Hoffmann, R. *The Conservation of Orbital Symmetry*; Verlag Chemie: Weinheim, Germany, 1970; pp 152–163.
234. Weiniges, K.; Klessing, K. *Chem. Ber.* **1974**, *107*, 1915.
235. Paquette, L.A.; Trova, M.P. *Tetrahedron Lett.* **1986**, *27*, 1895.
236. Paquette, L.A.; Trova, M.P. *J. Am. Chem. Soc.* **1988**, *110*, 8197.
237. Potter, S.E.; Sutherland, I.O. *J. Chem. Soc. Chem. Commun.* **1973**, 520.
238. Cooke, M.P. Jr. *J. Org. Chem.* **1981**, *46*, 1747.
239. Mitchell, R.H.; Boekelheide, V. *J. Am. Chem. Soc.* **1970**, *92*, 3510.
240. Mitchell, R.H.; Otsubo, T.; Boekelheide, V. *Tetrahedron Lett.* **1975**, 219.
241. Lai, Y.H.; Chen, P. *Tetrahedron Lett.* **1988**, *29*, 3483.
242. Bordwell, F.G.; Wilfinger, M.D. *J. Org. Chem.* **1974**, *39*, 2521.
243. Jones, S.L.; Stirling, C.J.M. *J. Chem. Soc., Chem. Commun.* **1988**, 1153.
244. Menichetti, S.; Stirling, C.J.M. *J. Chem. Soc., Perkin Trans. 2,* **1992**, 741.
245. Bates, R.B.; Siahaan, T.J.; Suvannachut, K. *J. Org. Chem.* **1990,** *55*, 1328.
246. Samuel, S.P.; Niu T.; Erickson, K.L. *J. Am. Chem. Soc.* **1989**, *111*, 1429.
247. Du, Z.; Haglund, M.J.; Pratt, L.A.; Erickson, K.L. *J. Org. Chem.* **1998**, *63*, 8880.
248. Buncel, E. *Carbanions: Mechanistic and Isotopic Aspects*; Elsevier: Amsterdam, 1975. Chapter 5; pp 161–166.
249. Hunter, D.H.; Stothers, J.B.; Warnhoff, E.W. In *Rearrangements in Ground and Excited States*; de Mayo, P., Ed.; Academic Press: New York, 1980; Vol. 1*;* Chapter 6, pp 410–436.
250. Janzen, A.F.; Smyrl, T.G. *Can. J. Chem.* **1972**, *50*, 1205.
251. Janzen, A.F.; Vaidya, O.C. *Can. J. Chem.* **1973**, *51*, 1136.
252. Dust, J.M.; Warren, C.S. *Water Qual. Res. J. Canada* **2001**, *36*, 589.
253. Williams, D.R.; McClymont, E.L. *Tetrahedron Lett.* **1993**, *48*, 7705.
254. Eskola, P.; Hirsch, J.A. *J. Org. Chem.* **1997**, *62*, 5732.
255. Gajewski, J.J. *Acc. Chem. Res.* **1980**, *13*, 142.
256. Burrows, C.J.; Carpenter, B.K. *J. Am. Chem. Soc.* **1981,** *103*, 6983.
257. Denmark, S.E.; Harmata, M.A. *J. Am. Chem. Soc.* **1982**, *104*, 4972.
258. Denmark, S.E.; Harmata, M.A. *Tetrahedron Lett.* **1984**, *25*, 1543.
259. Blechert, S. *Liebigs Ann. Chem.* **1985**, *754*, 673.
260. Danheiser, R.L.; Bronson, J.J.; Okano, S.M. *J. Am. Chem. Soc.* **1985**, *107*, 4579.
261. Denmark, S.E.; Marmata, M.A.; White, K.S. *J. Am. Chem. Soc.* **1989**, *111*, 8878.
262. Bowman, E.J.; Huges, G.B.; Grutzner, J.B. *J. Am. Chem. Soc.* **1976**, *98*, 8273.
263. Nemoto, H.; Suzuki, K.; Tsubuki, M.; Minemura, K.; Fukumoto, K.; Kametani, T.; Furuyama, H. *Tetrahedron* **1983**, *39*, 1123.
264. Banwell, M.G.; Onrust, R. *Tetrahedron Lett.* **1985**, *26*, 4543.

Author Index

Subject Index